ENCYCLOPEDIA OF

WEATHER AND CLIMATE

VOLUME I

ENCYCLOPEDIA OF WEATHER AND CLIMATE

VOLUME I

A–L

MICHAEL ALLABY

Facts On File, Inc.

For Ailsa
—M.A.

To my late wife, Jen, who gave me
inspiration and support for almost 30 years.
—R.G.

Encyclopedia of Weather and Climate

Facts On File, Inc.
132 West 31st Street
New York NY 10001

Library of Congress Cataloging-in-Publication Data

Allaby, Michael
Encyclopedia of weather and climate / Michael Allaby
p. cm.
Includes bibliographical references and index.
ISBN 0-8160-4801-0 (Volume I)
ISBN 0-8160-4802-9 (Volume II)
ISBN 0-8160-4071-0 (set) (alk. paper)
1. Meteorology—Encyclopedias. 2. Climatology—Encyclopedias. I. Title.
QC854 .A45 2001
551.6'03—dc21
2001023103

Text design by Joan M. Toro
Cover design by Cathy Rincon
Illustrations by Richard Garratt

Printed in the United States of America.

VB FOF 10 9 8 7 6 5 4 3 2 1

This book is printed on acid-free paper.

Contents

INTRODUCTION

The weather affects us all. At the most immediate level, an unexpected shower of rain will soak those who venture outdoors without a coat or umbrella, so before going outside we decide whether rain seems likely. Misjudgments can be more serious, of course. Snowfalls can make roads impassable, marooning people in their cars, where the low temperature can kill.

There are also less immediate ways in which the weather affects us. Fine growing weather across the farmlands of the nation, with rain when it is needed and sunshine to ripen the crops, produces heavy crop yields. Food is abundant, and when a commodity is abundant, its price falls, so fine weather can make food cheaper. In the same way, bad weather can lead to low yields and higher prices. Changes in food prices may make the difference between relative prosperity and hardship for the poorest members of our society, and in some parts of the world bad weather may lead to famine, in which people die.

Mild winters reduce our heating bills. We do not need so much energy to warm our homes as we do when the winter is hard. This makes a difference to our living costs, and it also has environmental consequences. When we burn fuels to generate electrical power or use them directly to heat areas, water, and food, the by-products of combustion are released into the air. Some of these, such as sulfur dioxide, carbon monoxide, and unburned hydrocarbons, cause pollution. Carbon dioxide is not a pollutant in the usual sense, because it causes no injury, but many scientists believe it has the capacity to change the climate of the world.

We all like to know what the weather will be like in the next few hours or days, but some people need to know. Fishermen must know whether it will be safe for them to put to sea. Sailors of all kinds need to know whether they are sailing into a severe storm and, if so, how to avoid it. Pilots need to know the speed and direction of the winds along the routes they plan to fly. These determine the time the journey will take and the amount of fuel the aircraft will consume.

We all have an interest in the weather, and in recent years that interest has intensified. Climates always change over long periods. The weather we experience today is different from that of the Little Ice Age of the 17th century and of the Middle Ages, when England was a major wine producer. Today, though, there are fears that the climate may be changing faster than it has for thousands of years and that the gases we release into the air, from our cars, factories, domestic fires, power stations, farming, and forest clearance, may be accelerating that change. This concern is now driving climatic research.

If we are to make informed decisions about climate change and the possibility that our activities may be partly responsible for it, we need to know something of the way the atmosphere works, of how our weather is produced. The *Encyclopedia of Weather and Climate* aims to address this need.

The *Encyclopedia* contains nearly three thousand entries. These vary in length. Some are simple definitions, no more than a few words long. Others are short essays. Some include tables, and nearly three hundred of the entries are illustrated. The length of entries is determined by the complexity of the ideas they seek to explain, and illustrations are used only where they may help clarify the entry. The entries are fully cross-referenced, so synonyms and acronyms are listed with a direction to the full entry.

The entries describe processes such as cloud formation, atmospheric phenomena such as rainbows, some of the techniques and instruments that are used to study the atmosphere, and the units of measurement that scientists use. They also explain the classification systems that are used for climate types, winds, and clouds.

Our understanding of the atmosphere did not begin suddenly in its present form. It has a history, and the *Encyclopedia* attempts to give a flavor of that history. It includes biographical entries, many of them with pictures of their subjects, outlining the lives and work of a few of the individuals who have contributed to the advancement of the atmospheric sciences.

The weather we experience is local. It may be raining on one side of a hill and fine on the other side. Winds, in particular, acquire local names. *Chinook, Santa Ana, mistral, harmattan, bora,* and *sirocco* are just a few of the local names for winds that people in certain places welcome or dread. The *Encyclopedia* lists some of these and explains what type of wind each one is and where it occurs. Before there were weather stations, orbiting satellites, and powerful computers to produce weather forecasts people had to rely on their experience and the signs they could read, or thought they could read, in the sky and in the natural world around them. Over centuries these experiences accumulated as weather lore, comprising sayings, rhymes, and references to clouds, plants, and animals. Some of these are also included here.

My own interest in weather and climate began many years ago. For a short time I was a military pilot, and so I was compelled to observe the weather and to respect it. I learned then that bad weather can kill and that it pays to listen to the weather forecast (we called it the "met briefing"). More recently my interest has developed from my studies of the environmental sciences. These include the atmospheric sciences of climatology and meteorology as well as the historical disciplines, such as paleoclimatology, the study of the climates of the distant past.

Most of what we know we learn from the books and articles we read. I am no exception, but reading does not usually teach us to think for ourselves and to try to figure out how things work. For that we need teachers, and here I must pay tribute to the stimulus I received from the many conversations I have had with my friend the chemist and proponent of the Gaia hypothesis Jim Lovelock. He and I collaborated in writing two books in which we had to work out the climatic consequences of certain events.

I am also very grateful to my friends at Facts On File. In particular I would like to thank my editor, Frank K. Darmstadt, for his patience and wisdom and for the kindly interest he always shows in the well-being of his author and the current location of the text he was supposed to have received weeks or months ago. The project was originally conceived and planned in collaboration with Eleanora von Dehsen. The *Encyclopedia* might not exist at all without her enthusiasm and encouragement in the very early stages.

This work is not a compilation of entries written by contributors. I wrote all of the entries myself. The result of that is that any mistakes that have escaped Frank's eagle eye are entirely my own work. So are the facts I got right, of course.

I hope you enjoy wending your way through the highways and byways of atmospheric science as you explore the processes that generate our weather. Writing the *Encyclopedia* was fun. I hope it is fun to use.

—Michael Allaby
Tighnabruaich
Argyll, Scotland

Entries A-L

A *See* AMPERE.

AABW *See* ANTARCTIC BOTTOM WATER.

AAC *See* ANTARCTIC POLAR FRONT.

Abbe, Cleveland (1838–1916) American *Meteorologist* The American meteorologist who was the first person to issue regular daily weather bulletins and forecasts and who is sometimes called the "father of the Weather Bureau." He was also influential in the establishment of standardized time zones across the United States.

Born and educated in New York City, he studied astronomy at the University of Michigan and privately with Benjamin Apthorp Gould (1824–96) at Cambridge, Massachusetts. He taught at the University of Michigan for several years. Afterward for two years, 1864–6, he completed his astronomical studies at the Pulkovo Observatory in Russia. When he returned to the United States, he was appointed director of the Cincinnati Observatory.

Like JOSEPH HENRY, Abbe received telegraphic reports of storms and used them to plot their location and timing across the country. These compilations provided him with the information he issued from Cincinnati Observatory in his daily reports, the *Weather Bulletin,* the first of which appeared on September 1, 1869. On September 22 of the same year, he published his first weather forecast.

The government was immediately interested, and on February 2, 1870, Rep. Halbert E. Paine introduced a Joint Congressional Resolution requiring the secretary of war to establish a national meteorological service. President Ulysses S. Grant signed the resolution on February 9, 1870, and the weather bureau began operations in November 1870. It formed part of the recently instituted Division of Telegrams and Reports for the Benefit of Commerce of the Army Signal Service and was headed by an army general. Abbe was appointed his scientific assistant, joined the Signal Service, and started work in 1871; he issued the first of his three-day weather forecasts, based on probabilities, on February 19, 1871. The first "cautionary storm signal" (for the Great Lakes region) was issued on November 8, 1871.

In 1891, the national bureau was renamed the United States Weather Bureau and was transferred from the Army Signal Corps to the Department of Agriculture (it was transferred to the Department of Commerce in 1940). Abbe was the meteorologist in charge and retained this position for the rest of his life. At the same time, he conducted research and taught meteorology at Johns Hopkins University.

His advocacy of standardized time zones culminated in a report on the subject that he wrote in 1879. At that time, each community used its own local time, which was accurate enough but meant travelers had to adjust their watches continually as they moved east or west. The railroad companies had devised their own standardization for the purposes of scheduling services, and these were the basis of the Abbe proposal. The government adopted the idea, and in 1883 the country was divided into four time zones. The same time was used throughout each zone and was based on an average value. This system was later extended to the entire world.

Cleveland Abbe died on October 28, 1916, in Chevy Chase, Maryland, where he and his wife are buried.

Abby A TYPHOON that struck Taiwan on September 19, 1986; it killed 13 people and caused damage estimated at $80 million.

Abe A TYPHOON that struck Zhejiang Province, China, on August 31, 1990; it killed 48 people.

ablation The removal of ice and snow from the ground surface by melting and also by the process of SUBLIMATION.

ablation zone An area in which the rate of loss of ice by ablation, and by iceberg calving from a glacier, exceeds the rate at which new snow and ice accumulates.

Abraham's tree A popular name for CIRRUS RADIATUS, when this cloud consists of long, parallel bands that seem to radiate from a particular point on the horizon.

absolute drought In Britain, a period of 15 consecutive days during which no rain falls.

absolute humidity The mass of the water vapor that is present in a given volume of air. It is usually expressed in grams per cubic meter (1 g m^{-3} = 0.046 ounce per cubic yard). This measure takes no account of the fact that changes in pressure and temperature alter the volume of the air, thus changing the absolute humidity without the addition or removal of any moisture. For this reason absolute humidity is not a very useful measure and is rarely used.

absolute instability The condition of air when the ENVIRONMENTAL LAPSE RATE (ELR) is greater than the DRY ADIABATIC LAPSE RATE (DALR). As it rises, a PARCEL OF AIR cools at the DALR, but because this is a slower rate of cooling than the ELR, it will always be warmer than the surrounding air. If the air contains water vapor, it may reach a height at which this starts to condense to form cloud. The release of LATENT HEAT of condensation warms the air, reducing the lapse rate from the DALR to the SATURATED ADIABATIC LAPSE RATE (SALR). The difference between the ELR and the SALR is greater than that between the ELR and DALR, so the instability of the air increases.

Being always warmer than the air immediately above it, the rising air will retain its buoyancy and will continue to rise. Its rise will be checked when it reaches a height at which the ELR decreases to less than the SALR or the DALR in the dry air above the cloud top or if no cloud has formed.

These conditions are likeliest to occur on very hot days. Air that is heated strongly from the ground then forms a layer that is at a much higher temperature than the air above it, producing a very steep ELR. Clouds that form in absolutely unstable air are seldom very deep, because the layer of very warm air that causes the instability is usually quite shallow and so the steep ELR does not extend very high. Consequently, absolutely unstable air does not usually produce storms or even showers.

absolute linear momentum *See* ABSOLUTE MOMENTUM.

absolute momentum (**absolute linear momentum**) The sum of the MOMENTUM of a particle in relation to the surface of the Earth and its momentum due to the rotation of the Earth. Both of these values are VECTOR QUANTITIES.

absolute pollen frequency (**APF**) The actual number of pollen grains that are counted in a unit volume of a sediment and, where the rate of deposition is known, the number per unit time. The pollen grains are those of a particular species, genus, or family of plants. This measure is more useful than counting the RELATIVE POLLEN FREQUENCY where several sites are to be compared and different plants at each site produce the most abundant pollen. Measuring the pollen frequency allows the vegetation of a site to be determined at a time in the past that can be dated, for example, by RADIOMETRIC DATING, and the type of vegetation that was present is a good indication of the type of climate prevailing at that time.

absolute stability The condition of air when the ENVIRONMENTAL LAPSE RATE is lower than the SATURATED ADIABATIC LAPSE RATE (SALR). If a PARCEL OF AIR is made to rise, at first it will cool at the DRY ADIABATIC LAPSE RATE (DALR). This is higher than the SALR, and therefore it is also higher than the ELR. The rising air will quickly reach a level at which it is cooler than the surrounding air and so it will sink once more. Even if it is forced to rise high enough for its water vapor to start to condense, for example, by being carried across a mountain, it will still tend to sink. Condensation of its water vapor will release LATENT HEAT of condensation, warming the air and altering its lapse rate from the DALR to the SALR, but this is still greater than the ELR, so the rising air will always be cooler than the air around it.

absolute temperature The temperature measured on the KELVIN SCALE, which has no negative values (*see* ABSOLUTE ZERO). It is reported in KELVINS (K). To convert a temperature on the FAHRENHEIT TEMPERATURE SCALE to the equivalent absolute temperature, $K = [(°F - 32) \times 5 \div 9)] + 273.16$. The absolute temperature at which water freezes (32° F) is 273.16 K and the temperature at which it boils (212° F) is 373.16 K.

absolute vorticity The VORTICITY about a vertical axis that a mass of fluid possesses when it moves in relation to the surface of the Earth. Absolute vorticity is the sum of the PLANETARY VORTICITY (f) and RELATIVE VORTICITY (ζ) and in the absence of friction it remains constant, as a result of the conservation of ANGULAR MOMENTUM. It is the constancy of absolute vorticity that gives rise to ROSSBY WAVES and LEE TROUGHS.

absolute zero The temperature at which the KINETIC ENERGY of atoms and molecules is at a minimum. It is 0 on the KELVIN SCALE and equal to –459.67° F (–273.15° C). It is the lowest temperature possible (and unattainable according to the third law of thermodynamics); its existence was first implied in the work of Jacques CHARLES (*see* GAS LAWS.)

absorbate *See* ABSORPTION.

absorbent *See* ABSORPTION.

absorption A process by which one substance (called the *absorbent*) takes up and retains another (called the *absorbate*) to form a liquid or gaseous solution, or the transfer of energy from electromagnetic radiation to atoms or molecules that it strikes. Certain gases are absorbed by ACTIVATED CARBON, which is used to reduce air pollution. ULTRAVIOLET RADIATION is absorbed by atmospheric gas molecules.

absorption of radiation A response to exposure to electromagnetic radiation in which energy is transmitted to molecules, causing them to vibrate more vigorously or move faster. This converts radiation energy into KINETIC ENERGY, which is dissipated among the surrounding molecules and converted into heat. It is the absorption of solar radiation that warms the surface of the Earth and contact with the warmed surface that warms the atmosphere and drives the GENERAL CIRCULATION. Some radiation is also absorbed by components of the atmosphere. The proportion of radiation that a material absorbs is known as the *absorptivity* of that material, and it is the ratio of the amount of radiation absorbed to the amount that would be absorbed by a BLACKBODY. This is equal to the reciprocal of the value of the ALBEDO of that material.

Most surface materials absorb radiation at all wavelengths, but certain molecules absorb radiation only at particular wavelengths. Water vapor absorbs at 5.3–7.7 µm and beyond 20 µm, carbon dioxide at 13.1–16.9 µm, and ozone at 9.4–9.8 µm, for example.

Below a height of about 44 miles (70 km) materials are in thermodynamic equilibrium and according to KIRCHHOFF'S LAW, they both absorb and emit radiation at the same wavelength. The amount of radiation they emit depends on their EMISSIVITY.

absorption tower A structure in which a solid or liquid absorbs a gas or liquid that is passed through it. The device is commonly used to remove pollutants from a stream of waste gases before these are discharged into the outside air. Sulfur dioxide (SO_2) and sulfur trioxide (SO_3) are absorbed by water to form sulfuric acid (H_2SO_4) that can be recovered, for example. Volatile organic compounds, including xylenes, that are released when ships are loaded with petroleum can also be recovered in absorption towers. This reduces pollution and the xylenes can be used, because they are raw materials for several industrial processes.

absorptivity *See* ABSORPTION OF RADIATION.

ABW *See* ARCTIC BOTTOM WATER.

Ac *See* ALTOCUMULUS.

Accademia del Cimento *See* FERDINAND II and RENALDINI, CARLO.

acceleration A rate of change of speed or VELOCITY that is measured in units of distance multiplied by the square of a unit of time, such as feet per second per second (ft s^{-2}) or meters per second per second (m s^{-2}). For a body that is moving in a straight line and accelerating at a constant rate from a speed u to a speed v, the acceleration (a) is given by $a = (v - u)/t$, where t is the time taken, and $a = (v^2 - u^2)/2s$, where s is the distance covered.

accessory cloud A small cloud that is seen in association with a much larger cloud of one of the cloud genera (*see* CLOUD CLASSIFICATION). The most common accessory clouds are called PILEUS, TUBA, and VELUM.

accidental drought *See* CONTINGENT DROUGHT.

acclimatization An adaptive, physiological response that allows an animal to tolerate a change in the climate of the area in which it lives. In addition to changes in temperature and precipitation, climatic change affects the availability of food and sometimes of nesting sites and materials. The most obvious examples of acclimatization occur as animals adjust to the changing seasons. As temperatures fall with the approach of winter, for example, in many animals the cells produce additional enzymes that help to compensate for the reduced activity of enzymes at low temperatures. Some animals are able to tolerate extremely low temperatures, provided the temperature falls slowly enough for them to adjust. There are insects, such as the parasitic wasp *Bracon cephi,* that convert glycogen to glycerol. This acts as "antifreeze" by lowering the freezing temperature of body fluids. *Bracon cephi* can survive at temperatures below –4° F (–20° C). Certain fish can produce trimethylamine, which has an effect similar to glycerol. Blood plasma in the Greenland or Labrador cod (*Gadus ogac*) freezes at 30.6° F (–0.8° C) in summer, but at 29.1° F (–1.6° C) in winter. Metabolic rate may also increase as the temperature falls. Tolerance of high temperatures also varies seasonally. The brown catfish, also called the brown bullhead (*Ictalurus nebulosus*), found from southern Canada through most of the United States, is likely to die if the water temperature exceeds 96° F (35.8° C) in August. In October it can tolerate only 88° F (31° C), and it may die in winter if it is exposed to temperatures higher than 84° F (29° C). Mammals living in high latitudes grow thicker fur, and in some species, such as the arctic fox (*Alopex lagopus*) and blue hare (*Lepus timidus*), the fur changes color from brown to white. Behavioral changes may also take place, as when animals prepare for and enter hibernation and later emerge from it. Acclimatization also occurs when an animal migrates from one region to another with a different climate. In humans moving to a warmer climate, the rate of sweating increases over several days until a new balance is struck that produces a comfortable level of cooling.

accretion The process by which an ICE CRYSTAL grows as it falls through a cloud containing many small, SUPERCOOLED water droplets. If the water droplets are very supercooled, so their temperature is well below freezing, they freeze immediately on contact with the ice crystal. New crystals are added one on top of another, with air trapped between them. This process produces a loose, spongy structure typical of GRAUPEL. If the water droplets are only slightly supercooled they may not freeze instantaneously. Instead they form a layer of liquid water that surrounds the ice crystal before freezing as clear ice. As more droplets are added the ice crystal grows into a HAILSTONE.

accumulated temperature The sum of the amount by which the air temperature is above or below a particular DATUM LEVEL over an extended period. The datum level is usually set at a value relevant to crop production or an ecological study. If, on a particular day, the mean temperature is m degrees above (or below, in which case it has a negative value) the datum level and it remains so for n hours (= $n/24$ days), then the accumulated temperature for that day is $mn/24$ degree-days. Adding the accumulated temperatures for each day yields the accumulated temperature for a week, month, season, or year. The concept is similar to that of the DAY DEGREE.

accumulation The extent by which the thickness of a layer of snow or ice increases over time through the addition of new snow or ice. It represents the amount of material added, minus the amount lost during the same period through ABLATION.

acicular ice (fibrous ice, satin ice) Ice that forms as long, pointed crystals and hollow tubes of varying shapes, with air held in the tubes and between the crystals. *Acicular* means "needle-shaped."

acid deposition The placing onto surfaces of airborne substances that are more acid than naturally occurring, clean rain, usually as a consequence of pollution from industrial or vehicle emissions. Although ACID RAIN is commonly thought to be the means by which acid deposition occurs, it is not the only one and is the least serious in its effects. DRY DEPOSITION is probably the most harmful type of acid deposition. Acid mist is harmful because it coats all exposed surfaces and keeps them wet, allowing ample time for chemical reactions to take place. Acid snow is harmful for a different reason. Snow accumulates and while it covers surfaces any acid it contains can have little effect, because the surfaces are exposed only to the layer in contact with them and reactions proceed very slowly, if at all, at temperatures below freezing. When the snow melts, however, its acid is released, often fairly slowly, onto surfaces and into the ground, where it can affect soil chemical characteristics.

acidity According to the theory published in 1923 by the Danish physical chemist Johannes Nicolaus Brønsted and the British chemist Thomas Lowry, a measure of the extent to which a substance releases hydrogen IONS (protons) when dissolved in water. Also in 1923, the American theoretical chemist Gilbert Newton Lewis defined acidity as the extent to which a substance acts as receptor for a pair of electrons from a base. The two theories describe different ways of looking at the same thing and do not contradict each other.

Acidity is measured on a scale of 0–14. This scale was introduced in 1909 by the Danish chemist Søren Peter Lauritz Sørensen. The acidity of a solution, measured at 25° C (77° F), is equal to $-\log_{10}c$, where c is the concentration of hydrogen ions in MOLES per liter. The scale measures the "potential of hydrogen," which is abbreviated to pH, so it is known as the pH scale. A neutral (neither acid nor alkaline) solution has a hydrogen-ion concentration of 10^{-7} mol 1^{-1}, so it has a pH of 7. A pH lower than 7 indicates an acid solution and one higher than 7 an alkaline solution. The scale is logarithmic, so a difference of one whole number in pH values indicates a 10-fold difference in acidity. A carbonated soft drink has an acidity of about pH 3, making it 100,000 times more acid than distilled water, pH 7, and ammonia, pH 12, is 100,000 times more alkaline than distilled water.

acid rain Rain that is more acidic than normal as a result of contamination by emissions from such sources as power plants, factories, vehicle exhausts, forest and bush fires, and volcanic eruptions. *Acid rain* is a blanket term that is used to describe the cause of all damage by acid pollution, but damage is less likely from acid rain than from other forms of ACID DEPOSITION. This is because rain runs off surfaces quickly, so they are exposed to the acid for only a very short time.

ACIDITY is measured on a pH scale where pH 7.0 is neutral, values below 7.0 are acid, and values above 7.0 are alkaline. Acid rain has a pH value of less than 5.0.

Ordinary, unpolluted rain has a pH of about 5.6. It is naturally acid, because it contains acid solutions of certain atmospheric gases. CARBON DIOXIDE (CO_2) dissolves into it to produce carbonic acid (H_2CO_3), nitrogen is oxidized by the energy of LIGHTNING and the oxides dissolve to form nitrous (HNO_2) and nitric (HNO_3) acids, and naturally occurring SULFUR DIOXIDE (SO_2) is oxidized and dissolved to form sulfuric acid (H_2SO_4).

Acidification as a result of pollution was first reported in 1852 in an area downwind from the industrial city of Manchester, in northwestern England. Cases were also well documented from copper smelters at Trail, British Columbia, Canada, from 1896 until 1930 and early in the 20th century at Anaconda, Montana.

Acid rain emerged again as a problem in the 1960s. This time, the pollution was experienced not close to its source, as it had been earlier, but over very much larger areas. Earlier attempts to reduce AIR POL-

LUTION from industrial sources had been based on dilution. Smokestacks had been made much taller and were modified to accelerate the emissions so they entered the air traveling at considerable speed. This carried them higher, with the idea that as they drifted downwind they would be greatly diluted by mixing with the surrounding air. Conditions improved near the pollution sources, but the improvement extended no more than 100 miles (160 km) downwind. Today, it is accepted that the contamination causing acid rain can be reduced effectively only by reducing the emissions responsible for it.

Acid rain caused serious erosion to limestone buildings and statues, and it was blamed for widespread damage in the forests of Central Europe, the phenomenon German environmentalists called WALDSTERBEN. It was also associated with the pollution of lakes, especially in Scandinavia.

Damage to buildings certainly occurred, but the effect on plants and lakes was found to be much smaller than had first been reported and the causes of it proved to be much more complicated. Drought and disease also affected forests; the proportion of damaged trees was revised from more than 50 percent to less than 20 percent when the method of measuring it was standardized; and the extent of damage varied with the type of soil. This does not mean there was no problem, however. Although the effect was smaller than had been feared, acid rain certainly contributed to a deterioration in the health of American and European forests.

Acid rain did cause harm, and measures have been taken in most countries to reduce the emissions of nitrogen and sulfur oxides that produce it. Recovery from its effects is slow, but it has begun and will continue provided emissions remain under control.

(For more information about acid rain and its effects on forests see Michael Allaby, *Ecosystem: Temperate Forests* [New York: Facts On File, 1999]. There is a more technical and detailed description in *Air Pollution's Toll on Forests and Crops,* edited by James J. MacKenzie and Mohamed T. El-Ashry [New Haven: Yale University Press, 1989].)

acid smut *See* ACID SOOT.

acid soot (acid smut) Particles of SOOT, approximately 0.04–0.12 in (1–3 mm) in diameter, that are bound together by water that has been acidified. Acidification is due to a reaction between water and sulfur trioxide (SO_3) present in the accompanying waste gases, to form sulfuric acid (H_2SO_4). Acid soot tends to cling to solid surfaces and is corrosive. It is a by-product of the inefficient burning of oil or coal with a high sulfur content.

Ac_{len} *See* LENTICULAR CLOUD.

actiniform An adjective that describes a cloud pattern in which lines of clouds radiate from a central point or branch from one another, like the branches of a tree. Actiniform clouds form by CONVECTION, and the pattern covers an area about 90–150 miles (145–240 km) in diameter. They commonly occur in groups over areas in which subsiding air, chilled from below by cold ocean currents, produces INVERSIONS. Actiniform clouds were recognized only when satellite images became available. These provided views over an area wide enough for the pattern to be seen.

actinometer *See* SUNSHINE RECORDER.

activated carbon (activated charcoal) Carbon that has been treated to make it highly absorbent to gases and some COLLOIDS. The carbon is obtained by heating plant material, lignite (brown coal), bituminous coal, or anthracite in the presence of a solution of a substance such as zinc chloride (ZnCl) or phosphoric acid (H_3PO_4) that dissolves the material and catalyzes the reaction. This process, known as *pyrolysis,* yields pellets of carbon. If the carbon is derived from coal, additional processing is needed. The carbon is then activated by heating it to 1,470°–1,830° F (800°–1,000° C) in a greatly reduced supply of air. It is heated by exposing it to steam or carbon dioxide. The activation process makes the surface of the pellets highly porous. Activated carbon is used in gas masks, in removal of odors from air, and in various devices for reducing air pollution.

activated carbon process A method that was invented in Japan for removing sulfur dioxide (SO_2) from FLUE gases and is now widely used. There are three ways the SO_2 can be removed: water washing, gas DESORPTION, and steam desorption. In the water washing process the gas is passed through ACTIVATED

CARBON and the SO_2 is absorbed. The activated carbon is then washed with water. This removes the SO_2 as sulfuric acid (H_2SO_4) or, if limestone or chalk (both are calcium carbonate, $CaCO_3$) is mixed with the carbon, as gypsum ($CaSO_4 \cdot 2H_2O$). In the gas desorption process the SO_2 is absorbed onto activated carbon and then released (desorbed) as SO_2. The steam desorption process is similar, but in it steam is used to desorb the SO_2.

activated charcoal *See* ACTIVATED CARBON.

active front A weather front (*see* FRONTOGENESIS) that is associated with appreciable amounts of cloud and precipitation. This implies that the RELATIVE HUMIDITY in the air behind the WARM FRONT is high, so cloud forms as it is forced to rise. If the air in the warm sector is unstable and there is a sharp difference in temperature on the sides of the front, a warm front may become extremely active, with CUMULUS and CUMULONIMBUS cloud accompanied by THUNDERSTORMS. This is uncommon and warm fronts are usually associated with STRATIFORM cloud. If the air in the warm sector is dry the warm front may produce no weather and pass unnoticed. It is then described as a "weak" front.

active glacier A GLACIER in which the ice is flowing. The mechanisms by which the ice flows vary, depending on whether the glacier is COLD, TEMPERATE, or COMPOSITE.

active instrument An instrument that sends out a signal that is reflected back to it. Instruments that use RADAR and LIDAR are active.

active layer The soil that lies above a layer of PERMAFROST and that thaws during the summer and freezes again in winter.

actual elevation The vertical distance between sea level and a weather station.

actual evapotranspiration (AE) The amount of water that is lost from the ground surface monthly by the combined effects of EVAPORATION and TRANSPIRATION. In the THORNTHWAITE CLIMATE CLASSIFICATION this can be compared with the POTENTIAL EVAPOTRANSPIRATION (PE) to determine the amount of the water surplus or deficit present in the soil. This provides farmers with a guide to the need for irrigation.

actual pressure The pressure recorded by a BAROMETER after it has been corrected for temperature, latitude, and any instrumental error, but before the reading has been reduced to the mean sea level pressure.

ACW *See* ANTARCTIC CIRCUMPOLAR WAVE.

adfreezing The process by which two objects stick to one another as the result of the freezing of a layer of water between them. The word is derived from the words *adhesion* and *freezing*. Objects can become frozen to the ground with a degree of firmness that is known as the *adfreezing strength*.

adhesion *See* SOIL MOISTURE.

adhesion tension *See* SURFACE TENSION.

adiabat The rate at which a PARCEL OF AIR cools as it rises (and warms as it descends). It is shown on a TEPHIGRAM as two lines, one representing the dry adiabat and the other the saturated adiabat. The dry adiabat is also a line of constant POTENTIAL TEMPERATURE (an isentrope).

adiabatic The adjective that describes a change of temperature that involves no addition or subtraction of heat from an external source. The word is from the Greek *adiabatos,* which means "impassable," suggesting that the substance in which adiabatic temperature changes occur is isolated from its surroundings.

Air close to the ground is subject to DIABATIC TEMPERATURE CHANGE, but air that is above this surface layer and moving vertically warms and cools adiabatically. The phenomenon is simple to demonstrate. When a bicycle tire is inflated vigorously by a hand pump the barrel of the pump and the valve on the tire become warm. This is because the air is being compressed inside the barrel and at the valve, and when air is compressed its temperature increases. If the valve of a bicycle tire is released, the air that rushes out feels cool. This is because the air is expanding and when air expands its temperature decreases. The air in the pump and tire has not been warmed or cooled from outside—the temperature change is adiabatic.

Adiabatic temperature change is a version of the first law of THERMODYNAMICS that can be stated as

temperature change = pressure change × a constant

Compression means a given number of molecules are forced to occupy a smaller volume. The molecules are packed more closely together as a result of external pressure. Energy is required to compress air—you must do work to pump up a tire—and some of this energy is absorbed by the molecules. Having more energy, they move faster and collide with one another more violently and, because they are closer together, more often. This change in the behavior of the molecules is what we measure as TEMPERATURE.

When the substance expands, its molecules expend energy in pushing each other aside so each of them occupies a larger volume. As they lose energy they slow down. Collisions between them become less frequent and less violent. This change is measured as a drop in temperature.

Air experiences large adiabatic temperature changes because it is very compressible and it is also a poor conductor of heat. AIR PRESSURE decreases with height, so a PARCEL OF AIR that moves vertically experiences a constant change in pressure and its compressibility allows it to expand or contract accordingly. Because it is a poor conductor of heat, there is little exchange of heat between a moving parcel of air and the larger body of air through which it passes, so the parcel tends to retain its thermal characteristics.

The rate at which the temperature of air changes adiabatically with height is a constant, known as the LAPSE RATE.

All fluids are subject to adiabatic temperature changes, but those in liquids are very much smaller than those in gases, because gases are very much more compressible than liquids. In the oceans, the adiabatic temperature change below the surface layer of well mixed water is usually less than 0.1° F for every 1,000 feet (0.2° C per kilometer).

adiabatic atmosphere A theoretical atmosphere in which the temperature decreases at the DRY ADIABATIC LAPSE RATE throughout the whole of its vertical extent.

adiabatic chart *See* STÜVE CHART.

adiabatic condensation temperature *See* CONDENSATION TEMPERATURE.

adiabatic equilibrium *See* CONVECTIVE EQUILIBRIUM.

adiabatic saturation temperature *See* CONDENSATION TEMPERATURE.

adret Sloping ground that faces in the direction of the equator and therefore is sunny. In the Northern Hemisphere an adret is a south-facing slope. The opposite slope is called an UBAC.

adsorbate *See* ADSORPTION.

adsorbent *See* ADSORPTION.

adsorption The chemical or physical bonding of molecules to the surface of a solid object or, less commonly, of a liquid. The adsorbed molecules form a layer on the surface. If they are attached by chemical bonds the process is known as *chemisorption;* if they are held physically, by van der Waals' forces, the process is *physisorption.* The substance that is absorbed is the *adsorbate* and the substance holding it is the *adsorbent.* Adsorbents are often used to remove pollutants from industrial waste gases.

Advanced Very High Resolution Radiometer (AVHRR) An instrument carried by weather satellites that senses clouds and surface temperatures. It stores its data on magnetic tape and transmits it on command to surface receiving stations. It also transmits both low- and high-resolution images in real time. The first AVHRR was launched in October 1978 on the *TIROS-N* satellite. It transmitted on four channels. Other four-channel AVHRRs were carried on *NOAA 6* and other even-numbered satellites in the NOAA series. The first five-channel AVHRR was launched in June 1981 on *NOAA 7* and others on subsequent odd-numbered NOAA satellites. The five channels comprise 0.58–0.68 μm (visible part of the spectrum); 0.725–1.10 μm (near-infrared); 3.55–3.93 μm (intermediate infrared); 10.3–11.3 μm and 11.5–12.5 μm (thermal infrared on *NOAA 7* and *9*); and 10.5–11.5 μm and 11.5–12.5 μm (thermal infrared on *NOAA 11*).

advection A change in temperature that is caused by the movement, usually horizontal, of air or water. A warm breeze that raises the temperature on what had been a cool day is an example of a heat transfer by advection and the movement of warm air over cold ground can produce ADVECTION FOG. Winds of the FÖHN type also transfer heat by advection. The transfer of heat by warm and cool ocean currents is also an example of advection, in which the GULF STREAM and KUROSHIO CURRENT are especially important.

In addition to the transfer of sensible heat—heat that can be felt and measured as a change in temperature—latent heat can be transported by advection. Water vapor that condenses out of warm air that is chilled by crossing a cold surface releases the latent heat of condensation.

As air moves horizontally its characteristics are modified by the surfaces with which it comes into contact. These advective effects on the moving air are of three types, known as the *clothesline effect, leading-edge effect,* and *oasis effect.*

The clothesline effect occurs when warm, dry air enters and flows through vegetation, such as a forest or farm crop. Near the edge the moving air raises the temperature and the rate of evaporation. This has a drying effect on the soil. Farther into the vegetation stand, the air cools, raising its RELATIVE HUMIDITY.

As moving air encounters new surface conditions, the air that is in immediate contact with the surface is affected, but the air behind or above this boundary layer is not. This produces a leading-edge effect, in which the altered boundary layer spreads downwind with only its lower part fully adjusted to the new conditions. Above this lowest layer the air is partly changed by the new conditions, but above the boundary layer the characteristics of the air are determined by the air above the moving air, not by the surface below. Because the leading edge extends downwind over a distance or FETCH, this is sometimes called the *fetch effect.* A leading-edge effect always occurs where air moves from one surface to another that is markedly different, such as between land and water or dry and irrigated farmland.

The oasis effect occurs because moist ground is always cooler than adjacent dry ground, a phenomenon that is most clearly observed in a desert oasis. Water evaporates into the dry air, absorbing the latent heat of vaporization from the surface and thus cooling it. Although this is obviously the case in an oasis, it also occurs elsewhere, for example, where irrigated cropland is adjacent to unirrigated dry ground, in a city park surrounded by streets and buildings, or over a lake in a dry region. Air moving across the warmer, dry ground is warmed by it and carries that warmth over the cooler surface, warming it by advection. Downwind of the "oasis," the air is moister and cooler and cools the ground it crosses, completing the oasis effect.

advectional inversion A temperature INVERSION that is produced when cold air moves by ADVECTION across a warm surface, undercutting air that had previously been warmed by contact with the surface. The warm air then lies above the cold air.

advection fog FOG that forms when warm, moist air is carried horizontally across a cold surface by a wind of about 6–20 mph (10–32 km h^{-1}). This can happen in summer, when the land has been warming through the day much faster than water nearby and air moves over both. It also happens when air crossing relatively warm sea encounters a current carrying cold water. Air crossing the Pacific Ocean encounters the cold CALIFORNIA CURRENT off Cape Disappointment, Washington, making this the foggiest place in the United States. The warm, moist air is chilled when it comes into contact with the cold surface. This increases its RELATIVE HUMIDITY to 100 percent and its water vapor starts to condense, producing the fog. Turbulence in the wind carrying the air causes mixing. This carries cool air to a greater height, thus extending the height at which fog forms, and it also carries the fog itself to greater heights. Advection fogs can be 2,000 feet (610 m) deep.

advective thunderstorm A THUNDERSTORM that is triggered by the ADVECTION of warm air across a cold surface or of cold air above a layer of warmer air at a high level. Warm air is cooled when it crosses a cold surface. This causes its water vapor to condense, releasing LATENT HEAT. This may make the air sufficiently unstable to produce CUMULONIMBUS cloud. Cold air moving above warm air may also cause INSTABILITY as it sinks beneath the warmer air, thus raising it and causing the warm air to cool ADIABATICALLY and its water vapor to condense.

aeolian *See* EOLIAN.

aerial plankton BACTERIA, SPORES, and other minute organisms that are blown from the ground, carried aloft by rising air currents, and can be transported long distances.

aeroallergen An airborne particle or substance to which sensitive people are allergic. Aeroallergens make such persons sick, and therefore they benefit from the regular monitoring and reporting of allergen concentrations. These vary according to the weather conditions. Rain washes aeroallergens from the air, so concentrations increase in dry weather. They also increase when air is trapped beneath an INVERSION. The principal allergens are plant POLLEN grains. These are released when the wind-pollinated source plants are in flower and are absent at other times of the year. Those pollens causing the majority of adverse reactions are from grasses (family Poaceae), ragweed (*Artemisia* species), hazel (*Corylus* species), cypress (*Cupressus* species), alder (*Alnus* species), birch (*Betula* species), and hornbeam (*Carpinus* species). Fungal spores, most of which are released in the fall, can also cause allergic reactions.

(You can learn more about aeroallergens from www.allernet.com/default.asp and www.usatoday.com/weather/health/pollen.wpusa.htm. The Italian Association of Aerobiology coordinates an Italian Aeroallergen Network. You can learn about that at www.isao.bo.cnr.it/~aerobio/aia/AIANET.html.)

aerobiology The scientific study of airborne particles that are of biological origin. It includes studying the sources of such particles, the way they disperse, the distances they travel and time they remain aloft, and the surfaces on which they are deposited. Interest in the subject began in the 1960s, and there are now national aerobiological associations, most of which are affiliated to an international organization.

(You can find out about the Pan-American Aerobiology Association at www.geocities.com/ResearchTriangle/Campus/9792/PAAA.html and about the International Association for Aerobiology at www.isao.bo.cnr.it/~aerobio/iaa/.)

aerodynamic roughness Irregularities in a surface that impede the passage of air and significantly reduce the wind speed. Close to the surface the size, shape, and distribution of the irregularities determine the wind speed. With the exception of water when there are no waves and very little wind, all surfaces are aerodynamically rough to a greater or lesser extent.

Surface roughness reduces the wind speed from the surface up to a height equal to one to three times the height of the projecting elements causing the roughness; the magnitude of the effect decreases with height. If the reduction in wind speed from the top of the affected BOUNDARY LAYER is continued downward, a height is reached at which the wind speed is reduced to zero. This is known as the *roughness length* (usually represented as z_0). Provided the projecting elements do not bend in the wind and thereby reduce the friction, z_0 can be calculated from the height of the elements by

$$\log z_0 = a + b \log h$$

where h is the height of the elements in centimeters and a and b are constants. Estimates of the value of the constants vary, but commonly used values are $a = -1.385$ and $b = 1.417$. Using these values, in a fir forest where the trees are 555 cm (18 feet) tall, z_0 = 283 cm (9.3 feet); in a large city (in fact Tokyo), z_0 = 165 cm (5.4 feet); in grass about 6 cm (2.4 inches) tall z_0 = 0.75 cm (0.3 inch); and over a tarmac surface z_0 = 0.002 cm (0.0008 inch).

aerography *See* DESCRIPTIVE METEOROLOGY.

aerological diagram A THERMODYNAMIC DIAGRAM on which data from soundings of the upper atmosphere are plotted. The diagram usually shows ISOBARS, ISOTHERMS, and dry and saturated ADIABATS.

aerology The scientific study of the free atmosphere throughout its vertical extent, with particular reference to the chemical and physical reactions that occur at particular levels within it. Aerology forms one part of METEOROLOGY, and the two words are sometimes used synonymously, but unlike meteorology, aerology is not confined to studies of the lower atmosphere, where meteorological phenomena occur.

aeronomy The scientific study of the atmosphere and of the changes that occur within it as a consequence of internal or external influences. The study embraces the composition of the atmosphere, relative movements of air within it, the transport of energy, and the radiant energy that powers these processes. Aeronomical findings, especially those from the middle and upper atmosphere, are also applicable to the atmospheres of other planets and solar-system satellites.

aerosol A mixture of solid or liquid particles that are suspended in the air. Strictly speaking, a cloud, comprising water droplets, ice crystals, or a mixture of them suspended in air, is an aerosol. The word is more usually applied to solid particles, however. Aerosol particles are so small that gravity has little effect on them. They fall naturally at about 4 inches (10 cm) a day but are removed much more quickly by being washed from the air by rain or snow. They consist of soil particles; dust, some of which enters from space; salt crystals from the evaporation of water from drops of sea spray; smoke; AERIAL PLANKTON; and organic substances. The concentration varies from about 1,500 to several million particles per cubic inch (10 per centimeter). The total mass of them in the column of air resting on one square yard (meter) of the Earth's surface is more than 10 million times smaller than the mass of the air itself. Aerosol particles range in size from around one-thousandth of a micrometer (10^{-3} μm) to about 10 μm; those between 0.1 μm and 10 μm are considered large. Particles less than 0.1 μm in size are known as *AITKEN NUCLEI*.

aerospace The whole of the atmosphere together with the adjacent region of space in which satellites orbit.

aerovane An instrument that measures both wind speed and WIND DIRECTION. It therefore combines the functions of a WIND VANE and an ANEMOMETER. The aerovane has a tapering body with two large fins at one end and a four-bladed propeller at the other. It is mounted horizontally on top of a vertical column that raises it to the standard height of 33 feet (10 m) above the ground or roof, if it is mounted on the roof of a building. It is free to turn on its column. The fins hold the propeller so it faces into the wind, thereby indicating the wind direction, and the speed at which the propeller spins indicates the wind speed. Both readings are converted into electrical impulses and are shown on dials inside the meteorological office.

aestivation *See* ESTIVATION.

African wave *See* EASTERLY WAVE.

Aerovane. **The fins hold the propeller so it always faces into the wind. This indicates the wind direction. The speed of the propeller is converted into the wind speed.**

afterglow A bright arch that appears in the west above the highest clouds just after sunset. It is caused by the SCATTERING of light by dust particles suspended in the upper TROPOSPHERE.

Aftonian interglacial An INTERGLACIAL period that occurred in North America after the end of the NEBRASKAN GLACIAL and that was followed by the KANSAN GLACIAL. The Aftonian began about 600,000 years ago and ended about 480,000 years ago. It is approximately equivalent to the DONAU–GÜNZ INTERGLACIAL of the European Alps. During the Aftonian, summers were milder and winters warmer than they are in North America today.

Agassiz, Jean Louis Rodolphe (1807–1873) Swiss–American *Naturalist* Louis Agassiz was born on May 28, 1807, at Motier, not far from Friborg, on the shore of Lake Morat, in Switzerland, where his father was the Protestant pastor. The family was originally French but was forced to flee from France after Louis XIV revoked the Edict of Nantes in 1685 and Protestants were no longer tolerated.

Louis's mother, Rose Mayor, taught him to love the natural world. His formal education began with four years at the gymnasium (high school) in Bienne (Biel), northwest of Bern, after which he attended a school in Lausanne. In 1824 he enrolled at the University of Zürich and moved from there to the University of Heidelberg, Germany, in 1826. He caught typhoid fever in

Jean Louis Rodolphe Agassiz.* The Swiss-American naturalist who discovered the existence of former ice ages. *(From the collections of the Library of Congress)

Heidelberg and had to return to Switzerland to recuperate. In 1827 he enrolled at the University of Munich, Germany. He qualified as a doctor of philosophy (Ph.D) at the University of Erlangen in 1829 and as a doctor of medicine at Munich in 1830.

Two distinguished naturalists from Munich, J. B. Spix and C. P. J. von Martius, who had spent 1819 and 1820 touring Brazil, returned to Germany with a large collection of fishes, most of them from the Amazon River. Spix set about classifying the collection, but in 1826 he died and Martius handed the task over to Agassiz. He completed the classification and it was published in 1829, when Agassiz was only 22 years old, as *Selecta Genera et Species Piscum* (Selection of fish genera and species). This set the course for his major research.

In November 1831 Agassiz continued his studies of fishes in Paris for a short time at the Natural History Museum under the supervision of Georges Cuvier (1769–1832), the eminent comparative anatomist. Cuvier, who had read and been greatly impressed by Agassiz's work on the Brazilian fishes, befriended the young man. ALEXANDER VON HUMBOLDT also helped him. After the death of Cuvier in May 1832, Agassiz moved to the University of Neuchâtel, Switzerland, where von Humboldt helped him secure the professorship of natural history. Between 1833 and 1844 Agassiz published *Recherches sur les poissons fossiles* (Research on fossil fish), in which he classified more than seventeen hundred species. He also published major works on fossil echinoderms and mollusks.

In 1836 his attention turned to a new question. Boulders that were scattered over the plain of eastern France and in the Jura Mountains were different in composition from the solid rock beneath the ground on which they lay. Some scientists thought these "erratics" might have been transported to their present positions by GLACIERS. If glaciers can push boulders ahead of them, it means they flow, and if they pushed boulders deep into France, it means that at one time the Swiss glaciers must have extended much farther than they do today.

With some friends, Agassiz spent his 1836 and 1837 summer vacations on the Aar Glacier. They built a hut on the ice and called it the "Hôtel des Neuchâtelois." From this base they observed the rocks piled to the sides and at the ends of this and other glaciers and the grooves that appeared to have been made by scouring as harder stones were dragged past them. Convinced that in the geologically recent past all of Switzerland had been covered by ice and that all those parts of Europe where erratic boulders and gravel were found had also lain beneath a great sheet of ice resembling that in Greenland, in 1840 he published the most important of all his works, *Études sur les glaciers* (Studies of glaciers).

Agassiz continued his investigations; in 1839 he found that a hut that had been built on the ice in 1827 had moved a mile (1.6 km) from its original position. He drove a line of stakes into the ice across a glacier and found that by 1841 they had moved and the straight line had changed to a U shape, because the stakes at the center had moved faster than the ones at the sides.

In 1846, with the help of a grant from King Friedrich Wilhelm IV of Prussia, Agassiz visited the United States, partly to continue his studies but immediately to deliver a series of lectures at the Lowell Institute in Boston. He followed these with other popular and technical lectures in various cities. The lectures were popular and he extended his stay, studying North American natural history at the same time. In 1848 he was appointed professor of zoology at Harvard University. Agassiz became an American citizen and remained in the United States for the rest of his life, most of the time at Harvard.

He found evidence that North America had also been covered by ice, and he traced the shoreline of a vast, vanished lake that once covered North Dakota, Minnesota, and Manitoba. It is now called *Lake Agassiz*. Agassiz clearly demonstrated that over Europe and North America there had once been what he called a *Great Ice Age*.

At Harvard in 1858 Louis Agassiz developed the Museum of Comparative Zoology to assist research and teaching. It was built around Agassiz's own collection, and he was its director from 1859 until his death. His scientific research was of great importance, but he was also one of the finest teachers of science America has ever known. He was devoted to his students and treated them as collaborators. During the second half of the 19th century every well-known and successful teacher of natural history in the country had at one time been a pupil either of Agassiz himself or of one of his former students.

Despite being one of the most knowledgeable biologists of his time, Agassiz remained steadfastly opposed to the Darwinian concept of evolution by natural selection.

He married twice. Cecile Braun, his first wife, died in 1848 in Baden, a few months after Agassiz had taken up his position at Harvard. In 1850 he married Elizabeth C. Cary, who became his valued scientific assistant.

Louis Agassiz died at Cambridge, Massachusetts, on December 12, 1873. He is buried at Mount Auburn, Cambridge, where his grave is marked by a boulder from the Aar glacial moraine. In 1915 Agassiz was elected to the Hall of Fame for Great Americans.

(You can read more about Louis Agassiz at research.amnh.org/ichthyology/neoich/collectors/agassiz.html, www.nceas.ucsb.edu/~alroy/lefa/Lagassiz.html, and at www.uinta6.k12.wy.us/WWW/MS/8grade/Info%20Access/SPANTLGY/agassiz .htm.)

age-of-air The time that elapses between the appearance of a given amount of a GREENHOUSE GAS in the TROPOSPHERE and the appearance of a similar abundance of the same gas in the STRATOSPHERE.

ageostrophic wind A wind that blows above the BOUNDARY LAYER at a speed that differs from that of the GEOSTROPHIC WIND predicted by the PRESSURE GRADIENT.

Agnes A HURRICANE, rated category 1 on the SAFFIR/SIMPSON SCALE, that struck Florida and New England in 1972; it caused $2.1 billion of damage. Agnes is also the name of two TYPHOONS. The first struck South Korea on September 1, 1981; it delivered 28 inches (711 mm) of rain in two days and left 120 people either dead or missing. The second Typhoon Agnes, rated category 5 on the Saffir/Simpson scale, struck the central Philippines in November 1984; it generated winds of 185 mph (297 km h^{-1}). At least 300 people were killed and 100,000 rendered homeless. The damage was estimated at $40 million.

agricultural drought A DROUGHT that causes a decrease in agricultural production. This is not the same as a METEOROLOGICAL DROUGHT.

agroclimatology The study of the ways in which agriculture is affected by climate.

agrometeorology The study of weather systems in terms of their effects on farming and horticulture and the provision of meteorological services for farmers and growers. Agricultural and horticultural weather forecasts provide information on the likely effect of forthcoming weather on particular crops.

Agulhas Current A current that flows in a southwesterly direction at the surface of the Indian Ocean, between the eastern coast of Africa and Madagascar, between latitudes 25° S and 40° S. Its speed is 0.7–2.0 feet per second (0.2–0.6 m s^{-1}).

air The mixture of gases that compose the atmosphere and in which solid and liquid particles are suspended. In ancient times air was thought to be one of the four "elements" (the others being fire, water, and earth) from which all substances are composed. It is not an element or even a compound in the modern sense, however, but a mixture of elements and com-

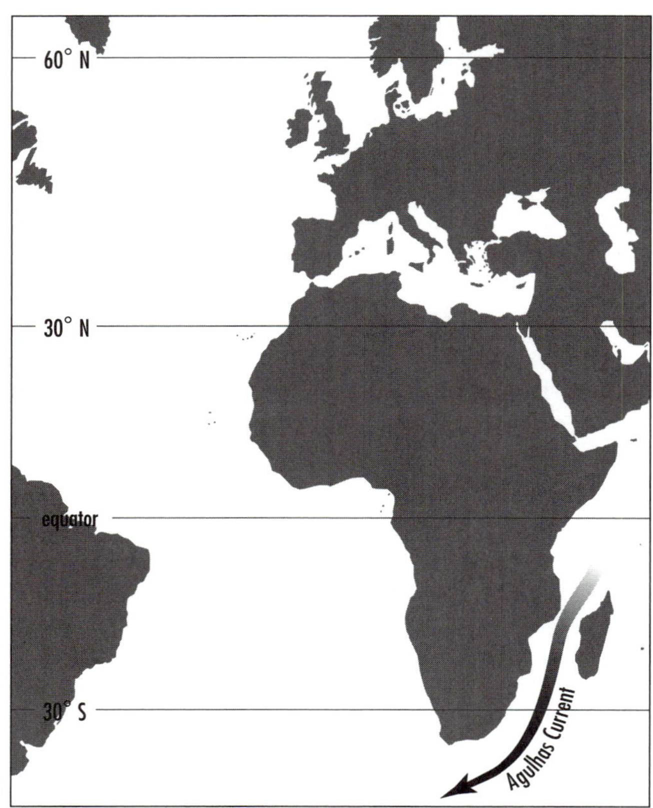

Agulhas Current. **An ocean current that flows in a southwesterly direction, parallel to the coast of East Africa.**

pounds, the relative proportions of which have changed over long periods (*see* ATMOSPHERIC COMPOSITION).

airborne dust analysis The sampling and categorization of dust particles that are seized from the air. Particles may be taken directly from the air or sucked from the surface of vegetation. They are then dried and graded according to their size, usually by shaking them through a series of increasingly fine sieves. They may also be classified by measuring their FALL SPEEDS.

aircraft ceiling The height of the CLOUD BASE when this has been measured by the pilot of an aircraft flying within 1.5 nautical miles (1.725 miles, 2.78 km) of the runway of the airport to which the ceiling refers. Airfield controllers often ask pilots to report the cloud ceiling in the vicinity of the field. Its ceiling is also the greatest altitude to which a particular type of aircraft is capable of climbing.

aircraft electrification The accumulation of an electric charge on the surface of an aircraft, or the separation of a surface electric charge into charges of opposite sign on different parts of the aircraft.

aircraft thermometry The measurement of air temperature by instruments that are mounted on aircraft.

air drainage The downhill flow of cold air under the influence of gravity. A drainage wind is also known as a *KATABATIC WIND*.

air frost The condition in which the air temperature is below freezing. Unlike GROUND FROST, air frost can cause damage to plants.

airglow (light of the night sky, night-sky light) A faint light that glows permanently in the night sky and is most clearly seen in middle and low latitudes. It is caused by the emission of light from molecules and atoms of oxygen, nitrogen, and sodium (from sea salt) that have absorbed photons from sunlight during the day, raising them to higher energy states from which they fall back at night, emitting photons as they do so.

airlight Light that is scattered toward an observer by AEROSOLS or air molecules lying between the observer and more distant objects. This light is visual NOISE that makes the more distant objects less clearly visible, thereby reducing VISIBILITY. The appearance of a cloudless sky in daytime is due entirely to airlight. At dawn the amount of airlight increases, and it is this that obscures the stars, rendering them invisible. At sunset, as the amount of airlight diminishes, the stars reappear.

air mass A body of air that covers a very large area of the Earth's surface and throughout which the physical characteristics of temperature, humidity, and LAPSE RATE are approximately constant at every height. The CONSTANT PRESSURE SURFACES at any height correspond to the ISOSTERIC SURFACES, and in a vertical section through the air mass the ISOBARS and ISOTHERMS are parallel at every height. Typically, an air mass covers a substantial part of a continent or ocean and extends from the surface to the TROPOPAUSE.

During the First World War, VILHELM BJERKNES and his colleagues at the Bergen Geophysical Institute studied meteorological data that were sent to them by observers located all over Scandinavia. When they plotted the distribution of temperature and humidity, the pattern that emerged showed that these atmospheric characteristics remained constant over large areas. They coined the term *air mass* to describe such homogenous bodies of air and *front* to describe the boundary between one air mass and another.

Air masses move, driven by the prevailing winds. In middle latitudes they travel from west to east. They are so large that they often take several days to pass a point on the surface. During the time it takes for an air mass to pass, the weather remains more or less unchanged except for local phenomena such as SHOWERS and THUNDERSTORMS. Constant weather associated with an air mass is known as *air mass weather.*

As an air mass travels, its characteristics are modified by the surface beneath it. Dry air accumulates moisture as it crosses an ocean, moist air loses moisture as it crosses a continent, and the temperature of the air changes as it crosses an extensive surface that is warmer or cooler. In this way one type of air mass is transformed into another type, with quite different characteristics. Air masses form in particular areas, called SOURCE REGIONS, and change by moving from one source region to another.

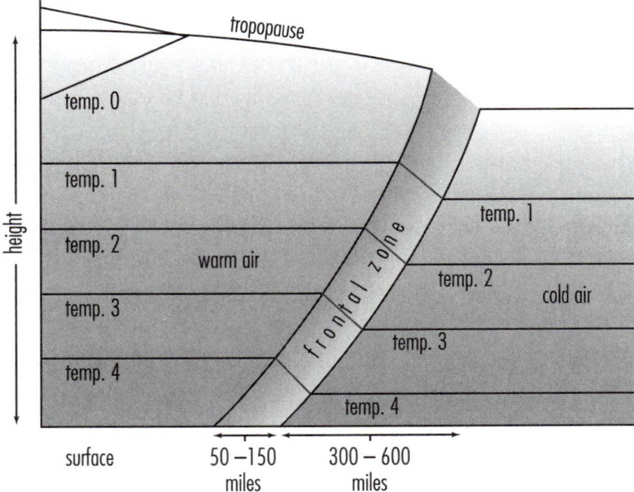

Air masses. **Two air masses, one warm and the other cold, are separated by a frontal zone. Throughout each air mass the temperature, humidity, pressure, and density of the air are constant.**

An air mass may be modified by contact with a surface, but not to such an extent that its characteristics change radically. If a mass of cold air crosses a warmer surface, for example, the air becomes unstable (*see* STABILITY OF AIR), and if warm air crosses a colder surface, it becomes more stable. Such changes as these produce secondary air masses that differ in only minor ways from the primary air masses that gave rise to them.

Types of air masses are classified according to the source regions in which they originate. The first division separates air masses into two types: CONTINENTAL AIR and MARITIME AIR. These are designated by the letters *c* and *m,* respectively. Air masses are further classified as ARCTIC AIR (A), POLAR AIR (P), TROPICAL AIR (T), and EQUATORIAL AIR (E). These are then combined to produce the six types of air mass:

continental arctic (cA)
continental polar (cP)
continental tropical (cT)
maritime tropical (mT)
maritime polar (mP)
maritime arctic (mA)
maritime equatorial (mE)

Combining the basic types can also produce continental equatorial air. This is not included, however, because most of the equatorial region is covered by ocean and continental equatorial air never occurs. MONSOON air is sometimes indicated separately, but its characteristics are no different from those of mT air.

Additional letters are sometimes used to designate secondary air masses. These indicate that the air is colder (k) or warmer (w) than the surface over which it is passing. If mT air crosses a continent in winter, for example, it is likely to be warmer than the surface, so it might be designated mTw. In summer, when the continental surface heats strongly, the mT air would be mTk. The designation *k* suggests air that produces gusty winds that quickly clear away air pollutants to bring clear, clean air. The designation *w* suggests very stable air, often with INVERSIONS that trap pollutants.

Air masses of the cA, cP, mP, cT, and mT types are responsible for North American weather.

air mass analysis A technique in weather forecasting that involves relating the characteristics of the AIR MASSES over a large area to the surface conditions illustrated on a SYNOPTIC CHART. The surface chart is studied in conjunction with a number of other charts and graphs, including vertical cross sections of the TROPOSPHERE, charts showing the winds at different heights, CONSTANT-PRESSURE CHARTS, and STÜVE CHARTS. The aim is to build up the most complete picture of the air masses possible in order to improve the reliability of predictions of their future behavior.

air mass climatology A type of SYNOPTIC CLIMATOLOGY in which the weather characteristic of a region is related to the characteristics of the AIR MASSES that affect it and the length of time each type of air mass remains over the region.

air mass modification The changes that take place in the characteristics of an AIR MASS as it moves away from its SOURCE REGION. For example, CONTINENTAL AIR, which is very dry, gathers moisture as it crosses the ocean and gradually becomes modified until it is MARITIME AIR. Air masses are modified by being heated or cooled by their passage over warmer or cooler surfaces. Heating from below tends to make the air unstable (*see* STABILITY OF AIR); cooling from below tends to make it more stable. EVAPORATION of water into the air and PRECIPITATION falling through an air mass from an air mass of a different type located above it on the upper side of a FRONTAL SLOPE absorb LATENT HEAT, which alters the LAPSE RATE. The CONVERGENCE of AIR STREAMS and OROGRAPHIC LIFTING also modify the characteristics of air.

air mass precipitation Precipitation that is wholly caused by the distribution of temperature and moisture within an AIR MASS and not by OROGRAPHIC LIFTING or the lifting of air in a FRONTAL SYSTEM.

air mass shower A SHOWER that falls from a convective cloud, such as CUMULUS or CUMULONIMBUS in unstable air (*see* STABILITY OF AIR).

air mass thunderstorm A THUNDERSTORM that results from CONVECTION in unstable air (*see* STABILITY OF AIR). The CUMULONIMBUS cloud producing the storm rises vertically and probably lacks an anvil (*see* INCUS) because there is no change in wind speed with height. The storm is produced entirely by conditions pertaining to the AIR MASS. This type of storm is frequent in the TROPICS. It is also the type of storm that occurs in middle latitudes late on summer afternoons.

air mass weather *See* AIR MASS.

air parcel *See* PARCEL OF AIR.

airpath *See* OPTICAL AIR MASS.

air pocket A downdraft that causes an airplane to drop briefly but suddenly with a motion that feels to the passengers like the descent of a fast elevator. In the early days of aviation it was supposed that the aircraft fell when it entered air that was insufficiently dense to support it. The air was likened to an empty pocket.

air pollution The release into the air of gases or AEROSOLS in amounts that may cause injury to living organisms. Certain pollutants can harm humans.

Pollution is not a new phenomenon. In the Middle Ages London air was so polluted by smoke from coal fires that in 1273 Edward I passed a law banning coal burning in an attempt to curb smoke emissions. In 1306 a Londoner was tried and executed for breaking this law. Despite this, pollution was not checked, and on one occasion in 1578 Elizabeth I refused to enter London because there was so much smoke in the air. Smoke killed vegetation and ruined clothes, and the acid in it corroded buildings.

Coal burning remained the most serious source of pollution until modern times. It caused the MEUSE VALLEY INCIDENT in 1930; a severe pollution episode at DONORA, PENNSYLVANIA, in 1948; and the LONDON SMOG INCIDENTS a few years later. These led to the introduction of legislation in many countries to reduce smoke emissions.

Certain products of combustion increase the acidity of precipitation, and some acids can be deposited on surfaces directly, from dry air (*see* ACID DEPOSITION). This causes the type of pollution known as ACID RAIN, which was first reported in 1852. The burning of those types of coal and oil that contain sulfur is discouraged in order to reduce the problems caused by acid deposition and acid rain.

PHOTOCHEMICAL SMOG occurs naturally in some rural areas, but it becomes a pollution problem where traffic fumes become trapped by an INVERSION and the sunlight is very intense. OZONE is one product of the chemical reactions among exhaust emissions that are driven by strong sunlight. It causes severe respiratory irritation in quite small concentrations. Where fuel is not completely burned, the oxidation of carbon remains incomplete and CARBON MONOXIDE is released. This is poisonous at high doses. LEAD POLLUTION, caused by the addition of tetraethyl lead to gasoline, is now decreasing as the use of lead in fuel is phased out.

At present, pollution from vehicle emissions can be dealt with only by improving the efficiency with which vehicle engines burn fuel and by reducing traffic density at critical times. A more effective remedy for the longer term will be the widespread introduction of new types of vehicle propulsion systems that do not burn gasoline or diesel. Diesel engines also emit very fine PARTICULATE MATTER that is believed to cause damage to lung tissue.

Other pollutants are not directly poisonous to any living organism and until recently were not considered to be pollutants at all. Their effects are subtle.

CARBON DIOXIDE is produced whenever a carbon-based fuel is burned, because combustion is the oxidation of carbon to carbon dioxide with the release of heat energy. Carbon dioxide is a natural constituent of the atmosphere, but its increasing concentration, which is believed to be due to the burning of FOSSIL FUELS, is suspected of causing GLOBAL WARMING. Methane, released when bacteria break down organic material, is also harmless in itself but implicated in undesired change as a GREENHOUSE GAS.

CFCs were introduced because they are so chemically inert that they are completely nontoxic (you can safely drink them and even inhale some of them) and nonflammable. Then it was found that they are broken down in the STRATOSPHERE by the action of sunlight, releasing chlorine that depletes the ozone in the OZONE LAYER.

Our understanding of air pollution has increased rapidly as scientists have learned more about the chemical characteristics of the atmosphere. At the same time, steps have been taken in many countries to reduce pollution. The air over the industrial cities of North America and the European Union is much cleaner now than it was half a century ago. Today the task facing us is to promote and encourage the economic and industrial development of the less-developed countries without reducing the quality of the air their people breathe.

air pressure The force that is exerted by the weight of overlying air. At sea level this averages 14.7 pounds per square inch (1 kg cm^{-2} or 1 BAR). Because air pressure is caused by gravity, it acts directly downward, so at sea level a surface area of 18 square inches (116 cm^2), for example, which is approximately the area of a hand, experiences a pressure of about 265 pounds (120 kg). The hand is not crushed, however, partly because of its structure, which can withstand the pressure, and partly because of the way the force is applied. Air pressure is exerted by the movement of air molecules. The more molecules there are, the greater the pressure they exert. You can picture them as rubber balls being thrown hard against a surface and bouncing back. Molecules move in all directions, however, so there are as many of them moving upward and sideways as there are moving downward. Consequently, the pressure applied to one side of the hand is precisely balanced by the pressure applied to the other side and to the edges. The true magnitude of air pressure can be seen when two surfaces are joined so closely that no air remains between them. It is then difficult to separate them. Air pressure varies with the temperature and density of the air, which are related to each other by the GAS LAWS. Air pressure decreases with height, because the distance to the top of the atmosphere decreases, and therefore so does the weight of overlying air.

air pressure tendency *See* BAROMETRIC TENDENCY.

air quality A measure of the extent to which air is polluted. If the concentration of pollutants is low, air quality is said to be high. Pollution levels may be judged aesthetically, for example by a bad smell made by a substance that is otherwise harmless, but more commonly they are related to the known damage they cause to human health, vegetation, or material structures such as buildings. The concept is not absolute, because pollution levels can change according to such factors as the wind direction, time of year, and the length of time during which a pollutant is released into the air.

Air Quality Act 1967 A federal law that empowers the Department of Health, Education and Welfare to define areas within which AIR QUALITY should be controlled, to set AMBIENT AIR STANDARDS, to specify the methods and technologies to be used to reduce pollution, and to prosecute offenders if local agencies fail to do so.

(You can learn more about U.S. clean air laws and regulations at www.legal.gsa.gov/legal14air.htm.)

air shed The geographic area that is associated with a particular source of air. The concept is analogous to that of a watershed and is used in estimating the likelihood that the area will be exposed to AIR POLLUTION from elsewhere.

airspace The air, from the surface to the uppermost limit of the atmosphere, above the surface area of a territory or nation. A nation exercises sovereignty over its airspace (*see also* AEROSPACE), allowing it to permit or forbid aircraft to enter it and to control aircraft movements within it. This concept of sovereign airspace was first defined in the Paris Convention on the Regulation of Aerial Navigation (1919) and was restated in 1944 in the Chicago Convention on International Civil Aviation. The powers exercised by the civil authority within national airspace are equivalent to those that apply in territorial waters as set out in the Geneva Convention on the High Seas (1958).

airspeed *See* TAILWIND.

airstream A large-scale movement of air across a continent or an ocean that is associated with the PREVAILING WINDS in that latitude. The airstream that most influences the climate of the western coast of North America arrives from the Pacific Ocean and carries maritime POLAR AIR. To the east of the Rocky Mountains the climate is dominated by airstreams of continental polar air and ARCTIC AIR.

air temperature *See* STEVENSON SCREEN.

air thermoscope The first THERMOMETER, which was invented by GALILEO in 1593. It consisted of a glass bulb connected to a narrow glass tube mounted vertically with its lower end immersed in colored water contained in a sealed vessel. Air in the bulb expands and contracts as the temperature rises and falls, pushing the water in the tube downward as it expands and drawing it upward as it contracts. The thermoscope was very sensitive to changes in temperature, but it made no allowance for changes in air pressure that also alter the volume of air, and so it was very inaccurate. The first reliable thermometer was invented in 1641 by FERDINAND II of Tuscany.

Air thermoscope. **The thermoscope that was invented by Galileo. As air in the bulb at the top expands and contracts colored water in the tube is pushed down or drawn up.**

Aitken, John (1839–1919) Scottish *Physicist* The scientist who discovered that the air contains large numbers of very small particles, now known as AITKEN NUCLEI, and invented the AITKEN NUCLEI COUNTER, an instrument for detecting them. He also discovered the part Aitken nuclei play, as CLOUD CONDENSATION NUCLEI, in the formation of clouds.

John Aitken was born and died at Falkirk, in Stirlingshire. Because of poor health, he was never able to hold any official position. Instead, he worked from a laboratory he made at his home. There, he constructed his own apparatus and conducted experiments. He described many of his findings in papers published in the journals of the Royal Society of Edinburgh, of which he was a member.

Aitken nuclei counter A laboratory device, invented by John AITKEN, that is used for estimating the concentration of CLOUD CONDENSATION NUCLEI in a sample of air. The sample of air is drawn into a chamber, where it is kept near SATURATION by the presence of sodden filter paper. A pump then makes the air expand rapidly. The expansion causes it to cool. Water droplets condense onto the nuclei present in the air, and some fall as a shower onto a graduated disk. A lens makes the droplets clearly visible, allowing them to be counted. Each droplet is assumed to represent one condensation nucleus.

Aitken nucleus AEROSOL particles that have diameters smaller than 0.4 μm, most of which are between 0.005 μm and 0.1 μm. Over dry land there are often about 28,000 nuclei in each cubic foot of air (100,000 per liter). The largest act as CLOUD CONDENSATION NUCLEI. Their existence was discovered by JOHN AITKEN.

Alaska Current A warm ocean BOUNDARY CURRENT that flows northwestward and then westward along the coast of Canada and southeastern Alaska. It results from the deflection of the NORTH PACIFIC CURRENT as it approaches the North American continent. It is sometimes called the ALEUTIAN CURRENT, although the two are usually regarded as distinct.

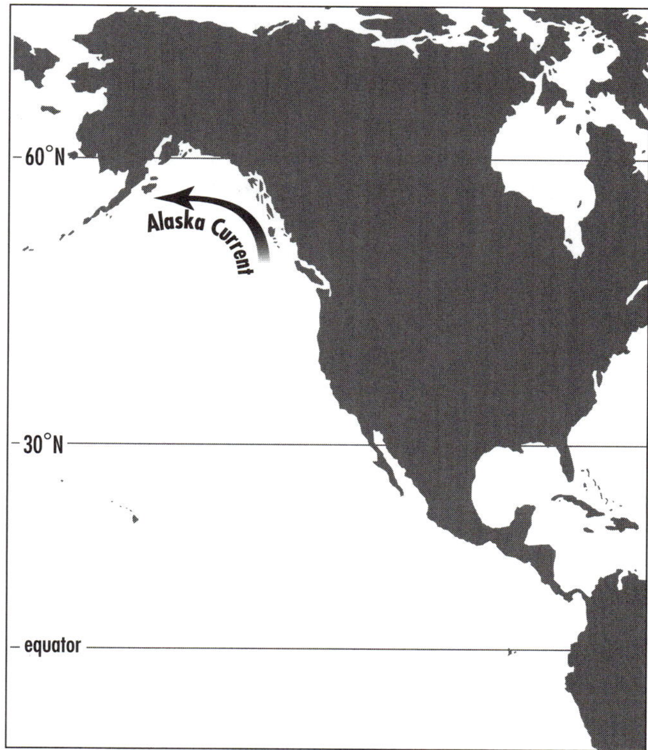

The branch of the North Pacific Current that carries warm water to the coasts of northwestern Canada and southeastern Alaska.

albedo A measure of the reflectivity of a surface, expressed as the proportion of the radiation falling on the surface that is reflected. If all of the radiation is reflected, the surface has an albedo of 1.0, or 100 percent. Radiation that is not reflected is absorbed (*see* ABSORPTION OF RADIATION) so the albedo of a surface strongly influences the extent to which anything beneath the surface is warmed by sunshine.

We take advantage of this property of materials. In summer we wear white or pale-colored clothes. Pale colors reflect light, and white reflects most of all, so these colors help to keep us cool. In winter we wear dark colors, which absorb more of the sunshine.

Unfortunately, it is not quite so simple, because albedo varies according to the wavelength of the radiation. Materials that have a high albedo in visible light, so we see them as pale, may absorb quite well at INFRARED wavelengths. Wearing garments of such a material will do little to protect against solar warmth. Solar radiation is most intense at about 0.5 μm, which

Comparative albedos

Surface	Green	Infrared
Dry sand	0.23	0.30
Wet sand	0.12	0.19
Clean ice	0.54	0.32
Dirty ice	0.33	0.19

is the wavelength we see as green visible light. Many surfaces have a different albedo at this wavelength than at an infrared wavelength of 0.8 μm.

Calculations of the amount of energy a surface absorbs must take account of the variation of albedo with wavelength. Regardless of wavelength, absorbed radiation is converted to heat, but a measure of the albedo of a surface in visible light does not give a reliable value from which to calculate the warming effect of sunlight. Whether it is wet or dry, sand is more highly reflective at infrared than at shorter wavelengths, and whether it is clean or dirty, ice absorbs more infrared than green radiation.

The amount of radiation absorbed by a surface is most closely related to the albedo at short wavelengths, of less than 4 μm. When values for albedos are given, unless stated otherwise they refer to the albedo at these wavelengths.

Albedo

Surface	Value
Fresh snow	0.75–0.95
Old snow	0.40–0.70
Cumuliform cloud	0.70–0.90
Stratiform cloud	0.59–0.84
Cirrostratus	0.44–0.50
Sea ice	0.30–0.40
Dry sand	0.35–0.45
Wet sand	0.20–0.30
Desert	0.25–0.30
Meadow	0.10–0.20
Field crops	0.15–0.25
Deciduous forest	0.10–0.20
Coniferous forest	0.05–0.15
Concrete	0.17–0.27
Black road	0.05–0.10

Albedo also varies with the ANGLE OF INCIDENCE of the solar radiation. This is seen most clearly over water, such as a lake or the sea. When the Sun is high in the sky the water appears dark. In temperate latitudes in summer the noonday Sun is at an elevation of about 50° and a water surface has an albedo of about 0.025. If the Sun is directly overhead, the albedo of water is even lower, at about 0.02. When the Sun is low in the sky, at an angle of 10°, the albedo increases to 0.35, and at dawn and sunset, when the Sun is touching the horizon and the angle of incidence is close to 0°, the albedo is greater than 0.99. Although this is most obvious in the case of water, to a greater or lesser extent it applies to all surfaces and most of all to level ones. Level sand and ice surfaces also have albedos approaching 1.0 when the angle of incidence is close to 0°.

Since albedo varies with latitude, season, and time of day, the values for particular surfaces are calculated as the global average through the year. Figures for albedo values therefore represent global averages of albedo at wavelengths of less than 4 μm, which is at the shortwave end of the wavelength of violet light.

Alberta low An area of low pressure that sometimes develops on the eastern slopes of the Rocky Mountains, in the Canadian province of Alberta. Air passing over the mountains develops a CYCLONIC circulation, and then the system moves eastward, carrying storms with heavy precipitation.

Aleutian Current (Sub-Arctic Current) A warm ocean current that flows in a westerly direction to the south of the Aleutian Islands. It runs parallel to the NORTH PACIFIC CURRENT, but to the north of it, and carries a mixture of warm water from the KUROSHIO CURRENT and cold water from the OYASHIO CURRENT. The ALASKA CURRENT is sometimes called the *Aleutian Current,* although the two are usually considered distinct.

Aleutian low
One of the two semipermanent areas of low pressure in the Northern Hemisphere (the other is the ICELANDIC LOW). It is centered over the Aleutian Islands, southwest of Alaska in the North Pacific at about 50° N, and covers a large area. It is described as "semipermanent" because although it forms, dissipates, and reforms, it is present for most of the winter and moves very little. The intensity of the system varies: The lowest pressure occurs when the atmospheric circulation is strong. Pressure is lowest in January, when it averages 1,002 mb. The Aleutian low is farther south than the Icelandic low. This is due to the presence of the Aleutian Islands, which restrict the northward movement of currents in the North Pacific Ocean, and to the GULF STREAM, which pushes the Icelandic low northward.

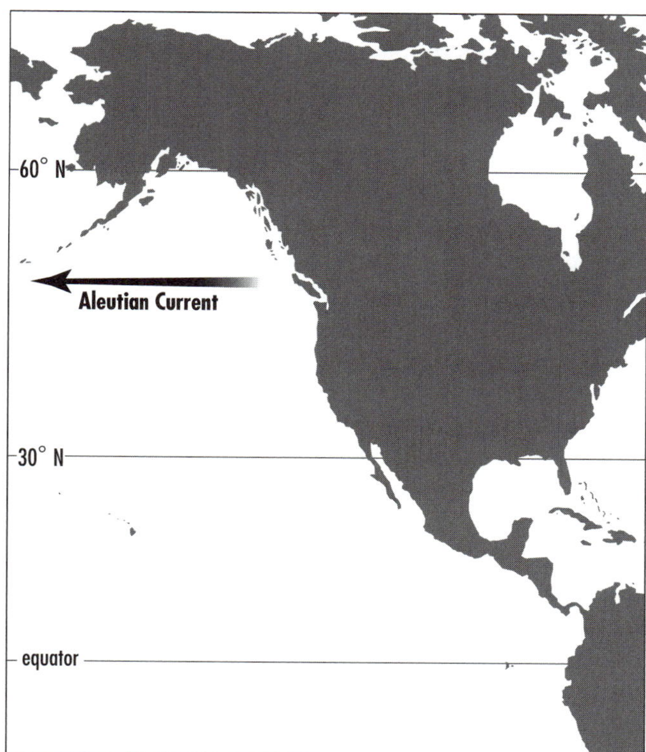

Aleutian Current. **The current flow in a westerly direction to the south of the Aleutian Islands.**

The Aleutian low generates many storms that travel eastward along the POLAR FRONT and tend to merge.

Alex A TYPHOON the struck Zhejiang Province, China, on July 28, 1987. It triggered a huge landslide. There was widespread damage and at least 38 people were killed.

algorithm In the atmospheric sciences, a set of equations that make it possible to infer one set of data, such as the concentration of ozone in the air, from another set, such as the intensity of radiation at certain wavelengths. This allows quantities that are difficult or impossible to measure to be calculated from quantities that can be measured simply and precisely. In this example, because ozone absorbs electromagnetic radiation at particular wavelengths, measuring the intensity of radiation at those wavelengths and comparing it with the intensity at other wavelengths reveal the amount of radiation being absorbed by ozone. From this the ozone concentration can be calculated.

In computing, an algorithm is a sequence of steps that, if followed, leads to the accomplishment of a specified task. Written as an ordered series of instructions in a language or code appropriate to the computer, an algorithm becomes a program.

d'Alibard, Thomas-François *See* FRANKLIN, BENJAMIN, and THUNDERBOLT.

Alicia A HURRICANE, rated category 4 on the SAFFIR/SIMPSON SCALE, that struck southern Texas on August 18, 1983. It generated winds of 115 mph (185 km h^{-1}) and caused extensive damage in Galveston and Houston. At least 17 people were killed and the damage was estimated at $1.6 billion.

Allen A HURRICANE, rated category 5 on the SAFFIR/SIMPSON SCALE, that struck islands in the Caribbean and the southeastern United States in August 1980. It had winds of 175 mph (280 km h^{-1}) gusting to 195 mph (314 km h^{-1}). Barbados, St. Lucia, Haiti, the Dominican Republic, Jamaica, and Cuba were affected. More than 270 people died, most of them in Haiti.

Allerød A place to the north of Copenhagen, Denmark, where clay is removed for making tiles and plant remains found at the site provide evidence of a period of warm, moist weather as the Weichselian glacial (DEVENSIAN GLACIAL) was drawing to a close. The lowest layer of clay contains evidence of the cold, OLDER DRYAS stade. Above this layer there is evidence of the warm Allerød INTERSTADE. It began about 11,800 years ago and lasted for about 800 years before temperatures fell once more with the start of the YOUNGER DRYAS. During the Allerød the tundra vegetation disappeared, the soil stabilized, and birch (*Betula* species) woodland covered the area, together with other warmth-loving plants such as meadowsweet (*Filipendula ulmaria*). The Allerød was followed by the Younger Dryas stade.

alpha decay A type of RADIOACTIVE DECAY in which the nuclei of the decaying atoms emit alpha particles. These consist of two protons and two neutrons and have a charge of +2. The loss of an alpha particle decreases the atomic number (the number of protons in the nucleus) by 2. Uranium-238 decays by alpha decay to thorium-234. A stream of alpha particles is known as *alpha radiation*. Alpha radiation is emitted naturally by elements present in the soil and rocks. IONIZING RADIATION ionizes air molecules with which it collides.

Alpine glacier *See* VALLEY GLACIER.

alpine glow A series of colors that are sometimes seen over mountains in the east, especially if they are covered with snow, as the Sun is setting in the west, and over mountains in the west as the Sun is rising in the east. The phenomenon is caused by the SCATTERING of light reflected from the mountains. The colors change from yellow to orange to pink to purple at sunset and in the reverse order at sunrise.

altimeter An instrument that measures the altitude of the person or device carrying it. There are two types, one that relates altitude to atmospheric pressure and one that measures the time taken for electromagnetic radiation to be reflected. The pressure altimeter, which is the type most widely used in aircraft, consists of an ANEROID BAROMETER located in a PITOT TUBE and linked to a dial on the instrument panel. In addition to the pointers indicating altitude in feet or meters, the dial displays sea-level barometric pressure in millibars. This can be adjusted and is set to the current value. The instrument then computes altitude by comparing the sea-level pressure with the pressure detected by its aneroid capsules. This type of altimeter indicates height above sea level.

A hypsometer also measures atmospheric pressure but does so by its effects on the boiling point of a liquid. This varies inversely with pressure, so altitude can be calculated from the temperature at which the liquid boils. The instrument consists of a cylindrical vessel containing the liquid (which is usually water), which is surrounded by a jacket through which the vapor can circulate and that contains a thermometer.

A radar altimeter measures altitude above the ground surface. It transmits a radio signal that is reflected from the surface and measures the time that elapses between the transmission and receipt of the reflection. A laser altimeter works in the same way but uses a laser beam rather than a radio signal. The altitude (a) is calculated by the equation $a = ct/2$, where c is the speed of light and t is the time that elapses between transmission and reception.

Altithermal A period lasting from about 5,000 to 8,000 years ago during which the average temperature was up to about 5° F (2.8° C) warmer than that of today.

altocumulus (Ac) A genus of middle clouds (*see* CLOUD CLASSIFICATION) that is composed of water droplets. It is white, gray, or both white and gray and is made up of elements, each about 1° to 5° across, which is approximately the thickness of three fingers held at arm's length. Sometimes there is shading around the elements, sometimes not. The elements are arranged in lines or waves and may be so close together that their edges merge and they form a sheet of cloud. There are often IRISATIONS around the edges of the elements.

Altocumulus is difficult to describe, because its appearance is extremely variable. In summer it often forms late in the evening, lasts through the night, but disappears during the course of the following morning. At night, the lower part of the cloud absorbs heat radiated from the ground surface but loses heat by radiation from the upper surface. This produces a precarious balance between water droplets that are cooled at the top of the cloud and sink, and water droplets that are warmed at the base of the cloud and rise. The following morning the cloud continues to absorb radiation from below but also absorbs more direct sunlight from above than it radiates away and the entire cloud warms, evaporating its water droplets. Altocumulus often has a wavy or banded appearance caused by small vertical movements of this kind within it.

Altocumulus has little predictive value. Its variability is evident from the large number of species and varieties of it that can be seen. It occurs in the species CASTELLANUS, FLOCCUS, LENTICULARIS, and STRATIFORMIS, and in the varieties DUPLICATUS, LACUNOSUS, OPACUS, PERLUCIDUS, RADIATUS, TRANSLUCIDUS, and UNDULATUS.

altostratus (As) A genus of middle clouds (*see* CLOUD CLASSIFICATION) that is composed of water droplets. It appears as a fibrous or striated veil or a uniform sheet of cloud and is grayish or bluish in color. It does not cause HALOES. The Sun or Moon can sometimes be seen through it as though through ground glass, but it can also be thick enough to obscure them totally. It is similar to CIRROSTRATUS, with which it often merges imperceptibly, but it forms at a lower level.

Altostratus develops at WARM FRONTS, where warm, moist air rises over cooler air. Its appearance is usually an indication of approaching precipitation. A sky that is overcast with altostratus is often described as "watery."

There are no species of altostratus, but it occurs in the varieties DUPLICATUS, OPACUS, RADIATUS, TRANSLUCIDUS, and UNDULATUS.

ambient Surrounding. The ambient temperature is the temperature of the air surrounding the place where the measurement is made, and the ambient pressure is the pressure of the surrounding air. AMBIENT AIR STANDARDS set quality criteria for the surrounding air. The word is derived from the Latin verb *ambire,* which means "to go around."

ambient air standard A standard for the AIR QUALITY in a particular place that is defined in terms of pollution levels. Industries operating in or close to such an area are required by the AIR QUALITY ACT 1967 to limit their emissions to levels that will not reduce the air quality to below the standard that has been set by the federal authorities.

ammonium sulfate haze The first stage in the formation of CIRRUS cloud, which occurs when the RELATIVE HUMIDITY (RH) exceeds 75–80 percent. Ammonium sulfate and sea-salt AEROSOLS are soluble in water. Liquid water condenses onto them (and will condense onto sulfuric acid aerosols at any RH). The droplets form a thin HAZE. As the RH rises to above 100 percent, condensation accelerates and the haze droplets grow larger. When they are more than 2 μm across they are classed as cloud droplets and at this stage the cloud is visible. The temperature at the height where cirrus forms is below freezing, and many of the supercooled (*see* SUPERCOOLING) liquid water droplets contain solid particles that act as FREEZING NUCLEI. This allows ICE CRYSTALS to grow. Ice crystals form in the absence of freezing nuclei if the temperature is below –40° F (–40° C). The ice crystals grow rapidly until they are about 50 μm across, after which their growth slows.

AMO *See* ATLANTIC MULTIDECADAL OSCILLATION.

Amontons, Guillaume (1663–1705) French *Physicist* Guillaume Amontons, one of the most ingenious inventors of his age, was born in Paris on August 31, 1663. His father was a lawyer, originally from Normandy. Guillaume studied the physical sciences, celestial mechanics, and mathematics, as well as drawing, surveying, and architecture, but so far as is known he did not attend a university. He earned his living as a government employee, working on a range of public works.

While still in his teens, Amontons became profoundly deaf. Far from regarding this as a handicap, he considered it a blessing, because it allowed him to concentrate on his scientific work without distraction.

In 1687 he invented a new type of hygroscopic HYGROMETER, based on the expansion and contraction of a substance as it absorbs and loses atmospheric moisture (the modern hair hygrometer was invented in 1783 by HORACE DE SAUSSURE). The following year he devised an optical telegraph (*see* TELEGRAPHY), which he thought would be of help to deaf people. Messages were transmitted by means of a bright light that was visible to a person with a telescope at the next station. He demonstrated it to the king some time between 1688 and 1695, but it was never adopted.

In 1695 Amontons invented a BAROMETER that did not require a reservoir of mercury. This meant it could be used at sea, where mercury barometers gave unreliable readings because the level of the mercury in the reservoir oscillated with the motion of the ship.

The same year he improved on the AIR THERMOSCOPE that had been invented by GALILEO in 1593. Galileo's design used the expansion and contraction of the air in a tube to alter the level of water. The disadvantage of this was that the water was also affected by changes in air pressure. Galileo was unaware of the effect of air pressure, but Amontons knew how to remove it. Instead of water he used mercury, then adjusted the height of the mercury until the air filled a fixed volume. After that, changing the temperature of the air in the tube altered the pressure it exerted on the mercury and it was this changing pressure that the instrument measured. The Amontons thermometer was more accurate than Galileo's, and he was able to use it to show that, within the limits of his instrument, water always boiled at the same temperature, but it was not accurate enough for most scientific uses.

No one had yet devised a scale by which temperature could be measured. This lack prevented him from discovering CHARLES'S LAW, but this new thermometer allowed him to take the study of gases a step further than EDMÉ MARIOTTE had. He noticed that for a particular change in temperature, the volume occupied by a gas always changed by the same amount. This led to AMONTONS'S LAW, which he described in 1699 and which can be stated as $P_1T_2 = P_2T_1$. It also allowed Amontons to visualize a temperature at which gases contracted to a volume beyond which they could contract no further. This was the concept of ABSOLUTE ZERO.

Amontons also invented a type of clock called a *clepsydra,* operated by the flow of water. He proposed that his clepsydra could be used at sea, although it would not have been accurate enough to measure longitude.

In 1690 Amontons became a member of the Académie des Sciences. He published a number of papers and one book, *Remarques et expériences physiques sur la construction d'une nouvelle clepsydre, sur les baromètres, thermomètres, et hygromètres* (Observations and physical experiences on the construction of a new clepsydra, on barometers, thermometers and hygrometers), which appeared in 1695.

Amontons died in Paris on October 11, 1705.

Amontons's law *See* AMONTONS, GUILLAUME.

amorphous cloud Cloud such as NIMBOSTRATUS that forms a flat, featureless sheet covering most or all of the sky.

amorphous snow SNOW that has an irregular crystalline structure.

ampere (A) The SYSTÈME INTERNATIONAL D'UNITÉS (SI) UNIT of electric current, which is equal to the constant current that, if maintained in two straight, perfectly cylindrical, parallel conductors of infinite length and negligible cross section placed 1 meter apart in a vacuum, would produce between the two conductors a force of 2×10^{-7} NEWTON per meter of their length. The unit is named in honor of the French physicist and mathematician André Marie Ampère (1775–1836).

amphidromic point A geographical position around which seawater circulates in the course of the tidal flow. The TIDES produce no rise or fall of water at the amphidromic points themselves. The word *amphidromic* is derived from the Amphidromia, an Ancient Greek naming festival for a newly born child, at which friends carried the child around the hearth.

An amphidromic point occurs where water flows from the ocean into a partially landlocked sea. The resulting tidal stream is affected by the CORIOLIS EFFECT (CorF), which causes it to swing, so it ends by circulating around a point. The TIDAL RANGE increases with distance from an amphidromic point.

The North Sea is a good example of a sea with an amphidromic tidal flow. Water enters the North Sea from the Atlantic through the Straits of Dover in the south and around the north of Scotland. This produces three amphidromic points, midway between the bulge of eastern England and the Netherlands, in the eastern North Sea level with Denmark, and off southwestern Norway.

The times of high and low water follow a circle around an amphidromic point. Lines, called *cotidal lines,* can be drawn radiating from an amphidromic point to link the points at which high tide is reached at a particular time. Corange lines can also be drawn to link points at which the average tidal range is the same. These surround the amphidromic point, sometimes as a series of concentric circles.

amplitude The vertical distance between the mean water level and the bottom of a wave trough or top of a wave crest. This is equal to half the wave height, which is the vertical distance between the base of a trough and the top of the adjacent crest.

Amy A TYPHOON that struck southern China on July 20 and 21, 1991. It killed at least 35 people and injured 1,360.

anabaric (anallobaric) An adjective that is applied to a phenomenon associated with a rise in atmospheric pressure.

anabaric center *See* PRESSURE-RISE CENTER.

anabatic wind A wind that blows up the side of a hill (*see* VALLEY BREEZE). It occurs on warm afternoons on the sides of narrow valleys, especially those that are aligned north–south, so the valley sides and floor are

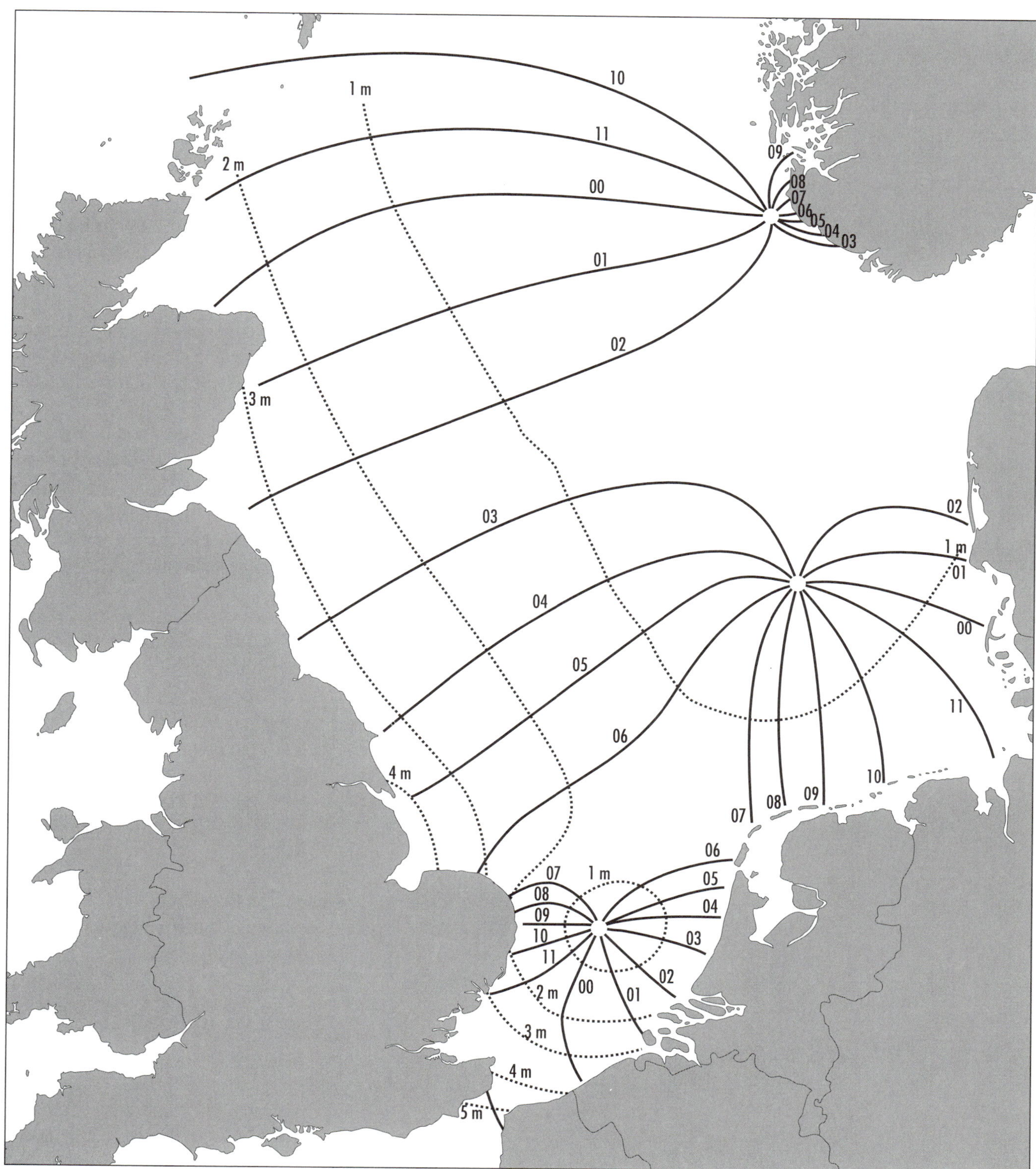

Amphidromic points. **The three amphidromic points in the North Sea, between eastern Britain and continental Europe. The solid lines are cotidal lines. They indicate the time of high water measured in lunar hours, of approximately one hour two minutes, after the Moon has passed the Greenwich meridian. The broken lines are corange lines, indicating the average tidal range.**

not shaded from the Sun. As the ground warms, the air in contact with it is also warmed, and so the air expands. The sides of the valley prevent the air from expanding to the sides, and therefore it expands upward, producing a gentle flow up the valley sides. At the same time, in many valleys the sides are warmed more strongly than the valley floor, so the warming of the air increases with height. This accelerates the upslope movement of air.

anafront A WARM FRONT or COLD FRONT at which air in the WARM SECTOR is rising. As the air rises it cools ADIABATICALLY and its water vapor condenses. Consequently, unless the air in the warm sector is unusually dry, anafronts are very active. They produce large amounts of cloud, mainly of a STRATIFORM type but sometimes including CUMULONIMBUS near the cold front and extensive, continuous, and often heavy precipitation.

anallobaric *See* ANABARIC.

analog model A climatic MODEL that aims to match present atmospheric conditions to times in the past when similar conditions prevailed and then to use the way the past conditions developed to predict the development of the present conditions. Analog models are used mainly in long-range weather forecasting and in climatological studies of ways the global climate may be changing. Concerns about GLOBAL WARMING, for example, have generated interest in a warm period that occurred in the early part of the Pliocene epoch, about 4.3–3.3 million years ago, during the EEMIAN INTERGLACIAL of northern Europe, about 130,000–72,000 years ago, and in the CLIMATIC OPTIMUM that occurred about 5,000 years ago. The HOLOCENE climatic optimum is the subject of the COOPERATIVE HOLOCENE MAPPING PROJECT.

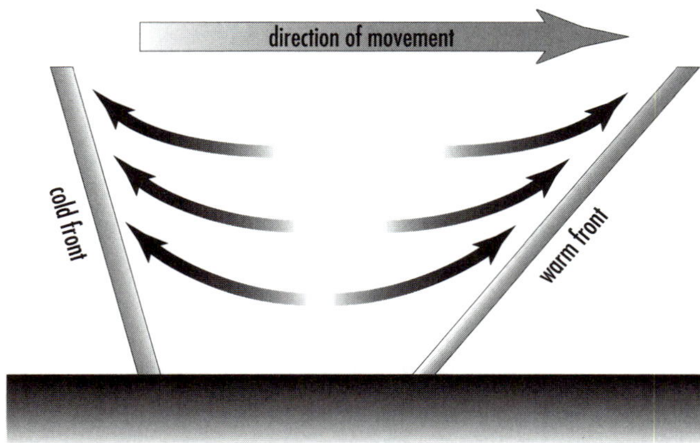

Anafront. **Air rises against the fronts.**

andhis The local name given in the northwestern part of the Indian subcontinent to a DUST STORM that accompanies a violent SQUALL. The squall is caused by vigorous CONVECTION.

Andrew A HURRICANE, rated as category 5 on the SAFFIR/SIMPSON SCALE, that struck the Bahamas, Florida, and Louisiana from August 23 to 26, 1992. It generated winds of up to 164 mph (264 kmh^{-1}). It was the costliest hurricane in U.S. history up to that time. In Florida, Homestead and Florida City were almost destroyed, 38 people were killed, 63,000 homes were destroyed, and damage was estimated at $20 billion. In Louisiana, 44,000 people were rendered homeless and damage was estimated at $300 million.

anemogram An instrument consisting of a pen linked to an ANEMOMETER and a rotating paper drum. The pen makes magnetic contact with the drum and the drum is driven by a motor linked to a clock. The anemograph makes a permanent record of the WIND SPEED.

anemograph A record of the WIND SPEED that is made by an ANEMOGRAM pen on a rotating paper drum.

anemology The scientific study of winds. The word is derived from the Greek *anemos,* which means "wind," and *logos,* which means "word."

anemometer An instrument that is used to measure the surface WIND SPEED. It is very difficult to measure wind speed directly (*see* ANEMOMETRY). To do so would necessitate labeling a small volume of air and tracking its movement over a measured distance. RADIOSONDE balloons are used to measure the speed and direction of the upper-level wind. The balloon moves with the air surrounding it.

The surface wind is measured by its effect on a physical object. The rotating cups anemometer is the device that is most widely used. It consists of three or four hemispherical or conical cups that are mounted on arms separated by 120° or 90°, depending on the number of cups, and attached to a vertical axis. The wind exerts more pressure on the concave surfaces than on the convex surfaces. This causes the cups to turn about the axis, and the rotational speed of the axis is converted into the wind speed and displayed on a panel attached below the cups or remotely, by a needle on a dial. A rotating cups anemometer tends to overestimate the speed of wind GUSTS, because the cups accelerate more quickly than they slow down.

The anemometer measures only the speed, not the direction of the wind, but it is sometimes mounted at one end of a horizontal arm and a WIND VANE is mounted at the other end. In this design the direction indicated by the wind vane is also shown by a needle on a dial.

A swinging-plate anemometer (also called a *pressure-plate anemometer*) is better for measuring the speed of gusts. It consists of a flat plate that swings freely at the end of a horizontal arm. A wind vane is fixed to the other end of the arm and the arm is free to rotate about a vertical axis. The vane ensures that the plate is always at right angles to the direction of the wind. Air pressure makes the plate swing inward, and the distance it swings is converted into a wind speed that is read from a scale.

A bridled anemometer resembles a rotating cups anemometer but has more cups: commonly 32. The vertical axis is "bridled," which is to say its movement is checked by a spring. The rotation of the axis exerts tension on the spring and this is shown as wind speed on a dial. This type of anemometer is often used on ships.

A four-bladed propeller spins if it is held at right angles to the wind. This is a propeller anemometer or AEROVANE.

A pressure anemometer also measures the pressure exerted by the wind. It uses a PITOT TUBE that is attached to a vane so it always faces directly into the wind.

The SPEED OF SOUND is also used to measure wind speed. The sonic anemometer has three approximately U-shaped arms mounted at right angles to each other. The two tips of the arms each carry an acoustic transducer, and the instrument measures the speed at which an acoustic signal travels from one tip to the other and

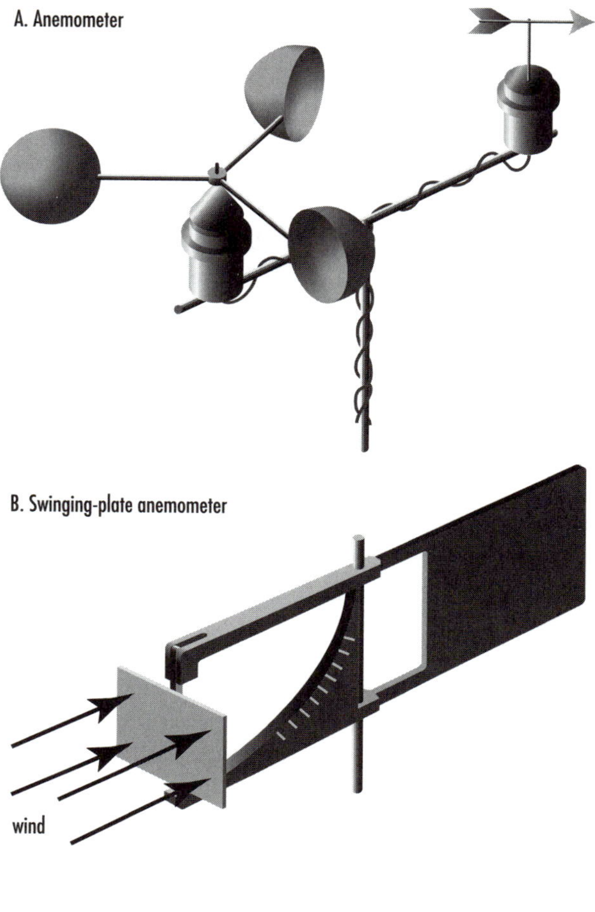

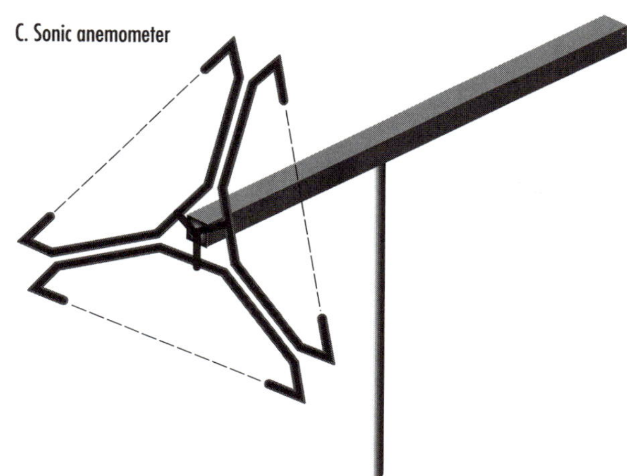

Anemometers. **The rotating cups anemometer measures the speed with which its cups spin around their vertical axis. It is often linked to a wind vane, so the combined instrument measures wind direction as well as speed. The swinging-plate anemometer measures the pressure the wind exerts on its flat plate. The sonic anemometer measures the speed of sound through moving air.**

then back again. A sound wave moves through the air. Consequently, if the air itself is moving, the speed of sound through the moving air differs from the speed of sound in still air. If the sound wave propagates in the same direction as the wind, its speed increases to the sum of the speed of sound and the speed of the wind. If the sound propagates in the opposite direction to the wind, the speed of sound is reduced by an amount equal to the wind speed. By measuring the speed in both directions between each pair of transducers the sonic anemometer corrects for the temperature and humidity of the air, which also affect the speed of sound. Because of their reliability and accuracy, sonic anemometers are now used at many weather stations.

WIND PROFILERS are used to measure wind speeds throughout the TROPOSPHERE and lower STRATOSPHERE.

Anemometers must be sited in the open, well clear of obstructions that deflect and slow the wind. They are usually fixed on tall poles for this reason.

anemometry The scientific measurement of WIND SPEED. Speed is the distance a body travels in a unit of time. It is not practicable, however, to label a small volume of air in order to make it visible and then to time its movement over a measured distance. Instead, it is necessary to measure the speed indirectly, by the effect of the wind at different speeds on visible objects. The first successful method was the one devised by Admiral Sir FRANCIS BEAUFORT. The advantage of the BEAUFORT WIND SCALE is that it is based on the response of commonplace objects, such as smoke, flags, trees, and umbrellas. Consequently, it requires no instruments. The Beaufort scale is still used. Weather stations use ANEMOMETERS to measure wind speed.

anemophily The pollination of plants by the wind. Wind pollination is an unreliable method, and wind-pollinated plants produce very large amounts of POLLEN to increase the chance that enough pollen grains will reach female flowers to allow the plants to reproduce. Wind-pollinated plants have small flowers, usually without colored petals. People who suffer from hay fever are allergic to pollen grains and so, despite being an entirely natural AEROSOL, pollen is often regarded as an atmospheric pollutant. All grasses are wind pollinated, but although grass pollen is the most widespread cause of hay fever, other pollen also affects sufferers.

aneroid barometer A BAROMETER that measures the effect of air pressure on a small metal box from which most of the air has been removed. The box is corrugated and acts as a bellows. It partly collapses when the air pressure increases, but a spring inside the box prevents it from collapsing completely. When the pressure decreases, the box expands. The position of the surface of the box may be measured electrically, but in most aneroid barometers the surface is linked to a spring and the spring to levers that move the needle on a dial. Aneroid barometers are less accurate than mercury barometers, but because they do not contain a reservoir of mercury they are more convenient to use. Most barometers that people hang on the walls of their homes are of the aneroid type. Their dials are often labeled *change, rain, much rain, stormy, fair, set fair,* and *very dry.* These annotations were introduced in the 17th century by ROBERT HOOKE. Pressure ALTIMETERS and BAROGRAPHS are modified aneroid barometers.

aneroidograph *See* BAROGRAPH.

Angela Two TYPHOONS, the first of which struck the Philippines in October 1989. It killed at least 50 people. The second Typhoon Angela, rated at category 4 on the SAFFIR/SIMPSON SCALE, struck the eastern Philippines on November 3, 1995, generating winds of 140 mph (225 kmh^{-1}). It killed more than 700 people, ren-

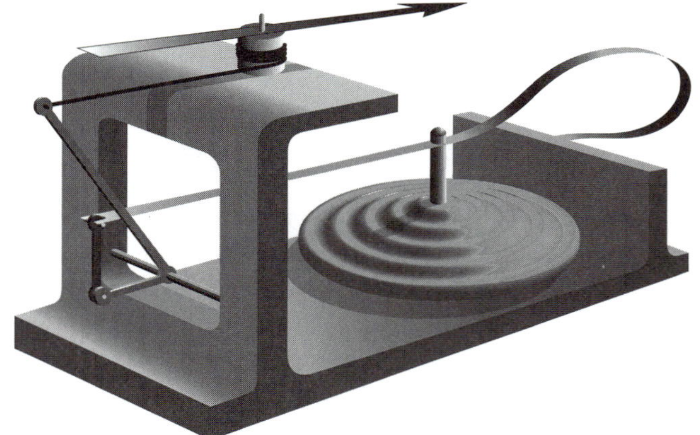

Aneroid barometer. **The corrugated metal box expands and contracts with changes in air pressure. Movements of its surface are transferred by a spring and levers to a needle that moves on a dial.**

dered more than 200,000 homeless, and destroyed 15,000 homes. Damage to crops, roads, and bridges was estimated at $77 million.

angiosperm *See* POLLEN.

angle of incidence The angle at which solar radiation strikes the surface of the Earth. The angle measured is that between the incident radiation and a tangent at the surface. Because the Earth is almost spherical, at noon at each EQUINOX the angle of incidence is 90° at the equator, and elsewhere it is equal to 90° minus the latitude. At latitude 40°, for example, the angle of incidence is 50° and at latitude 70° it is 20°.

The angle of incidence also changes with the SEASONS. In latitudes between those of the TROPICS and the Arctic and Antarctic Circles, the maximum seasonal variation, at the SOLSTICES, is 47°. In latitudes lower than that of the Tropics and higher than that of the Arctic and Antarctic Circles, the maximum variation is less than 47°. At the equator, the angle of incidence is never less than 66.5° and the maximum variation is 23.5°. At the Poles, the angle of incidence never exceeds 23.5°.

The angle of incidence determines the intensity of the solar radiation received at the surface. When the angle of incidence is high, and the Sun is high in the sky, a beam of radiation of a given diameter illuminates a smaller area of surface than would a beam of similar diameter with a low angle of incidence, and the Sun low in the sky. The intensity of the radiation received varies inversely with the area the beam illuminates and is therefore directly proportional to the angle of incidence.

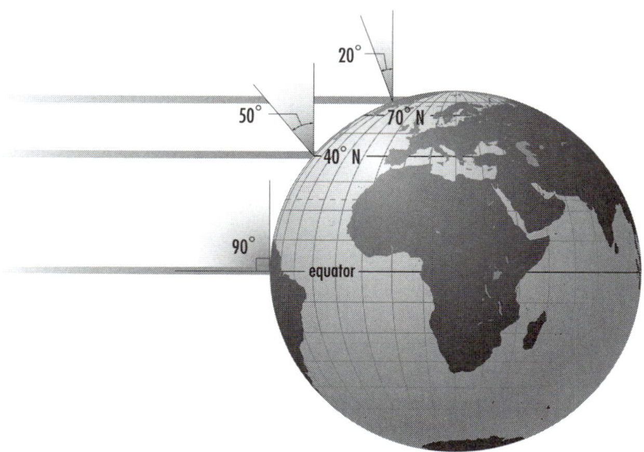

Angle of incidence. **The angle at which solar radiation strikes the Earth's surface. This varies with latitude.**

angle of indraft The angle between the direction of a steady wind and the ISOBARS. This indicates the extent to which the wind departs from the GEOSTROPHIC WIND. The angle of indraft is said to be positive when the wind direction is toward the area of low pressure (as it almost invariably is). The angle is greater over land than over the sea, because FRICTION is greater over land. Over the sea the angle is sometimes zero and occasionally negative when a strong THERMAL WIND blows toward a high-pressure region across a shallow PRESSURE GRADIENT.

angle of reflection The angle at which light is reflected (*see* REFLECTION) from a surface. This is always equal to the ANGLE OF INCIDENCE.

angle of refraction The angle through which light is refracted (*see* REFRACTION) when it passes from one transparent medium to another in which its speed changes. This angle varies according to the angle at which the light strikes the boundary between the two media and the difference in their refractive indices. The smaller the incident angle, the greater is the angle of refraction, and when the incident angle is 90° light is not refracted at all, although its speed changes. The greater the difference in the refractive indices of the two media, the greater is the angle of refraction.

Anglian glacial A GLACIAL PERIOD in Britain that lasted from approximately 350,000 years ago until

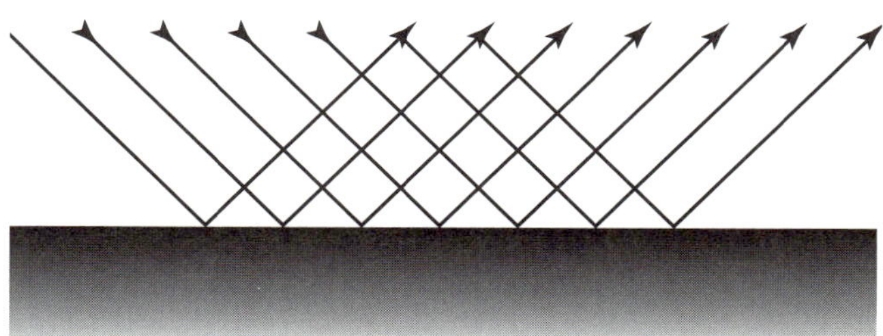

Angle of reflection. **When light strikes and is reflected from a surface, the angle between the reflected light and the surface is identical to the angle between the incident light and the surface.**

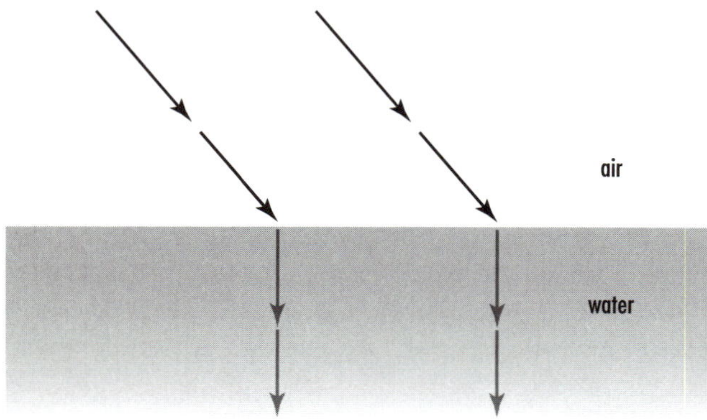

Angle of refraction. **Light is bent when it passes from one transparent medium to another with a different refractive index. The angle of refraction is the angle by which it is bent.**

about 250,000 years ago. It is equivalent to the Elsterian glacial of northern Europe and the MINDEL GLACIAL of the European Alps and partly coincides with the KANSAN GLACIAL of North America. GLACIERS advanced and retreated several times during the Anglian. It was preceded by the CROMERIAN INTERGLACIAL and followed by the HOXNIAN INTERGLACIAL.

ångström (Å) A unit of length that was formerly used to measure very small distances, such as those between molecules and the wavelengths of electromagnetic radiation. It was devised by the Swedish spectroscopist Anders Jonas Ångström (1814–74) and is equal to 10^{-10} m. It has been replaced by the Système International d'Unités (SI) unit the nanometer (1 Å = 0.1 nm). *See* SYSTÈME INTERNATIONAL D'UNITÉS.

angular momentum The MOMENTUM of a body that is following a curved path. Angular momentum is conserved. This means that once a body possesses angular momentum, this momentum remains constant provided no external force acts to accelerate or slow it.

The consequence of the conservation of angular momentum is seen at its most dramatic in the intense speeds generated near the center of a TROPICAL CYCLONE or TORNADO. Close to the center of these systems, air is spiraling upward. This produces a low-level region of low pressure, and this low-pressure area draws in a flow from the surrounding air. As it spirals toward the center, the approaching air turns in a progressively smaller circle and accelerates because of the conservation of its angular momentum.

Three factors affect the motion of a body that is turning in a circular path about an axis: the mass of the body (M), the radius of its circle of turn (r), and the speed with which it is turning (V). Its speed of turn is known as its ANGULAR VELOCITY. For any rotating body, the product of these is a constant:

$$MVr = \text{a constant}$$

The constant, MVr, is the angular momentum. Once the body is rotating, this constant is preserved. This means that if one or more of the three factors change, then one or more of the others will also change automatically, in order to preserve the constant.

Suppose a body of air with a mass of 1 unit (M = 1) is rotating with an angular velocity (V_1) of 5 units in a circle with a radius (r_1) of 20 units.

$$MV_1r_1 = 1 \times 5 \times 20 = 100$$

Now suppose the same air turns in a much tighter circle, with a radius (r_2) of 5 units. The angular momentum must be conserved, but the mass remains unchanged, so if the radius decreases, the angular velocity must increase (and vice versa). This can be expressed as

$$MV_2r_2 = 100$$

$$\therefore V_2 = 100 \div (Mr_2) = 20$$

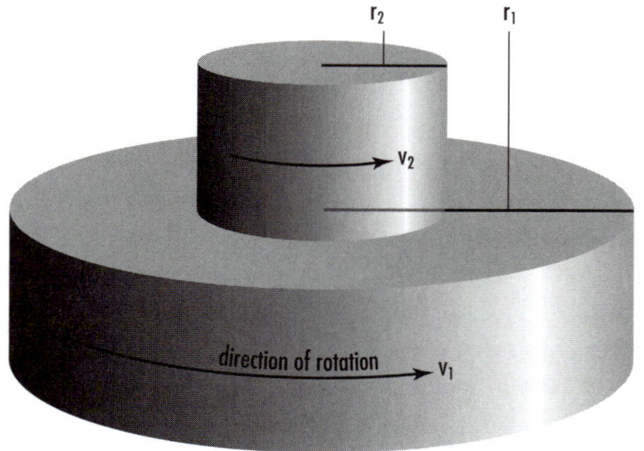

Angular velocity. **If air that is rotating in a large circle, radius r_1 at an angular velocity V_1, starts turning in a much tighter circle, radius r_2, it accelerates, so its new velocity V_2 is greater than V_1.**

Dividing the radius by 4 multiplies the angular velocity by 4.

A pirouetting ice skater can exploit the conservation of angular momentum if he or she begins to turn with the arms fully extended and then draws them toward the body. This alters the radius of rotation, and, therefore, with no further effort (no additional work is needed), the skater spins faster. At the end of the spin he or she slows by extending the arms once more.

The Earth is a rotating body and the atmosphere rotates with it. Within the total atmosphere, every mass of air possesses angular momentum. If a body of air moves toward or away from the equator its radius of rotation alters, because that radius is the distance between the air and the rotational axis of the Earth. Like that of the pirouetting skater, its angular velocity increases as it moves away from the equator and its rotational radius grows smaller. The change in its angular velocity produces a movement toward the east, in the direction of the Earth's rotation. If a body of air that was stationary over the equator moved to latitude 20°, it would then be moving eastward at about 66 mph (106 kmh^{-1}). Friction with the surface and air movements induced by pressure differences mask much of this effect, but the overall motion of the atmosphere is influenced by it.

angular velocity The speed of a body that is moving along a curved path. It is usually expressed in RADIANS per second (rad s^{-1}) and described by the symbol Ω. The circumference of a circle is equal to 2π radians. It follows, therefore, that $\Omega = 2\pi \div T$, where T is the time taken to complete one revolution. The tangential velocity (V)—the velocity in a straight line that can be measured in miles or kilometers per hour—is given by $V = \Omega r$, where r is the radius of the circle.

angular wave number (hemispheric wave number) The circumference of the Earth at a given latitude divided by the WAVELENGTH of the long waves (*see* ROSSBY WAVES) associated with a particular weather pattern. It gives the number of waves of that wavelength that are required to encircle the Earth and, therefore, the number of times the weather pattern repeats around the world.

anion An ION that carries a negative charge, because the number of electrons surrounding its nucleus is greater than the number of protons in its nucleus.

anisotropic Having properties that change in a particular direction. The properties of an anisotropic substance vary according to the direction from which the substance is approached. On a large scale the TROPOSPHERE is strongly anisotropic. Air temperature and pressure decrease with height and air pressure decreases or increases along a horizontal gradient toward centers of low or high pressure. On a smaller scale the atmosphere is much more ISOTROPIC.

annual snow line *See* FIRN LINE.

anomaly The deviation of some meteorological feature from the averages that are associated with a particular set of atmospheric conditions. Anomalies can sometimes be predicted. A low ZONAL INDEX in middle latitudes often leads, for example, to BLOCKING, which is the anomalous persistence of weather patterns for much longer than is usual. Varying patterns of SEA-SURFACE TEMPERATURE affect the circulation of air and the location of the JET STREAM over the oceans. In this way they can produce weather anomalies over land.

Antarctic air AIR MASSES inside the ANTARCTIC CIRCLE are of ARCTIC AIR and POLAR AIR. The continent of Antarctica, including the Antarctic Peninsula, is covered by CONTINENTAL arctic air (cA) in both winter and summer. In winter there are MARITIME polar air (mP) over the Southern Ocean adjacent to the South Atlantic (between South America and Africa) and Indian Oceans and maritime arctic air (mA) over the Ross Sea, opposite the Pacific Ocean. In summer, mP air covers the whole of the Southern Ocean.

Antarctic bottom water (AABW) A mass of dense water that forms in the Ross and Weddell Seas by the same mechanism that produces the NORTH ATLANTIC DEEP WATER. The freezing of surface water to produce sea ice removes freshwater, increasing the salinity of the adjacent seawater and therefore its density. The temperature of the surface water is slightly above freezing, at which the density of water reaches a maximum. This dense water sinks to the bottom of the Southern Ocean and then moves in an easterly direction around Antarctica, driven by the WEST WIND DRIFT. The temperature of the AABW is 28°–31° F (–2°-0.4° C) and its salinity is 34.66 parts per thousand (‰).

Antarctic Circle At latitude 66°30' S, the line that marks the boundary of a region within which there are a period in winter when the Sun does not rise above the horizon and a period in summer when it does not sink below it. At the Antarctic Circle, this period is confined to the winter and summer SOLSTICES. At higher latitudes, the period of continuous darkness and continuous daylight lasts longer. At the summer solstice, the Sun is no higher than 47° above the horizon anywhere inside the Antarctic Circle. The variation in the length of daylight is due to the inclination of the Earth's rotational axis with respect to the PLANE OF THE ECLIPTIC.

Antarctic Circumpolar Current *See* WEST WIND DRIFT.

Antarctic Circumpolar Wave (ACW) A set of two atmospheric and oceanic waves that travel through the Southern Ocean from west to east on a track that takes them all the way around the continent of Antarctica. The waves move at 2.4–3.1 inches per second (6–8 cm s^{-1}) and have a PERIOD of 3–5 years. It takes them 8–10 years to complete one circuit of the continent. As they pass, the waves affect wind speeds, the atmospheric pressure at sea level by up to 8 MILLIBARS, the sea-surface temperature by about 2.9° F (1.6° C), and the location of the edge of the sea ice by 217 miles (350 km). The cause of the waves is not yet known, but they are believed to arise from instabilities in the relative motions of the WEST WIND DRIFT and the air above it. The waves affect the climate in latitudes south of about 25° S.

(You can learn more about Antarctic Circumpolar Waves in a short abstract, B. Qiu and F. F. Jin, "Antarctic Circumpolar Waves: An Indication of Ocean–Atmosphere Coupling in the Extratropics," at www.agu.org/pubs/abs/gl/97GL02694/97GL02694.html; Chung-Chieng Aaron Lai and Zhen Huang, "Antarctic Circumpolar Wave and El Niño" at www.ees.lanl.gov/staff/cal/acen.html; and "An Antarctic Circumpolar Wave in Surface Pressure, Wind, Temperature, and Sea Ice Extent" at jedac.ucsd.edu/wbwhite/ray/paper.html.)

Antarctic convergence (AAC) *See* ANTARCTIC POLAR FRONT.

Antarctic front The FRONT that marks the boundary between ARCTIC AIR and POLAR AIR over the Southern Ocean. The Antarctic front is almost permanent and almost continuous around the continent of Antarctica. In summer the front lies close to the coast of Antarctica, separating CONTINENTAL arctic air (cA) from MARITIME polar air (mP). In winter the front moves farther north over the Ross Sea, where it separates maritime arctic (mA) from mP air but remains in the same position around the remainder of the continent.

Antarctic intermediate water (AIW) A mass of water that forms at the surface of the Southern Ocean, at the ANTARCTIC POLAR FRONT close to latitude 50° S. It then moves northward. Its salinity of 33.8 parts per thousand and low temperature, of 36° F (2.2° C), cause it to sink beneath the warmer, less dense water that it encounters. It sinks to about 2,950 feet (900 m) and continues to flow northward. It can be detected as far as 25° N in the North Atlantic.

Antarctic Oscillation A periodic change in the distribution of atmospheric pressure that is believed to occur naturally in the Southern Hemisphere, over the South Pole, and over latitude 55° S. It is the counterpart to the ARCTIC OSCILLATION.

Antarctic ozone hole *See* OZONE LAYER.

Antarctic Polar Current An ocean current that flows from east to west, parallel to the coast of Antarctica. It is driven by winds that blow from an easterly direction off the ice cap. The current affects only surface waters. This current is not to be confused with the WEST WIND DRIFT, which is also known as the *Antarctic Circumpolar Current*.

Antarctic polar front (Antarctic convergence; AAC) A boundary along the edge of the Southern Ocean between latitudes 50° S and 60° S where cold Antarctic water sinks beneath the warmer water in higher latitudes and forms the ANTARCTIC INTERMEDIATE WATER.

antecedent precipitation index A summary of the amount of precipitation that falls each day in a particular area, weighted so it can be used to estimate SOIL MOISTURE.

anthelion A spot of bright light that is occasionally seen in the sky at the same altitude as the Sun, but at the opposite AZIMUTH. The phenomenon is caused by the REFLECTION and REFRACTION of light by hexagonal ICE PRISMS with vertical axes.

Anthropogene *See* PLEISTOGENE.

anthropogenic An adjective that is applied to substances or processes that are produced by humans or that result from human activities. Strictly, this is an incorrect use of the word, which is derived from *anthropogenesis,* which is the study of human origins. The word is from the Greek *anthropos,* which means "human being," and *gen-,* which means "be produced."

anticyclogenesis The stages by which an ANTICYCLONE forms or is intensified. There are several ways anticyclones can form. In middle latitudes, where families of DEPRESSIONS travel one behind another from west to east, there is often an incursion of POLAR AIR behind the last member of the family. This cold, dense air establishes an anticyclone that dissipates with the arrival of the next batch of depressions. Both the depressions and their accompanying anticyclones are related to RIDGES and TROUGHS in the JET STREAM. The flow of air in the jet stream imparts an ANTICYCLONIC motion to the air on its southern side. This is associated with CONVERGENCE at high-level, subsiding air (*see* SUBSIDENCE) and DIVERGENCE at the surface.

Anticyclones also form in the final stage of the INDEX CYCLE. These can remain stationary for weeks on end. They are then known as *BLOCKING anticyclones.* Air that is subsiding on the poleward side of the HADLEY CELL circulation produces anticyclones called *SUBTROPICAL HIGHS.* Persistent anticyclones also form in winter over the interior of northern Canada and Asia, caused by the subsidence of cold, dry air.

anticyclolysis The weakening and final disappearance of an ANTICYCLONE or RIDGE as air flows outward from it.

anticyclone A region in which the atmospheric pressure is higher than it is in the surrounding air. Pressure is highest at the center and decreases with distance from the center. Air flows outward, from the region of high pressure to one of lower pressure at a speed proportional to the PRESSURE GRADIENT. As it moves, the air is subject to the CORIOLIS EFFECT (CorF), which swings it to the right in the Northern Hemisphere and to the left in the Southern Hemisphere. The balance between the CorF and the PRESSURE-GRADIENT FORCE produces an ANTICYCLONIC CIRCULATION. There are several ways in which anticyclones can form (*see* ANTICYCLOGENESIS).

The high pressure at the center of an anticyclone is produced by the SUBSIDENCE of air. Subsiding air is stable (*see* STABILITY OF AIR). It produces generally clear skies, except where the air is rendered unstable by moving across a warm surface, but the stable air can also produce INVERSIONS and ANTICYCLONIC GLOOM. Winds are usually light and change direction as the anticyclone passes. In winter, anticyclones in middle and high latitudes are usually associated with POLAR AIR and bring fine, cold weather. These are called COLD

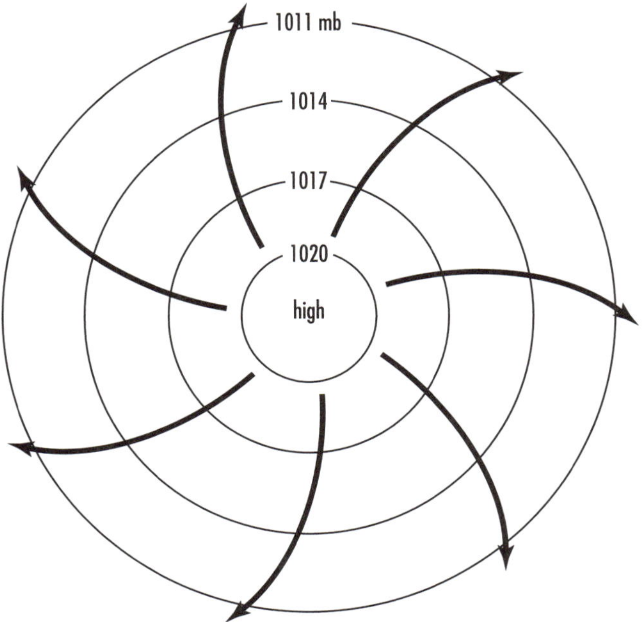

Anticyclone. **Highest pressure is at the center and decreases with distance from the center, as indicated by the isobars. Winds blow outward from the center, moving clockwise in the Northern Hemisphere and counterclockwise in the Southern Hemisphere.**

ANTICYCLONES. Summer anticyclones produce fine, warm weather, though with a risk of showers and thunderstorms if the air is moist.

Anticyclones range in size from a few hundred to as much as 2,000 miles (3,200 km) in diameter. They move more slowly and more erratically than CYCLONES.

anticyclonic The adjective that describes the direction in which the air flows around an ANTICYCLONE. The direction is opposite to that of the Earth's rotation as seen from directly above the North and South Poles, being clockwise in the Northern Hemisphere and counterclockwise in the Southern Hemisphere. Air also flows anticyclonically around a RIDGE. At a height of 33 feet (10 m), which is the standard height for observing surface winds, the winds cross the ISOBARS at an angle of 10° to 30°, depending on local topographical features and the wind speed. Above the BOUNDARY LAYER the winds flow almost parallel to the isobars at speeds proportional to the PRESSURE GRADIENT.

anticyclonic gloom Dull conditions that can develop when an ANTICYCLONE remains stationary for more than a few days. There are HAZE that reduces VISIBILITY and sometimes a layer of STRATOCUMULUS cloud that is just thick enough to hide the Sun.

The gloom is the product of SUBSIDENCE. Air in an anticyclone sinks at a rate of about 3,300 ft (1 km) a day and, if it is unsaturated (it usually is), it warms at the DRY ADIABATIC LAPSE RATE of about 17.6° F (9.8° C) a day. This sometimes produces a layer of air that is warmer than the air beneath it, forming a type of INVERSION known as a *subsidence inversion*. Air trapped beneath the inversion contains all the particles and substances entering from the surface. These accumulate, reducing visibility. At the same time, enough water vapor may condense to form a thin layer of cloud immediately below the inversion level. Anticyclonic gloom is best seen from an aircraft flying above and to the side of the affected area.

In middle latitudes, anticyclonic gloom is more likely to form in winter than in summer. In summer, strong solar heating of the ground generates CONVECTION currents that break through the inversion, although the strong sunlight can also supply the energy for reactions among substances held in the inversion to form PHOTOCHEMICAL SMOG. This type of smog frequently occurs over Los Angeles, Mexico, Athens, and other cities with warm climates. In the subtropics, where anticyclones are more or less permanent (*see* SUBTROPICAL HIGH), the ingredients that make the gloom are often able to accumulate over a lengthy period.

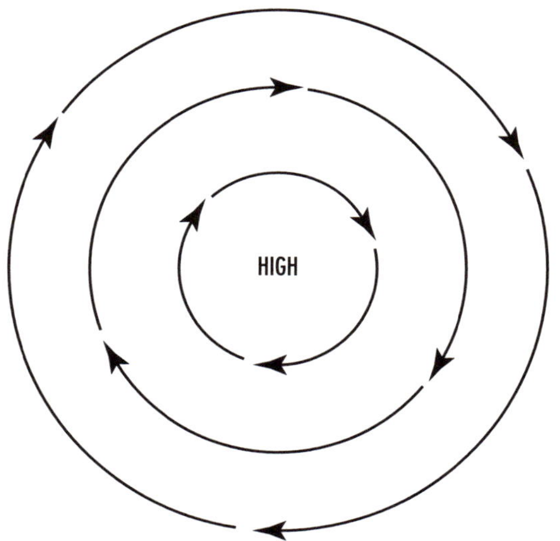

Anticyclonic. **Well clear of the surface, the winds flow almost parallel to the isobars, around the center of high pressure. Anticyclonic flow is clockwise in the Northern Hemisphere and counterclockwise in the Southern Hemisphere.**

anticyclonic shear Horizontal WIND SHEAR that produces an ANTICYCLONIC flow in the air to one side of it.

Antilles Current An ocean current that branches from the NORTH EQUATORIAL CURRENT and carries warm water along the northern coasts of the Great Antilles, in the Caribbean.

antisolar point The position in the sky that is directly opposite to the Sun. It is in the direction in which shadows point.

antitrade A wind that blows at a high level in the Tropics as part of the HADLEY CELL circulation. Its direction is opposite to that of the TRADE WINDS, so it blows from the southwest in the Northern Hemisphere and from the northwest in the Southern Hemisphere. Air that has been carried away from the equator by the

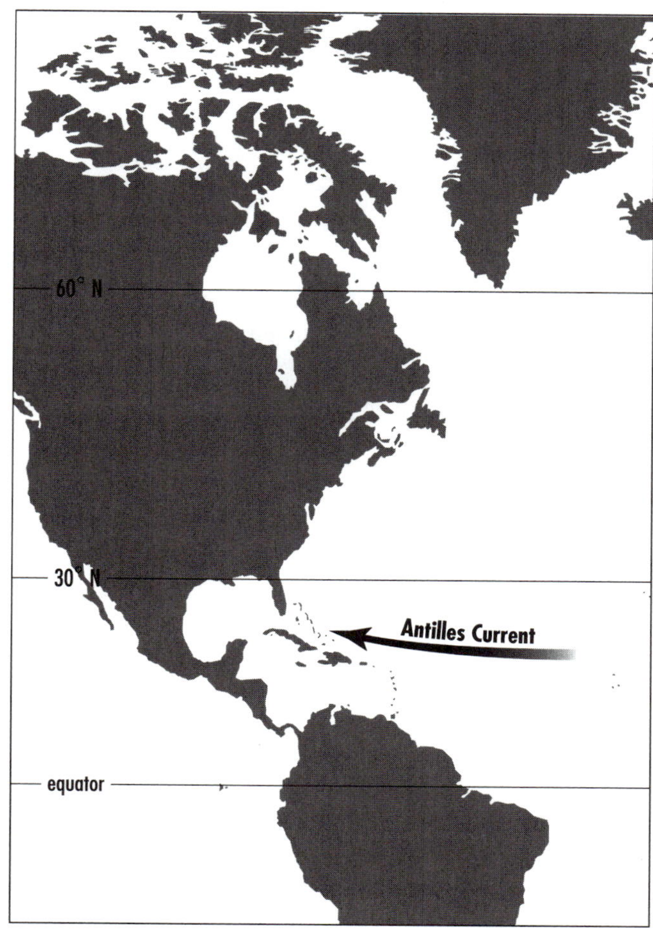

Antilles Current. **A warm ocean current that flows past the Antilles, in the Caribbean.**

antitrades subsides over the TROPICS to produce the SUBTROPICAL HIGHS.

antitriptic wind A wind that occurs locally or on a small scale and that is caused by differences in pressure or temperature. ANABATIC WINDS, KATABATIC WINDS, LAND AND SEA BREEZES, MOUNTAIN BREEZES, and VALLEY BREEZES are examples of antitriptic winds.

antitwilight arch A pink or lilac band that is seen at TWILIGHT as an arch rising to about 3° above the horizon at the ANTISOLAR POINT.

anvil See INCUS.

AO *See* ARCTIC OSCILLATION.

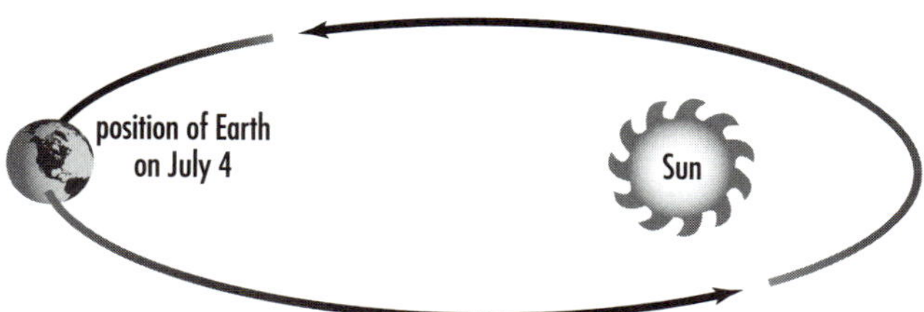

Aphelion. **The position in its elliptical solar orbit at which a body is farthest from the Sun.**

APF *See* ABSOLUTE POLLEN FREQUENCY.

aphelion The point in the eccentric solar orbit of a planet or other body when it is farthest from the Sun. At present, Earth is at aphelion on about July 4, but the dates of aphelion and PERIHELION change over a cycle of about 21,000 years (*see* MILANKOVICH CYCLES and PRECESSION OF THE EQUINOXES). The Earth receives 7 percent less solar radiation at aphelion than it does at perihelion.

Appleton layer A region of the atmosphere at a height of about 125 miles (200 km), where the IONIZATION of atoms and molecules causes the REFRACTION and REFLECTION of radio waves. It is named after the Scottish physicist Sir Edward Appleton (1892–1965), the inventor of RADAR.

applied climatology The use of climatological information and concepts to help in solving economic, social, and environmental problems. In some cases, the climatological contribution is so great that it has generated a scientific specialization. The application of climatological studies to agricultural planning, for example, produced AGROCLIMATOLOGY. An understanding of climate is also relevant to fisheries, the management of water resources, energy requirements (the length and severity of winters), the design of buildings, pollution control, aviation, and many other areas of life.

applied meteorology The preparation of weather reports and forecasts for specified groups of users, such as farmers, fishermen, and aircraft pilots. Specialized reports and forecasts are also prepared for climbers, backpackers, skiers, and others planning outdoor activities.

April showers The changeable weather, with showers of rain, that occurs in the month of April in many places in middle latitudes. During winter, deep DEPRESSIONS tend to follow one another, dominating the weather pattern and producing persistent PRECIPITATION and leaden skies. In early spring, the difference in temperature between the sea surface and land is at a maximum, as the sea is much colder than the land, which is beginning to warm as the sunshine grows stronger. Air crossing the sea is cold, and therefore its capacity for holding moisture is low. It warms as it crosses land. This increases its capacity for holding water vapor. Water evaporates into it and condenses again in the rising air to produce CUMULIFORM clouds and showers. April showers are usually light, because the air is still too cold to hold much moisture. Summer showers are much heavier. Despite its reputation, therefore, April is often drier than the summer months. In New York City, for example, an average 3.2 inches (81 mm) of rain falls in April, but 4.2 inches (107 mm) falls in July. London, England, receives an average 1.5 inches (37 mm) of rain in April, 1.8 inches (46 mm) in May, and 2.2 inches (57 mm) in July.

APT *See* AUTOMATIC PICTURE TRANSMISSION.

aquiclude (aquifuge) A rock that may be saturated with water, but that is almost impermeable to the flow of water through it. An aquiclude may form a boundary to an AQUIFER.

aquifer A layer of porous, permeable material (*see* PERMEABILITY) lying below the ground surface that is saturated with water and through which water flows. The water flowing through an aquifer is GROUNDWATER and the uppermost limit of the saturated zone is the water table. The aquifer lies above a layer of impermeable material, such as solid bedrock or compacted clay.

Water drains downward through the soil until it encounters the layer of impermeable material. It can then descend no farther, and so it accumulates above the impermeable layer, filling all the spaces between the mineral particles. The groundwater then flows laterally. Rock strata are rarely horizontal and the water flows down the slope. Immediately above the water table, where the soil is unsaturated, water is drawn upward by CAPILLARITY. The layer affected by capillarity is called the *capillary layer* or *capillary zone*.

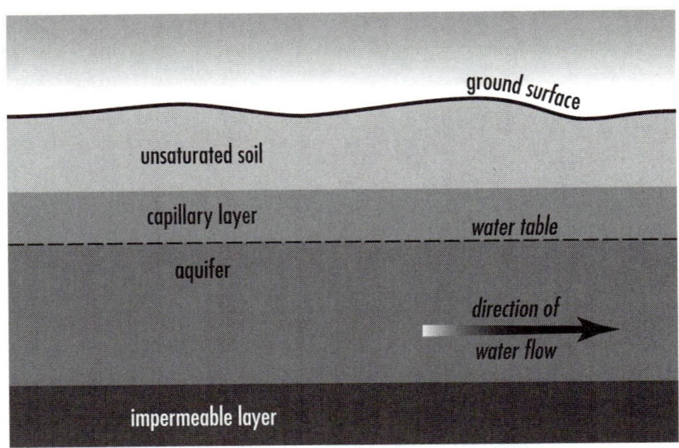

Aquifer. **Water drains downward through the soil until it encounters a layer of impermeable material. It accumulates above this layer. The saturated material is the aquifer.**

An aquifer may be composed of loose material such as sand or gravel, or of consolidated rock such as sandstone, siltstone, or shale. Sandstone is made from sand grains that have been packed and cemented together. Sand grains range in diameter from 62.5 μm to 2,000 μm (0.002–0.08 inch), siltstone is made from particles less than 4–62.5 μm (0.00016–0.002 inch) across, and shale is made from clay particles, which are less than 4 μm (0.00016 inch) across. More solid rocks can also form aquifers if they contain fissures and cavities through which water can move. Limestone, for example, is eroded by acids that are present naturally in water that has percolated through the soil. Over a long period this can produce caverns and tunnels carrying underground streams.

Faulting of the rocks may raise the underlying impermeable layer, causing the upper part of an aquifer to intersect the surface. Where this happens, water being carried by the aquifer emerges above the ground as a spring or seep.

If there is no impermeable layer above the aquifer, the aquifer is said to be *unconfined* and water can enter it by draining vertically from the surface. A confined aquifer lies beneath an impermeable layer, so it is sandwiched between two layers of impermeable material. Water cannot enter such an aquifer from directly above but flows in from the side that is at the higher elevation. A PERCHED AQUIFER may lie above a confined aquifer.

aquifuge *See* AQUICLUDE.

aquitard A rock through which water is able to pass slowly and with difficulty. An aquitard may form a partial boundary to an AQUIFER by slowing the movement of water, but without halting it altogether.

Ar *See* ARGON.

Aratus *See* WEATHER LORE.

arch cloud A stationary WAVE CLOUD, usually ALTOSTRATUS, that extends for a considerable distance along a mountain range with a wind blowing beneath it. The cloud is shaped like an arch and the wind blows down the mountainside as a FÖHN WIND. When seen from a distance, the arch indicates the approach of the wind. The Chinook Arch indicates the approach of a CHINOOK wind in the Rocky Mountains of North America, and the Southern Arch indicates the approach of a similar wind in the Southern Alps of New Zealand.

Archimedes' principle The physical principle that is used to calculate the BUOYANCY of a PARCEL OF AIR, or of any body that is immersed in a fluid. The principle states that when a body is immersed in a fluid, it displaces a quantity of fluid equal to its own volume. Consequently, its weight is reduced by the weight of the displaced fluid.

The Greek mathematician and inventor Archimedes (c. 287–212 B.C.E.) discovered this while seeking the solution to a puzzle presented to him by King Hieron II of Syracuse. Hieron had ordered a new crown from a local goldsmith, specifying that it was to be made of the purest gold. When it was delivered Hieron began to suspect that although the crown was golden in color the metal was alloyed with silver. Silver was less costly than gold, but the king had paid for gold and so, if his suspicion was correct, he had been cheated. Hieron asked Archimedes to determine whether or not the crown was made from pure gold. There was one condition: Archimedes must not damage the crown in any way.

Archimedes pondered this for some time. Then, as he was stepping into his bath, some of the water overflowed onto the floor. He realized that his foot and leg had displaced their own volume of water and this gave him a way to solve the problem of the crown. He was so delighted he ran naked down the street shouting, "Eureka; eureka" (I've found it; I've found it).

He weighed the crown very carefully, then borrowed from a trustworthy goldsmith a piece of pure gold of exactly the same weight. He immersed the gold in water, marking on the side of the vessel the height to which the water rose when he did so. Then he repeated the procedure with the crown and found the water rose a little farther. Silver weighs less than gold, so the volume of a given weight of an alloy of silver and gold is greater than the volume of the same weight of pure gold. Archimedes had proved that the crown was not made from pure gold. According to the story, Hieron had the dishonest goldsmith executed.

arcsecond See SECOND.

arctic air Air that is very cold and dry. The AIR MASS originates in the high-pressure areas of the Arctic and Antarctic (*see* SOURCE REGION). In winter, when the Arctic Ocean is completely covered by ice, continental arctic (cA) air forms over both the Arctic and over the continent of Antarctica. MARITIME arctic (mA) air forms off the coast of Antarctica. In summer, mA air forms over the Arctic, but the mA disappears from the Antarctic, where there is only cA air. Differences between cA and CONTINENTAL POLAR (cP) air are most noticeable in the middle and upper TROPOSPHERE, where cA air is the colder. In North America, cA air that forms in winter over the Arctic Basin and the GREENLAND ICE SHEET can bring COLD WAVES characterized by extremely cold, dry, and very stable air.

arctic bottom water (ABW) Very cold, dense water that sinks to a depth of about 20,000 feet (6,000 m) in the Greenland and Norwegian Seas, between Greenland and Norway. It fills the basins of those seas and from time to time spills through narrow channels in the ridge that lies between Scotland, Iceland, and Greenland.

Arctic Circle At latitude 66°30' N, the line that marks the boundary of a region within which there are a period in winter when the Sun does not rise above the horizon and a period in summer when it does not sink below it. At the Arctic Circle, this period is confined to the winter and summer SOLSTICES. At higher latitudes, the period of continuous darkness and continuous daylight lasts longer. At the summer solstice, the Sun is no higher than 47° above the horizon anywhere inside the

Arctic Circle. The variation in the length of daylight is due to the inclination of the Earth's rotational axis with respect to the PLANE OF THE ECLIPTIC.

arctic front A boundary that exists for most of the time in northern latitudes between ARCTIC AIR and more temperate AIR MASSES to the south. It extends across the whole of Eurasia and lies close to the ARCTIC CIRCLE.

A front that advances southward across North America and has arctic air behind it is also known as an *arctic front*. It usually yields bitterly cold weather.

A third type of arctic front forms in winter over snow and ice when the wind is weak or blows parallel to the edge of the SEA ICE. This situation produces a large difference in temperature between air that is chilled by its contact with the snow and ice and air that are warmed by contact with the sea. The resulting front is shallow, but it can trigger the formation of a POLAR LOW.

arctic haze A reduction in horizontal, but not vertical, VISIBILITY that sometimes occurs over the Arctic. It extends to a height of more than 30,000 feet (9,150 m).

arctic high An area of high surface atmospheric pressure that is located over the Arctic Basin. Maritime arctic (mA) air (*see* MARITIME AIR and ARCTIC AIR) covers the basin in summer, and continental arctic (cA) air in winter (*see* CONTINENTAL AIR), when ice and snow cover the entire surface. The lower layers of air are cooled by contact with the snow- and ice-covered surface. As a result of the low temperature the air contains little moisture; in winter the MIXING RATIO near the surface is 0.1–1.5 g kg^{-1} (0.0016–0.7 ounce of water vapor per pound of dry air). Because of the dryness of the air there is little cloud. Strong cooling from below makes the air very stable (*see* STABILITY OF AIR), and there is often a temperature INVERSION from the surface to about the 850–mb level. The high persists through the summer, but it is weaker and the cold AIR MASS is shallower than in winter.

arctic mist A very thin ICE FOG that sometimes occurs in high latitudes when the air temperature is well below freezing. It reduces VISIBILITY only slightly.

Arctic Oscillation (AO) A periodic change in the distribution of pressure over the North Pole and over a circle at about 55° N, passing through southern Alaska and central Europe. When pressure is high over the Pole it is low farther south, and vice versa. The AO is said to be positive when pressure is low over the Pole. High pressure farther south strengthens the westerly winds there and storms travel farther north, affecting Scandinavia and Alaska, but producing dry conditions in the Mediterranean region and California. Warm air is carried across Eurasia. When the AO is negative, California and the Mediterranean region experience wet weather and the interior of Eurasia is cold. Some climatologists suspect the NORTH ATLANTIC OSCILLATION may be part of the much larger AO. The AO has been positive for several decades and may be responsible for the recent warming of the high-latitude Northern Hemisphere.

(*See* Richard A. Kerr, "A New Force in High-Latitude Climate," *Science,* vol. 284, April 9, 1999, pp. 241–242.)

Arctic sea smoke FOG that forms when very cold air that has crossed ICE SHEETS and GLACIERS moves from the land across a sea surface that is 36°–54° F (20°–30° C) warmer. Because the air is very cold, it contains little water vapor. In the layer of air immediately adjacent to the much warmer water, the amount of water vapor needed to cause SATURATION is much greater than the water vapor content of the air. Consequently, water evaporates rapidly into the air and then rises by CONVECTION. This carries it into the cold air, where it condenses again to produce cloud. The "smoke" is often very dense, but it usually extends to a height of no more than about 35 feet (10 m).

arcus A supplementary feature of cumulonimbus clouds in which the lower, darkest part of the cloud is arched. This feature occurs most commonly in clouds that form along squall lines. *Arcus* is a Latin word that means "bow" or "curve."

area forecast A weather forecast that is prepared for a specified geographic area.

Argo A project that will use robots to measure the temperature of the oceans at a depth of 4,500–6,000 feet (1,370–1,830 m). A total of 3,000 robots will be used. Up to 150 of them will be in the water at any time, shared among the world's oceans. Each robot will be an upright cylinder, 3 feet (90 cm) long, with a 3-ft (90-cm) antenna on top. It will contain instruments to measure and record temperature and salinity, a battery-powered pump, and a bladder. The robot will allow its bladder to fill with water. This will cause it to sink to its operating depth, where it will remain for 10 days, drifting with the ocean current. At the end of this period it will pump the water from its bladder and rise to the surface. At the surface it will radio its data to a receiving station via a satellite link. Most data on ocean temperature and salinity are obtained from ships. The Argo project will allow changes in temperature and salinity to be monitored away from the main shipping lanes. The project is sponsored by the UK METEOROLOGICAL OFFICE, the Proudman Oceanographic Laboratory, Southampton Oceanographic Centre, and the Hydrographic Office. It plans to launch the first robots by the end of 2001 and for the project to be fully operational by 2003.

argon (Ar) A colorless, odorless gas that constitutes 0.93 percent of the atmosphere by volume (*see* ATMOSPHERIC COMPOSITION) and that is considered one of the major atmospheric constituents. It is a NOBLE GAS and has no true compounds. It was discovered in 1894 by LORD RAYLEIGH and Sir William Ramsay. Argon has atomic number 18, relative atomic mass 39.948, and density (at sea-level pressure and 32° F [0° C]) 0.001 ounce per cubic inch (0.00178 g cm^{-3}). It melts at –308.2° F (–189° C) and boils at –301° F (–185° C).

arid climate In the THORNTHWAITE CLIMATE CLASSIFICATION, a climate in which the MOISTURE INDEX is between –100 and –67 and the POTENTIAL EVAPOTRANSPIRATION is 5.6–11.2 inches (14.2–28.5 cm). It is designated *E*.

aridity Dryness; the extent to which its lack of PRECIPITATION makes a climate incapable of supporting plant growth.

aridity index A measure of the extent to which the amount of water available to plants falls short of the amount needed for healthy growth. The term was introduced in 1948 by the American climatologist C. W. THORNTHWAITE and is calculated as $100W_d/PE$, where W_d is the WATER DEFICIT and PE is the POTENTIAL EVAPOTRANSPIRATION.

Aristotle (384–322 B.C.E.) Greek *Philosopher, Scientist* Aristotle was born in 384 B.C.E. at Stagirus, a Greek colony on the coast of Macedonia. Both his parents were Greek. His father, Nichomachus, the personal physician to the king of Macedonia, Amyntas III, died when Aristotle was still a boy. Aristotle was then brought up by a guardian, Proxenus, who in about 367 B.C.E., when Aristotle was 17, sent him to the Academy in Athens, which was led by the philosopher Plato (428 or 427–348 or 347 B.C.E.). Aristotle remained at the Academy for 20 years, first as a pupil and then as a teacher.

In 347 B.C.E. Athens was at war with Macedonia. Amyntas had died and the new king was his son, Philip II. Then Plato died. Perhaps for political reasons, because he was sympathetic to the Macedonian cause, or perhaps because of the change in the leadership of the Academy, Aristotle left Athens. He settled first on the coast of Anatolia, in what is now Turkey; then on the island of Lesbos, where he lived from 345 B.C.E. to 343 B.C.E. Finally he returned to Macedonia. In the course of his travels he married Pythias. She was the daughter or niece of Hermias, the ruler of the land where Aristotle first arrived after leaving Athens. Back in Macedonia, Philip appointed Aristotle to supervise the education of his son, the 13-year-old Alexander, who later became known as Alexander the Great. Later in his life Aristotle was very wealthy, possibly from the money he was paid for teaching Alexander.

Aristotle returned to Athens in about 335 B.C.E. and for the next 12 years taught at the Lyceum, one of the three most famous schools in the city. There he began to assemble a library of books and maps and a museum of natural history. He also established a zoo with animals that were captured during Alexander's campaigns in Asia.

Alexander died in 323 B.C.E. and the opponents of the Macedonians became powerful in Athens. Aristotle was associated with a renowned Macedonian general and he was charged with impiety. Rather than face trial, and possibly death, he moved to Chalcis (now called *Khalkis*) on the Greek island of Euboea, north of Athens. The following year he fell ill and died. He was 62 years old.

Aristotle was one of the most original thinkers who ever lived. Every subject interested him. He wrote about logic, ethics, politics, biology, physics, astronomy, and many other topics. His writings are contained in 47 surviving works. Some of these comprise several volumes, and others are very short.

One of his works is called *Meteorologica.* The title means "account [*logos*] of lofty things [*meteoros*]"; from it we derive our word *meteorology.*

In *Meteorologica,* Aristotle set out his own explanations for the weather. He had studied Egyptian ideas on the subject and the methods for classifying winds that had been devised by the Babylonians, and in addition to these he drew on a wide variety of sources.

Aristotle. **The Greek philosopher who attempted to explain the weather and who gave us the word *meteorology. (From the collections of the Library of Congress)***

He maintained that the weather is confined to the region between the Earth and the Moon. This region is composed of four elements: earth, air, fire, and water. Earth and water are heavy and sink, whereas air and fire are light and rise. Aristotle proposed theories to explain the formation of clouds, rain, hail, wind, thunder, lightning, and storms. He argued, for example, that some of the water vapor formed during the day does not rise very high because the ratio of the fire that is raising it to the water being raised is too small. At night the water cools and descends and is then called *dew* or *hoar frost* if it freezes before it has condensed to water. It is dew, he said, when the vapor has condensed to water and the heat is not so great as to dry up the moisture that has been raised, nor the cold sufficient for the water to freeze. He observed that although hailstones are made from ice, hailstorms are most common in spring and autumn. They are rare in winter and happen when the weather is mild. He suggested that in warm weather the cold forms discrete areas within the surrounding heat, which can cause water vapor to condense rapidly. That is why raindrops are bigger in warm weather than in cold. If the cold is very concentrated, however, it can freeze the raindrops as they form, producing hail.

Aristotle had no instruments to measure temperature, pressure, or humidity. Nor did he have the facilities to compile a picture of weather conditions over a large area. Lacking any means to validate his ideas, Aristotle was not in a position to develop an accurate understanding of atmospheric processes. The importance of his contribution to scientific thinking arises from his insistence on basing theories on observed facts rather than tradition or unsupported opinion.

arithmetic mean *See* MEAN.

Arrhenius, Svante August (1859–1927) Swedish *Physical chemist* Svante Arrhenius was born on February 19, 1859, on the estate of Vik, near Uppsala, Sweden, which was owned by the University of Uppsala. In 1860 the family moved to Uppsala.

Svante began his education at the cathedral school in Uppsala. He was brilliantly clever, especially at mathematics, and was accepted as a student at Uppsala University when he was only 17. He studied chemistry, physics, and mathematics; he graduated in 1878, then stayed on to start working for his doctorate. After a time he grew dissatisfied with the quality of the teaching in physics, and in 1881 he moved to Stockholm to study under the physicist Erik Edlund (1819–88).

He completed his doctoral thesis in 1884 and submitted it to Uppsala University. He wrote it in French: *Recherches sur la conductibilité galvanique des électrolytes* (Investigations on the galvanic conductivity of electrolytes). An electrolyte is a solution of a chemical in water that conducts electricity. The first part of his thesis dealt with ways to measure the electrical conductivity of very weak solutions and the second with the reason the solution is conductive. This, he proposed, is because the dissolved substance dissociates into charged ions (for example, common salt, which is sodium chloride [NaCl], dissociates into sodium [Na^+] and chloride [Cl^-] ions) and the ions move through the solution.

Arrhenius presented his thesis to his professor, a man he greatly admired. According to his own account, he said, "I have a new theory of electrical conductivity as a cause of chemical reactions." The professor replied, "This is very interesting. Good-bye." The professor knew that many theories are formed and almost all of them turn out to be wrong and soon disappear. He concluded, therefore, that Arrhenius's theory was most probably mistaken. Like most of his colleagues, the professor believed in experimentation and Arrhenius had performed no experiments in the course of developing his idea. The thesis was accepted, but it was awarded only fourth class, the lowest grade. The thesis later earned a Nobel Prize.

Undeterred by the lack of interest at Uppsala, Arrhenius sent copies of his thesis to several of the most eminent chemists of the time. This led to the offer of a job at the University of Riga, Latvia, and the offer persuaded the Uppsala authorities to reconsider their opinion. Toward the end of 1884 he was offered a post at Uppsala, and later a traveling fellowship that allowed him to meet other scientists working in the same field.

In 1891 Arrhenius was offered a professorship at the University of Giessen, Germany, but he declined it, because he preferred to remain in Sweden. Instead, he accepted a lectureship at the Stockholms Högskola (high school), where in 1895 he became professor of physics. From 1897 until 1905 Arrhenius was also rector. The high school was equivalent to the science faculty of a university, but it was not empowered to award degrees or accept doctoral theses. It became the University of Stockholm in 1960.

Svante August Arrhenius. **The Swedish physical chemist who was the first to calculate the climatic effect of an increase in the atmospheric concentration of carbon dioxide.** ***(From the collections of the Library of Congress)***

He retired from the professorship in 1905 and refused another invitation from Germany, this time to become a professor at the University of Berlin. In 1905 the Swedish Academy of Sciences decided to establish a Nobel Institute for Physical Chemistry and appointed Arrhenius director. He remained in this position until shortly before his death.

Arrhenius was a man of wide interests. He applied his knowledge of chemical reactions to the effects on the body of toxins and antitoxins, which he described in a series of lectures he delivered in 1904 at the University of California. He was interested in immunology and the origin of life on Earth. It was Arrhenius who first proposed the idea of *panspermia,* according to which life arrived on Earth in the form of spores that had drifted through space. He wrote about astronomy, especially comets and the possibility of life on Mars.

He also studied what is now called the GREENHOUSE EFFECT. In 1896 he published a paper, "On the Influence of Carbonic Acid in the Air upon the Temperature of the Ground," (*Philosophical Magazine,* vol. 41, pp. 237–271). He was not the first scientist to consider the absorption of energy by carbon dioxide, and it was the French mathematical physicist Jean Baptiste Joseph Fourier (1768–1830) who suggested in 1827 that the atmosphere acts in the same way as the glass of a greenhouse, allowing light in but preventing heat from leaving. Arrhenius turned earlier speculations into hard numbers.

He calculated the effect carbon dioxide would have if the atmospheric concentration of it were altered. He worked out the resulting change in mean temperature for 13 belts of latitude, each 10 degrees, from 70° N to 60° S, for the four seasons of the year and the mean for the year. For each of these belts of latitude and seasons, and for the whole year, he worked out what the temperature would be if the carbon dioxide concentration were 67 percent, 150 percent, 200 percent, 250 percent, and 300 percent of the concentration that actually existed in the late 19th century. He calculated that a doubling of atmospheric carbon dioxide would increase the mean annual temperature by 4.95° C at the equator and by 6.05° C at 60° N. The task involved thousands upon thousands of calculations, all of which he performed by hand.

For his work on electrolytes, Arrhenius received the 1903 Nobel Prize in chemistry. He also received many other awards and honorary degrees. He married twice, first in 1894 to Sofia Rudbeck, by whom he had a son, and in 1905 to Maria Johansson, by whom he had one son and two daughters.

Arrhenius was a happy, contented, genial man who made many friends and delighted in meeting them. During World War I he worked successfully to obtain the release of German and Austrian scientists who were prisoners of war.

He was also a popular lecturer and author. In his later years he was in constant demand and traveled widely to attend meetings and deliver lectures. His incessant hard work may have weakened his health, because he was only 68 when he died in Stockholm on October 2, 1927. He is buried in Uppsala.

(You can learn more about Arrhenius at www.nobel.se/chemistry/laureates/1903/arrhenius-bio.html and there is a summary of his paper on the greenhouse effect at maple.lemoyne.edu/~giunta/Arrhenius.html.)

As *See* ALTOSTRATUS.

ASOS *See* AUTOMATED SURFACE OBSERVING SYSTEMS.

aspect The direction that sloping ground faces. This determines its exposure to direct sunlight (*see* ADRET and UBAC).

aspirated hygrometer A PSYCHROMETER in which the necessary flow of air over the thermometers is provided by placing them inside a tube through which air is blown.

astronomical twilight The dim daylight that illuminates areas inside the ARCTIC CIRCLE and ANTARCTIC CIRCLE during the early and late part of the winter. The Sun is below the horizon, but when it is less than 18° below the horizon the SCATTERING and REFRACTION of light allow some sunlight to reach the surface.

Atlantic conveyor A system of ocean currents that conveys cold water away from the edge of the Arctic Circle in the North Atlantic and warm water from the Pacific through the Indian Ocean and into the Atlantic. The existence of the conveyor was discovered in the 1980s by an American geochemist, Wallace S. Broecker, the Newberry Professor of Geology at Columbia University.

Since the conveyor forms a closed loop, a description of it can begin anywhere, but it is driven by the formation of the NORTH ATLANTIC DEEP WATER (NADW). This sinks to the floor of the North Atlantic and flows south, across the equator and to the edge of the Antarctic Circle, where it joins the WEST WIND DRIFT, or Circumpolar Current, flowing from west to east. Part of the current turns north into the Indian Ocean, past the eastern coasts of Africa and Madagascar, and turns south again to the south of Sri Lanka, rejoining the main current.

This diverges from the West Wind Drift to the south of New Zealand, turning northward into the Pacific Ocean and rising to become an intermediate current, flowing about 3,500 feet (1,070 m) below the surface. It crosses the equator; makes a clockwise loop in the North Pacific; then travels westward through the

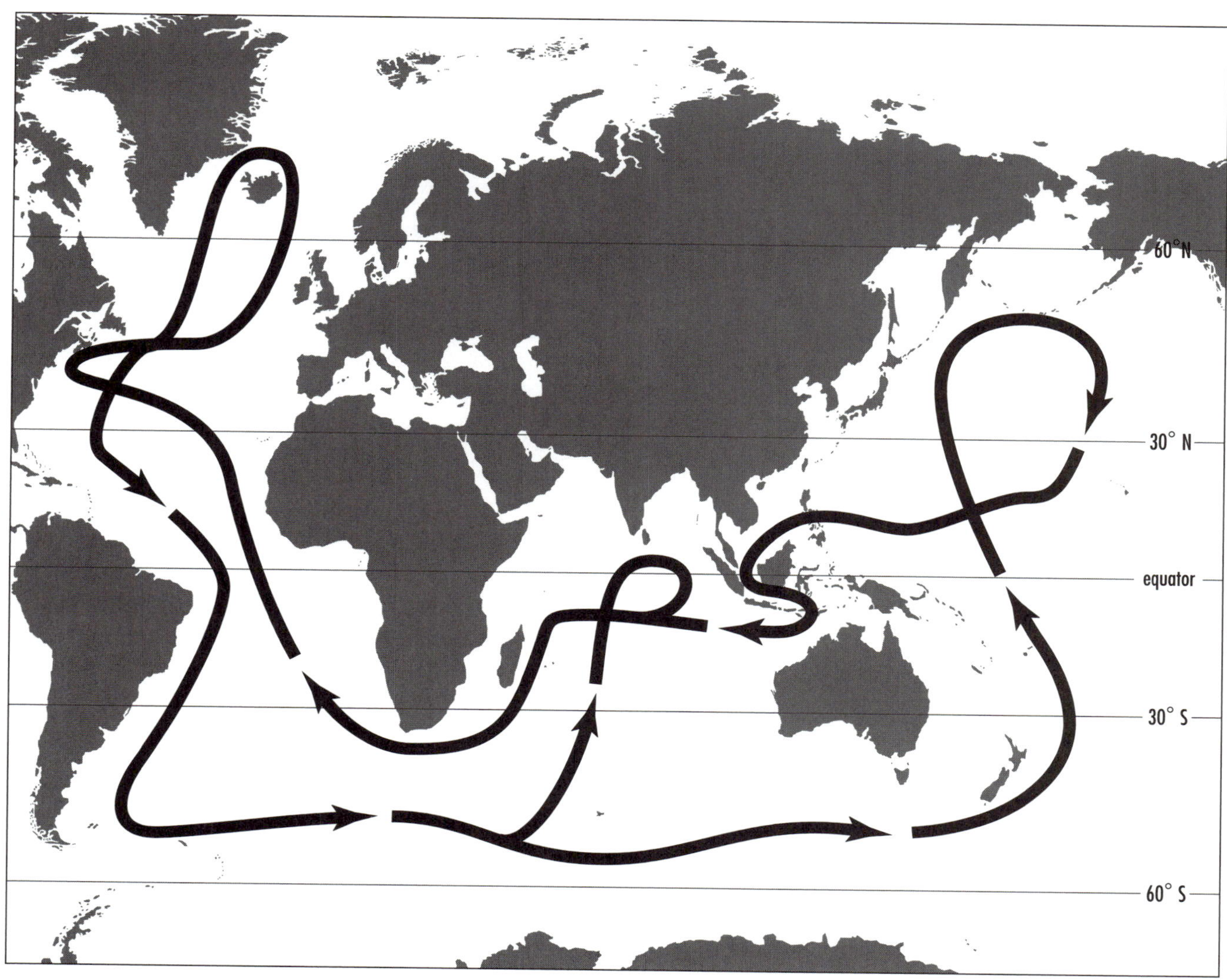

Atlantic conveyor. **A system of ocean currents carries cold water toward the equator and warm water toward the poles, strongly influencing climate.**

islands of Indonesia, where it crosses the equator once more; across the Indian Ocean; around Africa; and then northward, crossing the equator for the fourth time and returning to the North Atlantic.

The NADW is cold and removes cold water from the North Atlantic. This water remains cold until it rises in the South Pacific. During its progress through equatorial and tropical regions the water warms, and it returns as warm water to the area near Greenland, where it replaces the NADW that is sinking and moving south.

This circulation is of major importance in regulating the climates of the world. It is believed to have failed several times in the past. When it did so, the world experienced climates very different from those of today, when the conveyor is active (*see* YOUNGER DRYAS). Bill Gray, an American meteorologist, has suggested links between the conveyor and climate over the past 125 years. When the conveyor flows strongly there are an increase in the number of HURRICANES, heavy rainfall in the SAHEL region along the southern edge of the Sahara Desert, few ENSO events, and a general decrease in global mean temperatures. These conditions occurred between 1870 and 1899, and between 1943 and 1967. When the conveyor flows weakly, as it did between

1900 and 1942, and 1968 and 1993, there are fewer hurricanes and more ENSO events, rainfall in the Sahel region is average or below average, and global mean temperatures are higher. Some scientists fear that a general increase in global temperatures may affect the formation of NADW and trigger a change in the conveyor, causing it to flow more weakly or even to shut down. Were this to happen, it is possible that GLOBAL WARMING might induce ice age conditions in northern Europe.

Atlantic high An ANTICYCLONE that covers a large part of the subtropical North Atlantic Ocean. There is a similar anticyclone over the South Atlantic. The Atlantic highs are SOURCE REGIONS for maritime AIR MASSES. The North Atlantic and PACIFIC HIGH together cover one-quarter of the Northern Hemisphere and for six months of each year cover almost 60 percent of it.

Atlantic Multidecadal Oscillation (AMO) A change in the climate over the North Atlantic Ocean that occurs over a cycle of about 50–70 years. This has been detected over several centuries. Early in the 20th century it took the climate from unusually cold to unusually warm and back again. It was first noted in 1964 by JACOB BJERKNES, who suggested the slow warming of surface waters in the 1910s and 1920s might have been caused by a surge of warm water carried by the GULF STREAM. This produced high global air temperatures by the 1940s. These were followed by a sharp cooling, which ended in the 1980s. The temperature changes are of several tenths of a degree Celsius to either side of the mean. It is thought that the AMO is linked to the NORTH ATLANTIC OSCILLATION and ARCTIC OSCILLATION, but in ways that are not understood.

Atlantic period A time during the present, FLANDRIAN INTERGLACIAL, that lasted from about 7,500 to 5,000 years ago, after the BOREAL PERIOD. During the Atlantic period the climate of northwestern Europe was moist and warmer than it has been at any time since.

atmometer An instrument that measures EVAPORATION. It consists of a calibrated glass tube with one end that is open to allow water to evaporate.

atmophile An adjective that describes one of the chemical elements that are concentrated in the atmosphere and that together typify its composition (*see* ATMOSPHERIC COMPOSITION). Atmophile elements may be uncombined, for example, oxygen (O_2) and nitrogen (N_2), or combined, for example, carbon and oxygen in carbon dioxide (CO_2), hydrogen and oxygen in water vapor (H_2O), and carbon and hydrogen in methane (CH_4).

atmosphere *See* STANDARD ATMOSPHERE.

atmosphere, evolution of The air we breathe today has a very different composition from the atmosphere that existed on Earth billions of years ago. The present atmosphere consists of nitrogen, oxygen, and a number of trace gases (*see* ATMOSPHERIC COMPOSITION). This composition evolved long ago and is now stable.

When the Earth first formed, about 4.5 billion (4.5×10^9) years ago, it may have had an atmosphere consisting mainly of hydrogen and helium. These gases are very common throughout the universe and were major constituents of the cloud of gas and dust from which the solar system condensed. The gases are also very light and were soon swept away, because Earth does not exert sufficient gravitational attraction to retain them.

As the first atmosphere was lost, a second was already replacing it. This was composed of gases released from the many volcanoes on the early Earth and from the solid bodies that bombarded the Earth from space. No one knows the composition of that atmosphere, but one point is certain: It exerted a strong GREENHOUSE EFFECT.

When the Earth first formed, the Sun emitted about 30 percent less warmth than it does today. Had there been no greenhouse warming, all the water on Earth would have been frozen. Yet there are sedimentary rocks that have been dated to about 4 billion years old. These formed from sediments that were deposited on the seabed; clearly, the world was not entirely frozen. At that time, the sky would have looked very different. The Earth was spinning faster, so days were about 14 hours long, and the Moon was much closer and would have looked very much bigger than it does now. The sky itself was probably white, or perhaps pale yellow. It was certainly not blue, because that is the color of oxygen and the air contained no more than 0.1 percent oxygen by volume.

At one time, scientists thought carbon dioxide might have been the principal constituent of the atmosphere; there may have been 300 to 1,000 times more of it than there is in the present atmosphere. The carbon dioxide would have been released from volcanoes, and it would have exerted a powerful greenhouse effect. It would also have reacted with iron present on the surface to form iron carbonate. Unfortunately, rocks of the appropriate age are not rich in iron carbonate, so it seems unlikely that the air was mainly carbon dioxide. Perhaps, then, ammonia was the greenhouse gas? It could have formed by chemical reactions among the compounds dissolved in surface water and it is a strong greenhouse gas. Unfortunately, it breaks down in bright sunlight, so it could survive only if some other gas, such as methane, formed a protective haze above it. By shielding the ammonia atmosphere from sunlight, however, the haze itself would have had a cooling effect that might have completely offset the greenhouse warming.

Life was present by this time. There are chemical indications of it in rocks 3.8 billion years old and PHOTOSYNTHESIS had begun by 2.7 billion years ago. Alongside the photosynthesizing organisms, there were others that released methane as a by-product of their metabolism. Some scientists now suspect that methane, produced by living cells, may have accumulated in the atmosphere and produced the necessary amount of greenhouse warming. Methane reacts with oxygen to produce carbon dioxide and water. Today a molecule of methane survives in air for only about 12 years before being oxidized, but in the oxygen-free early atmosphere it might have survived 20,000 years.

Oxygen is released as a by-product of photosynthesis. At first, it reacted with volcanic gases and iron exposed at the surface. Eventually all the exposed iron had been oxidized and volcanic eruptions became less frequent. Oxygen began to accumulate, and between 2.2 and 1.8 billion years ago the atmospheric content of oxygen increased rapidly to 10–15 percent of its present concentration. The oxygen would have destroyed the methane, and some scientists suggest this ended the greenhouse effect, causing an ICE AGE that covered the entire planet. A second rapid increase in the oxygen concentration, possibly linked to a decrease in the concentration of carbon dioxide, occurred about 600 million years ago and also triggered a sharp fall in temperatures.

The early atmosphere probably contained less nitrogen than the present atmosphere; its concentration increased once life became established and the NITROGEN CYCLE began.

Once photosynthesis was widespread and the nitrogen cycle fully functional, oxygen and nitrogen accumulated in the atmosphere and carbon dioxide was removed from it to be incorporated in living organisms and eventually to be buried in the form of carbonate rocks. By about 500 million years ago the atmospheric gases had reached approximately their present proportions.

(For more information see James Lovelock, *The Ages of Gaia,* [New York: Oxford University Press, 1989], and Tyler Volk, *Gaia's Body: Toward a Physiology of Earth,* [New York: Springer-Verlag, 1998].)

atmospheric boundary layer *See* PLANETARY BOUNDARY LAYER.

atmospheric cell A PARCEL OF AIR inside which the air is moving. In the circulation of the atmosphere, HADLEY CELLS, FERREL CELLS, and POLAR CELLS are atmospheric cells.

atmospheric chemistry The scientific study of the chemical composition of the atmosphere (*see* ATMOSPHERIC COMPOSITION) and of the chemical processes that occur within it. Much of our knowledge of gases and the process of combustion was obtained by scientists who were studying the gases present in the air. The atmosphere has not always had its present composition, and atmospheric chemists have studied the stages by which that composition evolved (*see* ATMOSPHERE, EVOLUTION OF). Today, atmospheric chemists are especially concerned with the fate of pollutants released into the air (*see* AIR POLLUTION, OZONE DEPLETION POTENTIAL, and OZONE LAYER) and with the chemical behavior of gases that absorb radiation (*see* GLOBAL WARMING POTENTIAL).

atmospheric composition The air we breathe is a mixture of gases. That is to say, it consists of a number of gases that are thoroughly mixed, but that remain distinct so each gas can be considered separately (*see* PARTIAL PRESSURE) and can be separated from the mixture. When we breathe we inhale the complete mixture, but our lungs extract oxygen from the mixture and add carbon dioxide, the by-product of respiration, to it. We are able to do this because oxygen is present in a mixture. In carbon dioxide (CO_2) oxygen is present in a compound, and separating it is much more difficult. The air also contains AEROSOL particles.

The proportions of the principal gases are fairly constant throughout the lower layers of the atmosphere that the HOMOSPHERE comprises. Nitrogen is the most abundant gas, followed by oxygen; in all there are 18 gases present.

Composition of the Present Atmosphere

Gas	Chemical formula	Abundance
Major constituents		
nitrogen	N_2	78.08%
oxygen	O_2	20.95%
argon	Ar	0.93%
water vapor	H_2O	variable
Minor constituents		
carbon dioxide	CO_2	365 p.p.m.v.
neon	Ne	18 p.p.m.v.
helium	He	5 p.p.m.v.
methane	CH_4	2 p.p.m.v.
krypton	Kr	1 p.p.m.v.
hydrogen	H_2	0.5 p.p.m.v.
nitrous oxide	N_2O	0.3 p.p.m.v.
carbon monoxide	CO	0.05–0.2 p.p.m.v.
xenon	Xe	0.08 p.p.m.v.
ozone	O_3	variable
Trace constituents		
ammonia	NH_3	4 p.p.b.v.
nitrogen dioxide	NO_2	1 p.p.b.v.
sulfur dioxide	SO_2	1 p.p.b.v.
hydrogen sulfide	H_2S	0.05 p.p.b.v.

The atmosphere did not always have the composition it has now (*see* ATMOSPHERE, EVOLUTION OF). Oxygen is present as a by-product of PHOTOSYNTHESIS, for example, and nitrogen as a consequence of microbial activity. This indicates that prior to the emergence of living organisms the atmosphere consisted of a quite different gaseous mixture from the present one.

The table lists the constituents of the atmosphere. Nitrogen, oxygen, and argon together constitute 99.96 percent of the air and their proportions are given as percentages of the total. Water vapor and ozone are present in such widely variable amounts that proportions cannot be given. For the minor constituents, the amounts present are given in parts per million by volume (p.p.m.v.) and for the trace constituents in parts per billion by volume (p.p.b.v.). To compare these units of measurement, 1 p.p.m. = 0.0001 percent and 1 ppb. = 0.0000001 percent.

atmospheric dispersion The dilution of pollutants (*see* AIR POLLUTION) as they mix with a much larger volume of air. The rate of dispersion varies according to local atmospheric conditions. Pollution incidents occur when atmospheric dispersion fails to reduce pollutant concentrations to levels that are harmless. Atmospheric dispersion should not be confused with the DISPERSION of light.

atmospheric layer *See* ATMOSPHERIC SHELL.

Atmospheric Research and Environment Program *See* WORLD METEOROLOGICAL ORGANIZATION.

atmospheric shell (atmospheric layer) One of the layers of the atmosphere (*see* ATMOSPHERIC STRUCTURE).

atmospheric structure Conditions in the atmosphere do not remain constant or change at a constant rate all the way from the surface of the Earth to its uppermost limit. Rather, the atmosphere is layered, and each layer forms a shell around the Earth and is enclosed by the shell outside it.

There is no precise upper boundary to the atmosphere. About 90 percent of its total mass lies between the surface and a height of about 10 miles (16 km), and half of its mass is in the layer below about 3.5 miles (5.5 km). Above 10 miles, the remaining one-tenth of the atmospheric mass extends at least to a height of 350 miles (550 km), although the density of the air there is about one-million-millionth (one-trillionth) of the density at sea level. Beyond that height the atmosphere merges imperceptibly with the atoms and molecules of interplanetary space, and especially with the outer fringes of the Sun's atmosphere.

The lowest layer of the atmosphere is the TROPOSPHERE. In this layer the air temperature decreases steadily with height at a fairly constant rate known as the LAPSE RATE. It is the region in which all weather phenomena occur. The lowest part of the troposphere, where the movement of air is strongly affected by its contact with the surface, constitutes the PLANETARY BOUNDARY LAYER.

The upper boundary of the troposphere, called the *TROPOPAUSE,* is the height at which the temperature ceases to decrease with height. The tropopause is not always continuous over the entire Earth. There are sometimes breaks where one section ends and another begins. Such breaks mark the fronts between ATMOSPHERIC CELLS and are associated with the JET STREAMS.

The tropopause forms a very effective physical boundary. Tropospheric air rarely crosses it and only under special circumstances. Vigorous CUMULONIMBUS clouds can pierce it, but their tops do not penetrate far. Volcanic eruptions can project particles of ash and sulfate through the tropopause, but when they do so, a long time, sometimes several years, passes before the particles fall back into the troposphere.

In the lower STRATOSPHERE temperature remains constant with height. Then, at about 12 miles (20 km) the temperature begins to increase with height and at about 20 miles (32 km) it increases more rapidly and continues to do so until it reaches another region, where the temperature remains constant with height. This is the STRATOPAUSE and it forms the upper boundary to the stratosphere. The stratosphere contains the OZONE LAYER.

Only about 0.1 percent of the mass of the atmosphere is located above the stratopause. The layer above the stratopause is called the *MESOSPHERE.* Above a height of about 35 miles (56 km) the temperature falls rapidly with height all the way to the MESOPAUSE, at about 50 miles (80 km), above which lies the THERMOSPHERE. The temperature remains constant with height in the lower thermosphere, but at about 55 miles (88 km) it begins increasing rapidly with height. This does not mean the air at this height feels hot. Its temperature is high because atoms and molecules are exposed to intense sunlight, which they absorb; they move faster, but they are too widely scattered to have any appreciable warming effect on a body resting in this air or moving slowly through it.

At the thermopause, which is the upper boundary of the thermosphere, the temperature is sometimes above 1,830° F (1,000° C). The thermopause is between 310 and 620 miles (500–1,000 km) above the

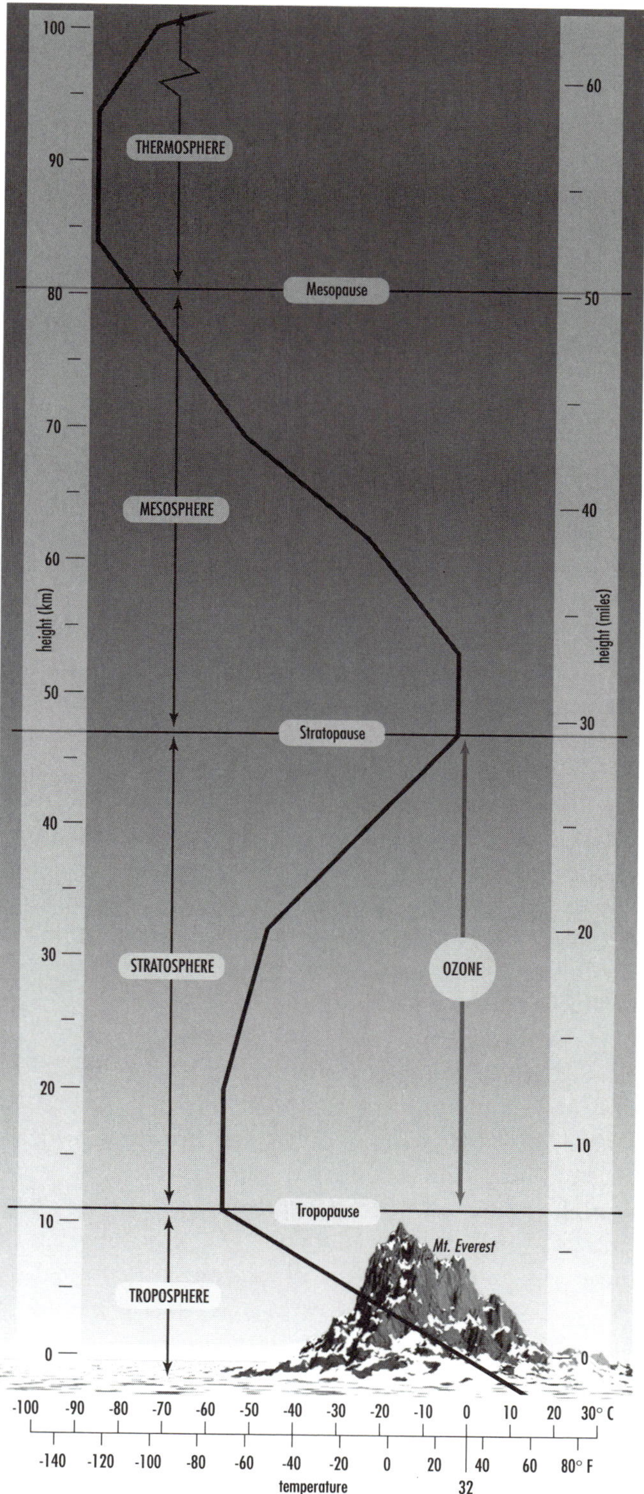

Atmospheric structure. **The atmosphere consists of a series of layers, one outside the other, and identified mainly by the way temperature changes with height within them.**

surface, its height varying with the intensity of the sunlight.

In the EXOSPHERE, which is the outermost layer of the atmosphere, the atoms and molecules are so widely dispersed they may not collide with one another.

(You can read more about the structure of the atmosphere in Michael Allaby, *Elements: Air, the Nature of Atmosphere and the Climate,* [New York: Facts On File, 1992].)

atmospheric tides TIDES that occur in the atmosphere. Like ocean tides, atmospheric tides are produced by the gravitational attraction of the Moon and Sun, but as a result of the low density of air the gravitational effect is small. The maximum value is at the equator, where the tidal variation never exceeds 0.4 lb ft^2 (20 Pa). It can be measured as a twice-daily variation in atmospheric pressure, averaged over a long period, that coincides with the lunar orbit.

The Sun exerts a much stronger effect, however, producing a maximum pressure every morning and evening between about 0900 and 1000 local time. Because the Sun is much farther from the Earth than the Moon, and gravity is subject to the INVERSE SQUARE LAW, the gravitational force exerted at the surface of the Earth by the Sun is only about 47 percent of that exerted by the Moon. Despite this, the pressure difference due to the solar atmospheric tides is almost 6.3 lb ft^2 (300 Pa), which is 15 times greater than that produced by the lunar tides.

The difference is due to the heating effect that is exerted by the Sun, but not by the Moon. This produces what are sometimes called *thermotidal oscillations.* Heating produces a strong pressure contrast between day and night that propagates as a wave around the world. The atmospheric wave moves vertically as well as horizontally. The combination of the horizontal and vertical expansion and contraction of the atmosphere produces a cycle with a period of half a day, just like that of the gravitational tides.

The AMPLITUDE of the atmospheric tides is greatest at the equator. Away from the equator it is reduced by the CORIOLIS EFFECT, and in middle and high latitudes the passage of ordinary weather systems produces such large pressure changes that the tides cannot be detected directly, but only by statistical analysis of pressure records over a long period.

The propagation of atmospheric tides is felt as a very light wind. It is light because the strength of a wind is proportional to the PRESSURE GRADIENT and the tides produce only a very small difference in pressure over a very large distance. At the surface, the tidal wind rarely blows at more than about 2 mph (1 m s^{-1}).

In the upper atmosphere, the tidal wind is stronger. As the expansion of the atmosphere causes the density of the air to decrease, the wind accelerates in order to conserve energy (*see* THERMODYNAMICS, LAWS OF). At a height of 30 miles (50 km) it can accelerate by more than 20 mph (9 m s^{-1}) and at 60 miles (100 km) by more than 110 mph (50 m s^{-1}).

atmospheric wave A vertical displacement in stable air (*see* STABILITY OF AIR) in the PLANETARY BOUNDARY LAYER at a height of 500–650 feet (150–200 m). It resembles the internal waves that occur beneath the surface of the sea. Atmospheric waves are detected by acoustic sounders. These are instruments that transmit sound waves with a WAVELENGTH of about 4 inches (10 cm) and detect the BACKSCATTER from layers of air with different densities.

Most atmospheric waves are gravity waves. These form when air is displaced upward by passing over a surface obstruction. The air then enters air that is less dense and therefore sinks by gravity but overshoots and enters denser air, causing it to rise again.

Other atmospheric waves are caused by WIND SHEAR and superficially resemble breaking waves on the sea surface. Different wind speeds on each side of a layer produce an instability that makes the layer oscillate up and down. This is the same mechanism that causes a flag to flap in the wind, but it operates in a horizontal rather than a vertical plane. The height of the waves is restricted by the stability of the air, which acts to restore the displaced air to its original level.

atmospheric window The 8.5- to 13.0-μm wave band within which there is no atmospheric gas that absorbs radiation. Consequently, outgoing infrared radiation from the atmosphere and surface of the Earth that has wavelengths within these limits escapes into space. Radiation at other wavelengths is absorbed by the various GREENHOUSE GASES present in the atmosphere.

attenuation The weakening of a signal with increasing distance from its source. It is caused by the absorption of part of its energy by the medium through which the signal travels and by divergence of the signal, which spreads its energy over an increasing area. Attenuation can be seen in the ripples that are produced on a pond by dropping a stone into the water. As the circle of ripples spreads, the length of the WAVE FRONT increases, because it is the circumference of the growing circle. No more energy is added to the waves and so the original energy is spread increasingly thinly. The consequence is that the AMPLITUDE of the waves decreases until, at a certain distance from the source, no waves can be seen.

Audrey A HURRICANE, rated category 3 on the SAFFIR/SIMPSON SCALE, that struck the U.S. Gulf Coast near the border between Louisiana and Texas on June 17, 1956. It killed nearly 400 people and generated winds of 100 mph (160 km h^{-1}) and a STORM SURGE of 12 feet (3.6 m).

aureole 1. A bright white or pale blue disk, surrounded by a brown ring, that is sometimes seen around the Sun or Moon. 2. A white area with no clearly defined boundary that sometimes surrounds the Sun in a clear sky is also called an *aureole.*

aurora Lights that are sometimes seen high in the sky in regions close to the North and South Magnetic Poles. Those occurring in the Northern Hemisphere are known as the *aurora borealis,* or "northern lights," and those in the Southern Hemisphere as *aurora australis,* or "southern lights." Occasionally, they appear in latitudes as low as 40°, but auroras are most often seen within oval-shaped areas in both hemispheres defined by latitude 67° at midnight and about latitude 76° at midday. These are regions that vary little in their position in relation to the Sun. These are geomagnetic latitudes—measured in relation to the magnetic poles, not the geographic poles—and people within the ovals often see two auroras in the same day, one in the morning and a second the same evening.

The lights may resemble curtains hanging vertically or appear as bands, patches, or arcs of light. Usually the lower part of the display is more clearly defined than the upper part. Auroras are mainly white, but often with parts that are pale green or red, and the sky around them may have a greenish tinge. Sometimes the display does not move, but at other times it may undulate gently, like a curtain stirred by a slight draft. Displays may last for several hours; when they end the lights move toward the magnetic pole, then fade.

Auroras are caused by the interaction of the SOLAR WIND with atoms of oxygen and nitrogen in the upper atmosphere. The solar wind consists of charged particles. When these encounter the Earth's magnetic field they compress it on the daylight side (facing the Sun) and stretch it into a long tail on the nighttime side. Solar-wind protons and electrons from inside the tail are caught and travel along the magnetic field lines. These descend to the surface at the magnetic poles and the particles descend with them. As they enter the upper atmosphere the particles start to collide with oxygen and nitrogen atoms. These collisions impart energy to the atoms, raising them to an excited state. Then, as they return to their ground state, the energy they absorbed is released as photons of light. It is this light that causes the auroras.

The occurrence of auroras is linked to climate, because the intensity of the solar wind varies with SUNSPOT activity, which in turn is linked to periods of warm or cool climatic conditions (*see* MAUNDER MINIMUM and SPÖRER MINIMUM). During the Maunder minimum between 1645 and 1715, Peter Dass, a Norwegian priest, described many aspects of Norwegian life but failed to mention auroras, although these would have been clearly visible had they occurred. More recently sunspot activity has increased, reaching a maximum in 1991, and the frequency of auroras has also increased.

(More information about auroras, with links to other sites, can be found at www.pfrr.alaska.edu/~pfrr/aurora. There are a short, simple explanation of them at www.uit.no/nordlyset/hvaernordlyset.en.html and a more detailed article by Frank Pettersen of the Northern Lights Planetarium, at the University of Tromsø, Norway, at www.uit.no/npt/nordlyset/waynorth/00-innhold.en.html.)

austru A cold, westerly wind that blows in winter across the low-lying plains on each side of the Danube, mainly in northern Serbia. The wind carries dry, clear air.

autan A wind, often strong, that blows from the southeast along the valley of the Garonne River in southwestern France. It is a wind of the SIROCCO type, carrying warm, moist air and rain, although it varies in strength and in the weather associated with it. Ordinarily it is a continuation of the MARIN after this wind has passed through a gap in the high ground near the town of Carcassonne. If, instead, it crosses the higher ground to the southeast of Toulouse, its descent sometimes gives it the character of a FÖHN WIND. Other winds in this part of southern France are sometimes called *autan,* although they are not related to the true wind of that name.

Automated Surface Observing Systems (ASOS) A network of about 1,000 stations across the United States comprising instruments that automatically make almost continuous measurements of surface pressure, temperature, wind speed and direction, runway VISIBILITY, the height of cloud ceilings, cloud types, and intensity of precipitation. ASOS units are designed to provide meteorological information for airports and are augmented by a separate system to provide warning of thunderstorms. ASOS are operated by the NATIONAL WEATHER SERVICE, and the thunderstorm facility is run in collaboration with the Federal Aviation Administration.

(More information on ASOS can be found at http://www.nws.noaa.gov/om/asos.htm)

automatic picture transmission (APT) A method used by satellites to transmit images to ground-based receivers. Usually, images have a resolution of 2.5 miles (4 km) and are transmitted at two lines per second. Images are transmitted as soon as they have been taken, rather than being stored for transmission later, as was necessary with earlier equipment.

automatic weather station A weather station that transmits its instrument readings to a receiving center at predetermined times without assistance. No personnel are required to operate it.

available water *See* SOIL MOISTURE.

avalanche A mass of material that is descending at high speed down a mountainside. The word usually refers to moving snow, but it can also be applied to ice, earth, or rock or to any mixture of these.

Snow will not accumulate on a steep slope or on a slope that is less steep but slippery because it is covered with grass. Its own weight makes the snow slide away harmlessly. On a very shallow slope snow can accumulate to a considerable depth, but because the underlying surface is almost horizontal the snow remains stable and will not slide. Avalanches are a risk on intermediate slopes, with a gradient of about 30° to 40°. This gradient is neither so steep as to prevent snow from accumulating nor so shallow as to allow it to accumulate safely.

An avalanche occurs when the layer of snow becomes unstable. There are three ways this can happen. The addition of more snow, due to a heavy fall of snow or the deposition of wind-driven snow, increases the weight of the layer. Alternatively, the snow layer may lose its adhesion to the underlying surface. This can happen if a sudden warming in spring or a FÖHN WIND melts some of the snow at a higher level, sending a stream of water beneath the snow farther down the slope. It can happen that the bonds between snow grains weaken naturally, so the cohesive strength of the mass of snow decreases. Once the layer is unstable the slightest vibration may dislodge it. A passing skier, a gunshot, or the loud snap of a breaking tree branch may be all that is needed.

There are two types of snow avalanches. In a point-release avalanche, the disturbance causes just a few snow grains to move. They dislodge other grains below them and the disturbance grows rapidly. The resulting avalanche has the shape of an inverted V and involves only the surface layer of snow.

The much more dangerous type of avalanche moves an entire slab of snow. A slab avalanche has an approximately rectangular shape and is up to half a mile (800 m) wide. It occurs when the mass of snow has already crept a little way down the slope. Its sides and base are under stress, and when a vibration disturbs it, that stress is released suddenly and the entire mass moves as a single unit.

As an avalanche slides and falls down the slope it gathers more snow and accelerates. The fastest speeds are attained when the snow is dry and powdery, because there is a great deal of air between the grains of powdery snow and it reduces friction between snow grains. Speeds of 80–100 mph (130–160 km h^{-1}) are common and they have been known to reach 145–190 mph (235–300 km h^{-1}). Wet snow moves more like mud, flowing down the slope rather bounding down it, and it travels more slowly, rarely exceeding about 55 mph (88 km h^{-1}).

An avalanche is very destructive. Even a small one can have an impact of 0.1–0.5 ton per square foot (900–4,500 kg m^{-2}). This is sufficient to destroy a timber chalet. A major avalanche can exert a force of more than 9 tons per square foot (8,000 kg m^{-2}). That will uproot trees and demolish solidly constructed buildings.

Small avalanches are sometimes triggered deliberately by explosives to prevent snow accumulation to a depth that could cause a major avalanche. Strong fences to hold the snow are built across slopes where avalanches are likely to begin.

avalanche wind The wind that is associated with an AVALANCHE. It is caused by air that is pushed ahead of the descending snow. A major slab avalanche can generate a wind of up to 185 mph (300 km h^{-1}). This can cause serious structural damage to buildings, so they are already weakened when the snow reaches them.

AVHRR *See* ADVANCED VERY HIGH RESOLUTION RADIOMETER.

aviation weather forecast A weather forecast that is prepared for aircrews. It includes information relevant to the operation of aircraft, such as cloud base, cloud type, wind speed and direction at various heights throughout the TROPOSPHERE, risk of ICING, and CLEAR AIR TURBULENCE.

Avogadro constant The value 6.02252×10^{23}. It is the number of atoms or molecules in one MOLE of a substance. It is derived from AVOGADRO'S LAW. It is expressed N_A or L.

Avogadro's law The law stating that equal volumes of all gases at the same temperature and pressure contain the same number of smallest particles. The smallest particles are atoms or molecules and the number per MOLE is known as the AVOGADRO CONSTANT. The law was proposed in 1811 by the Italian physicist Count Amedio Avogadro (1776–1856).

axial tilt The angle between the Earth's axis of rotation and a line passing through the center of the Earth that is at right angles to the PLANE OF THE ECLIPTIC. It is because of the axial tilt that the climates of the Earth experience SEASONS. The Earth's axial tilt varies over a cycle of about 41,000 years from a minimum of 22.1° to a maximum of 24.5°. At present the tilt is 23.45°. The axial tilt determines the location of the TROPICS, which are at 23.5° N and 23.5° S, and the height of the Sun above the horizon at the summer SOLSTICE over the Poles, which is also 23.5°. The contrast between summer and winter is greatest when the tilt is greatest, and if the axis were not tilted, there would be no seasons. The cyclic change in the axial tilt is one of the factors in the MILANKOVICH CYCLES that are linked to the onset and ending of ice ages.

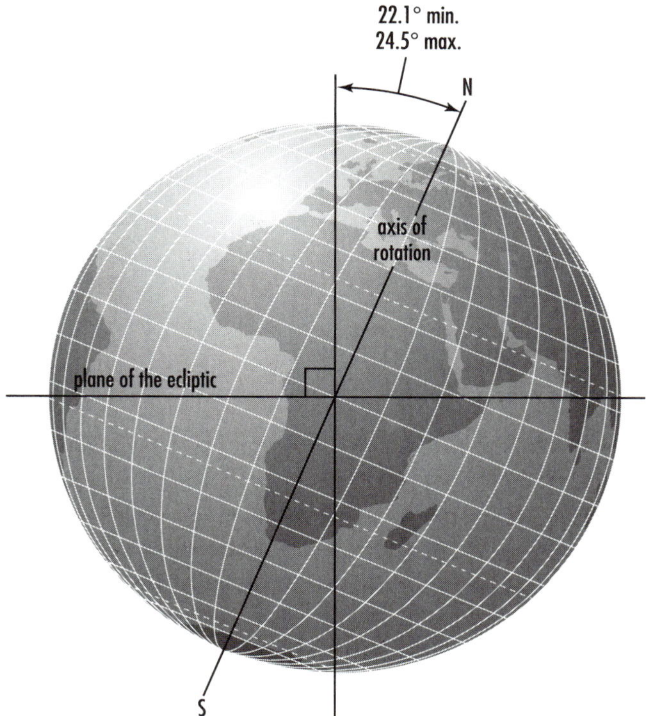

Axial tilt. **The angle between the Earth's rotational axis and a line passing through the center of the Earth at right angles to the plane of the ecliptic. The tilt varies from 22.1° to 24.5°.**

azimuth The angle between two vertical planes, one of which contains a celestial body or satellite and the other the meridian on which an observer is located. It is commonly used for reporting the position of satellites and is equal to the number of degrees from north (0°) counting in a clockwise direction. If the azimuth of a satellite is reported as, say, 170°, this means the satellite will be found by measuring 170° clockwise from north, so it will be 10° to the east of south.

Azores high Part of the ATLANTIC HIGH that is centered above the Azores, about 800 miles (1,290 km) to the west of Portugal. The ANTICYCLONE often extends westward as far as Bermuda; when it does it is known in North America as the *Bermuda high*. The difference in atmospheric pressure between the Azores high and the Icelandic low drives weather systems from west to east across the Atlantic. Periodic variations in this pressure difference are known as the *NORTH ATLANTIC OSCILLATION*.

B

Babs A TYPHOON that struck the Philippines in late October 1998. It caused floods and landslides in which at least 132 people died and about 320,000 people were rendered homeless.

back-bent occlusion (bent-back occlusion) An OCCLUSION that has reversed its direction of motion because a new CYCLONE has formed or the old cyclone has been shifted.

back-bent warm front (bent-back warm front) A WARM FRONT that curves around the CYCLONE. This happens when the COLD FRONT separates from the warm front and starts to move at right angles to it and at the same time the low-pressure center is moving rapidly. The strong winds ahead of the warm front then bend the front around the low center.

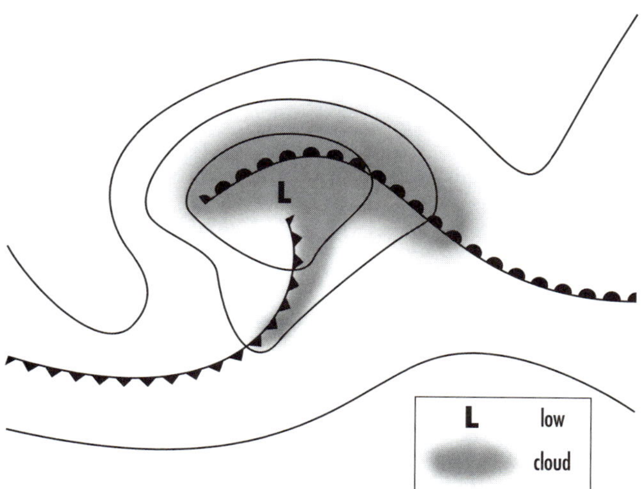

Back-bent warm front. **The cold and warm fronts have separated and the warm front is bent back around the low-pressure center.**

backdoor cold front A COLD FRONT that carries a cold AIR MASS southward along the Atlantic coast toward the south and southwest of the United States. The front develops where the sea-surface temperature is much lower than the air temperature over land. Warm and cold air then become sharply separated. The front arrives from the northeast, rather than from the west or southwest, the direction from which most weather systems arrive. Similar fronts occur along the eastern coasts of continents in the Southern Hemisphere, but there they travel northward.

backing A change in the wind direction that moves in a counterclockwise direction, for example from the northwest to the southwest. If the wind direction is given as the number of degrees from north, a backing wind decreases the number.

backscatter The proportion of a signal that is scattered from the surface at which the signal is directed. The signal may be an electromagnetic wave, such as RADAR, or acoustic. Acoustic signals are used to study ATMOSPHERIC WAVES, for example. The amount of backscatter can be calculated by comparing the strength of the transmitted signal with the strength of

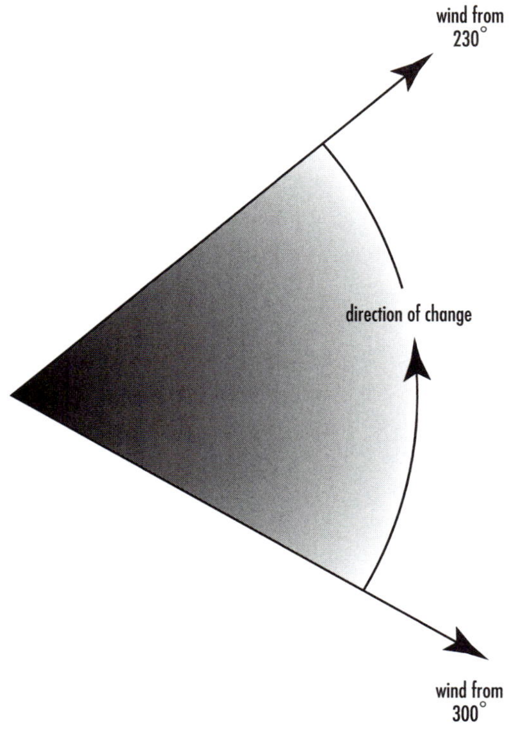

Backing. **A change in wind direction in a counterclockwise direction, in this case from 300° to 230°.**

the signal reflected from the surface to the receiver, after making allowance for ATTENUATION and absorption by the medium through which the signal travels.

bacteria Living organisms that are classified in the domain Bacteria. Most consist of a single cell. Bacteria are present almost everywhere, and some species play an important role in decomposing organic wastes and in the cycling of nutrients (*see* NITROGEN CYCLE). Decomposition involves the oxidation of carbon present in biological molecules to carbon dioxide that is released into the air between soil particles or that dissolves in water and eventually returns to the atmosphere. Many bacteria can survive long periods in a dormant state, as SPORES. Bacteria and their spores can be transported long distances as part of the AERIAL PLANKTON.

bag filter A device that is used to remove small particles from industrial waste gases. It consists of a tube-shaped bag made from woven or felted fabric, up to 33 feet (10 m) long and 3 feet (1 m) wide and closed at one end. The open end is attached to the pipe carrying the gases. Provided the gas is traveling fairly slowly, the filter can trap more than 99 percent of the particles. The material from which the filter is made must be suitable for the temperature of the gas. Natural fibers can be used at temperatures up to 194° F (90° C), nylon up to 392° F (200° C), and glass fibers up to 500° F (260° C).

bagyo The local name for a TROPICAL CYCLONE that forms in the vicinity of the Philippines or Indonesia. *Baguio* is the name of a town in Luzon, Philippines.

bai A yellow mist that sometimes forms in parts of China and Japan. Winds that blow during dry weather in the interior of China raise fine soil particles. These are of a loose type of silt, called *loess,* and are generally yellow. They are lifted to a great height and carried eastward, where they enter moister air. Water condenses onto them and they sink as a colored mist.

Bai-u The period of heaviest rainfall during the early part of the summer MONSOON season in Japan. It is associated with a series of shallow DEPRESSIONS that originate along FRONTS in central China and advance slowly across Japan in a northeasterly direction. The cloudy, humid, rainy weather they produce lasts from the middle of June until the middle of July; the name *Bai-u* means "plum rains." As the last of the depressions passes, the weather changes abruptly, becoming mainly fine, hot, and sultry. Rainfall increases again in late summer.

ball lightning A luminous sphere that is occasionally seen, often during a thunderstorm. Most ball lightnings are 4–8 inches (10–20 cm) across, but they have also been reported as small as 0.5 inch (1.3 cm) and as large as 3 feet (1 m). They are most commonly red, orange, or yellow and rather less bright than a 100-watt lamp. They move horizontally, quite slowly, sometimes pausing and remaining still for a few seconds. Some appear to spin and some bounce off objects and off the ground. Most seem to give off a pungent smell, like ozone or burning sulfur. Although people who have been close to them report no sensation of heat, ball lightnings have been known to scorch or burn wood and to make water boil when they entered it. A few have lasted longer than one minute, but most disappear after about 15 seconds, either fading away and vanishing or ending with a loud bang. Several attempts have been made to explain ball lightning. The most plausible is probably the one proposed in February 2000 by John Abrahamson and James Dinniss of the University of Canterbury, New Zealand. They suggested that when lightning strikes soil particles of silicon, silicon compounds are heated to about 3,000 K (5,924° F [3,273° C]) and thrown into the air. They then slowly oxidize, releasing their stored energy as heat and light.

(You can find the Abrahamson and Dinniss explanation in John Abrahamson and James Dinniss, "Ball Lightning Caused by Oxidation of Nanoparticle Networks from Normal Lightning Strikes on Soil," *Nature,* vol. 403, pp. 519–21, and a discussion of their idea in Graham K. Hubler, "Fluff balls of fire," *Nature,* vol. 403, pp. 487–88. There is a discussion about ball lightning at www-bprc.mps.ohio-state.edu/~bdasye/balligh.html.)

balloon ceiling A CEILING CLASSIFICATION that is used when the CEILING is determined by timing the ascent and disappearance of a balloon.

balloon drag A small balloon that is used to retard the first part of the ascent of a larger balloon in order to allow more time for making measurements. The drag balloon contains ballast to weight it and is inflated in such a way that it will burst at a predetermined height.

balloon sounding A measurement, or set of measurements, of atmospheric conditions that are made by a RADIOSONDE. The word *sounding* is a nautical term, from the French verb *sonder,* which is derived from Latin *subundare, sub-* meaning "under" and *unda* meaning "wave." Taking a sounding is measuring the depth of water.

banded precipitation (rainbands) PRECIPITATION that varies in intensity in the area of precipitation ahead of a WARM FRONT; the variations form a pattern of bands. The bands occur on a small scale and are probably due to local instabilities along the front.

banner cloud A WAVE CLOUD that extends downwind from a mountain peak and resembles a flag flying from the summit.

bar A unit of pressure that is equal to 10^5 newtons per square meter (= 10^6 dynes cm^{-2}). The unit was introduced by VILHELM BJERKNES in *Dynamic Meteorology and Hydrography* (Washington, D.C.: 1911). Meteorologists and climatologists now measure atmospheric pressure in pascals (1 Pa = 1 N m^{-2}), which is a much smaller unit (1 bar = 0.1 MPa), but weather reports and forecasts published in newspapers and broadcast on radio and TV still use the millibar (1 bar = 1,000 mb).

barat A strong westerly wind that blows on the northern coast of Sulawesi, Indonesia, during the north MONSOON. It yields SQUALLS and sometimes causes damage. Monsoon conditions move north and south through this part of southern Asia with the movements of the EQUATORIAL TROUGH. The north monsoon lasts from November until April and the barat winds are most frequent between December and February.

barb A line that is drawn at an angle at the end of a longer line to indicate a wind speed on a STATION MODEL. Barbs are used only for wind speeds up to 47 knots (54 mph [87 km h^{-1}]). PENNANTS are used to indicate higher speeds. The other end of the shaft to which the barb is attached joins the station circle; the angle at which it projects indicates the wind direction.

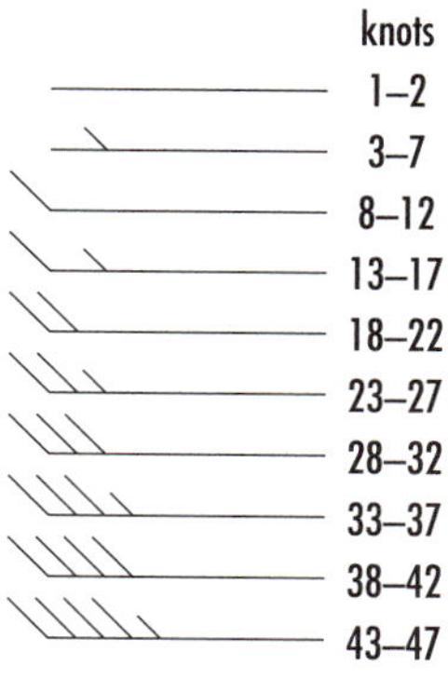

Barbs. **A short line drawn on the shaft indicating wind direction on a station model. The number of long and short barbs indicates the wind speed, which is given in knots.**

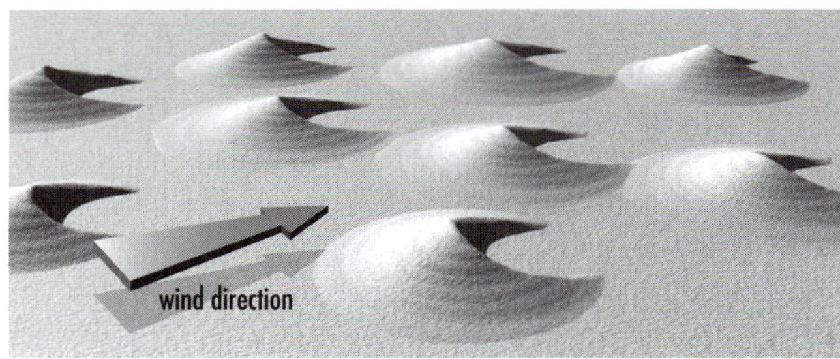

Barchan dune. **A crescent-shaped sand dune that is formed by wind that blows predominantly from a single direction.**

barchan dune A crescent-shaped sand dune that forms where the wind blows predominantly from one direction. Sand grains are blown up the shallow windward slope and fall from its crest down the steeper slope, which is at an angle of about 32°. The constant movement of sand from one side of the dune to the other causes the dune to move in the direction of the wind, typically by about 30–65 feet (10–20 m) a year.

barinés Westerly winds that occur in eastern Venezuela. They blow from the direction of the state of Barinas.

baroclinic An adjective that describes the very common condition of an atmosphere in which surfaces of constant pressure (ISOBARIC SURFACES) and constant air density (ISOPYCNIC SURFACES) intersect. The result is that the air density changes along each ISOBAR. Because the density of the air is related to its temperature, there are strong horizontal temperature gradients in a baroclinic atmosphere. These gradients produce strong THERMAL WINDS. In middle latitudes the atmosphere is usually strongly baroclinic. This explains the formation of FRONTS and JET STREAMS and the weather associated with them. A cross section of the atmosphere from Pole to Pole shows that the ISOPYCNALS or ISOTHERMS are nearly parallel to the isobars in the Tropics, where the atmosphere is BAROTROPIC, but cross them in middle latitudes.

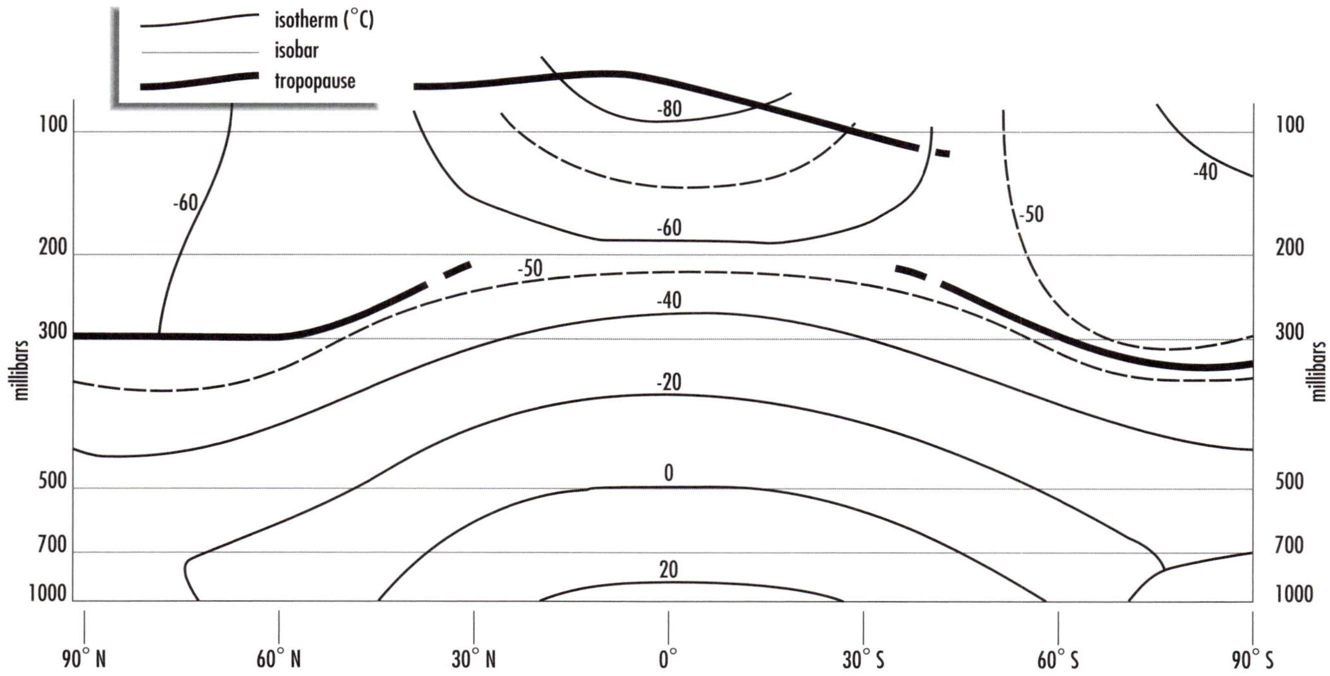

Baroclinic. **A cross section of the atmosphere from Pole to Pole, showing how the isotherms (isopycnals would follow the same curves) cross the isobars in middle latitudes, producing a baroclinic atmosphere. In the Tropics the isotherms and isobars are more nearly parallel and the atmosphere is barotropic.**

baroclinic disturbance **(baroclinic wave)** A change in the flow of air that results in an increase in the BAROCLINICITY. It occurs when a BAROTROPIC flow is diverted, for example when it travels around a mountain. Air then begins to flow across the isotherms (the flow becomes BAROCLINIC). This transports energy by ADVECTION, as cold air is carried into a region of warmer air and warm air is carried into a region of cooler air. The disturbance to the barotropic flow produces a wave pattern in the flow of air. BAROCLINIC INSTABILITY then causes the waves to grow in AMPLITUDE and also changes wind speed, because the air is accelerated as the flow becomes ANTICYCLONIC and decelerated when it becomes CYCLONIC. Although the disturbance begins by affecting the horizontal movement of air, it quickly generates vertical movements. Cold air sinks as it moves into a warmer region, and warm air rises as it moves into a cooler region.

baroclinic field A distribution of AIR PRESSURE and the mass of a given volume of air such that the DENSITY of the air is not a function only of atmospheric pressure.

baroclinic instability Atmospheric instability in the middle and upper TROPOSPHERE that is associated with the waves of a FRONTAL CYCLONE. If the change of wind speed with height (WIND SHEAR) exceeds about 3 feet per second for every 3,000 feet (1 m s^{-1} km^{-1}), waves with unstable wavelengths develop in the FRONT at these heights and their wavelengths increase as the wind shear increases. Wind shear is linked to the temperature gradient, so this type of instability is said to be BAROCLINIC.

baroclinicity The state of being BAROCLINIC.

baroclinic wave *See* BAROCLINIC DISTURBANCE.

barogram The chart of changing atmospheric pressure that is produced by a BAROGRAPH.

barograph **(aneroidograph)** An instrument that makes a continuous record of atmospheric pressure. It consists of a stack of partially evacuated boxes of the type used in ANEROID BAROMETERS. Instead of being linked to a needle on a dial, however, the surface of the boxes is linked to a pen that can move in the vertical plane. The pen inscribes a trace on a paper chart, graduated to show pressure vertically and time horizontally, that is wrapped around a rotating drum. The drum turns at a constant rate, so the chart records the changes in pressure over time.

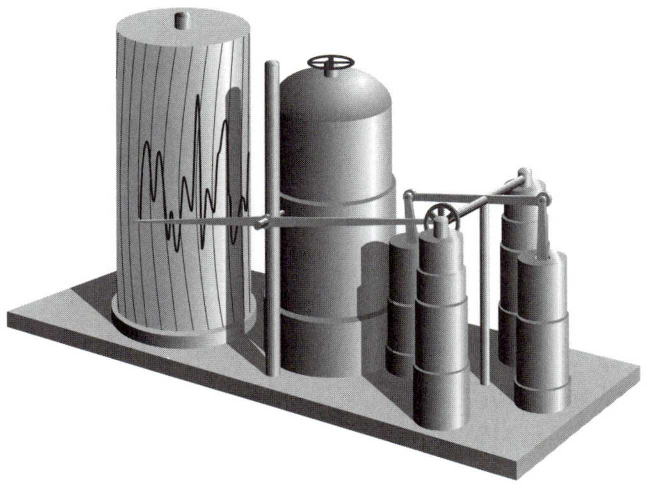

Barograph. **The barograph is an aneroid barometer linked to a pen that makes a continuous trace on a chart attached to a rotating drum.**

barometer An instrument that is used to measure AIR PRESSURE. There are four general types of barometer: the mercurial barometer, ANEROID BAROMETER, HYPSOMETER (hypsometric barometer), and piezoresistance barometer. All barometers measure the pressure that is exerted by the weight of the atmosphere on a surface of unit area (square inch or square centimeter).

Mercurial barometers measure the effect on mercury, a liquid of known density. This is the type of barometer that was invented in 1643 by EVANGELISTA TORRICELLI. His barometer consisted of an open-topped reservoir of mercury and a narrow tube that was sealed at one end and open at the other. He filled the tube with mercury, placed his finger over the open end to seal it, inverted the tube in the reservoir so it was upright, then removed his finger. The mercury in the tube fell, but the tube did not empty because the pressure exerted by the weight of the atmosphere on the open surface of the reservoir pushed mercury some distance into the tube. The upper end of the tube, above the level of the mercury, contained a vacuum (in

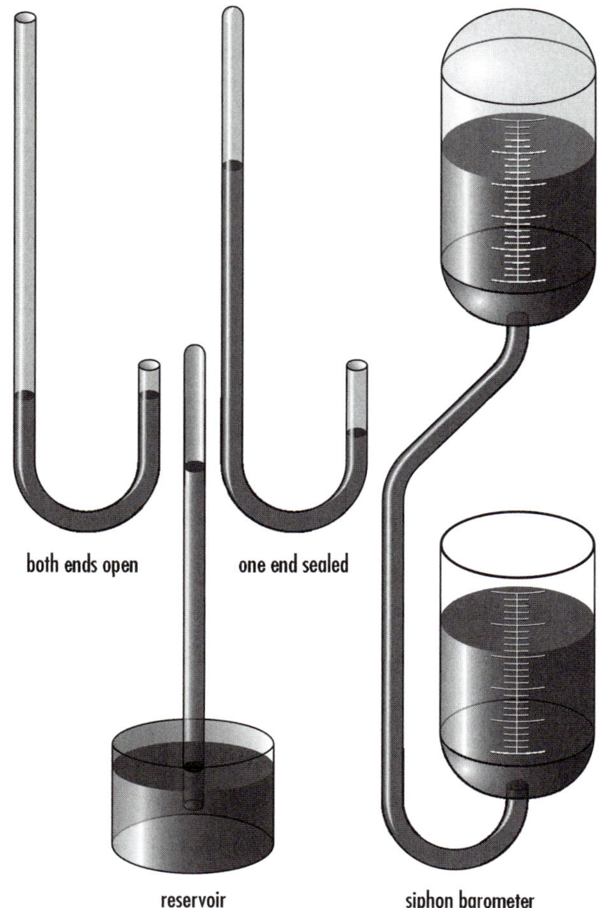

Barometer. **The simplest type of barometer comprises a reservoir containing mercury in which a tube stands upright. The reservoir can be dispensed with by using a U-shaped tube. If both ends of the tube are open, the mercury is at the same height in both arms of the U. If one end of the tube is sealed and there is a vacuum above the mercury, the mercury rises higher in the sealed arm than in the open arm. Pressure is calculated from the difference in the heights of the mercury in the two arms. The siphon barometer is a type of U-tube barometer.**

fact, it contains a small amount of mercury vapor that evaporates into it, but the effect is too small to be of importance).

Pressure (P) is calculated from the difference in height between the surface of the mercury in the reservoir and that of the mercury in the tube. This height difference (h) is multiplied by the density of mercury (d) and by the gravitational acceleration (g) acting on the mercury, so $P = hdg$. The density of mercury is 13.45 ounces per cubic foot (13.5 kg m^3) and g is 32.15 feet per second per second (9.8 m s^{-2}). If the mercury column is 29.92 inches (76 cm) tall, the air pressure is 14.7 pounds per square inch or 1,013.25 hPa or 1,013.25 millibars.

It is possible to dispense with the reservoir by using a U-shaped tube instead. If a tube that is open at both ends is partly filled with mercury, the level of mercury is the same in the two sides of the U. If one end is sealed, however, the mercury in the side with the open end acts as the reservoir. The air pressure can be calculated from the difference in heights of the two mercury columns.

There are several variations of the mercurial barometer. The siphon barometer is a U-shaped tube in which the two sides of the U are of unequal length. To make the instrument easier to read, the ends of the tube are much wider than the central part that connects them, and to save space the ends of the tube are arranged one directly above the other. The end of the shorter side is open and that of the longer side is sealed, with a vacuum above the mercury. The enlarged ends of both sides of the tube are calibrated and the pressure is calculated from the difference in the mercury level in the two sides of the tube.

A version of the siphon barometer is used to measure very small changes in air pressure. It uses a second liquid that is much less dense than mercury. Both liquids are held in the same U-shaped tube, but the part of one side of the tube that holds the lighter liquid is narrower than the part holding the mercury. A change in the level of the mercury produces a much bigger change in the level of the lighter liquid.

The disadvantage of all these types is that the air pressure cannot be read directly, because it depends on the difference between two levels of liquid. The problem is overcome in one type of siphon barometer. It has a flexible leather bag at the bottom of the open end of the tube and a screw that presses upon it. By adjusting the screw, and thus the thickness of the bag, the height of the mercury in the open tube can be moved to a zero point. Then the height of the mercury in the closed end of the tube always represents the difference in the two heights and it can be calibrated to read pressure directly. The Fortin and Kew barometers also eliminate the complication of measuring the difference in heights.

The piezoresistance barometer is based on elements that measure electrical resistance. Two of these are mounted at right angles to one another on the surface of a very thin silicon membrane. As changes in air pressure stretch the membrane the resistance also changes. Barometers of this type are accurate, as well as small, light, and robust; they are convenient to carry and to use on vehicles and RADIOSONDES. They are also inexpensive.

barometer elevation The vertical distance between sea level and the zero level of the mercury in the reservoir of a BAROMETER.

barometric law The law stating that the AIR PRESSURE balances the weight of all the air above the area being considered.

barometric tendency (**air pressure tendency, pressure tendency**) The direction in which the atmospheric pressure has changed at a weather station since the last time it was reported. This is shown on a STATION MODEL by a standard symbol to the right of the station circle.

barothermograph An instrument that records AIR PRESSURE and temperature simultaneously as a pen line on a chart fastened to a rotating drum. It is a combined BAROGRAPH and THERMOGRAPH.

barothermohygrograph An instrument that records AIR PRESSURE, RELATIVE HUMIDITY, and temperature simultaneously as a pen line on a chart fastened to a rotating drum. It is a combined BAROGRAPH, HYGROGRAPH, and THERMOGRAPH.

rising then falling
rising then steady or rising more slowly
rising steadily or unsteadily
falling or steady then rising or rising then rising faster
steady
falling then rising but still same or lower than before
falling then steady or falling more slowly
falling steadily or unsteadily
steady or rising then falling or falling then falling faster

Barometric tendency. **The symbols that are used on a station model to indicate the way atmospheric pressure has changed over the reporting period.**

barotropic An adjective that describes the condition of an atmosphere in which surfaces of constant pressure (ISOBARIC SURFACES) and constant air density (ISOPYCNIC SURFACES) are approximately parallel at all heights. Horizontal temperature gradients are low, there is little or no change in wind direction or speed with height, and atmospheric conditions tend to be uniform over large areas.

barotropic disturbance (**barotropic wave**) An alteration of a BAROTROPIC atmosphere that is caused by a horizontal WIND SHEAR. This converts part of the KINETIC ENERGY in the wind shear into a wave that propagates through the atmosphere.

barotropic field A distribution of AIR PRESSURE and the mass of a given volume of air such that the DENSITY of the air is a function of only atmospheric pressure.

barotropic model A MODEL that is used in NUMERICAL FORECASTING. At each level the model assumes the atmosphere to be BAROTROPIC. In a barotropic atmosphere the winds are GEOSTROPHIC and there is no CONVERGENCE or DIVERGENCE.

barotropic wave *See* BAROTROPIC DISTURBANCE.

bath plug vortex A way of describing an atmospheric VORTEX by means of a familiar metaphor. Water leaving a bath tub usually forms a vortex that can be used to illustrate several features of atmospheric systems. ANGULAR MOMENTUM is conserved. This can be seen by the acceleration of the water as it nears the center. The shape of the vortex resembles that of a TORNADO seen from above. The pressure surface is drawn down the center of the vortex, just as the TROPOPAUSE is drawn down to the surface in the eye of a TROPICAL CYCLONE.

beaded lightning *See* PEARL-NECKLACE LIGHTNING.

Beaufort, Francis (1774–1857) The British scientist who devised the scale of wind force that bears his name (*see* BEAUFORT WIND SCALE). He was born in county

Meath, Ireland, where his father, the Reverend Daniel Augustus Beaufort, was the rector of Navan. The family was of Huguenot origin. Francis's father was keenly interested in geography and topography—the art of drawing the natural and built features of a town or area of countryside—and from an early age Francis shared his enthusiasm.

In 1789, Francis joined the East India Company, the trading company that had been established to administer British commercial interests in India and eventually virtually governed that country. Francis stayed with the company for only a year before leaving to join the Royal Navy, with which he was to spend the rest of his working life. He was 16 years old and began as a cabin boy, but by the age of 22 he had risen to the rank of lieutenant and was serving on HMS *Phaeton*. In 1805 he was given his first command, of HMS *Woolwich*.

Within a very short time of going to sea, Francis recognized the importance of weather conditions and the value of recording them. He began to keep a journal, a habit he maintained for the rest of his life.

By the time of his first command he had become a hydrographer—a scientist who studies, describes, and charts river courses, coastlines, and the depth of the

Francis Beaufort. **The British naval officer who devised the scale of wind strengths named after him.** ***(By courtesy of the National Portrait Gallery, London)***

Beaufort wind force scale of 1831

Wind	Description	Sail
0. Calm		
1. Light air	Or just sufficient to give steerage way.	
2. Light breeze 3. Gentle breeze 4. Moderate breeze	Or that in which a man-of-war with all sail set, and clean full would go in smooth water from.	
5. Fresh breeze 6. Strong breeze	Or that to which a well-conditioned man-of-war could just carry in chase, full and by.	Royals, etc. Single-reefed topsails and top-gallant sail.
7. Moderate gale 8. Fresh gale 9. Strong gale		Double-reefed topsails, jib, etc. Treble-reefed topsails, etc. Close-reefed topsails and courses.
10. Whole gale	Or that with which she could scarcely bear close-reefed main topsail and reefed fore-sail.	
11. Storm	Or that which would reduce her to storm staysails.	
12. Hurricane	Or that which no canvas could withstand.	

sea—and his task with the *Woolwich* was to survey the Rio de la Plata region, in South America. In 1829, he was made the official hydrographer for the Admiralty. He carried out extensive surveying, for example around the Turkish coast in 1812, and was influential in sending out several important voyages of discovery. He also kept up his great interest in meteorology, especially those aspects that affected the operation of sailing ships at sea.

In 1806, Commander Beaufort, as he was then, drew up the chart of wind forces for which he is famous, his *Wind Force Scale and Weather Notation.* Its aim was to provide guidance for sailors, telling them how much sail they should set on a full-rigged warship according to the wind, and in its first version it included no information about the actual speed of the wind. These were not added until long after Beaufort had died. His original scale classified winds from force 0 to force 12—a wind "that no canvas could withstand."

It was in June 1812, during his surveying work in the eastern Mediterranean on HMS *Frederiksteen,* that Beaufort led the rescue of some of his men who had been attacked by forces commanded by the local rulers. Beaufort was seriously wounded in the encounter and at the end of the year he was ordered back to Britain. He never went to sea again.

Robert FITZROY, captain of the *Beagle,* on which Charles Darwin sailed, who used the Beaufort scale, spoke highly of it, and in 1838 it was introduced throughout the navy. Captains were required to include details of the wind in their daily logs. In 1874, modified to include details of the state of the sea and the visible effects of the wind on land, the scale was adopted for use in international meteorological telegraphy by the International Meteorological Committee.

Beaufort was knighted for his service in 1848, and, by the time he retired in 1855, he had reached the rank of rear admiral. Admiral Sir Francis Beaufort had served in the Royal Navy for 68 years; he died 2 years later.

(For more information about Beaufort see www.islandnet.com/~see/weather/history/beaufort.htm.)

Beaufort wind scale The classification of winds according to their speed and effects that was devised in 1805 by FRANCIS BEAUFORT, a British naval officer. Its

Beaufort Wind Scale

Force	Speed mph ($km\ h^{-1}$)	Name	Description
0.	0.1 (1.6) or less	Calm	Air feels still. Smoke rises vertically.
1.	1–3 (1.6–4.8)	Light air	Wind vanes and flags do not move, but rising smoke drifts.
2.	4–7 (6.4–11.2)	Light breeze	Drifting smoke indicates the wind direction.
3.	8–12 (12.8–19.3)	Gentle breeze	Leaves rustle, small twigs move, and flags made from lightweight material stir gently.
4.	13–18 (20.9–28.9)	Moderate breeze	Loose leaves and pieces of paper blow about.
5.	19–24 (30.5–38.6)	Fresh breeze	Small trees that are in full leaf sway in the wind.
6.	25–31 (40.2–49.8)	Strong breeze	It becomes difficult to use an open umbrella.
7.	32–38 (51.4–61.1)	Moderate gale	The wind exerts strong pressure on people walking into it.
8.	39–46 (62.7–74)	Fresh gale	Small twigs are torn from trees.
9.	47–54 (75.6–86.8)	Strong gale	Chimneys are blown down. Slates and tiles are torn from roofs.
10.	55–63 (88.4–101.3)	Whole gale	Trees are broken or uprooted.
11.	64–75 (102.9–120.6)	Storm	Trees are uprooted and blown some distance. Cars are overturned.
12.	more than 75 (120.6)	Hurricane	Devastation is widespread. Buildings are destroyed and many trees are uprooted.

purpose was to instruct the commanders of warships as to the amount of sail their ships should carry in winds of different strengths. In 1838 the Beaufort scale was introduced throughout the Royal Navy and commanders were required to record wind conditions in their daily logs.

The International Meteorological Committee adopted the scale in 1874 for use in sending weather information by telegraph. For this purpose the scale was expanded to include brief descriptions of the state of the sea or of conditions on land.

In its original form the scale made no direct reference to wind speed. The International Commission for Weather Telegraphers began in 1912 to calculate the wind speeds that would produce the effects in the scale, but their work was interrupted by the outbreak of World War I. It began again in 1921, when G. C. SIMPSON was asked to undertake the task. His equivalent wind speeds were accepted in 1926. In 1939, the International Meteorological Committee standardized the scale by asserting that the wind speeds are based on values that would be registered by an ANEMOMETER 20 feet (6 m) above the ground.

There are, therefore, two basic versions of the Beaufort scale. The first is as Beaufort prepared it; the second is the one that is used today and that refers to conditions on land. The modern version was subsequently extended by the addition of categories to describe hurricanes (*see* SAFFIR/SIMPSON HURRICANE SCALE).

Bebinca A TYPHOON that struck the Philippines in November 2000. It left 43 people dead and forced more than 630,000 people to leave their homes in metropolitan Manila and in 14 northern provinces.

Beckmann thermometer A THERMOMETER that is used for measuring very small changes in temperature. It is a mercury-in-glass thermometer with two bulbs. One bulb is located at the bottom of the thermometer tube, as in an ordinary thermometer. The top of the tube is shaped like an inverted U and the other, smaller bulb is at the end of one arm of the U. At the base of the upper bulb there is a second, upright, U-shaped tube. Mercury can be run from the upper bulb into the lower one. This alters the range of temperature the thermometer measures. The scale covers only about 5° C (9° F). The thermometer was invented by the German chemist Ernst Otto Beckmann (1853–1923).

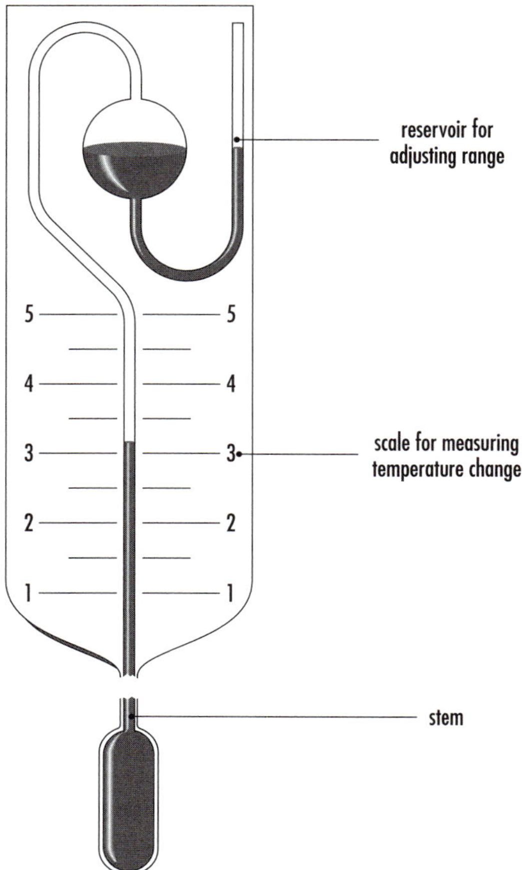

Beckmann thermometer. **The thermometer has two bulbs and a scale covering only about 5° C (9° F). It is used for measuring very small changes in temperature.**

becquerel (Bq) The Système International d'Unités (SI) unit of radiation. It is equal to an average of one transition of a radionuclide (one decay) per second. The unit is named in honor of the French physicist who discovered radioactivity, Antoine Henri Becquerel (1852–1908). (*See* SYSTÈME INTERNATIONAL D'UNITÉS.)

Beer's law A law stating that when light passes through a medium the amount that is absorbed and scattered varies according to the composition of the medium and the length of the path traveled by the light. The law assumes the medium to be homogeneous and the light to be of a particular WAVELENGTH, but the law holds fairly well for the absorption and scattering of light traveling through the atmosphere and the depth to which light penetrates water, snow, and ice. The law can be expressed as

$$K_z = K_0 e^{-az}$$

where K_z is the amount of light reaching a depth z, K_0 is the amount of light at the top of the medium (in the case of the atmosphere, the SOLAR CONSTANT), e is the base of natural logarithms (2.718), and a is the amount of radiation that is absorbed per meter (known as the *extinction coefficient*).

beetle analysis The use of the remains of beetles to infer past climatic conditions. The wing cases (elytra) of beetles are often preserved in the soil. The species can often be identified from the elytra, which can be dated (*see* RADIOCARBON DATING). The technique is possible because many species of beetles live only where the temperature is within certain fairly broad limits and beetles respond more rapidly than plants to changes in temperature. The temperature range each species tolerates is known from studies of living beetles of the same species. Where a number of such species occur together, all of their tolerable temperature ranges are compared. The temperature range in which all of them lived must be where the individual ranges overlap. This much narrower range is known as the *mutual climatic range* (MCR).

belat A strong northeasterly wind that sometimes blows across Yemen, in the south of the Arabian Peninsula. It produces hazy conditions because it carries dust from the desert.

below minimums Conditions of VISIBILITY, CLOUD BASE, and wind that are less than the minimum standards specified for aircraft to take off and land.

Benguela Current An ocean current that flows northward from the WEST WIND DRIFT and along the western coast of Africa, from about 35° S to 15° S. It carries cold water, with many UPWELLINGS, and flows fairly slowly, at less than 0.6 mph (0.9 km h^{-1}).

bent-back occlusion *See* BACK-BENT OCCLUSION.

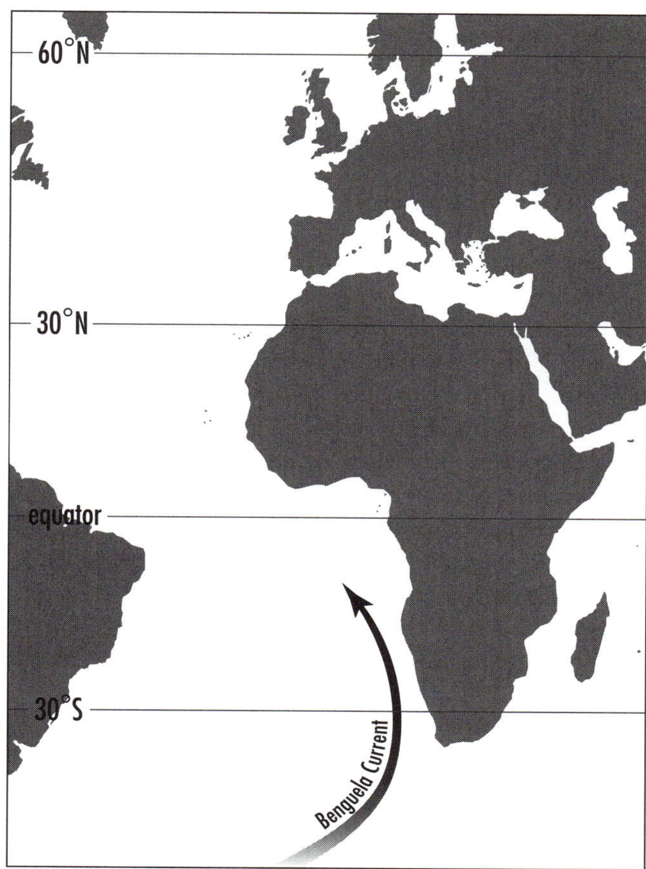

Benguela Current. **A cold current that flows parallel to the coast of southern Africa.**

Bentley, Wilson Alwyn (1865–1931) American *Photographer of snowflakes* For 50 years, Wilson A. Bentley studied and then photographed snowflakes. Eventually he accumulated an archive of more than 5,000 images and became widely known as "the Snowflake Man."

Bentley was born on February 9, 1865, at the family farm in the village of Jericho, Vermont. His mother had been a schoolteacher before her marriage, and she taught Wilson at home until he was 14. She used a small microscope as a teaching aid, and Wilson became fascinated by the world it revealed to him. In particular, he studied the shapes of SNOWFLAKES, dewdrops, FROST, and HAILSTONES. He recorded these by drawing what he saw, but this proved unsatisfactory and eventually he acquired the bellows camera and microscope objective that allowed him to photograph them. All of

Wilson Alwyn Bentley. **The American photographer who specialized in pictures of snowflakes.** ***(John Frederick Lewis Collection, Picture and Print Collection, The Free Library of Philadelphia)***

his photomicrographs were taken with this original camera.

Snowflakes consist of ICE CRYSTALS. These have a variety of shapes and can be arranged in an almost infinite number of ways, so that each snowflake is unique. What Bentley discovered, however, is that the temperature and pattern of air circulation within a cloud could be deduced from the form of the ice crystals that fell from it.

Each summer, Bentley turned his attention to the study of rain. He devised a method for measuring the size of RAINDROPS by exposing a dish containing a layer of sifted flour, about 1 inch (25 mm) thick. Raindrops formed the flour into little balls of dough. When these were dried they were approximately the same size as the drops that caused them. The method is still used. From the size of the raindrops he deduced how they had formed. Between 1898 and 1904 he made more than 300 measurements of raindrops.

In 1898, Bentley published his first magazine article, in *Popular Scientific Monthly.* After that, he wrote many popular articles and scientific papers; he described many of his most original ideas in the *Monthly Weather Review. Snow Crystals,* his only book, was published in 1931 with William J. Humphreys, the chief physicist at the U.S. Weather Bureau, who persuaded him to do it. Writing the book involved sifting through his collection of photomicrographs and selecting nearly twenty-five hundred of his favorites. *Snow Crystals* by Wilson A. Bentley and William J. Humphreys (New York: McGraw-Hill Book Co., 1931; republished New York: Dover Publications, 1962) contains about 10 pages of text and more than 200 pages of Bentley's photographs.

In 1924, he received the first research grant ever awarded by the American Meteorological Society. The amount was small, but it was a well-deserved recognition by the scientific community for the work Bentley had been doing for 40 years.

Despite his interest in the weather, Wilson Bentley remained a farmer. After the death of his father, Wilson and his brother worked the farm together and Bentley contributed his full share of the physical work. They succeeded and the farm prospered.

Bentley kept detailed meteorological records throughout most of his life, the last on December 7, 1931. Soon after that he fell ill and died of pneumonia at the farm on December 23, 1931.

(There is more information about Wilson Bentley in Duncan C. Blanchard, "The Snowflake Man," at www.snowflakebentley.com/sfman.htm.)

Benedict One of five CYCLONES that struck Madagascar between January and March 1982. The others were Frida, Electra, Gabriel, and Justine. Together they killed more than 100 people and rendered 117,000 homeless.

Bergen Geophysical Institute The institution that was founded in 1917 by VILHELM BJERKNES and his colleagues at Bergen, Norway. It had formerly been part of the Bergen Museum. Bjerknes and his team of meteorologists developed the institute as a center for meteorological research, and the ideas that were produced there in the 1920s and 1930s are often attributed to the "Bergen School" of meteorologists. These ideas included the theories of AIR MASSES and FRONTS. The Bergen Geophysical Institute is now part of the University of Bergen and specializes in meteorology and oceanography.

Bergen School *See* BERGEN GEOPHYSICAL INSTITUTE.

Bergeron–Findeisen mechanism A theory to explain how cloud droplets grow into raindrops in cold clouds (clouds in which the ambient temperature is below freezing). It was first proposed in 1935 by the Norwegian meteorologist TOR BERGERON and was demonstrated in large cloud chambers by the German meteorologist Walter Findeisen. Cold clouds contain both ice crystals and supercooled (*see* SUPERCOOLING) water droplets. The SATURATION VAPOR PRESSURE over an ice surface is lower than that over a liquid water surface. This is especially true at temperatures between about 5° F and –13° F (–15° C and –25° C), when the difference amounts to about 0.2 mb (20 Pa). Consequently, water will evaporate from the droplets and accumulate on the crystals by direct deposition from water vapor. The crystals grow, collide with one another, and stick together to form aggregations—SNOWFLAKES. This continues until the snowflakes are heavy enough to start falling. As they fall, the snowflakes collide with more supercooled droplets. The water freezes onto them, so the flakes keep growing. Some are broken apart by air currents, and the splinters of ice produced in this way act as FREEZING NUCLEI for the formation of more crystals that join into more snowflakes. Those that remain intact fall from the base of the cloud. If the temperature is above freezing in the air beneath the cloud, the snowflakes start to melt and some or all of them reach the ground as rain. In middle latitudes, rain showers often consist of melted snow, even in the middle of summer.

Bergeron, Tor Harold Percival (1891–1977) Swedish *Meteorologist* Tor Bergeron was born on August 15, 1891, at Godstone, near London, England. He was educated at the universities of Stockholm, Sweden, and Leipzig, Germany, and in 1928 he obtained his Ph.D. from the University of Oslo, Norway.

As part of his education, Bergeron spent the three years from 1918 until 1921 as a student and collaborator of VILHELM BJERKNES, the Norwegian meteorologist who in 1917 had established what became the most important meteorological research institute in the world, at the BERGEN GEOPHYSICAL INSTITUTE. After he qualified, Bergeron joined the staff at the institute.

In 1935, he proposed a mechanism for the formation of RAINDROPS in COLD CLOUDS. He calculated that ICE CRYSTALS would grow by gathering water at the expense of supercooled water droplets (*see* SUPERCOOLING). This is called the *Bergeron process*. The crystals would form SNOWFLAKES, and as the snowflakes fell from the base of the cloud into warmer air, they would melt and reach the surface as water droplets. This mechanism was later confirmed experimentally by the German meteorologist Walter Findeisen and is now known as the *BERGERON–FINDEISEN MECHANISM*.

From 1935 until 1945 Bergeron taught at the University of Stockholm, and in 1946 he moved to the University of Uppsala, also in Sweden. He was professor of meteorology at the University of Uppsala from 1947 until 1961.

Bergeron died at Stockholm on June 13, 1977.

Bergeron process *See* BERGERON, TOR HAROLD PERCIVAL.

berg wind A hot, dusty wind that blows across southern Africa, carrying continental air to the coast. Berg winds occur on the western coast on about 50 days of the year and are most frequent in winter. In southwestern Africa they are from the east, on the southern coast they are from the north, and in Natal they blow from the northwest. They last intermittently for two or three days and can create temperatures of 90° F (32° C) or higher. They tend to die down in the late afternoon, when sea breezes neutralize them.

Bering Current An ocean current that flows southward through the Bering Strait separating Alaska from eastern Siberia, moving cold water into the North Pacific.

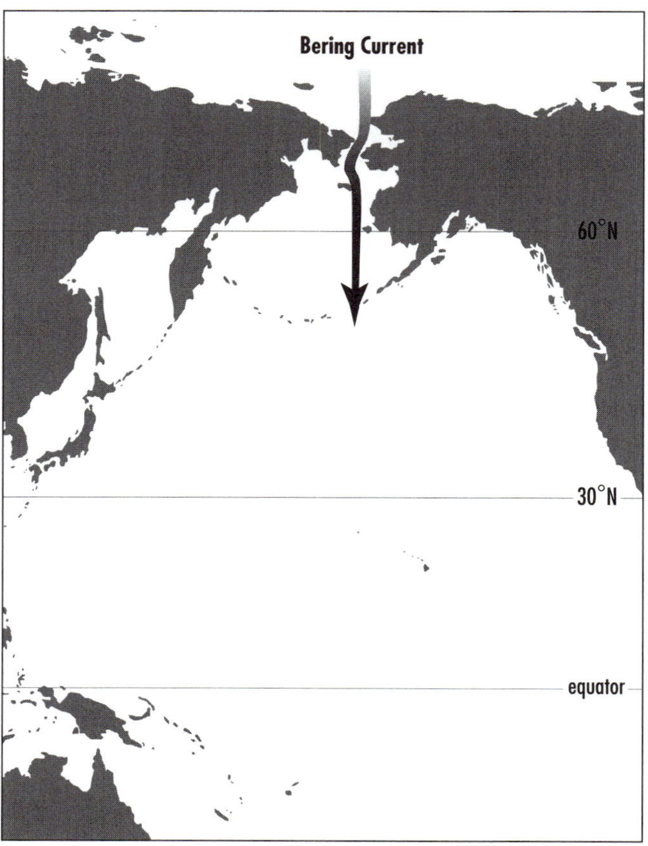

Bering Current. **A current that carries cold water from the Arctic into the North Pacific.**

Bermuda high *See* AZORES HIGH.

Bernoulli, Daniel (1700–1782) Swiss *Natural philosopher* Daniel Bernoulli was born at Groningen, the Netherlands, on February 9, 1700. His father, Johann (sometimes called Jean) Bernoulli (1667–1748), was trained as a physician but worked as a mathematician. Daniel had an older brother, Nikolaus (1695–1726). He studied law, and by the age of 27 he was professor of law at the University of Bern, Switzerland, but mathematics was his real passion.

At the time of Daniel's birth, Johann was a professor at the University of Groningen. Johann's older brother Jakob (sometimes called Jacques) Bernoulli (1654–1705) was also a distinguished mathematician and experimental physicist. From 1687 he was professor of mathematics at the University of Basel, Switzerland. After Jakob's death in 1705, Johann succeeded him at Basel and so, at the age of five, Daniel moved to Switzerland. His younger brother, Johann (1710–90), was born in Basel, and he, too, was a mathematician. In 1743 he succeeded his father as professor of mathematics at Basel.

That is where he was educated. He enrolled at the University of Basel when he was 13 and specialized in philosophy and logic, passing his baccalaureate examination (equivalent to graduation from high school) at the age of 15. When he was 16 he obtained his master's degree. His father wanted him to become a merchant, but commerce did not appeal to Daniel and he refused. He wanted to be a mathematician like the other members of his family, but his father insisted there was no money to be made in that profession. Eventually they compromised, and in 1717 Daniel began to study medicine at Basel. During his studies he also spent some time at the universities of Heidelberg in 1718 and Strasbourg in 1719 and returning to Basel in 1720. The thesis for which he was awarded his doctorate in 1721 was on the action of the lungs.

Having qualified as a doctor and satisfied his father, Daniel sought an academic position, but without success, and he moved to Venice to continue studying medicine. He intended to study at the famous medical school at Padua, but illness kept him in Venice, where he concentrated on mathematics. In 1724 he published the book *Exercitationes Mathematicae* (Mathematical exercises). He also designed an hourglass that could be used on ships, even in heavy seas. He submitted this to the French Academy of Sciences, and when he returned to Basel in 1725 he learned that it had won a prize.

Daniel also learned that his book had attracted wide attention, and he was offered a professorship in mathematics at the Russian Academy in St. Petersburg. His brother Nikolaus was also offered a professorship in mathematics there, and in 1725 the two men traveled to Russia together. In less than a year Nikolaus died of a fever. This greatly saddened Daniel and he did not like the Russian climate. Johann, his younger brother, joined him in 1731. Daniel was applying for posts at Basel, and in 1733 he became professor of anatomy and botany. He and Johann left Russia together; visited Danzig (now Gdansk), Hamburg, the Netherlands, and Paris; and finally reached Basel in 1734. In 1750 Daniel became professor of natural philosophy at Basel; he retained this post until he retired in 1777. The chair at St. Petersburg was later occupied by Daniel's pupil and nephew Jakob Bernoulli (1759–89).

The prize Daniel won for his hourglass was the first of 10 prizes he received from the French Academy for work on astronomy, magnetism, and a variety of nautical topics. He also wrote on political economy and probability.

Daniel Bernoulli. **The Swiss scientist and mathematician who discovered the Bernoulli effect. *(John Frederick Lewis Collection, Print and Picture Collection, The Free Library of Philadelphia)***

Daniel published his most important work, *Hydrodynamica* (Hydrodynamics), in 1738. In it he discussed the theoretical and practical aspects of pressure, velocity, and equilibrium in fluids and showed the link between the pressure of a fluid and its velocity. This is a consequence of the conservation of energy, although more than a century was to pass before that concept was formulated clearly (*see* THERMODYNAMICS, LAWS OF). The relationship between pressure and velocity is now known as the *Bernoulli principle,* and its consequences as the *BERNOULLI EFFECT.*

In *Hydrodynamica* Bernoulli also assumed that gases are composed of minute particles. This allowed him to produce an EQUATION OF STATE and to relate atmospheric pressure to altitude.

Daniel Bernoulli was predominantly a mathematician and a friend of many of the most eminent mathematicians of his day. The Swiss mathematician Leonard Euler (1707–83) traveled to St. Petersburg to work with him in 1727, and the French mathematician Jean le Rond d'Alembert (1717–83) was also a close friend.

Bernoulli received many honors and was elected to most of the scientific academies of Europe. He died in Basel on March 17, 1782.

(You can read more about Daniel Bernoulli at www.groups.dcs.st-and.ac.uk/~history/Mathematicians/Bernoulli_Daniel.html.)

Bernoulli effect The reduction in air pressure that occurs when a wind blows across a convex surface, such as a ridged roof or the wing of an airplane, provided the flow is laminar (*see* LAMINAR FLOW). This effect results from the Bernoulli principle, which states that the pressure within a fluid changes inversely with the speed of flow. The relationship can be expressed by

$$p + 1/2\rho V^2 = \text{a constant}$$

where p is the pressure, ρ is the density of the fluid, and V is the velocity. When air flows over a convex surface it must travel farther than adjacent air that does not flow over the surface, but it must do so in the same time, because the STREAMLINES must rejoin on the other side of the obstacle. This means the air crossing the convex surface must accelerate. The air must also possess the same amount of energy when it has passed the obstruction as it had before encountering it. Consequently, since its velocity increases as it passes the convex surface, the only other form of internal energy it possesses, its pressure, must decrease. In the equation, if V increases, p must decrease.

This is why a strong wind can lift the roof from a building. It does so not by blowing from beneath the roof, but by reducing the pressure above the roof. You may have experienced the same effect yourself when riding a bike. If a car passes you traveling fairly fast you are drawn toward the car, not (as you might expect) pushed away from it.

(You can read more about the Bernoulli effect in Michael Allaby, *Dangerous Weather: Hurricanes* [New York: Facts On File, 1997], and there is a description of several ways you can demonstrate the effect safely in Michael Allaby, *Dangerous Weather: A Chronology of Weather* [New York: Facts On File, 1998].)

Bernoulli principle *See* BERNOULLI, DANIEL.

Bertha A HURRICANE that struck the U.S. Virgin Islands on July 8, 1996, carrying torrential rain and winds of up to 103 mph (166 km h^{-1}) and triggering FLASH FLOODS and mudslides. It then moved to the British Virgin Islands, St. Kitts and Nevis, and Anguilla. On July 10, it reached the Bahamas, where it caused waves 20 feet (6 m) high and winds of 100 mph (160 km h^{-1}). On July 12 it crossed the Carolina coast, then traveled north, to Delaware and New Jersey. Bertha was initially rated as category 1 on the SAFFIR/SIMPSON SCALE but was then upgraded to 2 and on July 9 reached category 3. It was a large hurricane, with a diameter of 460 miles (740 km) and produced hurricane-force winds (*see* BEAUFORT WIND SCALE) 115 miles (185 km) from the eye. It killed seven people.

beta decay A type of RADIOACTIVE DECAY in which an unstable atomic nucleus changes into a nucleus with the same mass, but a different number of protons. There are two ways the decay can occur. A neutron may change into a proton with the emission of an elec-

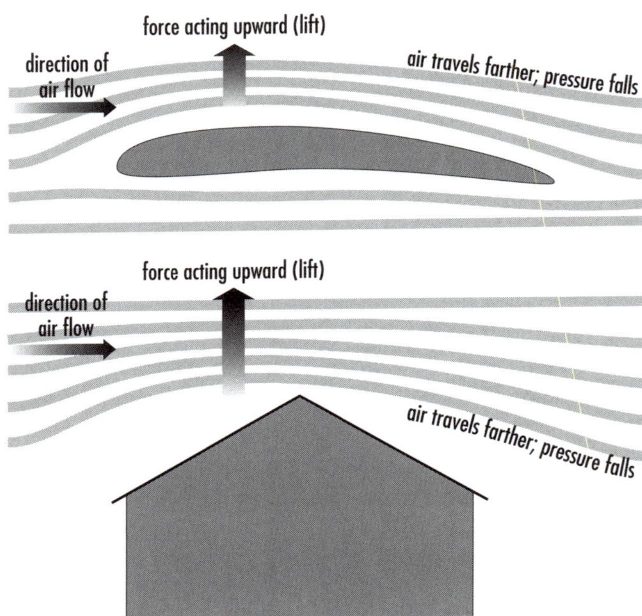

Bernoulli effect. **As the air passes over the upper surface of an airplane wing or a roof it must travel farther than the adjacent air. This accelerates the air and its pressure falls, generating an upward force (lift) on the wing or roof.**

tron and an antineutrino, or a proton may change into a neutron with the emission of a positron and a neutrino. Electrons and positrons (which are identical to electrons but carry a positive charge) emitted in this decay are known as *beta particles* and a stream of them is called *beta radiation.* This is the type of decay experienced by radioactive carbon-14. It is described by

$$^{14}_{6}C \rightarrow {}^{14}_{7}N + e^{-} + \bar{v}$$

Carbon-14 (^{14}C) has six protons ($_6C$). It changes into a nucleus with the same mass, but seven protons, which is nitrogen ($_7N$). The decay involves the emission of an electron (e^-) and an antineutrino ($\bar{v}$).

Bhopal A city in Madhya Pradesh, in central India, where one of the most serious industrial accidents in history occurred on December 3, 1984. It polluted the air and caused the deaths of about 2,000 people immediately and injuries to about 300,000. An estimated 8,000 people died later, making the final death toll about 10,000, although some organizations have estimated there were up to 16,000 deaths. Young children and elderly adults were especially susceptible. The accident occurred in the early morning, while most people were asleep, at a pesticide factory owned by the Indian subsidiary of Union Carbide Corporation. A series of mechanical failures and human errors caused a cloud of about 49.5 tons (45 tonnes) of methyl isocyanate to be released into the air. Heavier than air, the poisonous gas settled close to ground level and moved through the buildings nearby. The local authorities had no information about the toxic material being held inside the factory and no plans for dealing with an emergency. It was an hour before the alarm was raised; by that time many people were already dead or fatally ill. There was then a panic, and more people were injured as they tried to escape from the area, many suffering from eye and respiratory irritation caused by the poison.

(You can learn more about the accident at www.earthbase.org/home/timeline/1984/bhopal/text.html.)

Bilham screen A small container with louvered sides like a STEVENSON SCREEN that contains a WET-BULB THERMOMETER and dry-bulb thermometer mounted vertically and a MAXIMUM THERMOMETER and a MINIMUM THERMOMETER mounted horizontally.

Bilis A TYPHOON that struck Taiwan and the coast of mainland China in August 2000. It killed 11 people in Taiwan, including 7 farmers and a six-year-old girl who were buried in a mudslide and a woman who was killed by a falling power line. The typhoon had weakened by the time it reached the mainland province of Fujian, but it destroyed more than 1,000 homes.

billow clouds A type of UNDULATUS that consists of parallel rolls of cloud forming CLOUD BARS separated by clear sky. Billow clouds are produced in conditions of high RELATIVE HUMIDITY by the TURBULENT FLOW of air associated with CLEAR AIR TURBULENCE. The turbulence produces SHEAR WAVES and EDDIES, and the billow clouds mark the crests of the waves, where air has risen far enough for its water vapor to condense. Each billow cloud lasts for only a short time before it is dissipated by the turbulence.

bioclimatology The scientific study of the relationship between living organisms and the climates in which they live. Bioclimatology is a branch of biogeography, which is the study of the geographic distribution of plants and animals. Bioclimatic zones are distributed by latitude and distance from the ocean (*see* CLIMATE CLASSIFICATION), and they also exist on a smaller scale. A mountain, for example, has a number of distinct climates, each supporting its own community of organisms. At sea level, tropical rain forest covers the Andes of central Peru. This gives way to a more open type of forest at an elevation of about 3,300 feet (1,000 m) and to a still more open type in which the trees are smaller and covered in mosses and lichens, called *montane forest,* at about 6,600 feet (2,000 m). Above about 9,800 feet (3,000 m) the trees are smaller still. This is called *elfin woodland.* Beyond it, above about 13,000 feet (4,000 m), there is grassland, and above about 16,400 feet (5,000 m) the vegetation becomes increasingly sparse. There is continuous snow and ice above 19,700 feet (6,000 m). Mountains in other parts of the world are divided into similar altitudinal zones, but these vary in detail from one mountain range to another. Bioclimatology involves the study of climatic zones such as these and of the organisms that inhabit them. It also includes more specialized subdisciplines, such as AGROCLIMATOLOGY.

biometeorology The scientific study of the relationship between living organisms and the air around them. This includes the effect organisms have on the air, through PHOTOSYNTHESIS, RESPIRATION, and TRANSPIRATION, as well as the emission of gases and particulate material by plants (*see* ISOPRENE) and animals and as a consequence of human activity. Plants also affect the atmosphere by shading the ground beneath them and by intercepting PRECIPITATION. This produces distinct MICROCLIMATES in shaded and unshaded areas. The study also includes the response of plants and animals to atmospheric conditions. Plants wilt during a DROUGHT, for example, and many animals seek shade when the sky is clear and the sunshine intense. Biometeorology also includes specialized subdisciplines, such as AGROMETEOROLOGY.

bise The name given in Switzerland to a cold, dry, fierce, northeasterly wind that blows in winter and spring through the European Alps. It is often strongest in southwestern Switzerland. It can whip up water from the surface of Lake Geneva and blow it over plants on the shore, encasing them in thick ice. The wind can continue for several days, and in regions affected by it many of the houses do not have doors or windows on their northeastern sides. The bise occurs when there are an ANTICYCLONE in central and eastern Europe and a DEPRESSION in the western Mediterranean. This draws continental polar (cP) air (*see* CONTINENTAL AIR and POLAR AIR) through the mountains.

bishop's ring A faint, reddish brown CORONA around the Sun that is caused by the DIFFRACTION of light by dust particles. The clouds of dust are usually the result of a violent volcanic eruption. Bishop's rings were seen after the eruptions of KRAKATAU in 1883 and MOUNT KATMAI in 1912.

Bjerknes, Jacob Aall Bonnevie (1897–1975) Norwegian-American *Meteorologist* The son of VILHELM FRIMANN KOREN BJERKNES, Jacob was born in Stockholm and, during World War I, helped his father organize the network of weather stations that supplied the data they used to develop their theories of air masses and polar fronts. Jacob also discovered that DEPRESSIONS originate as waves on fronts. In 1939, he moved to the United States and in 1940 became professor of meteorology at the University of California, Los Angeles. He became a U.S. citizen in 1946. After World War II, Bjerknes conducted extensive studies of the upper atmosphere and JET STREAM; in 1952 he was among the first to use for this purpose photographs taken by high-altitude research rockets. Bjerknes also studied the climatic consequences of the interaction of the ocean and atmosphere in the tropical Pacific and in 1969 was the first to propose what became known as the *Bjerknes hypothesis*. This holds that ENSO arises from changes in sea-surface temperature, leading to changes in wind strength and direction, leading to changes in the ocean circulation, leading to further changes in sea-surface temperature. Bjerknes died in Los Angeles.

Bjerknes, Vilhelm Frimann Koren (1862–1951) Norwegian *Physicist and Meteorologist* A Norwegian scientist who was one of the founders of modern meteorology and scientific weather forecasting, he was born in Oslo. His father, Carl Anton Bjerknes, was professor of mathematics at Christiania (now Oslo) University, and Vilhelm helped him with some of his experiments in hydrodynamics before leaving to spend 1890 and 1891 in Germany working as an assistant to and collaborator with the physicist Heinrich Hertz. He then spent two years as a lecturer at the School of Engineering (Högskola) in Stockholm and in 1895 was appointed professor of applied mechanics and mathematical physics at the University of Stockholm.

In 1897, he developed a synthesis of hydrodynamics and thermodynamics that allowed him to propose a system for forecasting weather scientifically. In 1904, he published a scientific paper outlining a method of numerical forecasting. The Carnegie Institution supported this work, allowing Bjerknes to employ a long series of "Carnegie assistants" who joined the "schools" he founded at Leipzig and Bergen.

Bjerknes returned to Norway in 1907 as a professor at Kristiania (the spelling had been changed) University and, in 1910 and 1911, he and three of his assistants (the Swedish meteorologist Johan W. Sandström and the Norwegians Olaf D. Devik and T. Hesselberg) published *Dynamic Meteorology and Hydrography*. This book described their research to date and proposed many new techniques for weather forecasting as well as suggesting improvements to existing ones. In 1912, he was appointed professor of geophysics at the University of Leipzig. While there he founded the Leipzig Geophysical Institute.

Vilhelm Frimann Koren Bjerknes. **The Norwegian meteorologist who led the team that discovered air masses and fronts.** ***(University of Bergen, Norway)***

In 1917, during World War I, Bjerknes returned to Norway to found the Bergen Geophysical Institute as part of the Bergen museum. It is now part of the University of Bergen. He joined the staff of the University of Oslo in 1926 and remained there until his retirement in 1932.

It was while he was at the Bergen Institute that Bjerknes did his most important work. During World War I, he and his colleagues established a network of weather stations throughout Norway. These reported observations and measurements to Bergen, where they were assembled to produce general pictures of weather conditions at particular times over a wide area. Studying these pictures led Bjerknes and other members of the Bergen School to conclude that there exist AIR MASSES that differ from each other and that these masses are separated by distinct boundaries. Likening the masses to opposing armies, they called the boundaries between them *fronts* and developed a *frontal theory* to account for their development and disappearance and the weather associated with them.

The Bergen frontal theory also explained the way CYCLONES form over the Atlantic. Bjerknes described this work in 1921 in a book that became a classic: *On the Dynamics of the Circular Vortex with Applications to the Atmosphere and to Atmospheric Vortex and Wave Motion.* This formed the basis for the modern theory and practice of meteorology.

Vilhelm Bjerknes was an inspired and popular teacher who attracted talented workers and made sure they received full recognition for their work.

He died in Oslo on April 9, 1951.

(More information about Bjerknes can be found at http://www.mpae.gwdg.de/EGS/egs_info/bjerknes.htm.)

black blizzard A DUST STORM in which the dust consists mainly of dark-colored soil particles. Such a storm occurred in May 1934, during the DUST BOWL drought. The black blizzard covered an area of 1.35 million square miles (3.5 million km^2), extending from Canada to Texas and from Montana to Ohio, and the dust cloud was 3 miles (5 km) tall.

Black, Joseph (1728–1799) Scottish *Chemist* Joseph Black discovered the ways in which carbon dioxide can be released into the air naturally and therefore that this gas is a normal constituent of the air. He also discovered LATENT HEAT. He was born in Bordeaux, France, on April 16, 1728. His father was a wine merchant and the family was of Scottish descent, although his father was born in Belfast, Ireland.

In 1740 Joseph was sent to Belfast to be educated and from there he went to Glasgow University, where he studied medicine and natural sciences. His courses included chemistry, for which he displayed aptitude and enthusiasm; he became more an assistant to his teacher, William Cullen (1712–90), than a student. He moved to Edinburgh University in 1751 to complete his medical studies and in 1754 submitted a thesis for his doctor's degree.

The thesis described his research into the effect of heating "magnesia alba" (magnesium carbonate, $MgCO_3$). Black found that heating released a gas, which he detected by weighing, that was distinct from the ordinary air. In fact, this gas had been described more than a century earlier by JAN VAN HELMONT, but Black pursued his investigation further and published a fuller account of it in 1756, with the title *Experiments upon Magnesia Alba, Quicklime, and Some Other*

Alcaline Substances. His work showed that what he called "mild alkalis" (carbonates) are "causticized" (made more alkaline) when they lose this gas and that they become less causticized when they absorb it. This demonstrated that the gas is acid. Calcium carbonate ($CaCO_3$) was one of the "other alcaline substances" he studied. He found that when this is heated the gas is released and the solid substance is converted to quicklime, or calcium oxide (CaO), but that the quicklime can also recombine with the gas. Because it can be "fixed" by being absorbed into a solid substance Black called the gas "fixed air." This is the gas we now know as carbon dioxide, and the reaction he described would now be written as

$$CaCO_3 \leftrightarrow CaO + CO_2$$

Black used a balance to measure the change when $MgCO_3$ is heated and also measured the loss in weight when $CaCO_3$ loses CO_2. He measured how much $CaCO_3$ was needed to neutralize a measured amount of acid. This attention to measurement was new to chemistry and its importance was recognized some years later in the work of Antoine Lavoisier (1743–94). Investigating the properties of "fixed air," Black discovered that a candle would be extinguished if it was placed in an airtight container. He knew that heat released carbon dioxide and therefore suspected that this is what was extinguishing the flame, but when he added a substance that would absorb the gas, the candle still would not burn. He passed this problem on to one of his students, Daniel Rutherford (1749–1819), who discovered the gas that extinguished the flame is what he called "phlogisticated air" and what we know as nitrogen.

Joseph Black. **The Scottish chemist who discovered latent heat and recognized that carbon dioxide is a natural constituent of air.** ***(John Frederick Lewis Collection, Print and Picture Collection, The Free Library of Philadelphia)***

In 1756, the year his book was published, Black returned to Glasgow to succeed William Cullen as lecturer in chemistry and was appointed professor of anatomy at Edinburgh University, although he exchanged the post for that of professor of medicine. He was a practicing physician.

Around 1760 Joseph Black was becoming interested in a different problem. He found, by careful measurement, that when ice is warmed it melts slowly, but its temperature does not change. This led him to suppose that the intensity of heat is not the same as the quantity of heat, but that thermometers measure only heat intensity. As the ice melted, he concluded that it was absorbing a quantity of heat that must have combined with the particles of ice and become latent in its substance (*latent* meaning "hidden"). He called this *latent heat* and at the end of 1761 he verified its existence experimentally. He introduced the topic of latent heat into his lectures and described his work on it to a literary society in Glasgow in April 1762. In 1764 Black and his assistant, William Irvine (1743–97), measured the even larger amount of latent heat that is involved when water boils and water vapor condenses, although their measurements were not very accurate. Black never published any account of his work on latent heat, with the result that others were able to claim the credit. JEAN ANDRÉ DELUC also discovered latent heat, independently and at about the same time as Black.

Much of Black's research involved heating substances. In the course of it he noticed that equal masses of different substances require different amounts of heat to raise their temperatures by the same amount. This finding led to the concept of specific heat.

In 1766 he again succeeded his old teacher and friend William Cullen to become professor of chemistry at Edinburgh University. Joseph Black died in Edinburgh on November 10, 1799. He published very little during his lifetime, but after his death his friend John Robison (1739–1805) published lecture notes, with some additions from his pupils, together with a biographical preface by his friend. It appeared in 1803 as *Lectures on the Elements of Chemistry, Delivered in the University of Edinburgh.*

blackbody Any object (or body) that absorbs all of the radiant energy to which it is exposed and then radiates its acquired energy at the maximum rate possible for the temperature it has reached is known as a *blackbody.* The energy radiated by a blackbody is known as *blackbody radiation.*

The concept grew out of the theoretical and experimental work of physicists in the 19th century. The Swiss physicist Pierre Prévost (1751–1839) demonstrated that heat is not a substance and in 1791 pointed out that cold does not pass from snow to a person's hand; rather, heat flows from the hand to the snow. In 1824, the French physicist Nicolas Léonard Sadi Carnot (1796–1832) published *Réflexions sur la puissance motrice du feu et sur les machines propres à développer cette puissance* (translated into English as *On the Motive Power of Fire*). This related the efficiency of an engine that burns fuel to the difference between the maximum and minimum temperatures in that engine. This work attracted much attention, stimulating research in thermodynamics (*see* THERMODYNAMICS, LAWS OF), which is the scientific study of the laws governing the conversion of energy from one form into another, the direction in which heat flows, and the ability of energy to perform work.

Balfour Stewart (1828–87), a Scottish physicist, developed the ideas of Prévost and Carnot. These led him to identify the properties of a blackbody. These were also discovered independently by his contemporary, the much more famous German physicist Gustav Robert Kirchhoff (1824–87)—to whom the sole credit is often given.

The "body" is described as "black" because dark objects absorb energy, and, in principle, a perfectly black body would absorb all the energy falling on it. This cannot happen in the real world, because it is impossible to make a body that reflects no electromagnetic radiation at all. Nevertheless, the theoretical concept proved extremely valuable and Kirchhoff was able to explain its principle very simply by inviting us to suppose there is a box with blackened inside walls and only one tiny hole to provide access. Any radiation, of any wavelength, that enters the box through the hole will have only an infinitesimal chance of escaping again through the hole, so in effect it will have been absorbed. If the box is then heated until its interior is incandescent (until it glows), all wavelengths of light ought to emerge from the hole.

The Earth is not a perfect blackbody because some of the radiation falling on it is reflected (*see* ALBEDO), although the Sun is almost a perfect blackbody and the calculations based on the concept apply fairly precisely to the Sun and Earth.

The relationships between the amount of energy absorbed by a blackbody and the amount emitted as blackbody radiation were discovered toward the end of the 19th century by the German physicist Wilhelm Wien (1864–1928) and the Austrian physicists Josef Stefan (1835–93) and Ludwig Boltzmann (1844–1906). They are now known as *WIEN'S LAW* and the *STEFAN–BOLTZMANN LAW.* If the wavelength at which a blackbody radiates most intensely is known, the surface temperature of the body can be calculated from Wien's law. If the surface temperature is known, the

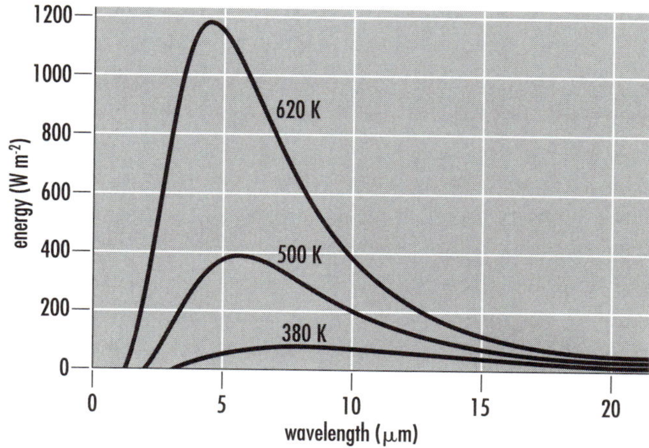

Blackbody radiation. **The graph relates the energy emitted by radiation at different wavelengths.**

amount of energy it radiates can be calculated from the Stefan–Boltzmann law.

Using Wien's law, the temperature at the visible surface (photosphere) of the Sun is 6000 K (*see* KELVIN SCALE). From this, the Stefan–Boltzmann law reveals that the Sun emits approximately 73.5 watts of energy from every square meter (W m^{-2}) of its photosphere. Multiplying this by the area of the photosphere shows that the total energy output of the Sun amounts to about 4.2×10^{20} MW: that is, 420 billion billion million watts. This is a very large number, but if the Sun has continued for the entire 4.6 billion years since it formed to emit radiation at this rate by converting matter into energy according to Einstein's equation $E = mc^2$, it will have consumed only one-thousandth of its mass.

Only a minute proportion of this energy falls upon the Earth. The Sun radiates in all directions and the Earth is a long way away and very small compared to its star. The amount of radiant energy reaching any object is inversely proportional to the square of the distance between that object and the source of the energy—in this case, about 93 million miles (150 million km). This means Earth intercepts no more than 0.0005 percent of the solar output. The actual amount of solar energy that arrives at the top of the Earth's atmosphere is known as the *SOLAR CONSTANT.*

blackbody radiation *See* BLACKBODY.

blackbody radiation curve A curve on a graph that shows the amount and wavelength of energy that is emitted by a BLACKBODY at a particular temperature. The graph shows radiation curves for several temperatures so they can be compared. Each curve is plotted by using WIEN'S LAW and the STEFAN–BOLTZMANN LAW.

black frost (hard frost) The type of FROST that leaves plants blackened, but with no ice crystals on external surfaces. It occurs when the air is very dry, so the temperature can fall far below freezing without causing SATURATION. No frost forms on exposed surfaces, but moisture freezes inside plant tissues. Many plants that grow naturally in high latitudes are able to tolerate freezing, and are said to be *hardy.* As water freezes in the intercellular spaces, osmotic pressure (*see* OSMOSIS) causes water to flow out of cells. This also increases the solute concentration inside the cells, lowering the freezing temperature of the cell contents and reducing the likelihood of ice formation. In plants that are not frost tolerant the loss of cell fluid leads to dehydration and may kill the affected cells. Hardy plants are believed to possess more flexible cell membranes and to allow their cells to shrink in size.

black ice A layer of ice that forms when rain that is close to freezing falls onto roads, the superstructure of ships, and similar surfaces that are below freezing. The water droplets spread on impact, forming a thin layer that then freezes. This results in a fairly even covering of ice. It is dark in color, hence the name *black ice,* and may be difficult to see. Black ice on roads produces extremely hazardous driving conditions. *Compare* GLAZE.

black smoke SMOKE that is produced when hydrocarbons are cracked (decomposed by heat) and then cooled suddenly. This releases particles of carbon, which are black. *Compare* BROWN SMOKE.

blizzard A wind that is accompanied by heavy snow and a low air temperature. The NATIONAL WEATHER SERVICE defines a blizzard as a wind of at least 35 mph (56 km h^{-1}), a temperature not above 20° F (–7° C), and snow that is either falling heavily enough to produce a layer at least 250 mm (10 inches) deep or that has been blown up from the surface and reduces visibility to less than 1/4 mile (400 m). In some areas the temperature requirement has been dropped.

Blizzards are extremely dangerous. Falling snow and snow lying on the ground combine to produce WHITEOUT conditions that are disorienting, and the cold wind has a strong WIND CHILL effect. When the temperature is 20° F (–7° C) in a 35-mph (56 km h^{-1}) wind, wind chill removes body heat at a rate equivalent to that of a temperature in still air of –20° F (–29° C).

Blizzards can occur anywhere. In February 1983, 47 people died in blizzards near Alayh, Lebanon. What were probably the worst blizzards to strike the United States in modern times occurred in 1888 and 1993. The 1888 blizzards lasted from January 11 to January 13 and were triggered by a COLD WAVE. They affected Montana, North and South Dakota, and Minnesota. Then, from March 11 to March 13 they struck the eastern states from Chesapeake Bay northward to

Maine, with winds gusting to 70 mph (113 km h^{-1}) and temperatures close to 0° F (–18° C). The East River froze in New York and snowdrifts almost 30 feet (9 m) deep lay in Herald Square, Manhattan. Fires could not be controlled, because fire engines could not reach them and tens of thousands of birds were killed when they were frozen solidly to trees. More than 400 people died. The 1993 blizzards, lasting from March 12 to March 15, affected the whole of eastern North America, killing an estimated 270 people in the United States and 4 in Canada.

blizzard warning A warning that is issued by the NATIONAL WEATHER SERVICE and broadcast on radio, television, and WEATHER RADIO. It alerts people in a particular area to the imminent arrival of strong wind and heavy snow. This is likely to produce deep snowdrifts. Visibility will be poor, possibly close to zero, and the wind will generate dangerously low WIND CHILL temperatures.

blob A signal on a RADAR screen that indicates a small-scale difference in temperature and HUMIDITY. It is produced by atmospheric turbulence (*see* TURBULENT FLOW).

blocking The situation in which a particular type of weather persists for much longer than usual because the movement of air that would ordinarily produce a change is obstructed or diverted. Blocking occurs in middle latitudes and can last for a month or more, although it usually lasts for about two weeks. It happens most often on the eastern side of the North Atlantic, rather less often on the eastern side of the North Pacific, and also over the Kara Sea, to the east of Novaya Zemlya off northern Siberia, and near Baffin Island, in northern Canada. In the Southern Hemisphere, blocking most often occurs near New Zealand, but also over the southern Indian Ocean and to the southeast of South America. In the Northern Hemisphere blocking is most common in winter and spring. In the Southern Hemisphere it most often occurs in winter and summer.

In the middle latitudes of both hemispheres the prevailing winds are from the west and most weather systems move from west to east. They are drawn in this direction by the POLAR FRONT JET STREAM, which also blows from west to east in both hemispheres. From time to time waves develop along the track of the jet stream. They grow more extreme until the jet stream breaks down and then resumes its approximately straight path. This sequence of events is called the *INDEX CYCLE*.

Changes in the jet stream affect the weather systems below it. During the final stage in the index cycle ANTICYCLONES, called *blocking highs,* often form with their centers between latitudes 50° and 70°. Often, there are also areas of low pressure, to each side of the blocking highs and to the south of them. While this situation lasts, the jet stream flows around each blocking high on the side nearer the pole. Sometimes the jet stream divides, with one branch diverted to each side of the blocking high.

Air circulates in an ANTICYCLONIC direction (clockwise in the Northern Hemisphere) around a blocking high. Consequently, it draws warm air from a low latitude into a higher one. This makes the core of the anticyclone warm. The lows, on the other hand, draw cool air into a lower latitude. The effect on the weather is more complicated, however.

FRONTAL SYSTEMS, with associated changes in temperature with the passing of fronts, together with precipitation and storms, slow as they approach the blocking high and are then diverted around it, following the path of the jet stream. Places to the north and south of the block are likely to experience more frontal weather than usual, but precipitation is much reduced in the area covered by the block and temperatures remain constant day after day.

Inside the blocking high and close to it the weather is drier and warmer than is usual for the time of year. Blocking is believed to have been responsible for the severe DROUGHTS that affected the Great Plains in the 1890s and 1930s, and it caused a drought in northwestern Europe that lasted from May 1975 to August 1976. To the west of the high, where the circulation carries air from a lower latitude, the weather is unseasonally warm. On the eastern side, where the same circulation draws air from a higher latitude, the weather is unusually cold. The extreme warm or cold weather lasts for as long as the blocking high remains in position.

Although blocking is well understood, it remains very difficult to predict. Scientists hope that more detailed information about changes in sea-surface temperatures will allow them to identify the conditions that trigger it before blocks develop. Of course, prediction can do nothing to protect people against the difficulties blocking sometimes causes.

blocking high *See* BLOCKING.

blood rain Rain that is red because it contains red dust particles that have been transported from a distant desert region. After it has fallen, exposed surfaces are left covered by a thin layer of the dust. Saharan dust often colors rain that falls in southern Europe and occasionally causes falls of blood rain as far north as Finland. Dust from the Australian desert has been known to fall as blood rain in New Zealand.

blossom showers *See* MANGO SHOWERS.

blowdown (windthrow) A WINDSTORM in which trees are broken or uprooted. Coniferous trees are more likely to be blown down than broad-leaved trees. This is because the roots of coniferous trees spread to the sides but do not descend very deeply into the soil, whereas many species of broad-leaved trees have taproots that penetrate vertically to a considerable depth and offer more secure anchorage. Conifers also tend to grow, and to be grown in plantations, on more exposed sites than broad-leaved trees. Broad-leaved trees can be blown down, of course; big, old trees with full crowns are more vulnerable than smaller trees, especially when they are in full leaf. When forest trees fall they often take down others and expose still more trees, so groups of trees often fall, leaving a gap in the forest. Blowdown that is caused by PREVAILING WINDS, known in Britain as *endemic windthrow,* can be anticipated and its effects in plantations minimized by the pattern of planting. Rare events cannot be predicted. These can prove devastating and are known in Britain as *catastrophic windthrows.*

blowing dust DUST that has been lifted from the ground by the wind and is transported through the air. Most airborne dust particles are less than 10 μm (0.004 inch) across. The local CONVERGENCE of air may raise a spiraling column of dust as a DUST DEVIL. Wind blowing over a larger area may produce a DUST STORM. Most dust consists of soil particles, and the removal of soil by wind is the most widespread form of EROSION, the most extreme example of which affected the region of the United States that came to be known as the *DUST BOWL.*

blowing sand Sand grains that have been lifted from the ground by wind and are transported through the air. A wind of 12 mph (20 km h^{-1}) can raise sand grains smaller than 0.01 inch (0.25 mm) in diameter provided they are dry. Strong local CONVERGENCE over a sandy desert produces WHIRLWINDS, and a strong wind blowing over a wider area produces a SAND STORM.

blowing snow SNOW grains that are lifted from the ground by the wind and transported through the air at a height of 6 feet (1.8 m) or more in amounts large enough to reduce VISIBILITY significantly. Snow that is not lifted to this height forms SNOW DRIFTS but does not reduce visibility. Blowing snow can cause a WHITE-OUT and forms one type of BLIZZARD.

blowing spray Water that is blown from the surface of the sea to form spray in an amount large enough to reduce VISIBILITY significantly.

blue Moon 1. A meteorological phenomenon in which the Moon appears blue. It occurs when the sky contains a large number of particles that are predominantly of one size. The Moon may appear green or orange, but the smaller the particles are, the farther the color is shifted toward blue. The cause of the coloration is believed to be the diffraction of light between the Moon and the observer. Suitable conditions occasionally follow DUST STORMS, forest fires, or large volcanic eruptions. The Sun also appears blue for the same reason—a blue Sun was seen after the 1883 eruption of Krakatau. 2. Astronomically, a term that was defined in 1946 in the magazine *Sky and Telescope* as the second full Moon to be seen in a single calendar month. In 1999, *Sky and Telescope* corrected this definition to the third full Moon in a season during which there are four and during a period of 12 SIDEREAL MONTHS in which there are 13.

blue-shift *See* DOPPLER EFFECT.

blue Sun *See* BLUE MOON.

Bob A HURRICANE that struck the eastern coast of the United States August 18–20, 1991. It killed at least 20 people.

bohorok A warm, dry wind that blows during the MONSOON season along the northeastern side of the mountains of Sumatra, Indonesia. It is a wind of the FÖHN type, and the drop in RELATIVE HUMIDITY that accompanies it sometimes damages crops.

boiling The change of PHASE that occurs when a liquid becomes a gas, absorbing LATENT HEAT to supply the energy required for the change. In the liquid phase, molecules form small groups that are constantly breaking and rejoining and that can slide easily past each other. At the surface of the liquid, even when it is cold, molecules are constantly escaping into the BOUNDARY LAYER of air immediately above the surface and molecules in the boundary layer are returning to the liquid. The application of an increasing amount of heat causes the molecules in the liquid to move faster—the motion that is measured as the TEMPERATURE—and a greater number of them to escape into the boundary layer. The molecules in the boundary layer exert a VAPOR PRESSURE; when the SATURATION VAPOR PRESSURE is reached molecules begin to escape from the boundary layer and into the air above it. If sufficient heat is applied to the liquid, groups of molecules break apart below the surface, forming bubbles of vapor (not air) that rise to the surface because the vapor is less dense than the liquid. This is boiling.

Boiling temperature °F	°C	Atmospheric pressure mb
392	200	15,536
320	160	6,176.8
248	120	1,984.9
212	100	1,013.25
140	60	199.33
68	20	23.38
32	0	6.11

In order for molecules to escape from the boundary layer, the vapor pressure they exert must exceed the atmospheric pressure. When the amount of heat applied to the liquid has raised its temperature to a level at which the vapor pressure exceeds the atmospheric pressure the liquid boils. Consequently, the temperature at which a liquid boils varies with the air pressure. At sea-level pressure of 1013.25 millibars (mb) pure water boils at 212° F (100° C). At any pressure below 6.11 mb water that is exposed to the air cannot remain in the liquid phase, because its boiling temperature is below its freezing temperature, which also varies with pressure. At a pressure of 6.11 mb and a temperature of 32.018° F (0.01° C), water exists simultaneously in all three of its phases: as liquid, with some ice floating on its surface, and water vapor above the surface. This is known as the *triple point* for water. The table gives the boiling temperature of water at a range of pressures.

Bølling A place in Jutland, Denmark, where there are sediments that were deposited on the bottom of a lake that existed toward the end of the Weichselian glacial (DEVENSIAN GLACIAL). Remains found in the sediments indicate there was a period of warmer conditions that lasted from about 13,000 years ago until about 12,200 years ago. Temperatures then were as high as those of today, or even higher, but fell toward the end of the INTERSTADE. The Bølling interstade was followed by the OLDER DRYAS STADE.

bolometer An instrument that is used to measure radiant energy. It was invented by S. P. LANGLEY in 1880 and works by measuring the rise in temperature of a blackened metal strip that is placed in one of the arms of a Wheatstone resistance bridge. This instrument compares the electrical resistance in an object placed in one of its arms with that of another arm, which is known. The change in temperature alters the electrical resistance in the strip. When it is linked to a galvanometer (an instrument that measures small electric currents), the deflection of the galvanometer needle is proportional to the intensity of the radiation. Modern bolometers use a metal strip that consists of strips of platinum made into four gratings. A bolometer can measure a temperature difference of 0.0018° F (0.0001° C). The distribution of the intensity of radiation through the spectrum is measured by a spectrum bolometer. This has a single metal strip set on its edge in one arm of a resistance bridge.

bolster eddy (roll vortex) An EDDY that sometimes forms along the foot of a steep slope or cliff on the upwind side. As the air approaches the slope most of it moves upward and over the top, accelerating as it does so (*see* BERNOULLI EFFECT). Air rising up the slope enters a region of lower atmospheric pressure. Its movement up the slope creates another area of reduced pressure near the base of the slope, and some of the air is deflected downward, towards this lower pressure. At the foot of the slope the air starts to move horizontally in the direction opposite to the main flow of air. It becomes caught in the main flow, which curls it around into an eddy that is approximately circular in cross section.

bomb A CYCLONE that develops very rapidly over the ocean. Its development differs from that of most cyclones. The COLD FRONT detaches from the WARM FRONT and starts moving at right angles to it. This means the cold front never catches up with the warm front. The center of the low moves rapidly and the warm front is left behind as a BACK-BENT WARM FRONT.

bora A north or northeasterly KATABATIC WIND that blows along the coast of the Adriatic Sea. It is cold, gusty, and usually dry, although it can carry rain or snow. Its average speed is about 24 mph (38 km h^{-1}) in summer and in winter 32 mph (52 km h^{-1}), although in winter it can be much stronger. At Trieste it blows on an average of about 40 days a year. It sometimes continues for several days. Its name is taken from the name of Boreas, the Greek god of the north wind.

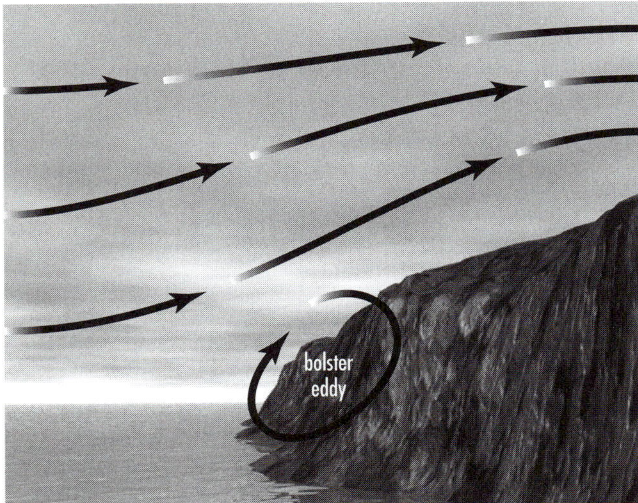

Bolster eddy. **Most of the air flows up the slope and over the top, but some is drawn down the slope and rolls over at the foot.**

bora fog A dense fog that forms when a strong BORA wind lifts clouds of spray from the sea.

boreal climate A climate that is associated with the belt of coniferous forest that lies across North America and Eurasia (in Russia it is known as the *taiga*). The name *boreal* refers to Boreas, the Greek god of the north wind. The region is bounded by the edge of the tundra in the north and extends southward to about 55° N in eastern North America and about 60° N in the west. In Eurasia it extends to about 50° N in the east and 65° N in the west.

The climate extends farther south in the east of the continents because it is within the midlatitude belt of westerly winds that carry AIR MASSES from west to east. Air masses entering North America and Eurasia from the Pacific and Atlantic, respectively, are maritime and therefore moist and relatively mild. As they cross the continents they become increasingly continental, making them drier and more extreme in temperature.

Summers are short, with average temperatures that reach about 60° F (15° C), but maximum temperatures that can be much higher. A temperature of 99° F (37° C) has been recorded at Fairbanks, Alaska, and Yakutsk, Siberia, has known 102° F (39° C). Winter is the dominant season, however. It lasts from six to nine months, during which the temperature remains below freezing. In many places away from coasts in up to four months the temperature remains below 0° F (–18° C). The Northern Hemisphere COLD POLE lies within this region. Heat is radiated away from the snow-covered ground so rapidly that in winter there is often a more or less permanent temperature INVERSION and the LAPSE RATE is negative (temperature increases with height) from ground level to a height of 3,300–5,000 feet (1,000–1,500 m). The range of summer and winter average temperatures is greater than in any other climate, typically of about 60° F (33° C).

Precipitation is generally light. Fairbanks has an average of about 12 inches (305 mm) a year and Yakutsk about 14 inches (356 mm). Coastal areas receive rather more. Archangel, in northern Russia, has about 21 inches (533 mm), for example, and Trondheim, Norway, has about 40 inches (1,016 mm).

Boreal period A time during the present FLANDRIAN INTERGLACIAL, which lasted from about 9,600 to 7,500 years ago, immediately prior to the ATLANTIC PERIOD. During the Boreal, the climate of northwestern Europe was drier and more CONTINENTAL than it is now, but it was also becoming steadily warmer. It was the last period since the end of the most recent GLACIAL PERIOD during which Britain was linked to mainland Europe by a land bridge across the Dover Strait.

bottom water The water that lies in the deepest part of the ocean. It is denser than water near the surface, and at a constant temperature of 34°–36° F (1°–2° C) in all oceans. The water flows very slowly and is driven by variations in density (*see* THERMOHALINE CIRCULATION).

Bouguer's halo A faint white arc of light, with a radius of about 39°, that is sometimes seen at the ANTISOLAR POINT. It is caused by REFLECTION and REFRACTION and is named after the French scientist Pierre Bouguer (1698–1758), who studied the refraction of light and the effect on light of its passage through the atmosphere.

boundary current An ocean current that flows close to the coast of a continent and parallel to it. Boundary currents flow in either a northerly or a southerly direction and are caused by the deflection of an east–west or west–east current where it meets the continental landmass. The boundary currents on the western sides of oceans in both hemispheres are deep, narrow, and fast-flowing and carry warm water. These currents are most prominent in the Atlantic and Pacific Oceans, as the GULF STREAM in the North Atlantic, the KUROSHIO CURRENT in the North Pacific, the BRAZIL CURRENT in the South Atlantic, and the AGULHAS CURRENT in the South Pacific. The currents on the eastern sides of the oceans are wide, shallow, and slow and carry cool water. These are the CANARIES CURRENT in the North Atlantic, the CALIFORNIA CURRENT in the North Pacific, the BENGUELA CURRENT in the South Atlantic, and the PERU CURRENT in the South Pacific. Currents in the Indian Ocean are more complicated.

boundary layer The layer of air that lies immediately adjacent to a surface and within which atmospheric conditions are strongly influenced by the proximity of the surface. A boundary layer may be very thin. The boundary layer above a water surface in which water molecules are constantly being exchanged with the liquid is about 0.04 inch (1 mm) deep. Other boundary layers are deeper. The PLANETARY BOUNDARY LAYER extends from the surface to an average height of 1,700 feet (519 m).

Bowen ratio The ratio of sensible to latent heat, which indicates how energy is apportioned at the Earth's surface. Solar energy is absorbed at the surface and is used to evaporate water. Evaporation (E) absorbs latent heat of vaporization (L) and some of the remaining energy is released from the surface as sensible heat (H)—heat that warms objects exposed to it. The temperature and water vapor content of the air at two levels above the surface can be measured, and from this the ratio (β) can be calculated as $\beta = H/LE$. Most surfaces tend to keep the ratio at a minimum. If β is greater than 1, more energy is being released into the atmosphere as heat than is being used for evaporation.

Boyle, Robert (1627–1691) English *Natural philosopher* Robert Boyle was an aristocrat. His father, Richard Boyle, was the earl of Cork and Robert was his 7th son and 14th child. He was born on January 25, 1627, at Lismore Castle, in Ireland. He learned to speak Latin and French while still a small child and was sent to Eton College, near London, at the age of eight. After three years, in 1638 he traveled abroad with a French tutor. In 1641 he arrived in Italy and spent the winter in Florence studying the work of GALILEO.

Boyle returned to England in 1644 and immediately devoted himself to a life of scientific inquiry. Drawn to others who shared his interests, Boyle soon joined a group of people who called themselves the "Invisible College." They held frequent meetings at Gresham College, in London, and some of the members also met in Oxford. Boyle moved to Oxford in 1654, and it was in that city that he carried out his most important scientific work.

In 1657 he read of an air pump that had been invented by the German physicist Otto von Guericke (1602–86). It was meant to evacuate the air from a chamber, and Boyle enlisted the help of ROBERT HOOKE to improve it. Boyle and Hooke became lifelong friends. The pump was finished by 1659 and Boyle began using it to experiment on the properties of air. He published the results of this work in 1660 with the title *New Experiments Physico-Mechanical Touching the Spring of Air and Its Effects.* Boyle had discovered that the volume occupied by a gas is inversely proportional to the pressure under which the gas is held. This relationship is known in English-speaking countries as *Boyle's law* (*see* GAS LAWS). He also found that the weight of a body varies according to the amount of BUOYANCY supplied by the atmosphere.

Boyle and Hooke also studied combustion. They found that neither charcoal nor sulfur burns when air is excluded, no matter how strongly the vessel containing them is heated, but they burst into flames as soon as air is allowed into the container. When either charcoal or sulfur is mixed with potassium nitrate (saltpeter), however, the mixture burns in a vacuum. Boyle concluded from this that both potassium nitrate and air contain some ingredient that is necessary for combustion. Boyle did not identify that ingredient, however. Joseph Priestley (1733–1804) isolated it in 1774, and in 1777 Antoine Laurent Lavoisier (1743–94) gave it the name *oxygen.*

In 1661 Boyle published another book, *The Sceptical Chymist,* in which he advanced the idea that matter is composed of "corpuscles." These are of various shapes and sizes and they are able to combine into groups. Each group of corpuscles comprises a chemical substance. Boyle was the first scientist to use the word *analysis* to describe the separation of a substance into its constituents. He invented a hydrometer for measuring the density of liquids and made the first match by coating a rough paper with phosphorus and placing a drop of sulfur on the tip of a small stick. The stick ignited when it was drawn along a crease in the paper. He made a portable camera obscura that could be extended or shortened like a telescope in order to focus an image on a piece of paper stretched across the back of the box, opposite the lens.

By a charter granted by King Charles II and passed on August 13, 1662, the Invisible College became the "Royal Society, for the improvement of natural knowledge by Experiment." The charter named Boyle as a member of the council. Boyle was elected president of the society in 1680 but declined because he was unwilling to take the necessary oath.

In addition to his scientific work, Boyle was deeply interested in theology. He learned Hebrew, Greek, and Syriac in order to be able to read scriptural texts in their original languages. His will provided for the founding of a series of lectures aimed at proving the Christian religion against the views of other religions, but with the proviso that disputes between Christians should not be mentioned.

In 1668 Boyle returned to live in London with his sister. He remained in London for the rest of his life. Boyle died there on December 30, 1691.

(You can read more about the life and work of Robert Boyle and the other scientists of his day in Lisa Jardine, *Ingenious Pursuits* ([Boston: Little, Brown and Co., 1999].)

Boyle's law *See* GAS LAWS.

Bq *See* BECQUEREL.

brave west winds The name given by sailors to the strong prevailing westerly winds that blow over the oceans between latitudes 40° N and 65° N and between 35° S and 65° S. They are more persistent in the Southern Hemisphere than in the Northern Hemisphere and are strongest in the roaring forties, in latitudes 40°–50° S. The winds are produced by the strong PRESSURE GRADIENT on the side nearer the equator of the frequent DEPRESSIONS that travel from west to east. Consequently, the generally westerly winds fluctuate between northwest and southwest.

Brazil Current An ocean current that carries warm water southward from the SOUTH EQUATORIAL CURRENT, along the eastern coast of South America, to join the WEST WIND DRIFT. The current moves very slowly and is no more than 330–660 feet (100–200 m) deep, and the salinity of its water is 36–37 parts per thousand (‰), which is saltier than the average for seawater.

break A sudden change in the weather. The term is usually applied to the ending of a prolonged period of settled dry, cold, or warm weather.

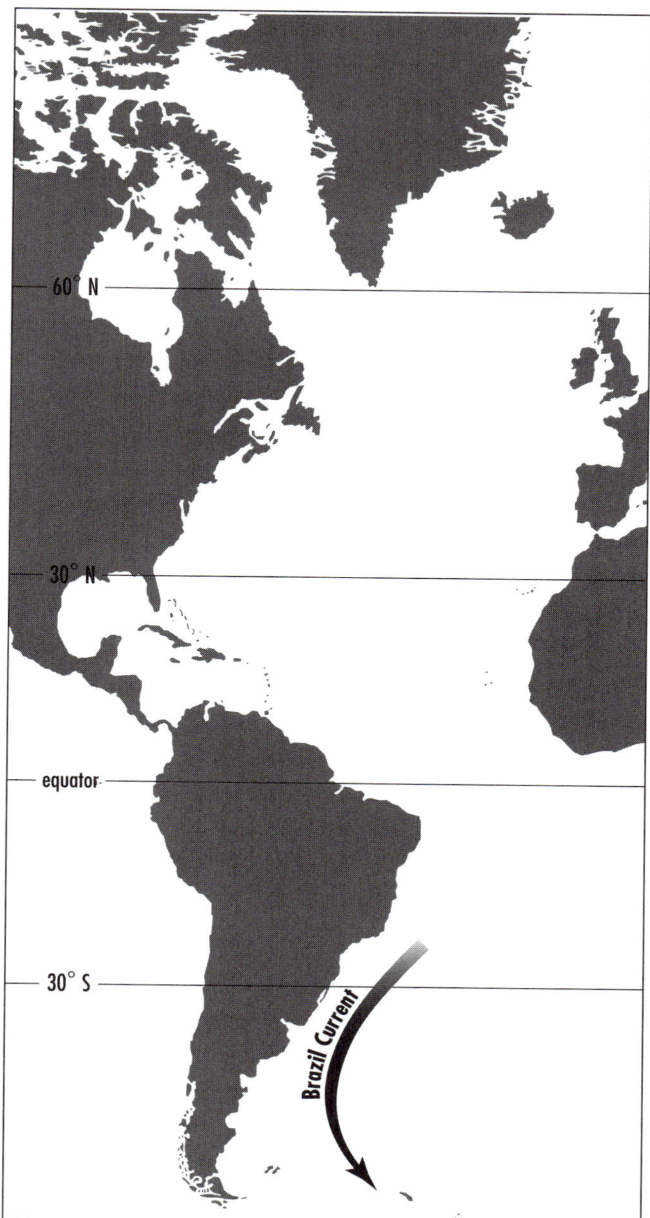

Brazil Current. **A warm current that flows parallel to the coast of Brazil.**

breakdown potential The potential gradient of the vertical electric field in the atmosphere at which LIGHTNING will occur in dry air. It is 3×10^6 V m^{-1}. This is 10 times greater than the largest field that is observed in CUMULONIMBUS clouds, showing that it is necessary for local processes to build charge between cloud droplets and ice crystals and initiate a flash leader to begin the discharge.

breaks in overcast The condition in which more than 90 percent but less than 100 percent of the sky is covered by cloud.

breeze A light wind. In the BEAUFORT WIND SCALE, a breeze is a wind blowing at 4–31 mph (6.4–49.8 km h^{-1}). Within this range, breezes are classified as LIGHT, GENTLE, MODERATE, FRESH, and STRONG BREEZES.

Brenda A TYPHOON that struck southern China in May 1989, killing 26 people.

Bret A TROPICAL STORM that struck Venezuela on August 8, 1993, causing floods and mudslides in which at least 100 people died. Bret is also the name of a HURRICANE that struck the coast of Texas in 1999. With winds of 125 mph (201 km h^{-1}), it hit Kenedy County at about 6:00 P.M. on August 22, traveling westward at about 7 mph (11.25 km h^{-1}). Its winds later weakened to 105 mph (169 km h^{-1}). The following day it was downgraded to a tropical storm and its winds abated. The area affected was sparsely populated, but four people lost their lives in floods caused by the hurricane.

brickfielder A hot, dry, dust-laden northerly wind that blows from the desert interior across southern Australia. In the north of Victoria the wind sometimes raises the temperature to above 100° F (38° C) and the hot conditions can last for several days. The name *brickfielder* originated in Sydney, where it was applied to SOUTHERLY BURSTERS that carried dust over the city from the brickfields to the south. The brickfields were the area where clay was dug from the ground and made into bricks. When gold mines opened in Victoria, miners recruited from Sydney gave the same name to the winds they experienced there, although these blow from the opposite direction and cross no brickfields.

bridled anemometer *See* ANEMOMETER.

bridled pressure plate A type of pressure-plate ANEMOMETER in which the movement of the plate is restrained by a spring (the plate is said to be *bridled*). The deflection of the plate is measured electronically.

bright band An enhanced RADAR echo that is seen in an image of a cloud where SNOWFLAKES are melting into RAINDROPS.

brightness temperature A unit that is used to color-code MICROWAVE images of snow- and ice-covered areas that are received from satellites. The measures correspond closely to the intensity of the microwave radiation, but they are given values in KELVINS. This reflects differences in EMISSIVITIES from different surfaces. At a microwave WAVELENGTH of 1.55 cm, for example, ice has a brightness temperature of 190 K or more, but water has a lower brightness temperature, of less than 160 K. The boundary between water and sea ice shows clearly on the resulting color-coded image. Brightness temperatures vary with the wavelength of the microwave radiation; by comparing brightness temperatures at two different wavelengths it is possible to calculate the depth of the snow covering an area.

bristlecone pine (***Pinus longaeva***) A species of pine tree that grows in the arid regions of California. Bristlecone pines survive to a remarkably old age. Some living specimens are 4,600 years old. Their age is measured by counting the TREE RINGS in cores drilled from the trunk of the tree (*see* DENDROCHRONOLOGY). When the ages of living trees are cross-dated with cores taken from dead trees in the same area, the time scale is extended to more than 8,200 years. A complete chronology covering 5,500 years has been developed for one group of bristlecone pines. This chronology is used as a standard reference against which other tree-ring sequences can be calibrated. It is also used to calibrate RADIOCARBON DATING methods, by measuring the $^{12}C:^{14}C$ ratio in individual rings and compiling a record of fluctuations in the ratio.

British thermal unit (**Btu**) A unit of work, energy, or heat that was first used by the English physicist James Prescott Joule (1818–89) in a paper on the relationship between heat and mechanical energy that he presented to a meeting of the British Association in 1843. The unit was given its name in 1876. One British thermal unit is the energy that is required to raise the temperature of one pound of water through one degree Fahrenheit. This varies according to the starting temperature, which is sometimes specified. A mean value for the Btu is given by dividing by 180 the energy needed to raise the temperature of one pound of water from 32° F to 212° F (from freezing to boiling). This Btu is equal to 1055.79 JOULES. The accepted international value for the Btu is 1055.06 joules.

Brocken specter A GLORY that is sometimes seen in the mist at the summit of the Brocken mountain, a peak in the Harz Mountains, Germany, and on some other mountains. It appears as a gigantic human figure. In fact, it is the shadow of the observer cast on the water droplets in the mist.

broken The condition when between 60 percent and 90 percent of the sky is covered by cloud.

Brørup interstadial An INTERSTADE that occurred about 59,000 years ago, during the Weichselian glacial (*see* DEVENSIAN GLACIAL). Evidence for it is taken from sediments that existed then in a lake beneath what is now a bog at Brørup, in West Jutland, Denmark. This lay beyond the limit of the ICE SHEET. During the interstade a forest consisting of coniferous trees with birch (*Betula* species) covered the area. The vegetation was similar to that found in England at about the same time during the CHELFORD INTERSTADIAL.

brown smoke SMOKE that contains particles of tar-like compounds. It is produced when coal is burned at a low temperature (*see also* BLACK SMOKE).

brown snow SNOW that is mixed with dust.

Brückner cycle A cyclical change in the weather that occurs over a period of about 35 years. The English scientist Francis Bacon (1561–1626) was the first person to suggest the existence of a 35-year weather cycle, shortly before he died in 1626, and in 1890 the German geographer and glaciologist EDUARD BRÜCKNER produced clear evidence for it. He based this on his detailed examination of weather records over many years. Brückner found that the temperature and precipitation in Europe vary over a period of 34.8 ± 0.7 years. Each cycle consists of a cool, moist half and a warm, dry half. In the course of each cycle the temperature varies by not more than 2° F (1.1° C) and the rainfall varies by 8–9 percent. The AMPLITUDE of the wave is small but is clearly recognizable when plotted on a graph. Modern climatologists consider the Brückner cycle to be of only minor importance in determining weather patterns.

Brückner, Eduard (1862–1921) German *Geographer and glaciologist* Eduard Brückner was born at Jena, in Saxony, Germany, on July 29, 1862, the son of Alexander, a teacher of Russian history, and Lucie Schiele. He was educated at the gymnasium (high school) in Karlsruhe and from 1881 until 1885 studied physics and meteorology at the University of Dorpat (now Tartu, Estonia). He then continued his studies in Dresden and Munich.

In 1885 Brückner joined the staff of the Deutsche Seewarte in Hamburg. Established in 1876, the Seewarte supplied weather information for ships using the port of Hamburg. It developed into the modern German weather service. From 1888 until 1904 he was a professor at the University of Bern, Switzerland, and from 1899 to 1900 he was rector of the university. He married Ernestine Stein in 1888. In 1904 he returned to Germany to become a professor at the University of Halle, and in 1906 he was appointed a professor at the University of Vienna, Austria, a post he held until his death.

Brückner was an authority on alpine GLACIERS, and he was especially interested in the effect of glaciers on surface features of the landscape. He was convinced that climate change is of great importance, with direct economic and social implications. He conducted extensive research and made many theoretical studies of changes that have occurred in the past. In the course of these he discovered the 35-year cycle that now bears his name (*see* BRÜCKNER CYCLE). His interest in the subject is indicated by the titles of some of the papers he published: "How Constant Is Today's Climate?" (1889), "Climate Change since 1700" (1890, and the paper in which his cycle was first mentioned), "Influence of Climate Variability on Harvest and Grain Prices in Europe" (1895), "An Inquiry about the 35 Year Periods of Climatic Variations" (1902), and "Climate Variability and Mass Migration" (1912).

Brückner died in Vienna on May 20, 1927.

(Some of his papers, including those listed here, are now available in English. You can find them at w3g.gkss.de/G/Mitarbeiter/storch/brueckner.html.)

Brunt–Väisälä frequency The frequency with which an atmospheric gravity wave (*see* ATMOSPHERIC WAVE) oscillates. If a PARCEL OF AIR is displaced upward it sinks again, overshooting and rising several times in a wave pattern about its level of neutral BUOYANCY. The frequency of this oscillation is the Brunt–Väisälä frequency and it is given by $N1/2\pi$, where $N = [(g/\theta)/\partial\theta/\partial\theta 0]^{1/2}$, where g is the gravitational acceleration, θ is the constant POTENTIAL TEMPERATURE of the parcel of air, and $\partial\theta/\partial z$ is the vertical gradient of potential temperature.

Btu *See* BRITISH THERMAL UNIT.

bubble policy *See* CLEAN AIR ACT.

bubnoff unit *See* EROSION.

Buchan, Alexander (1829–1907) Scottish *Meteorologist* The man who is acknowledged to have been the most eminent British meteorologist of the 19th century was born at Kinnesswood, Kinross, Scotland, on April 11, 1829. He became a schoolteacher, teaching all subjects. His favorite leisure pursuit was botany.

After a public meeting held in Edinburgh on July 11, 1855, a society was formed with the aim of establishing weather stations throughout Scotland. The society became the Scottish Meteorological Society and operated the weather stations from 1856 until 1920, when that task was taken over by the METEOROLOGICAL OFFICE. In December 1860 Buchan was appointed secretary to the society; he remained in the post until his death in 1907.

From 1864, he also edited the *Journal of the Scottish Meteorological Society* from its first issue and wrote a great deal of its material. During his editorship, the journal published Thomas Stevenson's description of his louvered screen (*see* STEVENSON SCREEN). The screen is still widely used.

In 1883 the society opened a meteorological observatory on Ben Nevis, the highest mountain in Britain. Buchan was closely involved with the establishment of the observatory and with the running of it, which continued until 1904.

Buchan established his reputation in 1867, when he published his *Handy Book of Meteorology*. This became a standard textbook and remained in use for many years. In 1869 he wrote the paper "The Mean

Pressure of the Atmosphere and the Prevailing Winds over the Globe" for the Royal Society of Edinburgh. He also wrote papers on the circulation of the atmosphere and on ocean circulation. It was in 1869 that he published his paper "Interruptions in the Regular Rise and Fall of Temperature in the Course of the Year" in the *Journal*, describing what came to be called BUCHAN SPELLS.

Buchan was made a member of the Meteorological Council in 1887 and was elected a fellow of the Royal Society in 1898. In 1902 he was the first person to be awarded the Symons Medal, the greatest honor meteorologists can bestow on one of their colleagues.

Alexander Buchan died in Edinburgh on May 13, 1907.

Buchan spells Periods in the year when the usual rise or fall of temperature with the seasons is halted or reversed. There are often a few days, or even a week or two, when the weather becomes colder in spring or warmer in the fall and people describe the weather as "unseasonal." The Scottish meteorologist ALEXANDER BUCHAN (1829–1907) suspected these periods might occur as a regular feature of the climate. To investigate this possibility, in 1869 he examined the temperature records for Edinburgh from 1857 to 1866. He discovered that cold departures from the temperature trend occurred on February 7–14, April 11–14, May 9–14, and June 29–July 4. The warmest weather was in July. After that, as the average temperature fell, there were warm periods on August 6–11, November 6–13, and December 3–14. It was popularly assumed that Buchan had found regular fluctuations that applied quite generally and could be anticipated. This is how they came to be called *Buchan spells*. In fact Buchan claimed no such thing. He made it clear that the periods he identified varied from year to year and that they applied only to southeastern Scotland. Similar spells of unseasonal weather do occur in most places, but their dates vary. They are now known as *SINGULARITIES*.

budget year The period of one year that commences with the start of SNOW accumulation at the FIRN LINE of a GLACIER. The year continues through the following summer, when snow is lost by ABLATION, and so a comparison of measurements taken at the start of successive budget years indicates the growth or diminution of the glacier.

Budyko classification A CLIMATE CLASSIFICATION that was proposed in 1956 by MIKHAIL I. BUDYKO. It is based on the net radiation that is available for the EVAPORATION of water from a wet surface (R_o) and the heat that would be required to evaporate the whole of the mean annual PRECIPITATION (Lr), where Lr is the LATENT HEAT of vaporization. The ratio of these two values is used to designate climate types. The drier the climate the larger is the ratio and unity (a ratio of 1.0) marks the boundary between dry and moist climates. The climate types used in the scheme are listed in the table.

R_0/Lr	Climate type
greater than 3.0	desert
2.0–3.0	semidesert
1.0–2.0	steppe grassland
0.33–1.0	forest
less than 0.33	tundra

Budyko, Mikhail Ivanovich (born 1920) Belorussian *Physicist and meteorologist* A Belorussian scientist who was the first to calculate the balance of heat received from the Sun and radiated from the Earth's surface, checking his calculations against observational data from all parts of the world. In 1956 he published his results in *Heat Balance of the Earth's Surface*. This work changed climatology from a qualitative discipline, based on measurement of climatic data from all over the world, into a more physical discipline. Professor Budyko became a pioneer of physical climatology, adding to his 1956 book an atlas, completed in 1963, that shows the Earth as viewed from space with all aspects of the Earth's heat balance displayed. Calculations of climate change are based on this atlas.

By 1960, Professor Budyko was already concerned about the possibility of a general rise in world temperatures caused by human activity. He suggested the day might come when it became necessary to scatter particles in the stratosphere in order to reflect solar radia-

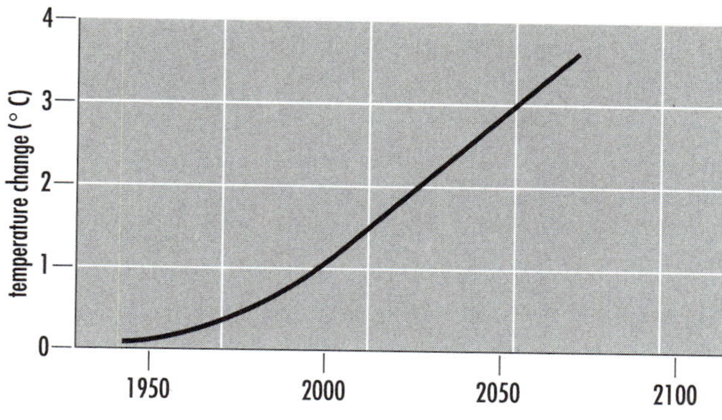

The rise in the average global air temperature as a consequence of the increasing atmospheric concentration of carbon dioxide predicted by Professor Mikhail Budyko in 1972.

tion and reduce the rate of temperature increase. In 1972, he was able to confirm a link between past climate changes and changes in the atmospheric concentration of carbon dioxide (*see also* ARRHENIUS, SVANTE). He warned then that his analysis indicated a general warming of the world's climates due to the rise in the carbon dioxide concentration brought about by the increasing consumption of fossil fuels. His 1972 calculations predicted a rise in temperature of about 6.3° F (3.5° C) from this cause between 1950 and about 2070.

His studies of the effects on climate of altering the composition of the atmosphere led him, in the early 1980s, to ponder the climatic consequences of a large-scale thermonuclear war. He suggested that such a war might inject such a huge quantity of AEROSOLS into the atmosphere that the entire world would be plunged into deep cold, a "nuclear winter" that might threaten human survival.

Mikhail Budyko was born on January 26, 1920, at Gomel, Belarus. He was educated in Leningrad and from 1942 until 1975 worked at the Main Geophysical Observatory, Leningrad, where he was the director from 1972 to 1975. He was then appointed head of the Division for Climate Change Research, at the State Hydrological Institute, St. Petersburg, the position he still holds. He was elected an Academician of the Russian Academy of Sciences in 1992.

He has been awarded many prizes, including the Lenin National Prize (1958), Gold Medal of the World Meteorological Organization (1987), A. A. Grigoryev Prize of the Russian Academy of Sciences (1995), and Blue Planet Prize (1999) for his contribution to environmental research.

(There is more information about Professor Budyko at http://www.af-info.or.jp/eng/whatnew/hot/enr-budyko.html.)

bulk modulus *See* SPEED OF SOUND.

buoyancy The upward force that is exerted on a body when it is immersed in a fluid. ARCHIMEDES' PRINCIPLE states that when a body is immersed in a fluid it displaces its own volume of the fluid. This reduces the weight of the body by the weight of the displaced fluid. This is why bodies weigh less in water and why very large animals, such as whales and hippopotamuses, are able to move freely and gracefully through water.

If the weight of the body is greater than the weight of its own volume of the fluid, the body sinks through the fluid. It then experiences negative buoyancy. If the body weighs less than the displaced fluid, it experiences positive buoyancy. This acts as an upward force, and the body rises.

Buoyancy occurs in air when a PARCEL OF AIR has a different DENSITY from the air surrounding it. This can be expressed as

$$F/M = g[(\rho' - \rho)/\rho]$$

where F is the buoyancy force, M is the mass of the air parcel, g is gravitational acceleration, ρ' is the density of the surrounding air, and ρ is the density of the parcel of air. Dividing F by M gives the buoyancy force per unit of mass, and $(\rho' - \rho)/\rho$ is the buoyancy, often designated by B. The force exerted by the buoyancy is therefore the buoyancy (B) multiplied by the gravitational acceleration (g), or gB.

buran A type of fierce BLIZZARD that occurs in winter on the open plains of southern Russia and throughout Siberia. In northern Siberia a very similar kind of storm is called the *PURGA*. The buran is a snowstorm driven by a wind of gale force or even hurricane force. As well as the falling snow, the air is filled with snow blown up from the ground, and visibility is reduced almost to zero. The air temperature is not especially low, but the combination of WIND CHILL and disorientation due to the poor visibility make the buran very dangerous.

burn-off The clearance of FOG, MIST, or low cloud during the course of the morning, as the sunshine intensifies and the temperature rises. As the air grows warmer the DEW POINT TEMPERATURE rises and suspended water droplets evaporate.

burst of monsoon The abrupt onset of the summer MONSOON, when the cool, dry weather of the winter monsoon gives way in a matter of a few hours to warm, humid air and heavy rain.

butterfly effect A metaphor that was invented by the American meteorologist Edward Lorenz (born 1917) to illustrate what is known formally as "sensitive dependence on initial conditions." On December 29, 1979, Lorenz presented a paper in Washington, D.C., at the annual meeting of the American Association for the Advancement of Science that had the title "Predictability: Does the Flap of a Butterfly's Wings in Brazil Set Off a Tornado in Texas?" Lorenz was then a research scientist at the Massachusetts Institute of Technology. While developing mathematical computer models of weather systems he found that weather patterns repeated themselves, but each time with differences that arose from extremely small variations in their starting conditions. Two apparently identical weather systems could develop entirely differently if one of their initial parameters differed by one part in a thousand—one millibar pressure, or a small fraction of a degree in temperature. This showed that weather is unpredictable for more than a few hours ahead and that changes in a weather system may be caused by factors arising within the system itself. Lorenz's work led to the development of the mathematical theory of CHAOS and had a strong influence on meteorological research.

(The butterfly effect is explained in James Gleick, *Chaos: Making a New Science,* London: William Heinemann, 1988.)

Buys Ballot, Christoph Hendrick Diderik (1817–1890) Dutch *Meteorologist* Christoph Buys Ballot was born at Kloetinge, Zeeland, the Netherlands, on October 10, 1817. In 1847 he was appointed professor of mathematics at the University of Utrecht; in 1854 he helped to found and was the first director of the Royal Netherlands Meteorological Institute. He remained in this post until his death.

Christoph Hendrick Diderik Buys Ballot. **The Dutch meteorologist who discovered the law named after him. *(Library KNMI Koninklijk Nederlands Meteorologisch Institut)***

In 1857 Buys Ballot described the wind circulation around areas of low and high atmospheric pressure. He based his description on his studies of meteorological records, and it quickly became known as a law attributed to him. Buys Ballot did not know that the American meteorologist WILLIAM FERREL had reached the same conclusion on theoretical grounds some months earlier. When he learned of this, Buys Ballot acknowledged Ferrel's prior claim to the discovery, but it was too late and what should be known as *Ferrel's law* is usually called *BUYS BALLOT'S LAW.*

Buys Ballot died on February 3, 1890.

Buys Ballot's law The rule that in the Northern Hemisphere if you stand with your back to the wind there is an area of low pressure on your left; in the

Southern Hemisphere, the area of low pressure is on your right. The law was deduced in 1857 on theoretical grounds by the American meteorologist WILLIAM FERREL and a few months later the Dutch meteorologist CHRISTOPH HENDRICK DIDERIK BUYS BALLOT announced his discovery of it, which was based on records of the wind circulation around midlatitude CYCLONES. Expressed a little more technically, the law states that the wind blows at 90° to the direction of the PRESSURE GRADIENT. This is true of the GEOSTROPHIC WIND in the FREE ATMOSPHERE, but it is not strictly true of the wind that blows in the PLANETARY BOUNDARY LAYER, where the wind is affected by FRICTION and the angle between the wind and the pressure gradient is less than 90°. The law does not apply close to the equator, where the CORIOLIS EFFECT is extremely small or, at the equator itself, where it does not exist.

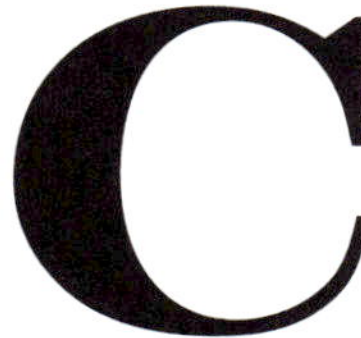

C

C *See* COULOMB.

cacimbo A heavy MIST or wet FOG associated with low STRATUS cloud and sometimes DRIZZLE that occurs along the coast of Angola during the dry season. The cacimbo usually forms in the morning and evening and may penetrate inland for some distance. It helps prevent extreme DROUGHT. The cacimbo is caused by onshore winds that carry warm air across the cold BENGUELA CURRENT.

Cainozoic *See* CENOZOIC.

cal *See* CALVUS.

calendar year *See* ORBIT PERIOD.

California Current A slow-moving, somewhat diffuse ocean current that conveys cold water southward parallel to the western coast of North America. In the latitude of Central America, it turns westward to become the NORTH EQUATORIAL CURRENT.

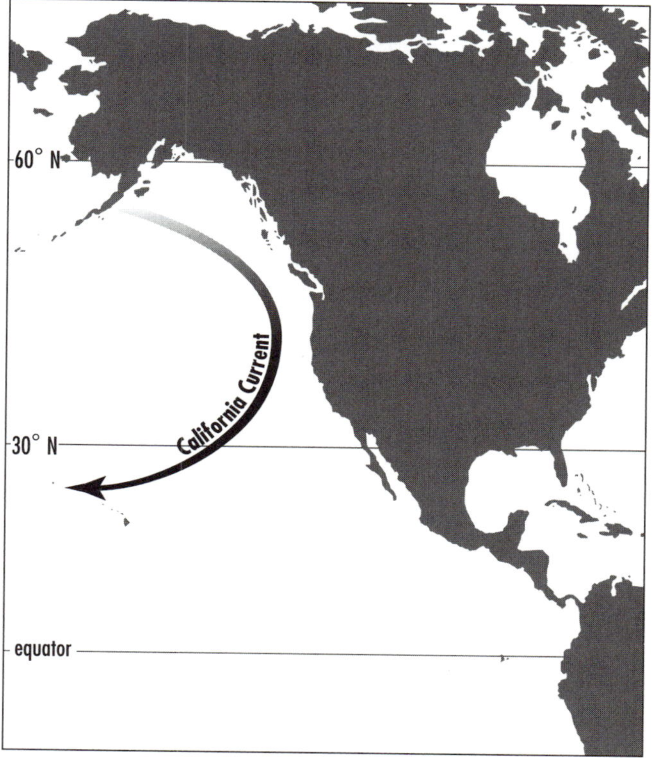

California Current. **A cold ocean current that flows parallel to the coast of northwestern North America.**

California fog FOG that affects the coastal regions of California and that drifts through the Golden Gate, at San Francisco, nearly every afternoon between May and October. It is an ADVECTION FOG driven by a sea breeze (*see* LAND AND SEA BREEZES). As the land warms during the day, air above it rises and cooler air is drawn in from over the sea. The air over the Pacific Ocean is relatively warm and moist, but as it approaches the coast the air crosses the cold CALIFORNIA CURRENT. The resulting drop in its temperature causes the condensation of some of the water vapor it carries, and it is this that produces the fog.

calina A HAZE that occurs in Spain and along parts of the coast of the Mediterranean Sea during the summer DROUGHT. It is caused by strong winds that pick up large amounts of DUST. The heat is intense and SHIMMER reduces VISIBILITY still further. The calina turns the clear blue sky to a drab gray.

calm In the BEAUFORT WIND SCALE, force 0, which is the condition in which the wind speed is 1 mph (1.6 km h^{-1}) or less. The air feels calm and smoke rises vertically.

calm belt One of the regions in which the winds are usually weak and the air is often still. These regions extend as latitudinal belts around the Earth. The calm belts occur in the HORSE LATITUDES. These are close to the TROPICS and are sometimes known as the *calms of Cancer* and the *calms of Capricorn.*

calms of Cancer *See* CALM BELT.

calms of Capricorn *See* CALM BELT.

calorie A centimeter–gram–second (C.G.S. SYSTEM) unit of heat that is equal to the amount of heat that is needed to raise the temperature of 1 gram of water by 1° C (= 1 K). This value varies with the temperature of the water, however, and so temperature had to be specified, with the result that there were eventually four separate calories in use. These were the International steam calorie (= 4.1868 J); the 15° C calorie (= 4.1855 J), which measured the heat needed to raise the temperature from 14.5° C to 15.5° C; the 4° C calorie (= 4.2045 J), from 3.5° C to 4.5° C; and the mean (0°–100° C) calorie (= 4.1897 J). The unit was introduced in 1880. Except for the kilocalorie, or Calorie, equal to 1,000 calories, which is sometimes still used in reporting food-energy values, in 1950 the calorie was replaced by the SYSTÈME INTERNATIONAL D'UNITÉS (SI) unit the JOULE: 1 cal = 4.1868 J (based on the value of the International steam calorie).

Calvin A HURRICANE that struck Mexico on July 6 and 7, 1993. It killed 28 people.

calving The breaking of a large mass of ice from the edge of a GLACIER where the glacier enters the sea, or from the edge of an ICE SHELF. The ice that breaks free moves away over the sea as an ICEBERG.

calvus (cal) A species of CUMULONIMBUS cloud (*see* CLOUD CLASSIFICATION) that lacks or is in the process of losing the billowing, cauliflowerlike structures and CIRRIFORM appendages from its upper part. *Calvus* is the Latin word for "bald."

CAM *See* PHOTOSYNTHESIS.

camanchaca *See* GARÚA.

Camille A HURRICANE, rated category 5 on the SAFFIR/SIMPSON SCALE, that struck Mississippi and Louisiana on August 17 and 18, 1969, killing about 250 people along the coast and causing $1.42 billion of damage. Camille then weakened and headed south and then east, over the Blue Ridge Mountains, Virginia, and was funneled through the narrow valleys of the Rockfish and Tye Rivers, where it encountered an advancing COLD FRONT with associated thunderstorms. This resulted in 18 inches (457 mm) of rain that flooded 471 square miles (1,220 km^2). FLASH FLOODS damaged or destroyed 185 miles (298 km) of road and 125 people died, either drowned or crushed by boulders. After 80 days the floods subsided.

Campbell–Stokes sunshine recorder An instrument that provides a daily record of the number of hours of sunshine. It comprises a spherical lens that acts as a burning glass, focusing the sunlight onto a card graduated with a time scale that partly encircles the lens. When the sky is clear the sunshine makes a scorch mark on the card. The position of the mark on the card is determined by the position of the Sun in the sky, and the graduation on the card interprets this as a time of day. The scorch mark ends whenever cloud obscures the Sun. The recorder produces a reading for as long as the Sun is more than about 3° above the horizon.

Canary Current A slow-moving ocean current that conveys cold water southward parallel to the coasts of Spain, Portugal, and West Africa. It is the cause of frequent SEA FOGS off northwestern Spain and Portugal.

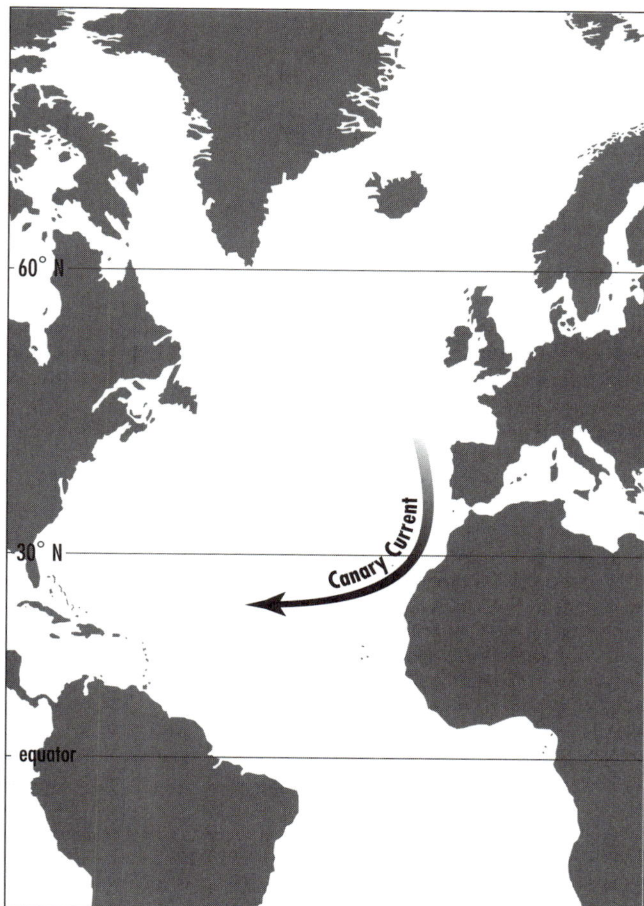

Canary Current. **A cold ocean current that flows parallel to the western coast of North Africa.**

candela (cd) The SYSTÈME INTERNATIONAL D'UNITÉS (SI) UNIT of luminous intensity, which is defined as the luminous intensity, measured perpendicularly, of a surface 1/600,000 m^2 in area of a BLACKBODY at the temperature of freezing platinum under a pressure of 101,325 PASCALS.

Candlemas February 2, the festival that is traditionally celebrated with lighted candles and the day on which Simeon recognized Jesus as "A light to lighten the Gentiles / And the glory of thy people Israel" (Luke 2:32). According to English weather lore, it is the day when the weather indicates whether winter has ended. It is an English equivalent of GROUNDHOG DAY, although it is also a day on which it is said that farmers should still have half of their hay and straw safe in the barn, because it will be some time before their animals can be grazed outdoors. The following is one of several versions of the belief:

If Candlemas be fair and bright,
Winter'll have another flight.
But if Candlemas Day be clouds and rain,
Winter is gone and will not come again.

Canterbury northwester A hot, enervating, northwesterly wind that blows across the Canterbury Plains of South Island, New Zealand. It is produced by depressions that travel from the southwest to northeast over the south of South Island or over the sea to the south. The wind carries tropical air from the interior of Australia, and it warms ADIABATICALLY as it descends on the eastern side of the Southern Alps, acquiring some of the characteristics of a FÖHN WIND.

canyon wind *See* MOUNTAIN-GAP WIND.

cap *See* CAPILLATUS.

capacitance A property of electrical conductors that allows them to store electric charge. The concept is most commonly applied to systems of conductors or semiconductors separated by insulators. Capacitance is measured in FARADS.

cap cloud A flat-topped, CUMULIFORM cloud that is seen blanketing a mountain peak. It is an OROGRAPHIC CLOUD that is also associated with a FÖHN WIND. The wind extends the cloud for some distance down the LEE side of the mountain, producing a FÖHN WALL. *See* ROTOR CLOUD.

CAPE *See* STABILITY INDICES.

Cape Hatteras low A deep depression that forms from time to time over the North Atlantic, off Cape Hatteras, North Carolina, and then moves northward. It carries strong northeasterly winds and storms to coastal areas from Virginia to the Maritime Provinces, often with flooding and damage to property. The storms, known as NOR'EASTERS, are most frequent between September and April.

cape scrub *See* CHAPARRAL.

capillarity The process by which WATER moves upward through a very narrow space, such as a tube or the spaces between soil particles. It occurs because the water molecule is polar (*see* POLAR MOLECULE); through this process water rises through unsaturated soil from the GROUNDWATER to within reach of plant roots.

Suppose that a narrow glass tube, open at both ends, is inserted vertically into a bath of water. One end of those molecules that are in contact with the sides of the tube is attracted to the opposite electric charge of molecules in the walls of the tube. This attraction draws the water molecules upward, along the sides of the tube. These water molecules are linked by HYDROGEN BONDS to the molecules behind and to the sides of them, so they are drawn behind the rising molecules. The liquid is drawn upward only at the sides of the tube, however, so the water at the sides rises higher than the water at the center. This causes the surface of the water in the tube to sag at the center, forming a concave shape.

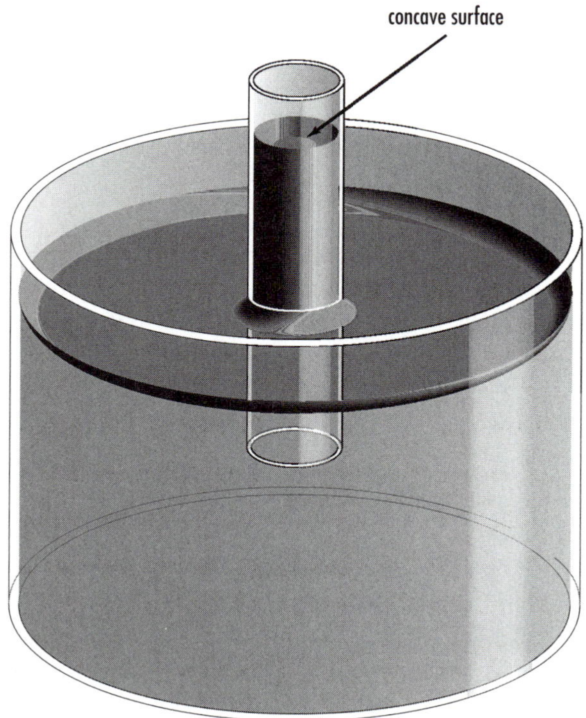

Capillarity. **Water is drawn up the narrow tube by the attraction between water molecules and the side of the tube and between the water molecules themselves.**

SURFACE TENSION acts on the surface molecules, pulling them toward the configuration that requires the least energy to maintain. This is a sphere and so surface tension seeks to make the surface resemble a sphere by pushing it upward at the center, into a convex (bulging) shape. More of the molecules close to the sides of the tube are then exposed to the attraction of opposite charges, so they move a little farther up the sides, drawing more molecules behind them. They leave the center behind, so it resumes its concave shape, which surface tension seeks to correct. In this way the water moves up the tube.

As soon as the surface of the water in the tube is higher than the surface of the water in the bath, gravitational force acts to restore them both to the same level. Water continues to rise up the tube for as long as the attractive force between the water molecules and the sides of the tube is stronger than the gravitational force. When the two are equal the water ceases to rise.

It follows from this that the distance water rises by capillarity depends on the width of the space. Wider spaces hold more water; therefore the column of water is heavier and the point at which the weight of the water column is equal to the attractive force is reached sooner. If the tube is wide, the weight of water in it exceeds the attractive force before the water is able to rise at all.

With even the narrowest tube, there is a limit to the height water rises by capillarity. This is the height at which the pressures acting on the water are in balance. In the bath, from which the water is rising, the pressure exerted by the water is greater at the bottom than at the top, because of the weight of the water. The water pressure decreases from the bottom to the top of the bath, and at the surface it is zero, because there is no water bearing down upon it. The water pressure continues to decrease with height in the capillary tube, but since it decreases from zero at the surface of the water in the bath, its value above the surface, in the tube, must be negative. The opposite of a pressure (pushing) is a tension (pulling), and in soils this is called *soil moisture tension.* The limit of capillarity is the height at which the negative value of the soil moisture tension is the same as the positive value of the water pressure at the bottom of the bath. At this point the two are in equilibrium. In soil, therefore, the depth of the groundwater determines the height to which water can be drawn by capillarity.

capillary layer *See* AQUIFER.

capillary wave A very small wave, produced on very still water by the slightest breeze, that causes "puckering" on the surface of water. The wave has a WAVELENGTH of less than 0.7 inch (1.7 cm) and the SURFACE TENSION of the water quickly restores the smooth surface.

capillatus (cap) A species of CUMULONIMBUS cloud (*see* CLOUD CLASSIFICATION) in which the uppermost part has a fibrous or striated, CIRRIFORM structure. The name of the species is derived from the Latin *capillus*, which means "hair."

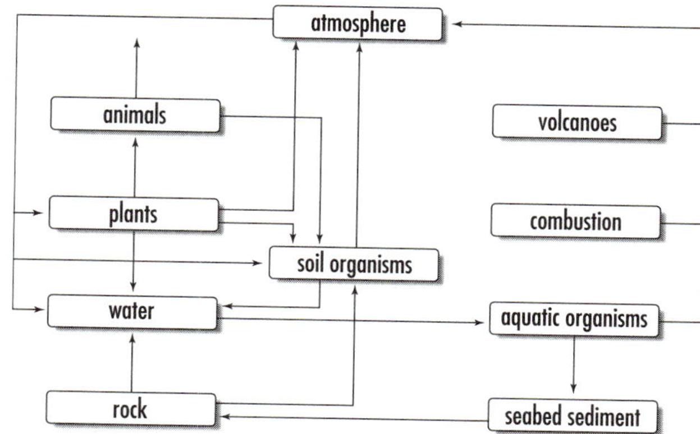

Carbon cycle. **Carbon moves through the air and is absorbed by plants. Plants and animals pass carbon to soil organisms and seabed sediments and their respiration returns it to the air.**

capping inversion An INVERSION that develops when a dry AIR MASS advances against a moist air mass more slowly at ground level than it does above the BOUNDARY LAYER, at about 1,000 feet (330 m). This is usually due to the retardation of low-level air by surface friction. The dry air overruns the moist air, preventing the development of convective clouds. Capping inversions are often associated with DRY LINES.

caracenet *See* CARCENET.

carbonate compensation depth *See* CARBON CYCLE.

carbon cycle The movement of the element carbon through the atmosphere, living organisms, soil, rocks, and water. Carbon enters the atmosphere initially in the form of carbon dioxide (CO_2) that is released from volcanoes.

Some atmospheric CO_2 dissolves in cloud droplets and falls to the surface as weak carbonic acid (H_2CO_3), and CO_2 also dissolves directly into surface waters. In water, H_2CO_3 dissociates into hydrogen (H^+) and bicarbonate (HCO_3^-) ions. Bicarbonate then dissociates further into carbonate (CO_3^{2-}) ions. These combine with positively charged ions, such as calcium (Ca^{2+}), to form salts, in this case, calcium carbonate ($CaCO_3$), which is insoluble in shallow water. At depths below about 2.5 miles (4 km) the low temperature of the water and the fact that the water is saturated with CO_2 cause $CaCO_3$ to dissolve. The depth at which this occurs is known as the *carbonate compensation depth* (CCD).

Some aquatic organisms exploit this reaction to make shells in which they live. When they die, the insoluble $CaCO_3$ sinks to the seabed to form part of the seabed sediment. Eventually this is transformed into carbonate rock, such as limestone and chalk, as a result of pressure and heating due to movements of the Earth's crust. Some sedimentary rocks are subducted beneath the crust (*see* PLATE TECTONICS) and become the source of the carbon that returns to the atmosphere from volcanoes. Other sedimentary rocks become exposed to the air as a result of tectonic processes. They are then subjected to WEATHERING, in the course of which $CaCO_3$ is broken down and CO_2 is returned to the air.

Plants absorb CO_2 directly from the air and use it in the production of carbohydrates by the process of PHOTOSYNTHESIS. Animals obtain the dietary carbohydrates they need by eating plants, and carnivorous animals eat herbivorous animals. Animals and plants obtain the energy that their bodies need by the process of RESPIRATION, in which carbohydrates are oxidized. This reaction releases energy, and the by-products are water (H_2O) and CO_2, both of which are returned to the air.

A hierarchy of organisms decompose dead plant and animal material. Each of these organisms utilizes compounds obtained from the decaying material for construction and repair of their own tissues and for respiration. By the time the soil organisms have completed the process of decomposition, all of the CO_2 that was absorbed during photosynthesis has been returned to the air by respiration.

At various times in the past decomposition has not been complete. Plant material fell into the soft, airless mud of tropical swamps, where it was buried and later compressed and heated to form coal. Some of the coal that is mined today consists of the remains of plants that grew during the Silurian period, about 400 million years ago, but most lived about 300 million years ago, during the Carboniferous period. Other organic remains were buried by sediments in river deltas. They became trapped between two layers of impermeable rock and were heated strongly under high pressure and in airless conditions. This "pressure cooking" converted them into petroleum and natural gas (mainly methane, CH_4).

Coal, gas, and petroleum are known as *fossil fuels.* The name reflects that they were formed a very long time ago; however, at one time anything dug from the ground was called a "fossil." The Latin *foss* means "dig."

Fossil fuels consist of carbon that has been removed from the air and stored. When we burn it, oxidizing the carbon to CO_2 and completing the process of decomposition, the carbon is returned to the air. This is one way in which our activities are affecting the carbon cycle.

It is the principal, but not the only way. We also accelerate the weathering of carbonate rocks. Calcium oxide (CaO), or lime, is used in the chemical process industries, and a suspension of calcium hydroxide ($Ca(OH)_2$), or slaked lime, in water is used to remove sulfur from the waste gases of industrial plants that burn fossil fuel, especially power plants, in order to combat ACID RAIN. These products are obtained by heating (called *kilning*) limestone. Heat breaks down calcium carbonate to produce lime (calcium oxide) and CO_2: $CaCO_3$ + heat → CaO + CO_2. Lime is converted to slaked lime by the addition of water: CaO + H_2O → $Ca(OH)_2$.

Many scientists fear that adding CO_2 to the atmosphere may seriously alter the climate (*see* GLOBAL WARMING and GREENHOUSE EFFECT).

carbon dioxide (CO_2) A gas formed by the complete oxidation of carbon (*see* CARBON MONOXIDE) that is a minor constituent of the atmosphere, at present constituting 365 parts per million by volume (p.p.m.v.), or 0.0365 percent. It is the most important GREENHOUSE GAS and climatic changes in the past have been associated with changes in the atmospheric concentration of CO_2. Data from ICE CORES obtained from the VOSTOK STATION cover four transitions from ICE AGES to warm periods that began about 335,000, 245,000, 135,000, and 18,000 years ago. In each case, the warming was associated with an increase in CO_2 concentration, from about 180 to 240–300 p.p.m.v. It is not certain, however, whether the rise in CO_2 concentration caused the warming or was a consequence of it. Some research suggests the last three warming episodes occurred 500–1,000 years before the rise in atmospheric CO_2.

(For details of the original research, see J. R. Petit et al. (there are 19 authors) "Climate and Atmospheric History of the Past 420,000 Years from the Vostok Ice Core, Antarctica," *Nature,* 399, June 3, 1999, pp. 429–36, and Hubertus Fischer et al. (there are 5 authors), "Ice Core Records of Atmospheric CO_2 around the Last Three Glacial Terminations," *Science* 283, March 12, 1999, pp. 1712–14.)

carbon isotopes The nuclei of all carbon atoms contain six protons, but they contain varying numbers of neutrons. The number of neutrons affects the mass of the nucleus, but not its chemical behavior, which is determined only by its protons. The sum of the number of protons and neutrons (nucleons) is called the *nucleon number* (or *mass number*) and is the way ISOTOPES are labeled. There are seven isotopes of carbon: ^{10}C, ^{11}C, ^{12}C, ^{13}C, ^{14}C, ^{15}C, and ^{16}C. The isotopes ^{12}C and ^{13}C are stable. The other isotopes are radioactive. Their HALF-LIVES are ^{10}C, 19.1 seconds; ^{11}C, 20.4 minutes; ^{14}C, 5,720 years (*see* RADIOCARBON DATING); ^{15}C, 2.4 seconds; and ^{16}C, 0.74 second.

carbon monoxide (CO) A gas formed by the partial oxidation of carbon (*see* CARBON DIOXIDE). It is emitted naturally by volcanoes and forest fires, and by the incomplete combustion of FOSSIL FUELS, especially in internal combustion engines. The amount of CO in the air is very small but varies greatly; the highest concentrations occur along busy main highways and city streets. When inhaled, CO forms a stable compound with blood hemoglobin, reducing its capacity to transport oxygen, and in high doses CO is lethal, although persons exposed to less than lethal doses recover fully. CO is chemically stable. It oxidizes to CO_2 and dissolves in the oceans, but it is also utilized by soil microorganisms, and this process is believed to be the way most of it is removed from the atmosphere.

carbon sequestration The long-term storage of carbon dioxide (CO_2) that is produced by processing and burning FOSSIL FUELS in order to prevent it from accumulating in the atmosphere. Natural gas is primarily methane (CH_4), but when it first emerges from its natural reservoir the methane is often mixed with CO_2. Most customers accept gas containing a maximum of 2.5 percent CO_2. Any CO_2 in excess of this proportion must be removed. Some of the gas produced in the North Sea contains up to 9 percent CO_2. Norway imposes a CO_2 tax, set in January 2000 at $38 for every ton of CO_2 released into the atmosphere. This encouraged the owners of the Sleipner oil and gas field in the Norwegian sector of the North Sea to install equipment to compress the CO_2 that is separated from the CH_4 and pump it under pressure into a sandstone formation beneath the seabed. Similar schemes are being planned at gas fields in other parts of the world, including the South China Sea and Barents Sea, as well as at the Alaskan oil fields. CO_2 from the burning of fuel can also be buried in this way, and there is a pilot gas-fired generating plant in Norway that pumps its CO_2 into underground reservoirs. CO_2 can be pumped into depleted oil or gas wells, coal beds that cannot be mined, salt domes that have been mined for their salt, and deep AQUIFERS that contain water that is too salty to be used. CO_2 can also be released directly into the sea, either frozen as dry ice or from a pipeline that is towed behind a ship or that runs from the surface to the seabed.

(You can learn more about carbon sequestration from Howard Herzog, Baldur Eliasson, and Olav Kaarstad "Capturing Greenhouse Gases," *Scientific American,* February 2000, pp. 54–61.)

carbon tax *See* GREEN TAXES.

carbon tetrachloride (tetrachloromethane, CCl_4) A clear, volatile liquid that was once widely used as a solvent, especially in dry cleaning, in fire extinguishers, and in the industrial preparation of other compounds. It is very toxic to humans if inhaled or swallowed; it also contributes to the enhanced GREENHOUSE EFFECT, having a GLOBAL WARMING POTENTIAL of about 1,550. It is also a source of free chlorine atoms that contribute to the depletion of stratospheric OZONE. At the fourth meeting of the signatories to the MONTREAL PROTOCOL ON SUBSTANCES THAT DEPLETE THE OZONE LAYER, held in Copenhagen in November 1992, it was agreed that the use of carbon tetrachloride should cease by January 1996.

carcenet (caracenet) A strong, cold wind that blows through mountain gorges in the eastern Pyrenees Mountains, on the border of southwestern France and northeastern Spain. The winds are accelerated by FUNNELING and are especially frequent in the upper part of the valley of the River Aude.

cardinal temperature The temperature below which almost no plant growth occurs. Cardinal temperatures are used in conjunction with ACCUMULATED TEMPERATURES to evaluate crop growth in the course of a growing season.

Caribbean Current An ocean current that flows westward through the Caribbean Sea. As it passes the coast of Florida it joins the FLORIDA CURRENT and then becomes part of the GULF STREAM. The Caribbean Current carries warm water and flows at an average of 0.85–0.96 mph (0.38–0.43 m s^{-1}).

Carlotta A HURRICANE that formed off the Mexican coast on June 19, 2000. It headed out to sea but generated heavy rain that caused mudslides, killing at least 6 people and causing more than 1,000 people to leave their homes.

cas *See* CASTELLANUS.

castellanus (cas) A species of cloud (*see* CLOUD CLASSIFICATION) that has many vertical protuberances that resemble small clouds arising from the main cloud. They are often shaped like the turrets of a castle and are most often seen on ALTOCUMULUS but also occur on clouds of the genera CIRRUS, CIRROCUMULUS, and STRATOCUMULUS. *Castellanus* is the Latin word for "castle."

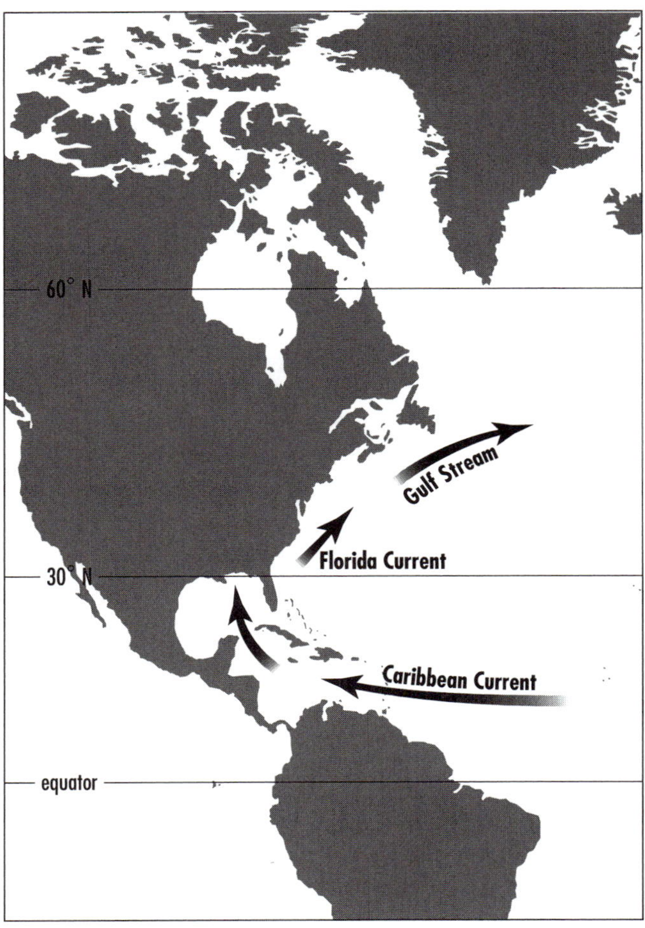

Caribbean Current. **The Caribbean Current flows westward through the Caribbean Sea, then joins the Florida Current before joining the Gulf Stream.**

CAT *See* CLEAR AIR TURBULENCE.

catastrophic windthrow *See* BLOWDOWN.

cation An ION that carries a positive charge, because the number of electrons surrounding its nucleus is smaller than the number of protons in its nucleus.

caustic scrubbing A process for the removal of SULFUR DIOXIDE (SO_2) from FLUE GASES that involves passing the gas stream through a solution of caustic soda (sodium hydroxide, NaOH). The SO_2 and NaOH react to form sodium sulfite (Na_2SO_3) and sodium hydrogen sulfite ($NaHSO_3$). The addition of calcium carbonate ($CaCO_3$) then causes the precipitation of insoluble gypsum (calcium sulfate, $CaSO_4$) and leaves the water enriched in sodium carbonate (Na_2CO_3), a harmless substance that is present in most mineral waters. After it has been diluted, the solution can be safely discharged into surface waters.

cavity *See* WAKE.

Cb *See* CUMULONIMBUS.

Cc *See* CIRROCUMULUS.

CCD *See* CARBON CYCLE.

CCN *See* CLOUD CONDENSATION NUCLEI.

cd *See* CANDELA.

Cecil Two TYPHOONS, the first of which struck South Korea in August 1982. It killed at least 35 people and caused damage estimated at more than $30 million. The second Typhoon Cecil struck Vietnam on May 25 and 26, 1989. It destroyed 36,000 homes and killed at least 140 people.

ceiling The height that is ascribed to the lowest layer of clouds or of anything else obscuring the sky, when the SKY COVER is described as *broken, overcast,* or *obscured* and the cover is not classified as *thin* or *partial.* If these conditions are not present, the ceiling is said to be *unlimited.*

ceiling classification A description of the way the CEILING was determined. It is included in weather reports for airports as a letter preceding the ceiling height. *A* means the ceiling was measured by an airplane, *B* means it was measured by a balloon, *E* that it was estimated, *M* that it was measured, *P* that there is precipitation, *W* that the ceiling is indefinite, and *V* that it is variable. *A25* would indicate that an airplane had measured the height of the ceiling at 2,500 feet (762.5 m).

ceiling light A small searchlight that projects a narrow beam of light, spreading by less than 3°, vertically upward onto the base of a dark cloud or onto a CLOUD BASE at night. The light illuminates a spot on the cloud.

An observer positioned a measured distance between 500 feet and 1,000 feet (150–300 m) from the ceiling light measures the elevation of the spot. The height of the cloud base (h) is then calculated by trigonometry: $h = a \tan \theta$, where a is the distance between the ceiling light and the observer and θ is the angle of elevation. This method has been used to measure the height of clouds up to 15,000 feet (4,575 m) with an accuracy of about 2,500 feet (750 m).

ceilometer A device that is used to measure the height of the CLOUD BASE from the ground. Its advantage over the CEILING LIGHT is that it can be used in daylight, when the sky is very much brighter than the area illuminated by a searchlight. The ceilometer consists of two lamps, focused by parabolic mirrors, that each shine through a shutter, which restricts the width of the beam. The mirrors, with the shuttered lamps above them, are arranged back-to-back and rotate at a given rate, so that the beam is transmitted as a series of pulses. The beams shine at an angle onto the base of the cloud. A detector consists of a photoelectric cell attached to electronic filters that allow it to respond only to a series of pulses at the predetermined frequency. The height of the cloud base is then calculated by trigonometry from the angles of the transmitted and reflected beams and the known distance between the projector and the detector. The signal from the photoelectric cell is then amplified. A ceilometer can measure cloud bases up to 10,000 ft (3,000 m) during the day and up to about 20,000 ft (6,000 m) at night.

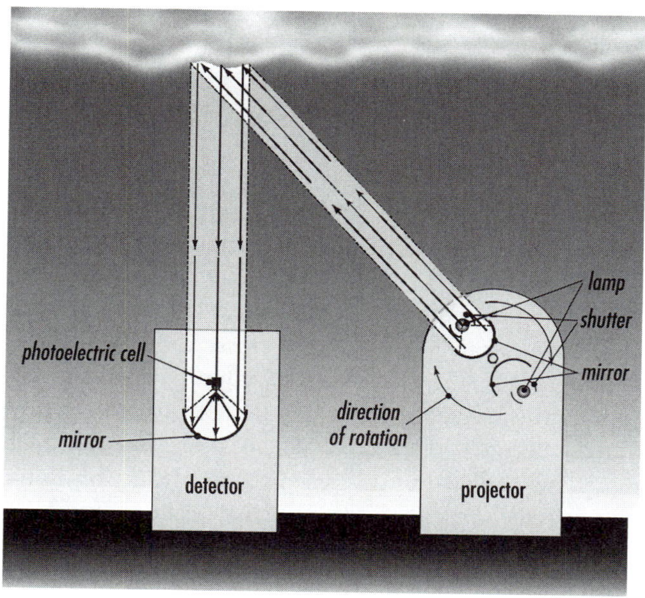

Ceilometer. **A beam of light, pulsating at a predetermined frequency, is shone onto the cloud base. A detector contains a photoelectric cell that will respond only to a signal at that frequency.**

celerity The VELOCITY with which a wave advances. Celerity (c) is proportional to the WAVELENGTH (λ) and FREQUENCY (f) of the wave, such that $c = \lambda f$. This applies to waves in either air or water. In deep water, $c = (g\lambda/2\pi)^{1/2}$, where g is the acceleration due to gravity (32.2 ft s^{-1} [9.81 m s^{-1}]). Taking this into account, $c = 1.25\sqrt{\lambda}$. In shallow water, $c = (gd)^{1/2}$, where d is the depth of water, and therefore $c = 3.13\sqrt{d}$.

celestial equator *See* DECLINATION and PRECESSION OF THE EQUINOXES.

Celsius, Anders (1701–1744) Swedish *Astronomer and physicist* Anders Celsius was born in Uppsala, Sweden, on November 27, 1701. His father, Nils Celsius, was professor of astronomy at the University of Uppsala, and both of his grandfathers were also professors at Uppsala. Magnus Celsius was professor of mathematics; his maternal grandfather, Anders Spole, preceded Nils as professor of astronomy. Several of his uncles were also scientists.

Anders was educated in Uppsala and in 1730 was appointed to succeed his father as professor of astronomy. There was no major observatory in Sweden at that time, and so soon after his appointment Celsius embarked on a tour of the leading European observatories. His tour lasted five years, and in the course of it he met many of the leading astronomers of the day. Between 1716 and 1732 Celsius and his companions made 316 observations of the AURORA Borealis. He published these in Nuremberg in 1733. It was Celsius and Olof Hiorter, his assistant, who discovered that the aurorae are magnetic phenomena.

While visiting Paris in 1734 Celsius met the French astronomer Pierre-Louis Maupertuis (1698–1759), who invited him to join an expedition to Torneå, in Lapland (today on the border between Sweden and Finland, but then in northern Sweden). The purpose of the expedition was to measure the length of one degree of latitude along a meridian (degree of longitude) close to the North Pole and to compare the result with a

Anders Celsius.* The Swedish astronomer and physicist who devised the temperature scale that bears his name. *(Circulating Collection, Print and Picture Collection, The Free Library of Philadelphia)

similar measurement taken in Peru (in a region that is now in Ecuador). The Lapland expedition, which took place in 1736–37, confirmed the hypothesis of Isaac Newton (1642–1727) that the Earth is flattened at the Poles.

His participation in this expedition made Celsius famous in his own country, and he was able to persuade the Swedish government to finance the building of an observatory at Uppsala equipped with instruments Celsius had bought during his European tour. The Celsius Observatory opened in 1741, with Celsius as its first director. Celsius made some of the earliest attempts to measure the magnitude of stars.

In the 18th century astronomy was not studied purely to obtain information about the stars and planets. Governments were busy delineating the borders of their territories, and to do this accurately their surveyors needed astronomical data to fix positions—inaccurate maps could and did lead to war. Accordingly, Celsius conducted many measurements that were used in the Swedish General Map. He may also have been the first person to observe that the Scandinavian landmass is slowly rising. We now know that this is due to the release of pressure that followed the melting of the FENNOSCANDIAN ICE SHEET (*see* GLACIOISOSTASY), but Celsius thought the sea level was falling because the sea was evaporating.

In 1742 Celsius presented a paper to the Royal Swedish Academy of Sciences in which he proposed that all scientific measurements of temperature should be made on a scale based on two fixed points that occur naturally. This led to the development of the temperature scale that bears his name (*see* CELSIUS TEMPERATURE SCALE).

Celsius published most of his scientific papers through the Royal Swedish Academy of Sciences and was its secretary from 1725 until 1744. He strongly favored the introduction of the Gregorian calendar. This had been tried in 1700 by omitting the leap days between 1700 and 1740, but 1704 and 1708 were declared leap years by mistake and in 1712 Sweden returned to the Julian calendar. Celsius and his supporters eventually succeeded, and the new calendar was introduced in 1753 and all 11 supernumerary days were dropped together.

By then Celsius was dead. He died in Uppsala on April 25, 1744.

(You can learn more about Anders Celsius at www.astro.uu.se/history/Celsius_eng.html and at me.in-berlin.de/~jd/himmel/astro/Celsius-e.html.)

Celsius temperature scale (**centigrade temperature scale**) The scale that is used to measure TEMPERATURE throughout most of the world, in which the freezing and boiling temperatures of water are separated by 100 degrees. In English-speaking countries the FAHRENHEIT TEMPERATURE SCALE is also used, although not for scientific measurements, for which the Celsius scale is preferred. Because there are 100 degrees, the scale is sometimes called the *centigrade scale,* from the Latin *centum,* meaning "hundred," and *gradus,* meaning "step." One degree on the KELVIN SCALE, which is often used in scientific publications, is equal to one degree on the Celsius scale, and measurements on this scale are made with a THERMOMETER graduated in degrees Celsius.

Any temperature scale must be based on the difference in temperature between two fixed points. This was recognized by the early 18th century and several eminent scientists, including Isaac Newton (1642–1727) and the French physicist and naturalist René Antoine Ferchault de Réamur (1683–1757), had proposed the freezing point of water for one of them. In 1742, the Swedish astronomer ANDERS CELSIUS published the paper "Observations on two persistent degrees on a thermometer" in the *Annals of the Royal Swedish Academy of Science*. He proposed that the two fixed points should be the temperature of melting snow or ice and the temperature of boiling water, and he described his reasons for choosing these two points. Obviously, water freezes and thaws at the same temperature, but it is more difficult to measure the point at which liquid water begins to freeze than it is to measure the temperature at which snow and ice melt. Celsius reported that he had used one of the thermometers made by Réamur to measure the temperature of melting snow. He repeated the measurement many times in the course of two winters and during all kinds of weather and at different atmospheric pressures. He even carried snow indoors and placed it in front of the fire in his room to measure its temperature as it melted. The temperature was invariably the same. Snow also melted at the same temperature in Paris and in Sweden at Uppsala (60° N) and Torneå (66° N). He was confident, therefore, in the first of his fixed points.

Measuring the temperature of boiling water was more complicated. Although the temperature of water will rise no further once it is boiling, Celsius thought the intensity of boiling might affect the thermometer, and he noticed that when he removed the thermometer from the boiling water the mercury rose before it began to fall. This, he suggested, happened because the glass tube contracted before the mercury began to cool. Daniel FAHRENHEIT had observed that the boiling temperature of water varies according to the atmospheric pressure. Celsius confirmed this but found a way to correlate the two, because the height of the mercury in the thermometer was always proportional to the height of the mercury in the BAROMETER.

Celsius then proposed a standard method for calibrating a thermometer. First, the bulb of the thermometer should be placed into thawing snow and the position of the mercury marked. Then the thermometer bulb should be placed into boiling water when the atmospheric pressure is approximately 1006.58 millibars (29.75 inches or 755 mm of mercury). The position of the mercury should be marked.

The distance between the two points should then be divided into 100 equal parts, or degrees, so that 0 degree corresponds to the boiling temperature of water and 100 degrees corresponds to its freezing temperature.

This was the Celsius scale. It was later reversed, so that 0 degree represents the freezing temperature of water and 100 degrees the boiling point. It is uncertain who made the change. It may have been Martin Strömer, a pupil of Celsius. It has also been suggested that it was Carl von Linné (Linnaeus). The most likely person, however, was the leading Swedish instrument maker of the time, Daniel Ekström.

(You can learn more about the Celsius temperature scale and read a copy of Celsius's original paper describing it [in Swedish] at www.santesson.com/engtemp.html.)

Cenozoic (Cainozoic, Kainozoic) The era of geological time that began about 65 million years ago and that extends to the present day. It includes the TERTIARY and QUATERNARY suberas (*see* GEOLOGICAL TIME SCALE).

center of action A center of high or low AIR PRESSURE that is located in a particular position more or less permanently. Centers of action are produced by the general circulation of the atmosphere, but changes in their shape, size, or intensity have widespread meteorological effects.

centigrade temperature scale *See* CELSIUS TEMPERATURE SCALE.

Central England climate record A continuous record of the mean monthly temperatures that have been experienced in central England since the year 1659 and a continuous record of daily temperatures in the same area since 1772. The area that is covered forms a triangle with the cities Preston, Bristol, and London approximately at its corners. Measurements taken since 1974 have been adjusted for the urban HEAT ISLAND effect. This is the longest continuous climate record, based on instrument readings, that exists anywhere in the world. The record is held at the Climate Data Monitoring section of the Hadley Centre of

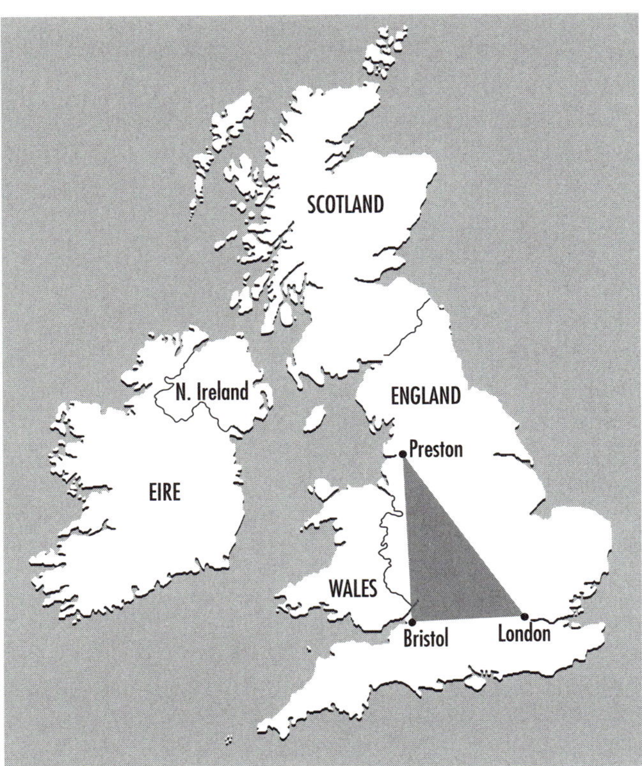

Central England climate record. **The record of temperatures from 1659 covers an area bounded by Bristol, Preston, and London.**

the United Kingdom METEOROLOGICAL OFFICE. It is not available to the general public.

central pressure The AIR PRESSURE at the center of an ANTICYCLONE or CYCLONE at a particular time. It is the highest pressure in a region of high pressure and the lowest pressure in a region of low pressure.

centrifugal force *See* CENTRIPETAL ACCELERATION.

centripetal acceleration The motion of a body that is following a curved path. Although the speed of the body may remain constant, its direction constantly changes. This means it is accelerating, because ACCELERATION is defined as the rate of change of VELOCITY, and velocity is a VECTOR QUANTITY that comprises both speed and direction. If the speed decreases, the effect is nevertheless an acceleration, in this case, a negative acceleration, although it is often called "deceleration" to prevent confusion.

According to Newton's first law of motion, *A body will continue in a state of rest or uniform motion along a straight path unless an external force is applied to it.* If such a force is applied, its effect will be to change either the speed at which the body is moving, or the direction in which it is moving, or both. This means it will change the velocity of the body, and a change in velocity is acceleration. Newton's second law states, *The acceleration of a body is proportional to and in the same direction as the force acting on that body.* It follows that if a body is moving along a curved path a force must be acting on it to accelerate the body toward the center of the curve. This is centripetal acceleration. Its magnitude is equal to mv^2/r, where m is the mass of the body, v its velocity, and r the radius of curvature of its path.

If you fasten a weight to the end of a string and swing it in a circle, the string will be taut and the weight will follow a curved path. A centripetal force acting along the string and toward your body accelerates the weight toward you. Should the string break, the weight will fly away, because once the centripetal force ceases to act the weight reverts to its motion in a straight path, according to the first law.

Centripetal acceleration can be observed and measured only by an observer who is in an external frame of reference that is in a state of inertia with respect to the body moving in a curved path. An observer inside the rotating frame of reference experiences conditions differently. When you swing the weighted string around your body, the force *feels* as though it is acting outward, not inward. If you imagine yourself riding in a car that is traveling fast around a tight corner you will feel yourself pushed away from the center of the turn. If there is a tennis ball lying on the flat shelf behind the rear seat, it will roll toward the outside of the turn. Sometimes people call this effect a "centrifugal force." It certainly feels real. Fighter pilots flying at high speed in a tight turn or pulling out of a steep dive experience it as a force pressing them down into their seats and draining the blood from their heads so they may lose first color vision (gray out) and then all vision (black out). Their loss of vision is genuine. They are not imagining it.

An observer in an INERTIAL REFERENCE FRAME would see more clearly what is really happening. That person would see that the tennis ball in the car, like the weight that breaks free from the string, does not expe-

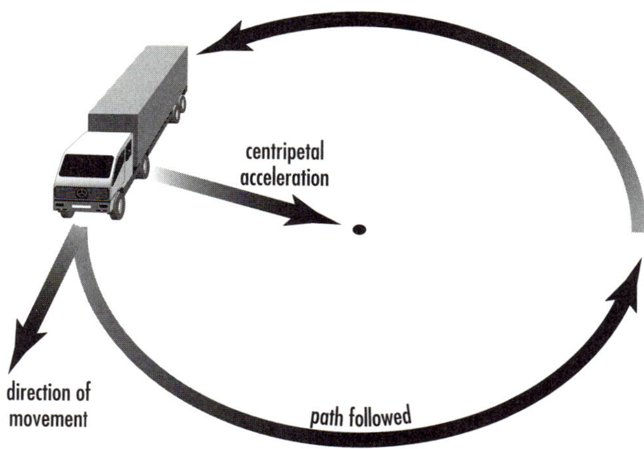

Centripetal acceleration. **As the truck follows its circular path, its motor propels it forward, at a tangent to the circle, but a countervailing force, exerted by the grip of its tires on the road surface, draws it toward the center of the circle. The two forces balance, allowing the truck to continue.**

rience a centripetal acceleration. Consequently it obeys Newton's first law and continues moving in a straight path. That is also what the bodies of passengers and fighter pilots are attempting to do. The pressures they experience are the combined effects of their inertia and their centripetal acceleration. The centrifugal force does not exist.

CFCs (chlorofluorocarbons) A range of chemical compounds in which chlorine and fluorine are bonded to carbon. Bromine and fluorine are bonded to carbon in a related series of compounds, with or without chlorine, and hydrogen is present in compounds known as *hydrochlorofluorocarbons* (HCFCs). Methyl chloroform (CH_3CCl_3) and carbon tetrachloride (CCl_4) are also grouped with the CFCs because their atmospheric effects are similar. Chlorine, fluorine, and bromine are known chemically as *halogens* and these compounds are also called *halocarbons*.

CFCs were invented in the 1930s by scientists working for DuPont de Nemours and Company and were given the trade name *Freon*. Their commercial value arose from their physical and chemical properties. Because they change between the liquid and gaseous phases at about room temperature, they can be used in freezers, refrigerators, and air-conditioning units. In these applications they lower the temperature inside the unit by absorbing the latent heat of vaporization from the surrounding air as the liquid expands and vaporizes. The heat is released outside, where a compressor causes the gas to condense. The same property made them useful as propellants in aerosol cans. CFCs are liquid while held under moderate pressure; when the pressure is released they vaporize and expand, spraying vapor from the nozzle of the can and carrying droplets of any substance that is mixed with them. They were also used as solvents and foaming agents in foam plastics.

Other compounds could also be used for these purposes, but CFCs were chemically highly stable. This means they are very reluctant to react with other substances, and that property in turn means they are nonflammable and completely nontoxic. The most likely alternatives to CFCs were ammonia, which can be used in refrigeration plants but is poisonous, and butane and propane, which are suitable propellants and foaming agents but are highly flammable. CFCs were very safe.

Their chemical stability also meant that once released into the air they would remain there for a long time before adhering to a solid surface and disappearing. In 1970, the Dutch chemist PAUL CRUTZEN showed that the OZONE LAYER was vulnerable to destruction by chemical reactions, and in 1974, the American chemists F. SHERWOOD ROWLAND and MARIO MOLINA warned that those reactions might involve CFCs. All three were awarded the 1995 Nobel Prize for Chemistry for these findings.

Because they are very stable, CFCs and other halocarbons survive long enough to enter the stratosphere. There they absorb ULTRAVIOLET RADIATION, which splits their molecules, releasing free atoms of chlorine. Chlorine atoms destroy ozone and atomic oxygen repeatedly by reactions that end by releasing the original chlorine atoms:

$$Cl + O_3 \rightarrow ClO + O_2$$

$$ClO + O \rightarrow Cl + O_2$$

Production and use of these halocarbons are now banned in many countries under the MONTREAL PROTOCOL ON SUBSTANCES THAT DEPLETE THE OZONE LAYER.

The principal compounds and the number of years they survive in the atmosphere are as follows:

Name	Formula	Lifetime
CFC-12 (Freon-12)	CCl_2F_2	100
CFC-11 (Freon-11)	CCl_3F	45
CFC-113	CCl_2FCClF_2	85
methyl chloroform	CH_3CCl_3	4.8
carbon tetrachloride	CCl_4	35
H-1301	$CBrF_3$	65
H-1211	$CBrClF_2$	16
HCFC-22	$CHCl_2F$	12
HCFC-142b	CH_3CClF_2	19
HCFC-141b	CH_3CCl_2F	9

c.g.s. system A system of units that was introduced for scientific use before the SYSTÈME INTERNATIONAL D'UNITÉS (SI) system was developed. The c.g.s. system was based on the centimeter, gram, and second, but it proved unsatisfactory and confusing when applied to electrical quantities and heat measurements (*see* CALORIE). Its use persisted for some time, but it has now been replaced by the SI system.

chain lightning *See* PEARL-NECKLACE LIGHTNING.

chain reaction *See* COMBUSTION.

Chandler wobble A periodic change in the position of the Earth's axis of rotation and, therefore, of the location of the geographic North and South Poles. The magnitude of the change is approximately 0.1 minute of arc and its PERIOD is about 14 months. Its effect is to alter all latitudes by that amount. This small change could produce much larger changes in the circulation of the atmosphere, with significant climatic effects. The cause of the wobble is uncertain. It is believed to be due to changes in the ANGULAR MOMENTUM of the solid Earth and atmosphere, combined with the effect on electrically charged water droplets of changes in the magnetic field, and possibly the positions of other planets in the solar system. The wobble was predicted in 1744 by the Swiss mathematician Leonhard Euler (1707–83), who calculated the period as precisely one year. In about 1881 the American geophysicist Seth Carlo Chandler (1846–1913) studied the phenomenon by using his own observations and by examining old records, especially those from the Greenwich observatory in England. He found that the actual period was about 14 months (428 ± 17 days). It is now known that the period would be one year if the Earth were completely rigid, but that the interior of the Earth is slightly elastic and it is this that increases the period.

chandui *See* CHANDUY.

chanduy (**chandui, charduy**) A cool, descending wind that blows over Guayaquil, Ecuador, during the dry season that lasts from July to November.

Channeled Scablands An area of about 13,000 square miles (33,670 km^2) between the valleys of the Columbia and Snake Rivers in Washington State, where the soil has been scoured from the surface and the land is dissected by deep canyons, called *coulees*. The coulees have steep, stepped sides and are approximately rectangular in cross section. Streams that flow through the coulees are much too small to have carved the canyons by erosion. The landscape is so harsh that scientists study it to help them understand the landscapes of Mars. In a series of papers in the *Journal of Geology*, the first of which was published in 1923, the American geologist J. Harlan Bretz proposed what is now the accepted explanation of how the landscape was formed.

During the last ice age, the WISCONSINIAN GLACIAL, glaciers formed a succession of dams at the edge of the ICE SHEET. Water accumulated behind the dams until there was a lake, now Lake Missoula, that had a surface area of about 3,000 square miles (7,770 km^2), in some places was 2,000 feet (610 m) deep, and that held 500 cubic miles (2,080 km^3) of water. About 15,000 years ago the climate warmed a little and the last of the dams broke. This released a wall of water 2,000 feet (610 m) high into the Clark and Flathead Rivers. The wave traveled at more than 50 mph (80 km h^{-1}) and carried up to 10 cubic miles (42 km^3) of water an hour. The torrent filled the valleys, cut new channels, and fell in huge waterfalls that cut deep plunge pools.

The landscape still bears the scars of this catastrophic event caused by a change in the climate.

(You can read more about the Channeled Scablands in "The Missoula Floods" at www.opb.org/ofg/1001/missoula/sitemap.htm; "Flood Evidence in Eastern Washington" at www.sentex.net/~tcc/scabland.html; "Boulders, Braids, and J. Harlan Bretx" at www.gi.alaska.edu/ScienceForum/ASF11/1160.html; "Channeled Scablands Theory" at www.spokaneoutdoors.com/scabland.htm; and in Michael Allaby, *Dangerous Weather: Blizzards* [New York: Facts On File, 1997].)

chaos A mathematical theory that describes dynamic systems governed by nonlinear equations. A nonlinear equation is one of the type $y = x^2$ that does not produce a straight line when plotted on a graph. If $y = x^2$, a small change in the value of x produces a much greater change in the value of y. Consequently, the development through time of a chaotic system is acutely sensitive to very small differences in the starting conditions. In the natural world such systems usually involve several equations, each of which is more complicated than the example given here. Because the sensitivity to the precise initial conditions is so acute, the behavior of the system is essentially unpredictable, because those initial conditions can never be known with sufficient accuracy. If the system is observed over time, it will appear to behave randomly—or chaotically. Systems that behave in this way are said to be *complex*.

In 1961, Edward Lorenz, a research meteorologist at the Massachusetts Institute of Technology, discovered that weather systems are complex in this sense. Computers model the atmosphere by constructing an imaginary three-dimensional grid, describing the state of the air at each intersection in the grid. When the model is run, an initial change in one part of the grid produces effects that ramify across the grid. These changes are calculated mathematically, in terms of pressure, temperature, and humidity, at each intersection of grid lines, and the calculations are repeated in a series of steps. With even the finest grid, conditions between the grid lines have to be assumed, so inevitably the initial data are somewhat approximate. Since the weather system is complex, these small discrepancies magnify at each step in the calculation, making a weather forecast based on the model increasingly inaccurate the longer it is run. If the model is run twice, with even the smallest differences in the initial conditions, the weather it describes in one run soon becomes vastly different from that in the other run. Lorenz discovered this when he started a run using initial data the computer had generated partway through a previous run, but to save time he entered numbers to three decimal places rather than six places stored in the computer memory. The second run should have duplicated the results of the first, but it diverged rapidly to describe an entirely different weather pattern.

This sensitive dependence on initial conditions came to be called the *butterfly effect*: the notion that a butterfly flapping its wings in China can affect the way a storm develops a month later in America. It implied that weather forecasts can be reliable for no more than a few days in advance and long-range weather forecasting was abandoned.

Despite being called chaotic, the behavior of complex systems is not random. Patterns emerge over time, offering the possibility that with better understanding prediction may become possible. In recent years, meteorologists have identified a number of cycles and oscillations that strongly influence the weather over large areas and allow very general predictions to be made, in some cases up to a year ahead.

(The first book to describe this theory at a popular level is James Gleick, *Chaos: Making a New Science* London: William Heinemann, 1988.)

chaparral A type of shrubland that is found on dry hillsides and ridges from southern Oregon to Baja California, but that is most widespread in Shasta County, California, and the area to the south and on the western side of the Sierra Nevada. Chaparral also occurs in discontinuous belts across Arizona. Similar types of vegetation around the Mediterranean are known as *maquis, machia,* or *garrigue;* in central Chile they are known as *matorral,* in southern Australia as *mallee scrub,* and on the southern tip of Africa as *fynbos* or *cape scrub.* Most of the perennial plants are evergreen shrubs and small trees. Many of them are broad-leaved, with leaves that are small, tough, and leathery as an adaptation to dry conditions. The climate is strongly seasonal, with hot, dry summers and mild winters (*see* MEDITERRANEAN CLIMATE). Rainfall is 12–36 inches (300–900 mm) a year, of which 65 percent falls between November and April in the Northern Hemisphere and between May and October in the Southern Hemisphere.

charduy *See* CHANDUY.

Charles, Jacques Alexandre César (1746–1823) French *Physicist and mathematician* The discoverer of CHARLES'S LAW was born at Beaugency, Loiret, on November 12, 1746. He worked as a clerk in the Ministry of Finance in Paris, and while there he became interested in science. Having heard about the experiments BENJAMIN FRANKLIN had conducted with electricity, Charles gave popular public lectures in which he popularized Franklin's discoveries, demonstrating them with apparatus he constructed himself.

In June 1783, the Montgolfier brothers made their first experiments with unmanned hot-air balloons at Annonay, in the south of France. When news of this reached Paris, the Academy of Sciences asked Jacques Charles to study the invention. He realized that hydrogen would be a much better lifting gas than hot air. With the help of two friends, Nicolas and Anne-Jean Robert, he successfully launched a hydrogen balloon in August 1783. On December 1, Charles and Nicolas Robert became the first people to ascend in a balloon. In later flights Charles reached a height of nearly 10,000 feet (3,000 m). This made him a popular hero and the king, Louis XVI, invited him to move his laboratory to the Louvre.

It was in about 1787 that Charles made his most important discovery. As long ago as 1699 GUILLAUME AMONTONS had published his finding that different gases expand by the same amount for a given rise in temperature. Using oxygen, nitrogen, and hydrogen, Charles repeated the experiments by which Amontons had reached this conclusion and was able to calculate the precise amount by which the gases expanded. He found that for every 1°-C (1.8°-F) rise in temperature their volume increased by 1/273 of the volume they had at 0° C (32° F). This meant that if the gas could be cooled to –273° C (–459.4° F) its volume would be zero. This came to be known as *ABSOLUTE ZERO*.

Jacques Alexandre César Charles. **The French physicist and mathematician who discovered the gas law that bears his name.** ***(From the collections of the Library of Congress)***

Charles did not publish the results of these experiments, but he did inform JOSEPH GAY-LUSSAC about them. Gay-Lussac repeated them and the resulting general rule came to be known in France as *Gay-Lussac's law;* outside France it is called *CHARLES'S LAW.*

In 1785 Charles was elected to the Academy of Sciences. Later he became professor of physics at the Paris Conservatoire des Arts et Métiers. He died in Paris on April 7, 1823.

Charles's law The second of the GAS LAWS, which was discovered by JACQUES CHARLES in about 1787 and is based on the observation that the volume of a gas changes in direct proportion to its temperature. The law states that at constant pressure the volume of a mass of gas is directly proportional to its temperature in KELVINS. In other words, if the pressure remains constant, an increase in the temperature of a gas leads to a proportional increase in volume. This can also be expressed as

$$V_1/V_2 = T_1/T_2$$

where V_1 and T_1 are the initial volume and temperature and V_2 and T_2 are the final volume and temperature.

Charley A HURRICANE that reached the British Isles on August 25, 1986. It caused at least 11 deaths.

Chelford interstadial An INTERSTADE that occurred in Britain about 60,000 years ago, during the DEVENSIAN GLACIAL. Evidence for it has been found at Chelford, Cheshire, in northwestern England. At that time a coniferous forest similar to that now found in Finland covered the region. Conditions were similar to those during the BRØRUP INTERSTADIAL.

chemical transport model (CTM) A three-dimensional model of the atmosphere that uses observed or analyzed winds, moisture, temperature, and other meteorological conditions to calculate the transport of chemical substances through the atmosphere and reactions among them as a function of time. A CTM includes the processes by which chemical species are converted to AEROSOLS and by which they are incorporated into rain and washed to the ground. The models can therefore be used to compute the way the distribution of aerosols varies from place to place and time to time.

chemisorbent *See* ADSORPTION.

chergui A hot, dry wind that blows from the east across Morocco, in North Africa. Because it has crossed a long stretch of desert, the wind has a strongly drying effect.

chili A hot, dry wind that blows from the south over Tunisia, in North Africa, most commonly in spring. It is a wind of the SIROCCO type.

chill wind factor An index that was developed by the Canadian army to help in relating the performance of equipment to that of personnel in the arctic winter. The index is equal to the wind speed measured in miles per hour minus the temperature in degrees Fahrenheit. For example, a wind speed of 25 mph at a temperature of –40° F would give an index of 25 – –40 = 25 + 40 = 65. This is in no way connected to WIND CHILL.

chimney height *See* STACK HEIGHT.

chimney plume The cloud that is emitted from a chimney or factory smokestack and that travels downwind in a CONING, FANNING, FUMIGATING, LOFTING, LOOPING, or TRAPPING pattern. The plume consists of gases, which are the chemical products of COMBUSTION. These are invisible. If the plume can be seen it is because water droplets and possibly solid particles of ash and SOOT are present. Water is also a product of the combustion of hydrocarbon fuels, through the oxidation of hydrogen. As the plume travels downwind, it mixes with the surrounding air and disperses. Dispersion is due to several processes. Heavier solid particles fall under gravity, as fallout. In air close to SATURATION, smaller particles act as CLOUD CONDENSATION NUCLEI and trigger the formation of cloud, a process known as *rainout.* If the cloud produces precipitation, the particles in it are carried to the surface as washout. Finally, where the plume encounters solid surfaces the molecules and particles in it adhere by a process known as *dry impaction.*

chinook A warm, dry, KATABATIC WIND that blows mainly in late winter and spring on the eastern side of the Rocky Mountains in Canada and the United States. It is of the same type as the FÖHN WIND of the European Alps. The chinook occurs when there is a strong

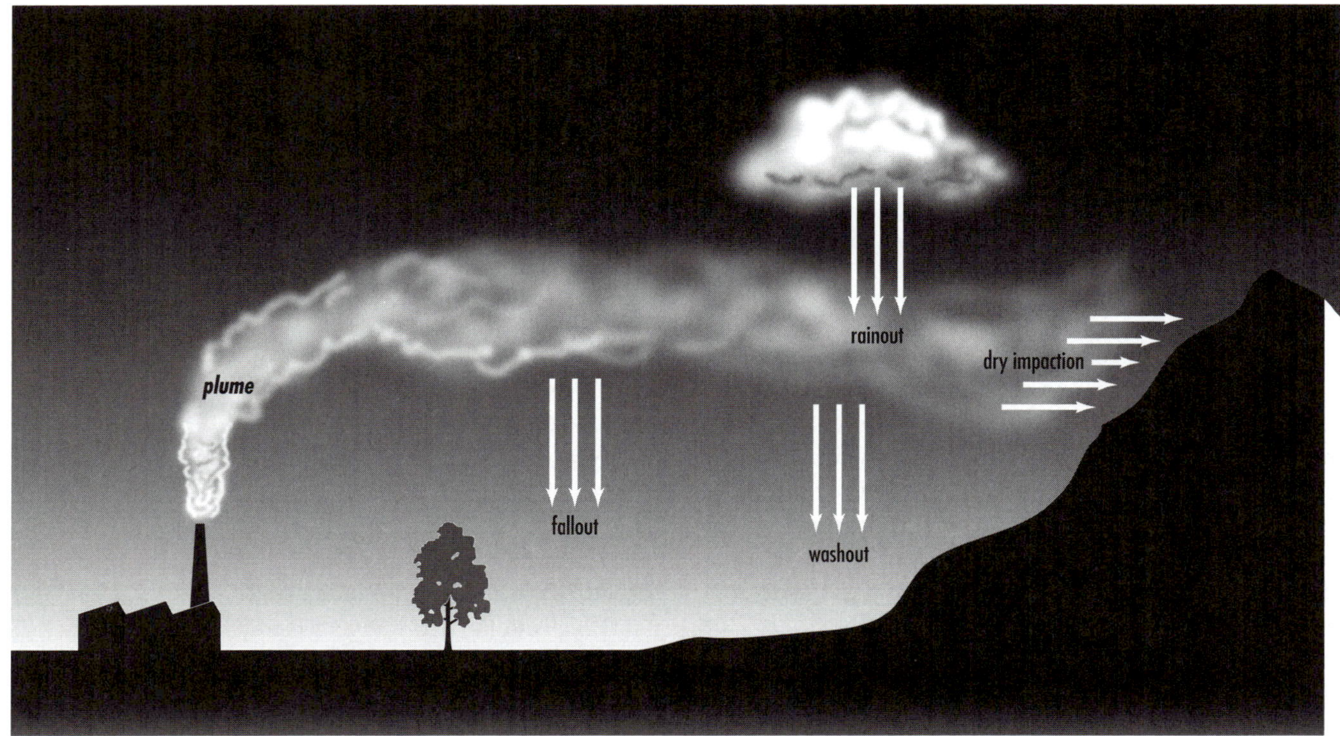

Chimney plume. **Gases and particles from the chimney are carried away by the wind, dispersing as they move farther from the source.**

flow of air across the mountains. A TROUGH of low pressure forms on the eastern (lee) side of the mountains and pulls air down the mountainsides. A cloud, known as the *chinook wall cloud,* forms over the mountains and is a reliable indicator of an approaching chinook. The air has lost its moisture on the western (windward) side of the mountains, so it descends as dry air and warms at the DRY ADIABATIC LAPSE RATE of 5.4° F per 1,000 feet (9.8° C per kilometer). When it reaches the plains to the east of the mountains it produces a large and rapid rise in temperature. The temperature sometimes rises by about 40° F (22° C) in less than five minutes. The wind can reach speeds of 100 mph (160 km h^{-1}), and because the air it carries is so dry it can SUBLIMATE large amounts of snow—often up to 6 inches (15 cm) a day. The RELATIVE HUMIDITY during a chinook usually falls to below 50 percent and has been known to reach 10 percent. This gives the wind its alternative name of "snow eater." The name *chinook* is taken from the Native American people of that name, whose lands are in the Northwest of North America.

Chinook Arch *See* ARCH CLOUD.

chinook wall cloud *See* CHINOOK.

Christmas Day According to WEATHER LORE, a day that is believed to indicate what the weather will be like in the following months. If Christmas Day is fine, the spring will be mild, there will be no frosts in May, and therefore there will be a good crop of tree fruit in the fall.

chromosphere The gaseous layer of the Sun (or any other star) that lies above the PHOTOSPHERE. The temperature rises through the chromosphere from about 4,000 K at the base to about 10,000 K at the top. When the Sun is hidden by a solar ECLIPSE the chromosphere appears as a pink glow; the color gives the layer its name *chromosphere,* or "colored sphere."

Chronology of Disasters *See* Appendix I.

Chronology of Discovery *See* Appendix II.

chubasco A heavy THUNDERSTORM, with strong SQUALLS, that occurs in the rainy season from May to October along the western coast of Central America, and especially in Nicaragua and Costa Rica.

Ci *See* CIRRUS.

circular vortex A VORTEX in which the STREAMLINES are parallel to one another around a common axis.

circulation The movement of air or water along a path that eventually returns it to its starting point. Circulation can take place in either the horizontal or the vertical plane. Air circulates vertically in a CONVECTION CELL, and the GENERAL CIRCULATION of the atmosphere involves a number of large-scale vertical cells (*see* HADLEY CELL, FERREL CELL, and POLAR CELL) that are combined in the THREE-CELL MODEL of the general circulation. The horizontal circulation of air may be CYCLONIC or ANTICYCLONIC, and the large-scale circulation may consist mainly of ZONAL FLOW or of MERIDIONAL FLOW. Horizontal flow is affected by periodic changes in the distribution of AIR PRESSURE, such as the NORTH ATLANTIC OSCILLATION, PACIFIC DECADAL OSCILLATION, and MADDEN–JULIAN OSCILLATION. Variations in the mean flow occur during ENSO events.

circulation flux FLUX that is associated with the overall movement of the atmosphere (its CIRCULATION) rather than EDDY motion.

circulation index A value that is ascribed to one of the major components of horizontal atmospheric CIRCULATION. There are two such indices, the MERIDIONAL INDEX and the ZONAL INDEX. ZONAL FLOW varies over a cycle known as the INDEX CYCLE.

circulation pattern The geometrical shape of the horizontal CIRCULATION of the atmosphere as this is shown on SYNOPTIC CHARTS, where it is indicated by the ISOBARS. The pattern may be ANTICYCLONIC or CYCLONIC.

circumhorizontal arc A brightly colored horizontal band of light that is seen at an elevation of less than 32° above the horizon when the Sun is a little more than 58° above the horizon. The light is caused by the REFLECTION and REFRACTION of light from ICE CRYSTALS that have vertical axes. The light enters the crystals through their vertical faces and leaves through their horizontal faces. The band of light displays the colors of the spectrum (*see* RAINBOW) with red at the top.

circumpolar vortex The CIRCULATION PATTERN formed by the westerly winds that blow around the North and South Poles. The winds circulate CYCLONICALLY around a persistent region of low pressure located in the TROPOSPHERE at an altitude of 6,500–33,000 feet (2–10 km).

circumscribed halo A HALO that is surrounded by a bright ring. When the Sun is high in the sky the outer ring touches the edge of the 22° halo at the top and bottom but is clearly outside it at each side, so it has an approximately elliptical shape. When the Sun is low in the sky the bottom of the outer ring sags below the halo. This rare phenomenon is caused by the REFRACTION of light through hexagonal ICE CRYSTALS that have horizontal axes.

circumzenithed arc A circular arc, brightly colored with red at the bottom, seen more than 58° above the Sun when the Sun is below 32°. It is caused by light rays entering the horizontal tops of hexagonal ice crystals with vertical axes and emerging from vertical sides.

cirque glacier *See* VALLEY GLACIER.

cirr- (or cirro- or cirri-) A prefix that is derived from the Latin word *cirrus*, which means "curl," such as a curl of hair. It is attached to cloud genera (*see* CLOUD CLASSIFICATION) that consist of wispy, fibrous cloud elements.

cirriform Stretched into long, fine, curling filaments that resemble the cloud genus CIRRUS (*see* CLOUD CLASSIFICATION). The Latin *cirrus* means "curl," as in curled hair.

cirrocumulus (Cc) A genus of high cloud (*see* CLOUD CLASSIFICATION) that is composed entirely of ICE CRYSTALS. It appears as small, white patches or sheets, or as more or less spherical masses, called *elements*, with no shading around or between them. Each of these elements has an apparent width of about 1°, which is approximately the width of a little finger held at arm's length. The elements are arranged in more or less regular patterns resembling the ripples seen in sand on the seashore or, less commonly, form groups or lines. Cirrocumulus is usually a degraded form of CIRRUS or CIRROSTRATUS, from which it retains a fibrous appearance. Cirrocumulus should not be confused with small ALTOCUMULUS, which is sometimes seen at the edge of sheets of altocumulus.

Cirrocumulus occurs as the species CASTELLANUS, FLOCCUS, LENTICULARIS, and STRATIFORMIS, and as the varieties LACUNOSUS and UNDULATUS. A MACKEREL SKY is produced by cirrocumulus.

cirrostratus (Cs) A genus of high cloud (*see* CLOUD CLASSIFICATION) that is composed entirely of ICE CRYSTALS. It appears as a thin white veil that does not blur the outlines of the Sun or Moon, although it often gives rise to HALOES. Sometimes it is so thin it does no more than give the sky a pale, milky appearance. At other times, it has a distinctly fibrous appearance, as though it consists of tangled filaments.

It often forms on a WARM FRONT, and if CIRRUS appears first, then thickens until it becomes cirrostratus, it is likely that an active DEPRESSION is approaching. This may become stationary and fill or may change direction, but if it continues its approach it will probably produce precipitation.

The species of cirrostratus are FIBRATUS and NEBULOSUS, and it occurs in the varieties DUPLICATUS and UNDULATUS. The supplementary feature VIRGA is sometimes seen below the base.

cirrus (Ci) A genus of high cloud (*see* CLOUD CLASSIFICATION) that is composed entirely of ICE CRYSTALS. It appears as long, wispy filaments; narrow bands; or white patches, always with a fibrous appearance. Cirrus is often seen ahead of an approaching WARM FRONT. It occurs as the species CASTELLANUS, FIBRATUS, FLOCCUS, SPISSATUS, and UNCINUS and the varieties DUPLICATUS, INTORTUS, RADIATUS, and VERTEBRATUS. *See also* MARES' TAILS.

CISK *See* CONDITIONAL INSTABILITY OF THE SECOND KIND.

CISOs *See* CLIMATOLOGICAL INTRA-SEASONAL OSCILLATIONS.

CLAES *See* CRYOGENIC LIMB ARRAY ETALON SPECTROMETER.

Clara A TYPHOON that struck Fujian Province, China, on September 21, 1981. It destroyed 130 square miles (337 km²) of rice crops.

Clarke orbit *See* GEOSTATIONARY ORBIT.

Clausius–Clapeyron equation An equation that relates the SATURATION VAPOR PRESSURE (e_s) to the ABSOLUTE TEMPERATURE (T). The equation is

$$d\,e_s/d\,T = L/T(\alpha_2 - \alpha_1)$$

where L is the LATENT HEAT of vaporization, α_2 is the SPECIFIC VOLUME of water vapor, and α_1 is the specific volume of liquid water. Since α_2 is usually very much larger than α_1, α_1 can be ignored.

Clean Air Act A law designed to improve air quality that was passed in 1956 in the United Kingdom; a law with the same name that was passed in 1963 in the United States. The British legislation, drafted in response to the LONDON SMOG INCIDENTS, empowered local governments to designate "smokeless zones" in which it became an offense to emit black smoke. As the act was implemented over the succeeding years this had the desired effect of eliminating the domestic burning of coal in most cities in favor of use of smokeless fuels such as coke. Industrial plants continued to burn coal but were required to fit devices to remove the smoke.

The U.S. act was introduced in order to strengthen the provisions of the AIR QUALITY ACT of 1960 and was revised in 1970; amendments were passed in 1977, 1987, and 1990. The Clean Air Act increased the powers of the Environmental Protection Agency, especially in stipulating the emissions permitted from particular industrial installations. This proved difficult in practice, because it would have forbidden any further industrial development in states where existing emissions exceeded the permitted limits. This difficulty was overcome by allowing companies to agree with the regulatory authority to accept stricter emission standards for one part of their operation in return for a relaxation in the standards for another part. This "bubble policy" increased the effectiveness of pollution control while reducing its cost.

Clean Development Mechanism A procedure that is included in the KYOTO PROTOCOL under which countries and companies are permitted to offset their own carbon emissions by paying for a project in another country that would reduce carbon emissions.

clear (*noun*) The condition in which cloud covers less than one-tenth of the sky. (*verb*) The dissipation of cloud that ends PRECIPITATION and leaves the sky largely cloudless.

clear air turbulence (CAT) Vertical air currents that occur in unstable air (*see* STABILITY OF AIR) that is not saturated and is therefore free of cloud. It can be caused in several ways. Unstable air rises by convection at fronts, where air is converging, and by being made to cross high ground. It can also do so where there is a sharp difference in surface temperature between two adjacent areas, such as the warm surface of a small island surrounded by the cool surface of water. Strong WIND SHEAR can also trigger instability, even in air that is stable when it is still. Wind shear associated with the JET STREAM is the most common cause of the clear air turbulence that occasionally affects aircraft.

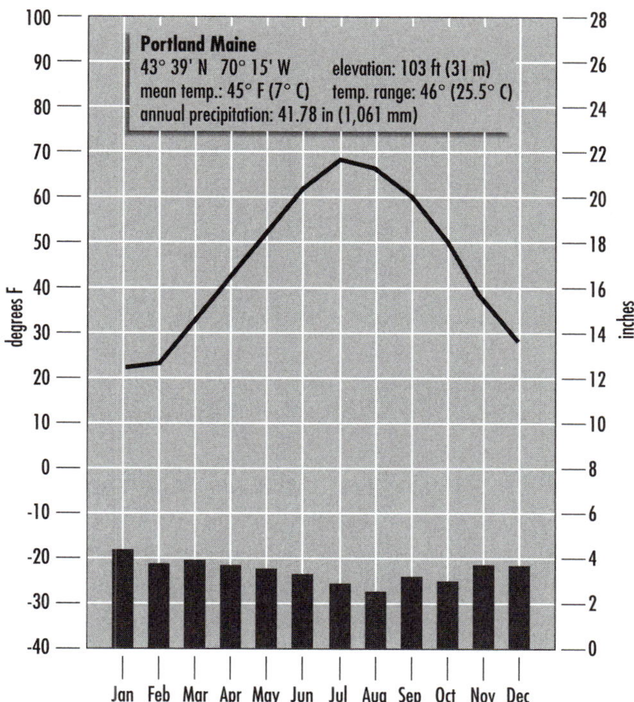

Climate. **The mean temperature for each month is shown by the line and the average precipitation by the bars. The diagram summarizes the climate for a particular place, in this case Portland, Maine.**

clear ice *See* GLAZE.

clepsydra *See* TOWER OF THE WINDS.

CLIMAP *See* CLIMATE–LEAF ANALYSIS MULTIVARIATE PROGRAM; *see also* CLIMATE: LONG-RANGE INVESTIGATION MAPPING AND PREDICTION.

climate The average weather conditions that are experienced in a particular place over a long period. This is contrasted with the WEATHER that is experienced from day to day. The climate of a place is determined principally by its latitude, its distance from the ocean, and its elevation above sea level. This makes it possible to define the different types of climate (*see* CLIMATE CLASSIFICATION).

climate classification The arrangement of climates according to their most important characteristics in order to provide each type with a short, unambiguous name or title by which it can be known. This is necessary if climates are to be compared, because without names that everyone understands, all the relevant features of each climate would have to be specified every time that climate was mentioned and discussions would become impossibly long-winded and confusing.

Before any group of things can be classified, their most important features must be identified. In the case of climates, the main features relate to temperature and precipitation. The earliest classifications were made in ancient Greece and took account of only temperature. The Greeks divided the Earth into three climatic regions: torrid, temperate, and frigid.

Attempts at more detailed classifications began in the 19th century. At first, they were made mainly by plant geographers, who were botanists and interested not so much in specific temperature and precipitation data as in their effects on plant communities. Consequently, they related climates to the types of vegetation associated with them, and some of the resulting names are still used. It is quite usual to speak of a savanna climate, for example, or a tropical rain forest, tundra, or boreal forest climate (the boreal forest is the predominantly coniferous forest that grows in the north of Canada and Eurasia). Other names, such as "penguin" climate, have been dropped, although they were once in use.

When the first attempts at classification were being made, the only reliable information available referred to temperature, precipitation, and vegetation types. Much more information has become available since then. It is now possible for a classification to include data concerning the climatic requirements for crop growth, for example, or the effect of climates on humans living in them, and data can be obtained from satellite readings and images as well as from surface observations. Modern computing power also makes it possible to classify climates by means of extremely complex statistical analyses of the data.

Most classifications are based on either a generic, or empirical, approach, or a genetic approach. Both are valid.

Generic classifications identify climates that are similar in their effects on plant growth. They rely on two primary criteria: aridity and warmth. Aridity takes account of both precipitation and temperature to determine the EFFECTIVE PRECIPITATION, which is what regulates plant growth. The most widely used of the generic classifications are those devised by WLADIMIR KÖPPEN and CHARLES W. THORNTHWAITE. The Russian climatologist MIKHAIL BUDYKO proposed a generic scheme in 1958 that is based on radiation and evaporation and introduces the concept of a RADIATIONAL INDEX OF DRYNESS.

Genetic classifications are based on those features of the general circulation of the atmosphere that cause particular climates to occur where they do. In other words, they relate climates to their physical causes rather than grouping similar types, as in generic systems. There are fewer genetic classifications than generic ones. One such scheme, the FLOHN CLASSIFICATION, was proposed in 1950, and in 1969 A. N. Strahler devised the STRAHLER CLIMATE CLASSIFICATION, which can be related to the Köppen classification.

No single scheme of climate classification is satisfactory for all purposes. Some provide no more than a convenient set of names. Many are intended mainly for agricultural application. Others relate to the distribution of natural vegetation and are of interest to geographers, botanists, and ecologists. Because classifications must serve so many different purposes, they tend to be more or less specialized, and consequently a large number of schemes have been proposed, reflecting the particular interests of the climatologists who devised them. Further proliferation has resulted from the many attempts that have been made to produce a single, comprehensive scheme that suits the requirements of all users. Each scheme has its advantages and its limitations. Through them all, however, the Köppen and Thornthwaite classifications have emerged as the ones that are most widely used.

climate controls The factors that determine the type of CLIMATE a particular place experiences. Latitude is the most important of these, because it defines the amount of solar radiation the area receives. The proportions of the surrounding area that have surfaces of land and water are also important, because of their different HEAT CAPACITIES. The geographical location of the place and the direction of the PREVAILING WINDS determine the AIR MASSES that reach it. MARITIME AIR, with its moderating influence on temperatures, is carried a long way inland, so a place on the WINDWARD side of a continent has a gentler climate than a place in the same latitude on the opposite side of the continent, where it receives CONTINENTAL AIR. The movement of air masses may be blocked by mountain ranges, however. This is the case in North America, where the moderating influence of maritime air from the Pacific does not extend very far inland because of the Rocky Mountains. Mountains also receive OROGRAPHIC RAIN on the windward side, and the LEE side has a dry climate. This effect is seen in all the continents. Ocean currents also influence climates on land. Western BOUNDARY CURRENTS carry warm water, which warms the air crossing over it, and eastern boundary currents carry cold water. As well as affecting the air temperature, boundary currents affect HUMIDITY. Eastern boundary currents tend to produce very arid coastal climates and western boundary currents produce moist climates. The Atacama and Namib Deserts are located along coasts exposed to eastern boundary currents (the PERU CURRENT and BENGUELA CURRENT). The mild climate of northwestern Europe is the result of the warm NORTH ATLANTIC DRIFT and Japan and the Aleutian Islands benefit from the KUROSHIO CURRENT.

climate diagram A diagram that shows the mean temperature and precipitation month by month for a particular place. There are several types, but the most widely used has a scale for temperature on one side, with a curve showing the temperature through the year, and a scale for the amount of precipitation on the opposite side. Precipitation is usually shown as a histogram. The diagram shows the name of the place to which the data refer, often with its latitude and longitude, and additional information may also be displayed, such as height above sea level, total annual precipitation, and temperature range.

Climate–Leaf Analysis Multivariate Program (CLIMAP) A program that uses plant leaves to help estimate past temperatures. Leaves of dicotyledonous plants (those with seeds that produce two seed leaves, or cotyledons) that are known to have been present at a site at a known time in the past are subjected to a statistical analysis of a suite of 29 characteristics. The technique is based on the strong correlation that has been observed between the warmth of the climate and the likelihood that dicotyledonous plants will have leaves with smooth edges.

Climate: Long-Range Investigation Mapping and Prediction (CLIMAP) A 10-year international scientific project, the aim of which was the reconstruction of the climates of the QUATERNARY subera. The project ran from 1971 until 1980 and was funded by the National Science Foundation and administered from Columbia University, New York. It involved the participation of scientists who specialize in Earth and atmospheric sciences and oceanography. CLIMAP has been succeeded by the COOPERATIVE HOLOCENE MAPPING PROJECT.

climate model A mathematical simulation of the processes that affect the atmosphere and produce local weather and the climates over large regions, or the entire world, and over extended periods. It is used as an aid to understanding those processes and predicting how changes to them may affect weather and climate. The model consists of a computer program in the form of a series of equations that allow the physical laws controlling the weather to be applied (*see* GAS LAWS). This is possible only to a limited extent, because not all climatic processes are fully understood, data are incomplete, and if the model is to describe conditions over a large region or long period the amount of computation involved may exceed the capacity of the available computers. Consequently, climate models are necessarily simplified.

Construction of most models begins with the imposition of a grid over the region to be studied. The grid may be two-dimensional or three-dimensional, like a cage made from several two-dimensional grids placed one above the other and linked vertically. Two-dimensional models may represent two horizontal dimensions, like the lines of latitude and longitude on a map, or, more commonly, the vertical and one horizontal dimension.

Initial data on such factors as pressure, temperature, humidity, cloud amount and type, and wind speed and direction are supplied to each intersection between grid lines, where all the calculations of physical effects are made. Conditions at every intersection affect those at every other because of FEEDBACK, which may be negative or positive, so a formidable amount of computation is needed to trace the development of a storm, for example, or a frontal system. When conditions change at one intersection, the results of that change supply the input data for adjacent intersections, where conditions are then recalculated. In this way the entire system evolves, usually in a "time-step" fashion.

The sensitivity of the model depends on the scale of the grid—the distance between grid lines—and the size of the time steps: The shorter these are the more accurate the model is likely to be.

Models must be tested before they can be used to estimate the consequences of change. Testing usually begins by supplying the model with data from the recent past then running it to see how well it simulates the weather conditions that were actually recorded. If this is successful the model is used to simulate conditions from the more distant past, partly as an aid to understanding how those conditions developed. Only if it passes these tests is the model used to estimate what may happen in the future.

Some models are one-dimensional and not based on a grid. Radiative convective models, for example, calculate the vigor of convection in a vertical column through the atmosphere. They do this by calculating the effects of incoming and outgoing radiation and computing the amount of convective motion needed to produce a known LAPSE RATE.

The most powerful supercomputers are used to construct and run the more complex models, and the reliability of models is directly proportion to the amount of computational power that is available to run them. Regardless of the amount of computational power, however, all models make assumptions about factors that are not well understood, such as the role of ocean currents in the transport of heat, and that occur on a scale much smaller than that of the grid, such as cloud formation.

There are several types of climate models, the most important of which are ENERGY BALANCE MODELS, GENERAL CIRCULATION MODELS, STATISTICALLY DYNAMICAL MODELS, and radiative convective models. *See also* ADVECTIVE MODEL, ANALOG MODEL, BAROCLINIC MODEL, BAROTROPIC MODEL, CHEMICAL TRANSPORT MODEL, CROP-YIELD MODEL, EQUIVALENT BAROTROPIC MODEL, HISTORICAL ANALOG MODEL, HYDROLOGICAL MODEL, STATION MODEL, STORM MODEL, SWAMP MODEL, SYNOPTIC MODEL, and THERMOTROPIC MODEL.

climate system The atmosphere, together with all the factors that affect it to produce the climates of the world. The system includes the oceans, lakes, rivers, and the HYDROLOGICAL CYCLE, the polar ICE SHEETS, the solid Earth, and living organisms. The energy driving the system is derived from the Sun; solar radiation, and periodic changes in it, is also included.

climatic divide The boundary between two or more regions that have markedly different types of CLIMATE.

climatic elements The factors that determine the WEATHER and CLIMATE. There are seven elements of major importance: TEMPERATURE, sunshine, AIR PRESSURE, wind direction and speed, HUMIDITY, cloudiness, and PRECIPITATION. VISIBILITY is also a climatic element, but of less general importance.

climatic forcing A perturbation of the balance between the amount of energy Earth receives from the Sun and the amount it reradiates into space that is imposed by some factor outside the climatic system but produces climatic effects. A change in the output of solar energy would have such a climatic forcing effect, for example, as would a change in the concentration of atmospheric CARBON DIOXIDE (CO_2). Climatic forcing is measured in watts per square meter; that due to the increase in CO_2 since preindustrial times amounts to about 1.5 W m^{-2}. Calculating the forcing due to different factors is essential in predicting future climates. It seems likely that natural changes in solar output may be the most important cause of climatic forcing, rather than the forecast changes in CO_2 concentration.

climatic geomorphology Geomorphology is the scientific study of landforms and how they develop. Climatic geomorphology is a branch of geomorphology that began toward the end of the 19th century and that concentrates on the effects of climate in shaping the surface of the planet. Climatic geomorphologists have defined 10 types of climatic processes that form landscapes and have produced a map of the world showing where each of them occurs.

climatic normal The mean values for TEMPERATURE, HUMIDITY, and PRECIPITATION at a specified place over a fixed period. In many countries, including the United States, the fixed period is of 30 years and the period changes every 10 years. Recent fixed periods lasted from 1951 until 1980, 1961 to 1990, and 1971 to 2000. The use of a 30-year mean ensures that short-term variations in climate are hidden and the regular updating ensures that the data depict the present climate reasonably accurately.

climatic optimum A period during which the climates over most or all of the world are warmer than the climates before or after. The MEDIEVAL WARM PERIOD was a climatic optimum, but the warmest period since the end of the most recent ice age occurred between about 7,000 and 5,000 years ago. At that time, summer temperatures in Antarctica and in Europe were about 4°–5° F (2°–3° C) warmer than they are today. This postglacial optimum did not reach Greenland or the northern part of North America until about 4,000 years ago. As the ice melted, sea levels rose. The rise began about 17,000 years ago, and by about 4,000 years ago sea levels may have been about 10 feet (3 m) higher than they are now.

climatic prediction An estimate of the possible climatic consequences of social or industrial changes that are already occurring and that are likely to continue. For example, as cities grow in size their heat production can be calculated fairly accurately (*see* HEAT ISLAND), and from this it is possible to estimate the effect on summer and winter temperatures and precipitation. MODELS that calculate the effect of releasing GREENHOUSE GASES into the atmosphere estimate the climatic consequences at various times in the future. These are also climatic predictions.

climatic province *See* CLIMATIC REGION.

climatic region (**climatic province**) A fairly large area over which the CLIMATIC NORMALS are fairly constant, so the normals from different stations can be grouped together. Attempts to define climatic regions more rigorously led to increasingly elaborate systems of CLIMATE CLASSIFICATION.

climatic snow line The altitude above which snow accumulates over a long period on a level surface that is fully exposed to sunshine, wind, and PRECIPITATION. Below this altitude, ABLATION between snowfalls is sufficient to prevent snow accumulation.

climatic zone A region of the Earth, defined by latitude, within which the climate is sufficiently constant to be characteristic of the region as a whole. The concept of the climatic zone was introduced by ARISTOTLE. He related the zones to changes in the length of daylight and called the zones *klimata,* from which we derive our word *climate.* He defined three *klimata*—"torrid," "temperate," and "frigid"—separated by the tropics of Cancer and Capricorn and the Arctic and Antarctic Circles. Climatic zones form the basis for some schemes of CLIMATE CLASSIFICATION (*see,* for example, KÖPPEN CLIMATE CLASSIFICATION). The principal climatic zones that are accepted today include the high polar (80°–90° N, 70°–90° S), subpolar (60°–80° N, 55°–70° S), temperate (40°–60° N, 35°–55° S), subtropical wet (30°–40° N, 30°–35° S), subtropical dry (20°–30° N and S), tropical seasonal (10°–20° N, 5°–20° S), and equatorial (10° N–5° S).

climatological forecast A forecast of the weather for a region that is based on its CLIMATE, rather than on a projection of the current SYNOPTIC situation, but with allowance made for such important features as fronts, pressure systems, and the location and strength of the JET STREAM.

Climatological Intra-Seasonal Oscillations (**CISOs**) A series of weather cycles that produce alternately wet and dry conditions in regions affected by the Northern Hemisphere summer MONSOON. Four CISO cycles occur between May and October. The first produces wet weather in the middle of May over the South China Sea and the Philippines, followed by dry weather in late May and early June. The wet weather of the second cycle occurs in the middle of June and marks the start of the monsoon over the western North Pacific. It is followed by dry weather in the first half of July. The third cycle causes wet weather in the western North Pacific in the middle of August, followed by dry weather, and the fourth cycle yields wet weather in the middle of October. The CISO propagates from the equator to the northern Philippines from May through July, then westward along latitude 15° N from 170° E as far as the Bay of Bengal during August and September.

(You can find more technical details about the CISO in Bin Wang and Xihua Xu, "Northern Hemisphere Summer Monsoon Singularities and Climatological Intraseasonal Oscillation," at www.soest.hawaii.edu/MET/Faculty/bwang/bw/pubs/52.html.)

climatological station An observing station where meteorological data are collected and stored over a long period for use in climatological studies (*see* CLIMATOLOGY).

climatological station elevation The datum level that is used as a reference for all records of AIR PRESSURE in a particular region. It is the vertical height above sea level of an identified point.

climatological station pressure The AIR PRESSURE that is calculated for the CLIMATOLOGICAL STATION ELEVATION. Its use allows all climate records to be compared, because they all refer to the same height above sea level. The climatological station pressure may differ from the STATION PRESSURE.

climatology The scientific study of CLIMATES. The word is derived from the Greek words *klima,* which means "slope," and *logos,* which means "account." The "slope" to which the name refers is the inclination of the Earth's axis of rotation to the PLANE OF THE ECLIPTIC. Although this inclination does not produce climates, it does produce the SEASONS, and the ancient

Greeks noticed the connection (*see* ARISTOTLE). Climatology encompasses every aspect of the physical state of the atmosphere over particular parts of the world and over extended periods.

The movements of and within AIR MASSES constitute DYNAMIC CLIMATOLOGY, and the relationship between regional climates and the circulation of the atmosphere is the subject of SYNOPTIC CLIMATOLOGY. The climates of particular regions are the subject of regional climatology, which involves CLIMATE CLASSIFICATION, the purpose of which is to facilitate comparisons between climates.

The scale of their subject matter also distinguishes branches of climatology. There are three such branches, called *MACROCLIMATOLOGY, MESOCLIMATOLOGY,* and *MICROCLIMATOLOGY,* in decreasing order of scale. There are also climatologists who specialize in the reconstruction of past climates. Their discipline is PALEOCLIMATOLOGY; one of the techniques available to paleoclimatologists involves the study of TREE RINGS: DENDROCLIMATOLOGY. Climatologists also attempt to estimate what the climates of the world will be like in years to come. They build CLIMATE MODELS to help them. Calculations of future GLOBAL WARMING are derived from climate models developed by climatologists.

Climate studies are directly relevant to a variety of everyday activities. Housing developments must take account of the climates in which people will live and design buildings accordingly. Climatologists also contribute to the planning of the use of natural resources, and the study of the effect of climate on farming is so important that it forms a branch of the overall discipline called AGROCLIMATOLOGY.

climatostratigraphy The study of traces of soil and living organisms that are found in sedimentary rocks that were formed during the QUATERNARY. These rocks can be dated (*see* RADIOMETRIC DATING), and the fossils and other materials found in them provide clues to the climatic conditions at the time the sediments were deposited.

close (oppressive, muggy, sticky, stuffy) A subjective feeling of discomfort that people sometimes experience when the air is still and warm (*see* COMFORT ZONE). The feeling can be experienced indoors or outdoors. It is caused by a combination of high temperature and a RELATIVE HUMIDITY high enough to inhibit the evaporation of sweat from the skin. The inability to cool the body by the evaporation of sweat produces an uncomfortably hot feeling, and sweat that fails to evaporate soaks into clothing, which then tends to stick to the skin.

clothesline effect *See* ADVECTION.

cloud A large concentration of liquid water droplets or ICE CRYSTALS that form by the CONDENSATION of WATER VAPOR in SATURATED AIR and that remain suspended in the air, clear of the surface. At any time about half the surface of the Earth is covered by clouds.

CLOUD DROPLETS and crystals range in size from less than 1 μm to about 50 μm (less than 0.0004–0.02 inch). CLOUD CONDENSATION NUCLEI must be present in order for water to condense out of saturated air. Inside a cloud, individual droplets and crystals usually survive for less than one hour before evaporating or subliming (*see* SUBLIMATION). Newly condensed droplets and newly frozen crystals immediately replace them. Close examination (using binoculars) of the edges of a CUMULIFORM cloud reveals what is happening. Fibers of cloud are constantly twisting away from the main mass and dissipating, but the cloud as a whole remains the same size or grows bigger. Cumuliform clouds are short-lived. A small cloud may take no more than half an hour to grow into a CUMULONIMBUS storm cloud that extends all the way to the TROPOPAUSE; an hour later that cloud may have dissipated completely. The ephemeral nature of such clouds is due to the fact that they form as isolated units in unsaturated air. Once liquid droplets and snowflakes falling from the top chill the rising warm air that sustains them, CONVECTION ceases and the cloud quickly evaporates into the surrounding air. STRATIFORM clouds form in stable air (*see* STABILITY OF AIR), and strong convection plays no part in their formation. Consequently, they last longer. Some can survive for a week or more, but even they do so by constant evaporation and condensation. An individual PARCEL OF AIR inside a stratiform cloud takes little more than one day to emerge into drier air, where its droplets evaporate.

There are many types of cloud, distinguished by their shape and the manner in which they form. This variety of cloud forms is described by an internationally accepted system of CLOUD CLASSIFICATION.

cloud amount The extent to which the sky is obscured by cloud. This information is included in reports from WEATHER STATIONS together with details of the cloud type. Cloud amount is measured from a reflection of the sky in a mirror that is marked with grid lines dividing it into equal areas. The observer counts the number of these areas that are filled with cloud. The total cloud amount is reported in tenths or in eighths, known as *oktas,* and it is indicated on a STATION MODEL by shading on the station circle. Symbols for cloud amount are in oktas, but these can be interpreted in tenths as follows: 1/8 = 1/10 or less; 2/8 = 2/10–3/10; 3/8 = 4/10; 4/8 = 5/10; 5/8 = 6/10; 6/8 = 7/10–8/10; 7/8 = 9/10 or overcast but with gaps in the cloud cover.

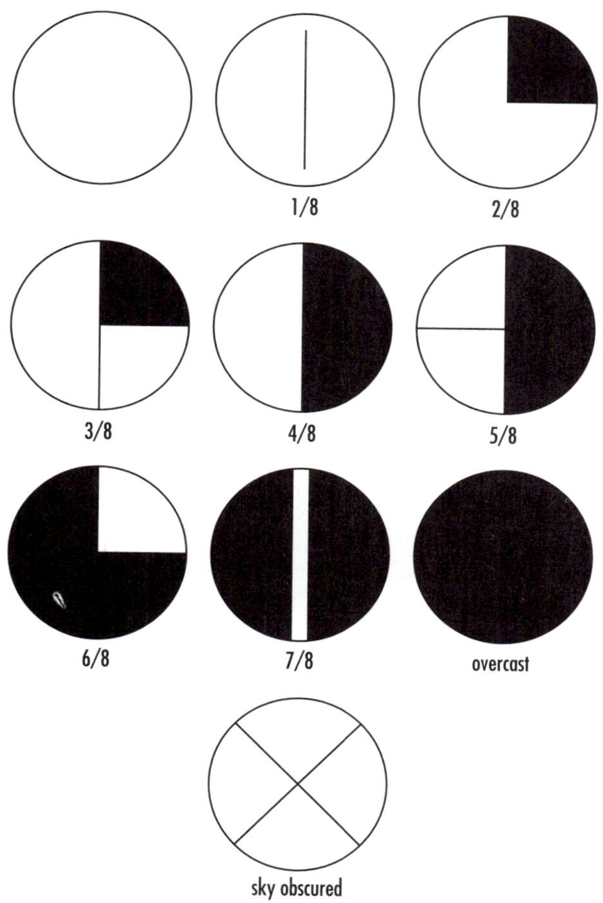

Cloud amount. **Shading of the station circle indicates the proportion of the sky that is covered by cloud. There are 10 possibilities. An open circle indicates no cloud, a fully shaded circle indicates complete cloud cover, a diagonal cross indicates that the sky is obscured, and the remaining symbols represent proportions from one-eighth to seven-eighths; these can also be interpreted in tenths.**

cloud band A linear formation of clouds that is 10–100 miles (16–160 km) wide and varies in length from tens to hundreds of miles. Spiral cloud bands are sometimes seen under the strongly BAROCLINIC conditions found behind an active COLD FRONT.

cloud bank A well defined mass of cloud that is seen from a distance. It extends across most of the horizon but does not cover the sky overhead.

cloud bar A long, narrow horizontal cloud that is clearly defined. It may be an element of a system of BILLOW CLOUDS or a LENTICULAR CLOUD. A cloud bar may also appear as a dark CLOUD BANK on the horizon that is the outermost edge of the clouds associated with a TROPICAL CYCLONE and that heralds the approach of the storm.

cloud base The height of the lowest part of an individual cloud or a layer of cloud, measured as the distance above sea level. Sea level is used as a datum because it is constant. Heights measured above ground level would vary according to the surface topographical characteristics, so although they would indicate the height of the cloud above the ground at a particular place, they could not be used reliably for any other place. Clouds are classified as high, medium, and low according to their bases, even though some, such as CUMULONIMBUS, may extend to a great height. *See* CLOUD CLASSIFICATION.

cloud bow An arc of light seen in the sky that is caused by the REFRACTION of light through spherical water droplets. It is similar to a FOGBOW.

cloudburst A sudden, very intense shower of rain that occurs when the mechanism sustaining a CUMULONIMBUS cloud fails and the cloud starts to dissipate. Individual cloudbursts are usually of brief duration, but they may be repeated, because as one cloud dissipates, another forms (*see* SQUALL LINE). Vertical currents inside a CUMULIFORM cloud carry water droplets upward, and in a large cumulonimbus the updrafts are often strong enough to hold a large amount of water airborne. Eventually downdrafts inside the cloud overlap the updrafts and suppress them. When this happens the water droplets are no longer supported and the cloud loses all its water at once. A big, fully developed cumulonimbus may hold 250,000 tons (275,000 tonnes) of water. If that amount of water falls on an area of 10 square miles (26 km^2), it delivers about 4 inches (10 cm) of rain. The resulting rainfall may cause FLASH FLOODING.

cloud chamber A device that is used to study CLOUD FORMATION and to detect the tracks of particles carrying an electric charge. It was invented in 1896 by the Scottish physicist Charles Thomson Rees Wilson (1869–1959), who worked at the Cavendish Laboratory of the University of Cambridge, England, and was especially interested in the way clouds form. Wilson shared (with Arthur Compton, 1892–1962) the 1927 Nobel Prize in Physics for this discovery. Between 1895 and 1899 Wilson showed that CLOUD DROPLETS can form in the absence of CLOUD CONDENSATION NUCLEI if the air is supersaturated (*see* SUPERSATURATION) and that droplets form more readily in very clean air if the air is exposed to X rays. This demonstrated that the droplets condense onto ionized (*see* IONIZATION) water molecules. In 1946, VINCENT JOSEPH SCHAEFER and BERNARD VONNEGUT used a cloud chamber to perform the experiments that led to techniques for CLOUD SEEDING. There is a simple home demonstration that shows how cloud forms when a drop in AIR PRESSURE reduces the temperature of moist air below its DEW POINT TEMPERATURE. In order to make your own cloud chamber you will need a glass bottle that holds about one gallon (3.8 liters), a stopper that will seal the bottle tightly, glass tubing, rubber or plastic tubing, and a strong source of light such as a slide projector. The stopper should be pierced by a hole that is just big enough for the glass tubing to fit snugly. Place about two inches of water in the bottom of the bottle. Insert the stopper with the glass tubing through it. Attach the rubber or plastic tubing to the other end of the glass tubing. Shake the bottle vigorously. This ensures the air and water in the bottle are at the same temperature throughout. Blow hard into the rubber or plastic tubing, then hold the tube tightly shut between your finger and thumb. Blowing into the bottle increases the air pressure inside. Keeping the tube sealed, shake the bottle again. This ensures that the air in the bottle is saturated. Place the bottle so the light is shining into it, watch carefully, and release the tube. The pressure inside the bottle will drop suddenly as it equalizes with the air pressure outside. This cools the air ADIABATICALLY and a thin mist will form inside the bottle. The mist is thin only because the cloud chamber is so small. If this happened on a large scale it would produce a big, dense cloud.

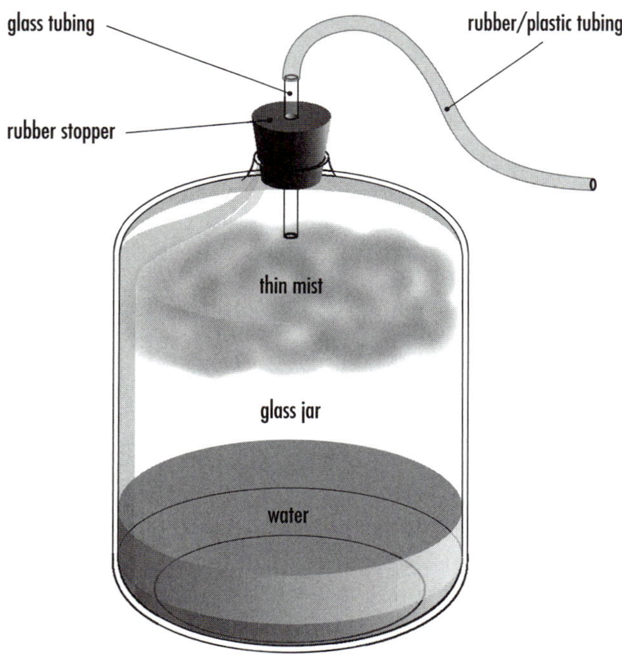

Cloud chamber. **Blowing through the tube into the bottle increases the internal pressure. When the pressure is released the air cools adiabatically and cloud forms.**

cloud classification Clouds vary greatly in color and shape and form at different heights above the surface. Throughout history people have sought ways to standardize the way in which clouds are described, so that a person in one place could send an unambiguous account of the appearance of the sky to someone who had not seen it. This is more difficult than it may seem. The Greek philosopher Theophrastus (371 or 370–288 or 287 B.C.E.) attempted to do so but could do no better than "clouds like fleeces of wool" and "streaks of cloud." If Theophrastus had told a friend in a far country that he had seen "streaks of cloud" the friend would have been little the wiser, because this might describe several quite different kinds of cloud.

The chevalier de Lamarck (1744–1829) also tried. Jean-Baptiste Pierre Antoine de Monet Lamarck was primarily a biologist, but one who specialized in classification, so classifying clouds must have seemed a similar task. He devised six categories. These were colorful, but somewhat vague: cloud sweepings, clouds in bars, dappled clouds, grouped or piled clouds, veiled clouds, and clouds in flocks.

Success was achieved finally in 1803, when LUKE HOWARD proposed a system that is the basis of the one in use today. In 1895, the International Meteorological Committee expanded the Howard system, and the resulting classification was revised several times before being adopted officially for use in the *INTERNATIONAL CLOUD ATLAS*.

In the modern system, clouds are classified first according to the height of the CLOUD BASE, as high, middle, and low. This division is made mainly for convenience, and it refers to the height at which the bases of clouds most commonly occur. They are not confined to these bands, however.

Then clouds are divided according to their appearance into 10 basic types, or genera, with names introduced by Howard based on the Latin names *cirrus, cumulus,* and *stratus;* these mean "curl," "pile," and "layer" (from *stratum*), respectively. (See table below.)

Cumulus and cumulonimbus clouds sometimes extend vertically to a considerable height. They are usually classified as "low" clouds by the height of their bases.

The 9 basic types are further divided into 14 species: CALVUS, CAPILLATUS, CASTELLANUS, CONGESTUS, FIBRATUS, FLOCCUS, FRACTUS, HUMILIS, LENTICULARIS, MEDIOCRIS, NEBULOSUS, SPISSATUS, STRATI- FORMIS, and UNCINUS.

There are also nine varieties: DUPLICATUS, INTORTUS, LACUNOSUS, OPACUS, PERLUCIDUS, RADIATUS, TRANSLUCIDUS, UNDULATUS, and VERTEBRATUS.

Clouds may also possess ACCESSORY CLOUDS, such as PILEUS, TUBA, and VELUM, and supplementary features such as ARCUS, INCUS, MAMMA, PRAECIPITATIO, and VIRGA.

cloud condensation nuclei (CCN) Small particles that are carried by the air and onto which water vapor condenses; the resulting droplets form clouds. In very clean air it is possible for the RELATIVE HUMIDITY to exceed 100 percent without water vapor condensation (*see* SUPERSATURATION). This is because water vapor condenses readily onto a surface but condenses only with difficulty if there is nothing on which droplets can form. Water vapor easily condenses onto plant surfaces as DEW and onto windows and other solid surfaces. In the air it condenses onto airborne particles—cloud condensation nuclei.

Airborne particles vary greatly in size, although all of them are microscopically small. Water does not condense onto the smallest particles, with diameters of about 0.002 µm because the SATURATION VAPOR PRESSURE is higher over a curved water surface than over a flat one and increases as the curvature increases. This *curvature effect* arises because the force that binds

Cloud level	Height of base					
	Polar regions		Temperate latitudes		Tropics	
	'000 feet	'000 meters	'000 feet	'000 meters	'000 feet	'000 meters
High cloud:	10–26	3–8	16–43	5–13	16–59	5–18
Types: CIRRUS, CIRROSTRATUS, CIRROCUMULUS						
Middle cloud:	6.5–13	2–4	6.5–23	2–7	6.5–26	2–8
Types: ALTOCUMULUS, ALTOSTRATUS, NIMBOSTRATUS						
Low cloud:	0–6.5	0–2	0–6.5	0–2	0–6.5	0–2
Types: STRATUS, STRATOCUMULUS, CUMULUS, CUMULONIMBUS						

water molecules to each other is strongest on a flat surface and weakens as curvature of the surface increases. Consequently, water evaporates much faster from a curved surface than from a flat one and small droplets evaporate much faster than big ones. Droplets that form on the smallest particles evaporate almost immediately.

Giant particles, more than 20 μm across, are so big they do not remain airborne long enough for water to condense onto them. The most effective particles are those between 0.2 μm and 2.0 μm in diameter. Air over land contains an average of about 80,000 to 100,000 of such cloud condensation nuclei in every cubic inch (5–6 million per liter), and air over the ocean, far from the nearest land, usually contains about 16,000 per cubic inch (1 million per liter).

As the relative humidity rises the HYGROSCOPIC NUCLEI are affected first. These are particles of a variety of substances, including dust, smoke, sulfate (SO_4), and salt (sodium chloride, NaCl). Water can start forming droplets around salt crystals when the relative humidity reaches 78 percent, and the other hygroscopic nuclei absorb water at somewhat higher humidity. When the relative humidity reaches about 90 percent, enough vapor may have condensed to form a fine haze that restricts visibility.

Hygroscopic nuclei dissolve in the water they absorb, leading to the *solute effect*. The droplets are not of pure water, but are solutions, and the saturation vapor pressure over any solution is lower than that over a surface of pure water. The smaller the droplet, the more concentrated the solution—because the hygroscopic nucleus has dissolved in less water—and the stronger the solute effect. It is because of the solute effect that the first droplets to form are very small and condense onto hygroscopic nuclei. As the relative humidity continues to rise, other nuclei become active, including the larger among the nonhygroscopic particles.

More water then condenses onto the droplets, increasing their size. This reduces the solute effect by weakening the solution, but at the same time it also reduces the curvature effect. Before long water vapor is condensing so rapidly that the relative humidity of the air between droplets starts falling. Air adjacent to droplets is no longer supersaturated. Once the relative humidity falls to 100 percent no more nuclei are activated. The result is to produce droplets of a size that is in equilibrium with the amount of water in the cloud. Supersaturation rarely exceeds 101 percent.

cloud crest *See* CREST CLOUD.

cloud deck The upper surface of a cloud layer.

cloud discharge A flash of LIGHTNING that occurs between areas of positive and negative charge within a single cloud. To an observer on the ground a cloud discharge appears as SHEET LIGHTNING.

cloud droplet A particle of liquid water that is held in suspension inside a CLOUD. Cloud droplets range in size from less than 1 μm to about 50 μm (less than 0.00004–0.002 inch) and usually last for less than one hour before they evaporate. Their size varies according to the size of the CLOUD CONDENSATION NUCLEI onto which the water vapor condenses. A typical cloud droplet is about 0.0004 inch (10 μm) in diameter, there are about 283 of them in every cubic foot of air (100,000 per liter), and they fall at about 0.4 inch per second (1 cm s^{-1}). Big cloud droplets are about 0.002 inch (50 μm) across, there are about three of them in every cubic foot of air (1,000 per liter), and they fall at about 11 inches per second (27 cm s^{-1}). Cloud droplets merge to form RAINDROPS by COALESCENCE or by the BERGERON–FINDEISEN MECHANISM and by the time it falls from the cloud a single raindrop consists of approximately 1 million cloud droplets.

cloud echo A RADAR signal that is reflected by CLOUD DROPLETS. The WAVELENGTH of the radar transmission determines the size of the objects that reflect it. The longer the wavelength, the bigger the objects it detects. Consequently, very short-wave radar beams are used to detect cloud droplets, which are extremely small. It is not possible to determine from a cloud echo whether the cloud is producing precipitation.

cloud formation An expression that has two meanings. It can be a particular pattern of CLOUDS with particular shapes—a formation of clouds. More usually, it refers to the processes by which clouds form.

When air rises its TEMPERATURE decreases in an ADIABATIC manner, and when it subsides its temperature increases adiabatically. Air can also be chilled or warmed by mixing with air at a different temperature.

The amount of WATER VAPOR a given volume of air can contain depends on the temperature of the air: Warm air can hold more water vapor than cold air. Consequently, as air rises and its temperature falls, or as it mixes with cooler air, its RELATIVE HUMIDITY (RH) increases. When its RH approaches 100 percent, water vapor begins to condense onto CLOUD CONDENSATION NUCLEI. The height at which rising air reaches its DEW POINT TEMPERATURE is called the *LIFTING CONDENSATION LEVEL*.

Immediately after air reaches its lifting condensation level, water vapor starts to condense into minute droplets. Cloud begins to form in a matter of a few seconds. CONDENSATION releases LATENT HEAT. This warms the air around the water droplets, causing it to continue rising and cooling, and further cooling leads to further condensation. The process ends when enough water vapor has been removed from the air for the dew point temperature to rise above the AMBIENT air temperature. The level at which this occurs marks the top of the cloud.

cloud height The vertical distance between the surface and the CLOUD BASE.

cloud layer STRATIFORM clouds or a number of clouds of the same or different types (*see* CLOUD CLASSIFICATION) that cover all or part of the sky and have a CLOUD BASE at approximately the same height everywhere. Cloud layers can occur at different levels, one above the other, with clear air between them.

cloud level One of the three groups into which clouds are classified by the height of the CLOUD BASE. The groups are HIGH CLOUD, MIDDLE CLOUD, and LOW CLOUD.

cloud particle A liquid CLOUD DROPLET or an ICE CRYSTAL that forms part of a CLOUD.

cloud physics The scientific study of the physical properties and behavior of CLOUDS. This branch of physics considers CONDENSATION and the other processes involved in CLOUD FORMATION; the formation of RAINDROPS, HAILSTONES, and SNOWFLAKES that fall as PRECIPITATION; the radiative properties of clouds, such as their ALBEDO and their absorption and emission of INFRARED RADIATION; and the transport of energy inside them. CLOUD DROPLETS grow into raindrops by collision and coalescence (*see* COLLISION THEORY), but if the cloud also contains ICE CRYSTALS raindrops may form by the BERGERON–FINDEISEN MECHANISM. Electric charge may separate inside large CUMULONIMBUS clouds. A cloud in which this happens may give rise to a THUNDERSTORM, and these electrical processes also form part of the subject matter of cloud physics.

cloud searchlight A powerful light that is used to measure the height of the CLOUD BASE. The light shines vertically upward, illuminating an area on the cloud directly overhead. An observer some known distance away measures the elevation of the center of the illuminated spot. The height of the cloud base can then be calculated by trigonometry. If the distance between the observer and the searchlight is y and the angle of elevation is θ, then the height of the cloud base is $y \tan \theta$. *See also* CEILING LIGHT and CEILOMETER.

cloud seeding Injecting material into supersaturated air (*see* SUPERSATURATION) in order to make water vapor condense into CLOUD DROPLETS or ICE CRYSTALS. The technique is used to make rain fall where it might not have fallen otherwise and to protect farm crops by inhibiting the formation of HAIL.

The advantages of reducing hail damage and of making rain fall where and when it is needed are obvious. For thousands of years people have dreamed of being able to control the weather in this way, but it was not until 1946 that scientists studying a different problem came across a way to do it. There had been earlier attempts. In 1891 the United States Congress appropriated $9,000 for experiments in which cannons were fired into clouds, and explosives were detonated inside low clouds from kites and balloons. The experiments were to check whether there was any truth in an old story that rain often fell after a major Civil War battle—the story is unfounded. Large, muzzle-loading mortars were also fired vertically upward into cloud to prevent hail formation. By 1899 there were thousands of such "hail cannons" in use in Europe. A very similar technique was tried in Russia in the 1960s, using rockets and artillery shells. Later there were attempts to make rain fall by throwing sand into clouds from airplanes. All these attempts failed.

In 1946, VINCENT JOSEPH SCHAEFER (1906–93) and BERNARD VONNEGUT (1914–97) were working as assistants to the Nobel Prize–winning chemist Irving Langmuir (1881–1957) at the General Electric Research Laboratory in Schenectady, New York, studying the problem of ICING on the wings of aircraft. They needed to know how icing was caused. Schaefer used a refrigerated box. The temperature inside the box was held at a constant –9.4° F (–23° C) and Schaefer added different kinds of particles to see what would cause ice crystals to form. There was a spell of very hot weather in July, and Schaefer found it difficult to keep the box cold enough for his purpose. To chill the air inside, on July 13 he dropped some crushed dry ice (solid carbon dioxide), at –109° F (–78° C), into the box. The instant the dry ice entered the box ice crystals formed and there was a miniature snowstorm. Shortly after that, Bernard Vonnegut found ice crystals formed when he burned SILVER IODIDE, allowing the smoke to enter the box. Dry ice was sharply lowering the temperature and causing ice crystals to form by HOMOGENOUS NUCLEATION. Silver iodide crystals triggered HETEROGENEOUS NUCLEATION.

On November 13, 1946, Schaefer dropped 6 pounds (2.7 kg) of dry ice pellets from an airplane into a cloud over Pittsfield, Massachusetts. This started a snowstorm. Further experiments followed, and by the 1950s commercial companies were offering cloud-seeding services. Silver iodide was much more convenient to use than dry ice, although dry ice is still used. Salt crystals are also used.

Silver iodide and dry ice are the most effective seeding agents where temperature inside the cloud is 5°–23° F (-5° to -15° C). Salt crystals that are larger than most cloud droplets (more than about 10 μm across) make bigger liquid droplets form. The bigger droplets, 30 percent to 60 percent larger than those present previously, are heavier, and so rain can be induced to fall from a cloud composed of droplets that are too small to fall. Other particles are also used. These include grains of volcanic dust and clays (such as kaolinite, which initiates ice-crystal formation at 15.8° F [9° C]), some proteins, and bacteria (such as *Pseudomonas syringae,* which initiates ice-crystal formation at 28.4° F [–2° C]). Others work best at temperatures between 5° F (–15° C) and 10° F (–12° C), but none of them is as effective or as simple to disperse as silver iodide.

Dry ice is administered by being dropped from an aircraft flying above the cloud. It is more effective when in the form of pellets about the size of peas. These produce a curtain of tiny ice crystals as they fall through the cloud. Silver iodide is released from aircraft or from the ground beneath the target cloud.

The addition of FREEZING NUCLEI can also reduce the size of HAILSTONES or even prevent hail formation entirely. An airplane flying at the base of a CUMULONIMBUS storm cloud releases particles into the updrafts. Increasing the number of freezing nuclei causes more small ice crystals to form. Freezing the supercooled water droplets onto a much larger number of much smaller crystals prevents the formation of big hailstones and may produce hailstones that are small enough to melt completely before reaching the ground.

For a long time it was impossible to tell whether cloud seeding really worked. Precipitation that fell from a cloud might have fallen in any case. The evidence now is that seeding can increase the amount of rain or snow that falls by at least 5 percent and sometimes by much more, but doubts remain and in Kansas some farmers oppose hail-suppression programs, believing these reduce rainfall.

There are weather modification programs based on cloud seeding in several states, including Kansas, Colorado, Texas, Oklahoma, and North Dakota.

(You can read more about cloud seeding in Daniel Pendick, "Cloud Dancers," in *Weather: What We Can and Can't Do about It,* [New York: Scientific American, 2000, pp. 622–669]. You can learn more about weather modification programs at users.pld.com/hailman/master.html.)

cloud shadow Shading of the ground by a cloud that is overhead. This is the primary cause of reduced sunshine. Because climate MODELS are unable to include details of processes that take place on the small scale of CLOUD FORMATION, the models must include broad assumptions about the effect of cloud shadow. Despite its importance, cloud shadow is one of the least understood phenomena in theories of climate change.

cloud street A row of small fair weather CUMULUS clouds that are aligned with the wind direction. Cloud streets most often form in the early morning and evening, and they require a wind speed of more than about 13 mph (6 m s^{-1}).

As the wind flows across the warm ground, small irregularities in the surface can trigger the development

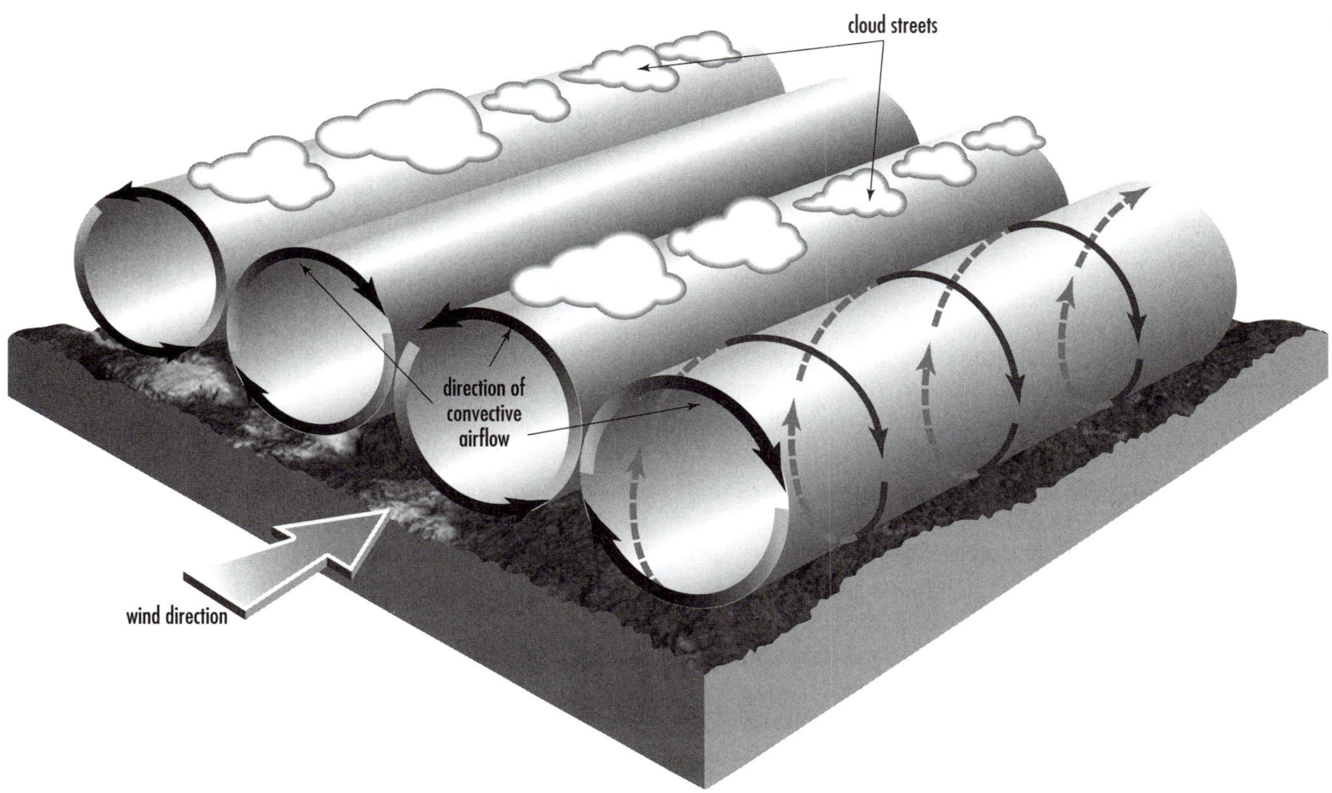

Cloud streets. **The small cumulus clouds are in rows parallel to the wind direction. This pattern is due to the spiraling airflow produced by the combination of convection and wind.**

of THERMALS that are carried downwind. The thermals form a series of CONVECTION CELLS, but in the early morning and evening, when the ground surface is cooler and CONVECTION is less vigorous, a breeze may cause the cells to merge into a series of convective spirals traveling downwind.

Air cools as it is carried upward in the spirals. If, at the top of each spiral, it cools to below its DEW POINT TEMPERATURE some of its water vapor condenses to form a CUMULIFORM cloud. Before the cloud has time to grow, the air has descended on the downward side of the spiral and its temperature has risen to above its dew point. Consequently, small clouds form at the top of each turn in the spiral, to produce a line of clouds parallel to the wind direction.

cloud symbols A set of ideograms that are used on a STATION MODEL to indicate the type of CLOUD that has been observed. There are symbols for all 10 cloud genera (*see* CLOUD CLASSIFICATION) as well as two more, for FRACTOSTRATUS and towering CUMULUS.

cirrus	cirrostratus	cirrocumulus
altostratus	altocumulus	nimbostratus
stratocumulus	stratus	fractostratus
cumulus	towering cumulus	cumulonimbus

Cloud symbols. **There are 12 symbols. These represent the 10 cloud genera, plus fractostratus and towering cumulus.**

cloud top The highest altitude at which there is a perceptible amount of a particular CLOUD or CLOUD LAYER. An aircraft flying above the cloud top is "in the clear," although there may be another cloud layer above.

cloudy The condition in which more than 70 percent of the sky, six OKTAS, remains covered by CLOUD for at least 24 hours.

cluster map A WEATHER MAP on which the weather conditions over a large area, such as the coterminous United States, are shown for a particular day. The map is prepared by a statistical technique that "clusters" data from many weather stations. This means that the reports of particular variables, such as temperature, pressure, wind, and cloud cover, from several stations tend to be similar. These results form a cluster around a mean value. The more stations with data that fit into a cluster the more confidence meteorologists have in the cluster. On the final map, the generalized areas are indicated by different colors or shading to indicate the conditions within them.

coalescence The merging of two or more cloud droplets into a single, larger droplet. Collision and coalescence is the mechanism by which RAINDROPS form in WARM CLOUDS. Colliding droplets may bounce away from each other. Droplets of similar size usually coalesce temporarily, oscillate, and then separate into two or more smaller droplets. Droplets of widely different sizes coalesce to form a stable droplet.

coalescence efficiency The proportion of colliding cloud droplets that merge to form larger drops. Coalescence efficiency is greatest where there is the greatest difference in size between the large and small droplets. It also varies with the relative velocities of the droplets and the angle at which they collide (*see* COLLISION EFFICIENCY). The higher the coalescence efficiency, the more rain the cloud will produce. Atmospheric electricity increases coalescence efficiency by placing opposite charges on the surfaces of droplets, so they are attracted to each other.

coastal desert A desert (*see* DESERT CLIMATE) that is adjacent to a coast. The desert of Baja California; Atacama Desert, in Chile; Namib Desert, in Namibia; and parts of the Sahara Desert, in North Africa, are coastal deserts. Air that approaches them from the ocean crosses a cold ocean BOUNDARY CURRENT with UPWELLING that flows parallel to the coast. The lowest layer of air is cooled by contact with the cold water. This increases its stability (*see* STABILITY OF AIR), producing a shallow INVERSION, and often chills it to below its DEW POINT TEMPERATURE. The effect is to lower the air temperature over the land and reduce the annual temperature range. It also produces frequent fog and low cloud. Although the air is moist, however, and the RELATIVE HUMIDITY often reaches 90 percent, the clouds rarely produce rain, because the inversion inhibits CONVECTION. These are the driest of all subtropical deserts. At Iquique, Chile, for example, the average relative humidity is 81 percent, but the average annual rainfall is 1.1 inches (28 mm). Over one five-year period it did not rain at all at Iquique during the first four years; in July of the fifth year there was a shower that delivered 0.6 inch (15.24 mm) of rain.

(You can read more about these deserts in Michael Allaby, *Ecosystem: Deserts* [New York: Facts On File, 2001].)

coastal upwelling *See* UPWELLING.

Cobra A TYPHOON, generating winds up to 130 mph (208 km h^{-1}) and waves up to 70 feet (21 m) high, that occurred in the Philippine Sea in December 1944. It struck a U.S. naval fleet, causing three destroyers to sink and destroying 150 carrier-borne aircraft. It caused the deaths of 790 sailors.

coefficient of haze (COH) A measure of the degree of AIR POLLUTION that is caused by small particles. It is calculated from the proportion of the total amount of light that is transmitted through a filter paper that has been exposed to the air for a specified length of time.

coefficient of viscosity The force per unit area, applied at a tangent, that is needed to maintain a unit relative velocity between two parallel planes set a unit distance apart in a fluid. It is measured in newtons per square meter per second (N m^{-2} s^{-1}) in the SYSTÈME INTERNATIONAL D'UNITÉS (SI) system and in dynes per square centimeter per second (dyn cm^{-2} s^{-1}) in the C.G.S. SYSTEM.

COH *See* COEFFICIENT OF HAZE.

cohesion *See* SOIL MOISTURE.

COHMAP *See* COOPERATIVE HOLOCENE MAPPING PROJECT.

col The area between two centers of high or low pressure where the pressure gradient is low. It is the region of highest pressure between two CYCLONES and of lowest pressure between two ANTICYCLONES. On a CONSTANT-PRESSURE chart a col is shaped like a saddle.

cold anticyclone (cold high) An ANTICYCLONE in which the air at the center is at a lower temperature than the surrounding air. The SIBERIAN HIGH that forms in winter is an anticyclone of this type. The high pressure at the surface weakens with increasing height and pressure is low in the upper air. A cold anticyclone rarely extends above about 8,000 feet (2,450 m).

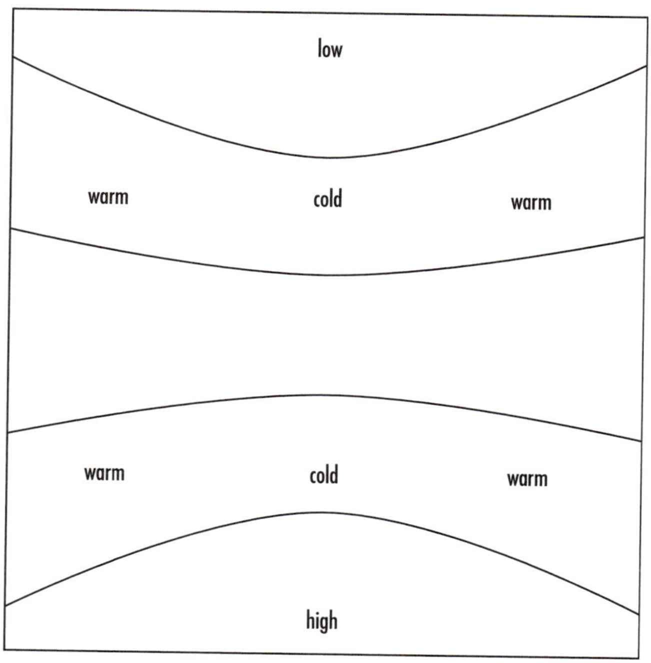

Cold anticyclone. **High pressure at the surface produced by a column of cold air draws air downward, causing the high pressure to weaken with increasing height until low pressure forms in the upper air.**

cold cloud A cloud in which the temperature is below freezing throughout. *See* BERGERON–FINDEISEN MECHANISM.

cold core The center of a DEPRESSION associated with fronts that have occluded (*see* OCCLUSION). The temperature difference between the warm air at the center and the cooler surrounding air has disappeared and the depression is located on the poleward side of the warm AIR MASS.

cold front A boundary between warm and cold air that advances with the warm air ahead of the cold air. Fronts are named *warm* or *cold* in respect to the air behind them, and the designation is relative. Air behind a cold front is cooler than the air ahead of the front, but *cold* implies no particular temperature. The cold air, being denser, forms a wedge beneath the warm air, lifting it. A cold front slopes at an angle of about 2°, the FRONTAL ZONE is 60–120 miles (100–200 km) wide, and the front advances at an average speed of 22 mph (35 km h^{-1}). A cold front is shown on a weather map as either a blue line or a black line with a row of black triangles along its leading edge.

The weather associated with a cold front varies, depending on whether it is an ANAFRONT or a KATAFRONT. In an anafront, which is the more active type, cloud occurs all the way to the top of the front, at the TROPOPAUSE. Much of the cloud is hidden from an observer on the ground by the complete cover of low-level cloud in the WARM SECTOR ahead of the front. Precipitation falls throughout a belt about 125 miles (200 km) wide.

cold front thunderstorm A THUNDERSTORM that is produced on a COLD FRONT. As cold air pushes beneath warm air, or warm air rises over cold air, moist air in the WARM SECTOR may become sufficiently unstable to generate CUMULONIMBUS clouds that are vigorous enough to cause storms.

cold glacier (polar glacier) A GLACIER in which the temperature of the ice at the base is well below the PRESSURE MELTING point at all times of year. Cold glaciers occur in parts of Antarctica. They flow as a consequence of internal deformation of the ice, which is "squeezed" outward by the weight of overlying ice.

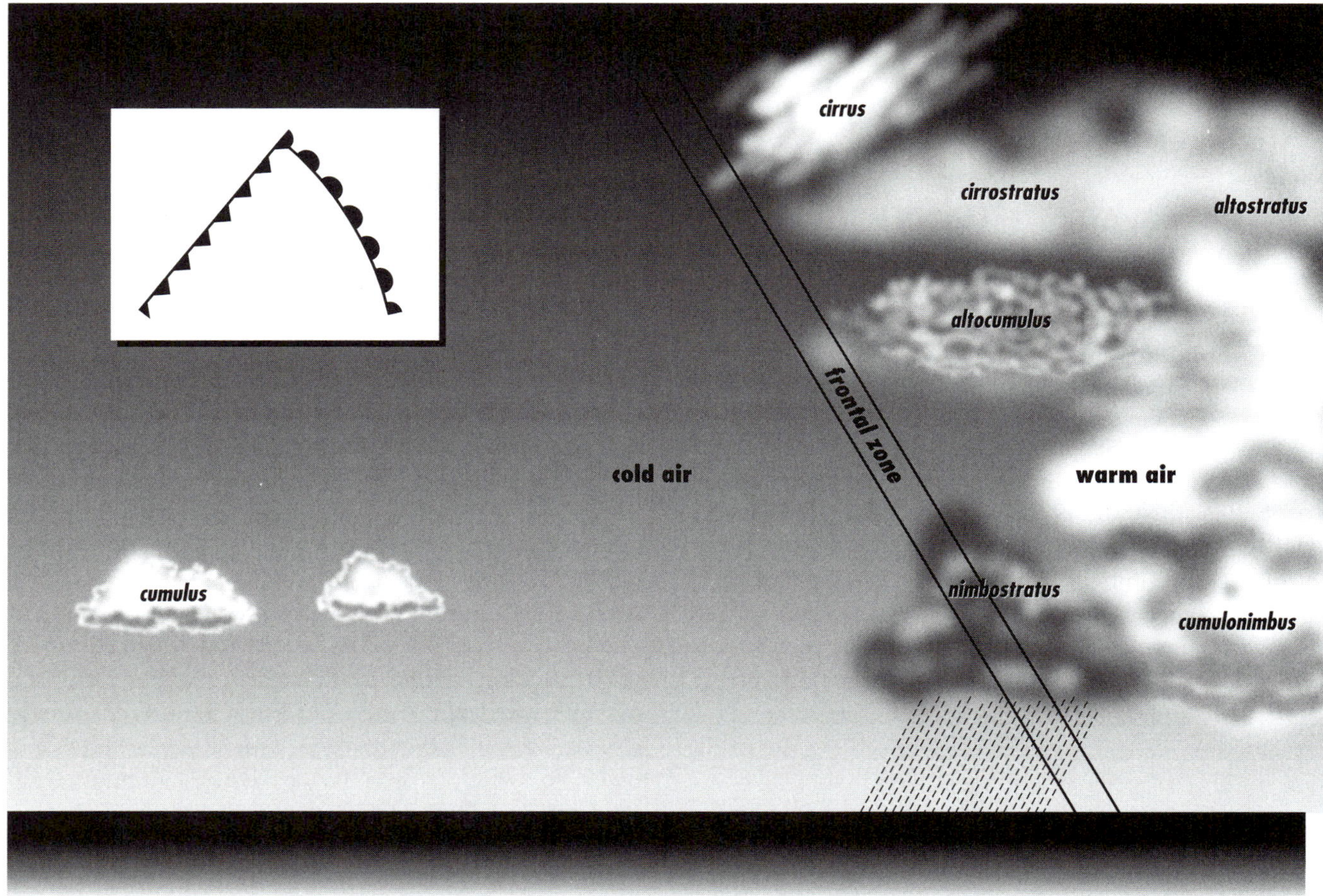

Cold front. **Cold air lies behind the front and warm air rises ahead of it, producing stratiform cloud. A cold front is shown on a weather map as a line with a row of triangles.**

cold high *See* COLD ANTICYCLONE.

cold lightning LIGHTNING that does not ignite forest fires, in contrast to HOT LIGHTNING. The current carried by the lightning stroke releases heat beneath the bark of trees. This can cause local explosions that strip away sections of the bark, but the current is not sustained into the RETURN STROKE.

cold low (cold pool) A DEPRESSION that consists of cold air surrounded by warmer air at a higher pressure. Cold lows often form in winter in the middle TROPOSPHERE over northeastern North America and northeastern Siberia, and they are usually persistent. They probably result from strong vertical movement and associated ADIABATIC cooling in OCCLUSIONS along the coast. Surface pressure may be high or low beneath the low. Cold lows produce extensive MIDDLE and HIGH CLOUD that reduces the rate at which the ground cools by radiation. They also form in lower latitudes when pools of POLAR AIR are isolated during the latter stages of the INDEX CYCLE. This type of cold low produces CUMULIFORM cloud with showers and THUNDERSTORMS in summer.

cold pole One of the places that experience the lowest mean temperatures on Earth. The cold poles do not coincide with the geographical North or South Poles. Water movements transport heat through sea water and through PACK ICE. Consequently, the climates of coastal areas and islands are strongly influenced by the adjacent sea. There is no land at the geographical North Pole, so its climate is of the MARITIME type. Winters are colder in places with CONTINENTAL CLIMATES, so, although summers are also warmer, in polar regions the annual mean temperature is likely to be lower over a continental landmass than over an island, coastal region, or ocean. The geographical South Pole lies close to the center of the large continent of Antarctica, and its climate is, therefore, continental, but the geographical and cold poles still do not coincide.

Surface atmospheric pressure is permanently high over both polar regions, but the centers of these high-pressure cells are some distance from the geographical poles and over landmasses. Air is subsiding and diverging and often there is a temperature INVERSION below a height of about 3,300 feet (1,000 m). Above the inversion air is diverging. The cells are not strong features, but they intensify the CONTINENTALITY, and it is this that produces extremes of temperature.

In the Southern Hemisphere, the cold pole is at the VOSTOK STATION, where the temperature on July 21, 1983, was –128.6° F (–89.2° C). This is the lowest surface temperature that has ever been recorded anywhere on Earth. It was exceptional, of course. August is usually very slightly colder than July at Vostok, with an average temperature of –89.6° F (–67.6° C). The annual mean temperature is –67.1° F (–55.1° C). The annual mean temperature at the geographical South Pole is –76° F (–60° C). There is also a wide temperature range. In January, which is the warmest month of the Antarctic summer, the average temperature at Vostok is –25.7° F (–32.1° C).

The Northern Hemisphere cold pole is at Verkhoyansk, Siberia, at 67°34'N 133°51'E. This is much farther from the geographical Pole than Vostok, because there is no continental landmass at the pole itself. Its mean annual temperature is 1.1° F (–17.2° C). January is the coldest month, when the mean temperature is –58.5° F (–50.3° C); the lowest temperature ever recorded there is –89° F (–67° C). Summers are warm, however. In the warmest month, July, the mean temperature is 56.5° F (13.6° C). This is the average of daytime and nighttime temperatures, and the mean daytime temperature is 66° F (19° C); the highest ever recorded was 98° F (37° C).

There are two continents in the Northern Hemisphere, Eurasia and North America. This means there is also a cold pole in North America. It is at Snag, Yukon, in northwestern Canada, at about 62°22'N 140°24'W. At Snag airport the temperature fell to –81° F (–63° C) in February 1947. This is the lowest temperature ever recorded in North America. January is usually the coldest month, when the average temperature is –18.5° F (–28.1° C). The mean annual temperature is 21.6° F (–5.8° C). July is the warmest month. Then the mean temperature is 57.0° F (13.9° C).

Elevation affects temperature. Vostok Station is 11,401 feet (3,475 m) above sea level, the elevation of Verkhoyansk is 328 feet (100 m), and that of Snag is 1,925 feet (587 m). When these differences are taken into account the temperature differences among the three cold poles becomes less marked.

cold pool *See* COLD LOW.

cold sector The cold air that partly surrounds the WARM SECTOR during the development of a FRONTAL SYSTEM. Once the system reaches the OCCLUSION stage and all of the air in the warm sector is lifted above the surface, the whole of the surface air composes the cold sector.

cold, snowy forest climate *See* HUMID CONTINENTAL CLIMATE and CONTINENTAL SUBARCTIC CLIMATE.

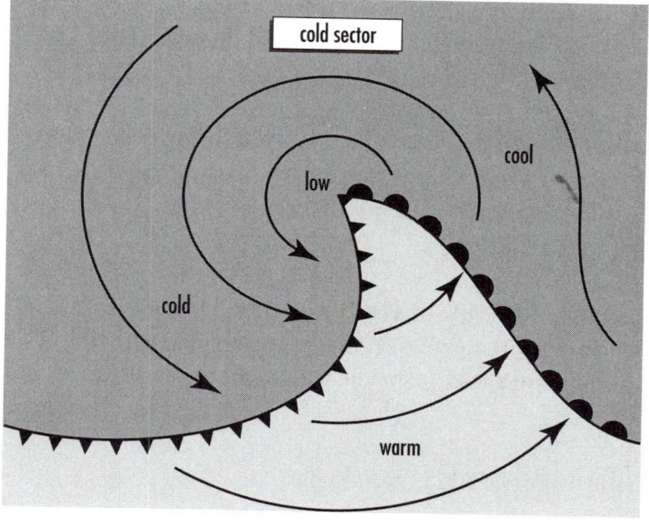

Cold sector. **As the cold front (on the left) advances into the warm air, cold air surrounds the wedge of warm air.**

cold tongue A narrow, tonguelike extension from a cold AIR MASS in the direction of the equator.

cold water sphere **(oceanic stratosphere)** That part of the ocean where the temperature is lower than 46° F (8° C). This is the oceanic equivalent of the STRATOSPHERE.

cold wave A sudden and large drop in temperature. Although cold waves happen throughout the middle latitudes, it is only in the United States that they are defined precisely. Over most of the United States a cold wave is defined as a temperature decrease of at least 20° F (11° C) that occurs over a period not exceeding 24 hours and reduces the temperature to 0° F (–18° C) or lower. In California, Florida, and the Gulf Coast states, the temperature drop must be of at least 16° F (9° C) to a temperature of 32° F (0° C) or lower. On average three or four cold waves affect the United States every winter. Cold waves cause more deaths than any other weather phenomenon. They are caused by the undulations in the POLAR FRONT JET STREAM that develop toward the end of the INDEX CYCLE. As the undulations extend southward, polar air follows, behind the POLAR FRONT. Pressure is low in the wave and the circulation around the edges of the trough is CYCLONIC. This draws air from the far north southward along the western side of the trough, making this the colder side. Cold waves can carry Canadian winter temperatures as far as the Gulf of Mexico. Cold waves are shallow, especially over the central and southern states, where they seldom extend higher than about 3,000 feet (900 m).

colla **(colla tempestade)** A southerly or southwesterly MODERATE GALE, blowing at up to 39 mph (63 km h^{-1}), with heavy rain and severe SQUALLS that occurs over the Philippines.

collada A MODERATE GALE that blows from the north or northwest over the northern part of the Gulf of California and from the northeast over the southern part.

colla tempestade *See* COLLA.

collision efficiency The proportion of droplets in a cloud that collide with other droplets. Although it might seem that water droplets are crowded closely together inside a cloud, in fact they are widely separated in relation to their own size and so collisions are by no means inevitable. A large droplet has a higher TERMINAL VELOCITY than a small one and therefore falls faster. A WARM CLOUD therefore consists of relatively large droplets falling through smaller ones. In order to collide, the small droplets must be very close to the center of the path followed by the large ones. If they are not, the displacement of air by the large droplets sweeps the small ones to the sides and away and collisions do not occur. Collision efficiency increases with the size of the large droplets. Droplets smaller than 20 μm in diameter are swept aside without colliding. Droplets more than about 40 μm across collide with most of the small droplets in their path. The higher the collision efficiency, the more rain the cloud produces. Collision does not necessarily mean the droplets will coalesce (*see* COALESCENCE EFFICIENCY).

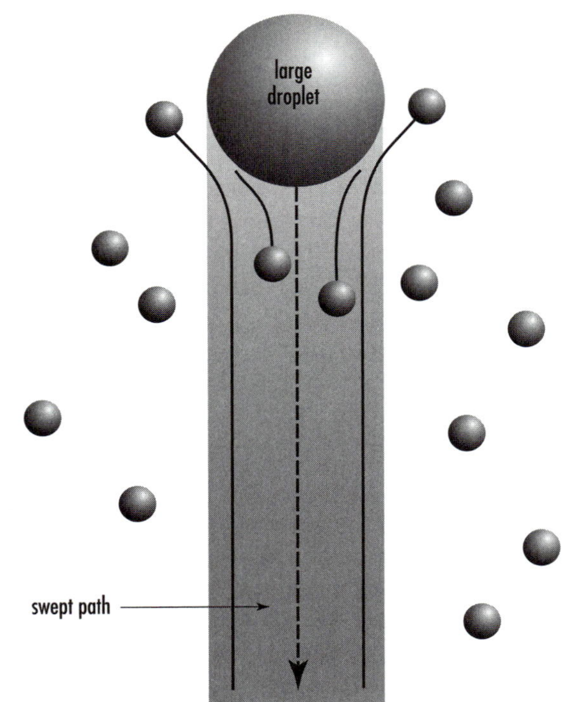

Collision efficiency. **As it falls, the large droplet collides with the smaller droplets in its path. Only those small droplets that are close to the center of the path swept by the large droplet collide with it. Those farther away are carried to the sides by the airflow.**

collision theory The theory that describes the way RAINDROPS form in WARM CLOUDS, where no ICE CRYSTALS are available and therefore the BERGERON–FINDEISEN MECHANISM does not apply. Warm clouds contain droplets of varying sizes. The larger ones fall through the cloud at a TERMINAL VELOCITY that varies with their size, so that large droplets fall faster than smaller ones. As they fall, the large droplets collide and coalesce with smaller droplets in their path. The rate at which collisions and coalescence occurs is measured by the COLLISION EFFICIENCY and COALESCENCE EFFICIENCY.

colloid Two homogeneous substances in different phases (solid and liquid, solid and gas, or liquid and gas) that are thoroughly mixed. A cloud is a colloid consisting of water droplets (liquid) distributed in air (gas).

colloidal instability The property of clouds, which are COLLOIDS, that causes water droplets to aggregate into larger drops (*see* COLLISION THEORY).

Colorado low An area of low pressure that sometimes develops on the eastern slopes of the Rocky Mountains, in the state of Colorado. Air passing over the mountains develops a CYCLONIC circulation, and then the system moves eastward, producing storms with heavy precipitation. It is similar to the ALBERTA LOW.

colpus *See* EXINE.

combustion A chemical reaction in which an element combines rapidly with oxygen and energy is released in the form of heat, light, or both. Certain elements, such as sodium and uranium, are oxidized spontaneously as soon as they are exposed to the oxygen in air and burn with a brilliant flame. Other substances must be heated before they ignite. All of the materials we use as fuel are of this type. They include petroleum, natural gas (primarily methane), coal, peat, wood, and other plant material. The key ingredients in all of them are hydrocarbons, which are compounds of carbon (C) and hydrogen (H). These are oxidized to carbon dioxide (CO_2) and water vapor (H_2O) with the release of energy:

$$C_xH_y + O_2 \rightarrow CO_2 + H_2O + \text{energy}$$

When a fuel is heated, for example by applying a match to it, some of the hydrocarbons are vaporized. Among the molecules of the hydrocarbon vapor there are some that react with oxygen molecules. That reaction releases energy, some of which is absorbed by other hydrocarbon molecules. The additional energy makes the molecules move faster and increases the violence of the impact when two molecules collide. Some of the collision impacts are violent enough to break a molecule into smaller molecules or atoms, producing IONS or free radicals, which are groups of atoms with unpaired electrons. Ions and free radicals are highly reactive. They combine rapidly with gas molecules. This facilitates the oxidation of those molecules, releasing more ions and free radicals. The result is a chain reaction, in which the oxidation of one molecule accelerates the oxidation of others. Once the reaction has started it advances very rapidly as a wave, moving in all directions.

As atoms absorb the energy that is released by oxidation, some of their electrons are excited; that means they jump to higher energy levels (orbitals). Then they fall back to their previous level. As it drops to a lower orbital, each electron emits the energy it absorbed as a photon of light. When large numbers of photons are emitted, they are visible as a flame. The WAVELENGTH of the light the electrons emit is proportional to the amount of energy being absorbed and released. Consequently, the color of a flame is an indication of its TEMPERATURE.

Hydrocarbons burn only if they are mixed with air in the correct proportions. Gas may not burn at all at the center of a flame, because there is insufficient oxygen to sustain combustion. The edge of a flame marks the boundary where there is too little fuel to sustain combustion. One consequence of this is that not all the fuel is consumed by combustion. Except in very efficient industrial incinerators, the combustion of fuel releases unburned hydrocarbons into the air. These may condense to form SOOT particles, which consist mainly of carbon, or participate in further reactions that lead to the formation of PHOTOCHEMICAL SMOG.

Although carbon dioxide and water are the only products of the combustion of hydrocarbons, hydrocarbons in fuels are attached to a large variety of other substances, some of which burn. Sulfur, for example, is oxidized to SULFUR DIOXIDE. If combustion generates high enough temperatures, as it does in modern car engines, atmospheric NITROGEN is oxidized, releasing NITROGEN OXIDES. Other components of fuel do not burn. They are vaporized, but the vapor quickly condenses into fine, solid particles of ash that may contain traces of many elements, including metals.

Although the combustion of hydrocarbons is a convenient way to obtain energy, it is also by far the largest cause of AIR POLLUTION. Natural FIRE also causes serious pollution.

(You can read more about combustion in Michael Allaby, *Elements: Fire* [New York: Facts On File, 1993].)

combustion nucleus A CLOUD CONDENSATION NUCLEUS that was released by COMBUSTION, such as a particle of ash, or condensed from a gas released as a product of combustion, such as sulfate (SO_4) formed by the oxidation of SULFUR DIOXIDE (SO_2).

comfort index *See* TEMPERATURE–HUMIDITY INDEX.

comfort zone The range of temperatures within which humans feel comfortable. For most people this is between about 65° F (18° C) and 75° F (24° C). Adjustments have to be made at temperatures higher or lower than these. People add or remove layers of clothing, light fires, sweat, or shiver.

Wind and RELATIVE HUMIDITY can make the air feel warmer or colder than it really is. Wind carries warm air away from the body and causes WIND CHILL, which can be harmful in cold weather.

When the air is warm, a high relative humidity can make the temperature feel higher than it is. This is because the high humidity reduces the rate at which sweat can evaporate from the skin. LATENT HEAT for the vaporization of sweat is taken from the skin surface. This cools the skin and helps maintain a constant body temperature. If evaporation is restricted in hot weather, people feel uncomfortable and talk of the air's feeling "close," "sticky," or "oppressive." At very high temperature and humidity, this can be dangerous to health.

When the humidity is low, the air can feel warmer than it really is in winter and cooler than it really is in summer. This can also be dangerous. A fine, sunny day can feel pleasantly warm despite a temperature far below freezing. Frostbite can occur unless the face, ears, fingers, and toes are kept adequately protected.

temp.	relative humidity (%)									
°F (°C)	10	20	30	40	50	60	70	80	90	100
80 (27)	75	77	78	79	81	82	85	86	88	91
85 (29)	80	82	84	86	88	90	93	97	102	108
90 (32)	85	87	90	93	96	100	106	113	122	
95 (35)	90	93	96	101	107	114	124	136		
100 (38)	95	99	104	110	120	132	144			
105 (40)	100	105	113	123	135	149				
110 (43)	105	112	123	137	150					
115 (46)	111	120	135	151						

Comfort zone. **The chart shows whether a particular combination of temperature and humidity will produce physical discomfort. Read the actual temperature in the left column and follow that row to the right to the column directly below the relative humidity, indicated in the row along the top of the chart. The number in the box is the *apparent temperature.* Below 80° F (27° C), there is no risk to most people. At 80°–90° F (27°–32° C), caution should be exercised. At 90°–106° F (32°–41° C), extreme caution should be exercised. The range 106°–130° F (41°–54° C) is dangerous; 130° F (54° C) is extremely dangerous.**

Similarly, on what feels like a pleasantly cool day in summer the temperature may be high enough to cause heat stroke in people who remain exposed to the heat for too long.

In 1979, R. G. Steadman of the National Weather Service calculated the effect of humidity on apparent temperature and used this to produce an index of heat stress. This sets out the relationships among the actual temperature, relative humidity, and apparent temperature. When the actual temperature is 95° F (35° C) and the relative humidity is 50 percent, for example, the temperature feels as though it is a very uncomfortable 107° F (42° C). If the relative humidity is only 10 percent, on the other hand, the apparent temperature is 90° F (32° C) and there is a danger people might remain in the open rather longer than they would if the air were moister, and that additional exposure could be harmful to them.

The National Weather Service categorizes apparent temperatures according to the heat stress they impose. It recognizes four categories: caution, extreme caution, danger, and extreme danger.

Below an apparent temperature of 80° F (27° C), there is no risk to most people.

At an apparent temperature of 80°–90° F (27°–32° C), caution should be exercised. Prolonged exposure to this range of apparent temperature combined with physical activity may cause fatigue in some people.

At an apparent temperature of 90°–106° F (32°–41° C), extreme caution should be exercised. Prolonged exposure combined with physical activity can cause sunstroke, heat cramps, and heat exhaustion.

An apparent temperature of 106°–130° F (41°–54° C) is dangerous; prolonged exposure combined with physical activity is likely to cause sunstroke, heat cramps, or heat exhaustion, and heat stroke may occur.

An apparent temperature in excess of 130° F (54° C) is extremely dangerous. Sunstroke and heat stroke are likely after quite short exposure.

complex low A region of low AIR PRESSURE within which there is more than one low-pressure center.

complex system *See* CHAOS.

composite glacier (subpolar glacier) A GLACIER in which the edges are at a temperature well below the PRESSURE MELTING point throughout the year, so they are cold or polar glaciers, but in the central part the ice temperature is above the pressure melting point except during winter, so they are also temperate or warm glaciers. A composite glacier moves by pressure melting at the center and by internal deformation ("squeezing") at the edges. Glaciers in Spitzbergen are composite.

compressional warming The mechanism by which a fluid warms when it is compressed. Compression means the volume of the body of fluid is reduced and therefore the molecules the fluid comprises move closer together. This happens because they have been pushed into a small space by molecules in the surrounding fluid. When molecules collide, KINETIC ENERGY is transferred from one to another. If the colliding molecules remain in the same body of fluid and at a constant temperature the amount of kinetic energy remains unchanged, because no energy is lost in the transfer of energy between molecules. In a contracting fluid, however, the molecules in one body of fluid (in the surrounding fluid) transfer kinetic energy to molecules in a different body (the contracting body). Consequently, energy leaves the surrounding fluid and is absorbed in the body of fluid undergoing compression. The gain of kinetic energy causes the molecules to move more rapidly. When they collide with a surface, such as that of the bulb of a thermometer or of heat sensors in human skin, the impact is more violent than it was formerly. This is measured by the thermometer and felt by the skin as a rise in temperature. The opposite happens when a fluid expands. This is known as EXPANSIONAL COOLING.

computer Literally, any device that performs mathematical calculations (i.e., computations). An electronic computer accepts data fed into it in a prescribed form, processes the data, and supplies the results of its computations either by displaying them on a screen or printing them or by sending them directly to another machine or feeding them into another process. Because electronic computers operate at great speed they are widely used by meteorologists to perform the many calculations that are required to prepare a weather forecast. Climatologists use computers to construct the CLIMATE MODELS of the atmosphere with which they study past, present, and future climates.

con *See* CONGESTUS.

concrete minimum temperature The lowest TEMPERATURE that is registered by a MINIMUM THERMOMETER that remains in contact with a concrete surface for a specified period. It is used as an alternative to the GRASS MINIMUM TEMPERATURE, because it gives more uniform results and provides a better indication of the likelihood of ice formation on road surfaces.

condensation The change of PHASE in which a gas is transformed into a liquid. Gases condense when they saturate the air (*see* SATURATION). The RELATIVE HUMIDITY (RH) of the air is then 100 percent with respect to the vapor. Water is the only common substance that exists in all three phases (gas, liquid, and solid) at the TEMPERATURES found at the surface of the Earth and in the TROPOSPHERE.

RH varies with temperature. In warm air, water molecules have more energy to move freely through the air, and so the PARTIAL PRESSURE of water vapor is higher than it is in cold air, where molecules have less energy. As the air temperature falls, the partial pressure approaches the SATURATION VAPOR PRESSURE.

WATER VAPOR condenses when the RH approaches 100 percent provided there is a surface onto which it can do so. In the phase change water molecules join together to form small groups in which the individual molecules are linked by HYDROGEN BONDS. If the vapor condenses onto the ground or surface vegetation the liquid is called *DEW*, or *FROST* if it then freezes or changes directly from the gas to the solid by DEPOSITION.

Above the surface, the condensation of water produces CLOUD DROPLETS. Droplets form around very small particles, called *CLOUD CONDENSATION NUCLEI*. The first droplets to form are very small. They consist of water that merges with HYGROSCOPIC NUCLEI or that condenses onto particles with surfaces that attract water, called *wettable* AEROSOLS. Tiny droplets then grow as more water condenses onto them and as droplets collide and coalesce.

In the absence of a suitable surface, water vapor does not condense until the RH reaches about 101 percent. The air is then said to be *1 percent supersaturated* (*see* SUPERSATURATION.)

Condensation releases LATENT HEAT. This warms the air and if the condensation is occurring in air that is rising by CONVECTION and cooling in an ADIABATIC process, the release of latent heat sustains the convection. This is the mechanism by which CUMULIFORM clouds grow. It is important in the development of THUNDERSTORMS and TROPICAL CYCLONES.

condensation level *See* CONVECTIVE CONDENSATION LEVEL, LIFTING CONDENSATION LEVEL, and MIXING CONDENSATION LEVEL.

condensation temperature (adiabatic condensation temperature, adiabatic saturation temperature) The temperature at which a PARCEL OF AIR reaches SATURATION if it cools at the DRY ADIABATIC LAPSE RATE.

condensation trail *See* CONTRAIL.

conditional instability The condition of air when the ENVIRONMENTAL LAPSE RATE (ELR) is greater than the SATURATED ADIABATIC LAPSE RATE (SALR), but lower than the DRY ADIABATIC LAPSE RATE (DALR). Unsaturated air does not rise unless it is forced to do so, for example by crossing a mountain, because if it does, it cools at the DALR. This is greater than the ELR, and so the rising air immediately becomes cooler than the surrounding air and tends to sink again. While it remains unsaturated the air is stable.

Should it be forced to rise high enough to reach its LIFTING CONDENSATION LEVEL, the water vapor it carries starts to condense into cloud droplets. This releases LATENT HEAT of condensation, warming the air and slowing its lapse rate from the DALR to the SALR. This is lower than the ELR, so the rising air is always warmer than the surrounding air because it is cooling more slowly. Being warmer, it remains buoyant and continues to rise. The air is then unstable.

Some outside force must compel the air to rise before it changes from being stable to being unstable. This force is the condition needed to trigger instability, which is why the air is said to be "conditionally" unstable.

conditional instability of the second kind (CISK) The type of CONDITIONAL INSTABILITY that leads to the formation of large, long-lived clusters of CUMULONIMBUS cloud over tropical oceans. Ordinarily, conditional instability produces CUMULIFORM clouds that are isolated because each cloud is built by a CONVECTION CELL that utilizes all the moisture and energy in its immediate vicinity, thus preventing the formation of another cloud nearby.

In the Tropics it is possible for conditional instability to be intensified by horizontal air movements. This happens when there is an area of low pressure covering a surface area about 600 miles (1,000 km) across and the air is conditionally unstable. Air flows toward the low-pressure region from all sides (*see* CONVERGENCE). Friction with the surface slows it, reducing the magnitude of the CORIOLIS EFFECT so the PRESSURE-GRADIENT FORCE becomes dominant. The converging air enters the low-pressure region and rises at the center. Alternatively, an ATMOSPHERIC WAVE may move across the area. Such waves often produce convergence in some places and DIVERGENCE in others.

Convergence triggers CONVECTION. At the same time, the converging air is warm and moist, so it feeds moisture into the convection cells. It also pushes the developing clouds closer together. LATENT HEAT is released as water vapor condenses in the rising air. This warms the air, further fueling convection. Divergence above the clouds reduces the amount of air beneath and therefore reduces the surface pressure. This intensifies and sustains the low-pressure system.

This alternative type of conditional instability is involved in the formation of TROPICAL CYCLONES. The difference between it and ordinary conditional instability was first recognized in 1964 by the meteorologists Jule Charney and Arnt Eliassen. Katsuyuki Ooyama also described it; he called it *conditional instability of the second kind,* the name it has retained. Ordinary conditional instability is sometimes called *conditional instability of the first kind.*

conduction The transmission of heat through a substance from a region that is relatively warm to one that is relatively cool until both are at the same temperature. In a gas or liquid the heat is transferred by collisions between atoms or molecules. When molecules with higher KINETIC ENERGY collide with molecules having lower kinetic energy a proportion of the energy is transferred; the molecule with lower kinetic energy gains energy and the molecule with higher kinetic energy loses energy. In this way energy is exchanged until all the molecules possess the same energy. TEMPERATURE is a measure of the kinetic energy of molecules or atoms, and so a transfer of kinetic energy implies a distribution of temperature. Conduction is one of the three ways in which heat is transmitted. The others are CONVECTION and radiation.

conductive equilibrium *See* ISOTHERMAL EQUILIBRIUM.

confined aquifer *See* AQUIFER.

confluence A flow of air in which two or more STREAMLINES approach one another. This accelerates the air, because a narrowing stream must carry the flow, but unlike CONVERGENCE, it does not lead to an accumulation of air or to any vertical movement.

congestus (con) A species of CUMULUS clouds (*see* CLOUD CLASSIFICATION) that is large and growing rapidly, usually by the development of towering, billowing structures in the upper parts of the cloud. A cumulus congestus (Cu_{con}) cloud looks like a cauliflower. The name of the species is derived from the Latin verb *congere,* which means "to bring together."

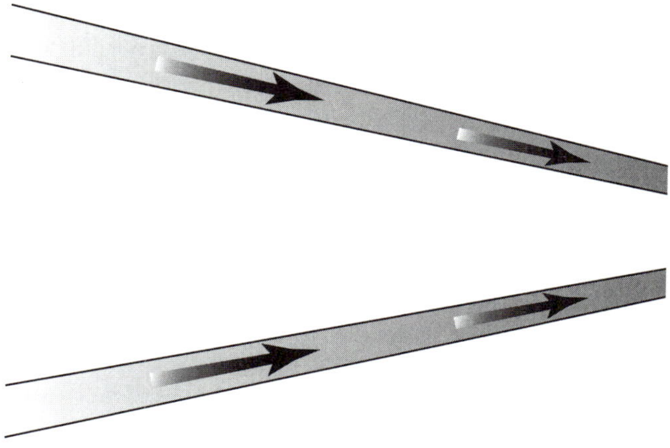

Confluence. **Two airstreams are approaching one another and accelerating.**

coning One of the patterns a CHIMNEY PLUME may make as it moves away downwind. The plume of gases and particles widens with increasing distance from the smokestack. A line through the center of the plume is horizontal. Coning occurs when the wind is fairly strong and the plume is moving through stable air. In the layer of air extending from the surface to beyond the height of the plume the DRY ADIABATIC LAPSE RATE is greater than the ENVIRONMENTAL LAPSE RATE.

connate water *See* JUVENILE WATER.

Coning. **The plume of gas and particles from a factory smokestack travels horizontally, spiraling away downwind.**

Connie A HURRICANE that struck the United States in August 1957, producing rain that saturated the ground, shortly before the arrival of Hurricane DIANE.

conservation of energy, law of *See* THERMODYNAMICS, LAWS OF.

constant gas A constituent atmospheric gas that is present in the same proportion by volume to an altitude of about 50 miles (80 km). The most abundant constant gases are NITROGEN (78.1 percent), OXYGEN (20.9 percent), and ARGON (0.9 percent). The atmosphere also contains VARIABLE GASES.

constant-height chart A SYNOPTIC CHART on which the meteorological conditions at a particular altitude, based on RADIOSONDE data, are plotted. The heights most often used are 5,000, 10,000, and 20,000 feet (1,525, 3,050, and 6,100 m). Constant-height charts help in the identification of AIR MASSES and the boundaries between them.

constant-level balloon A weather balloon that is designed to rise to a predetermined altitude and then remain there. The balloon is contained within an inelastic cover, so its volume cannot exceed a certain value. This prevents the gas in the balloon from expanding until it becomes less dense than the air at the desired height. The altitude of the balloon cannot be controlled precisely, however, because it is affected by vertical air movements and by changes of temperature and therefore volume and gas density caused by moving into and out of direct sunshine.

constant-level chart A SYNOPTIC CHART on which meteorological conditions at a particular level are plotted. The level may be defined as the altitude above sea level, in which case the chart is a CONSTANT-HEIGHT CHART, or the level at which the atmospheric pressure remains constant, in which case the chart is a CONSTANT-PRESSURE CHART.

constant-pressure chart A map that shows the distribution of atmospheric pressure, but not by means of ISOBARS. Instead, the map assumes the existence of a level surface where the pressure is constant throughout. It then shows pressure contours that indicate heights above or below this imaginary surface. The resulting chart resembles an isobaric map and can be interpreted in the same way, but its lines represent heights, not pressures. It is often a more convenient way to present the data.

constant-pressure surface (isobaric surface) A surface across which the atmospheric pressure is everywhere the same. Although such a surface is level with respect to pressure, it is not level with respect to height above the land or sea surface. Consequently the height of the constant-pressure surface can be shown by contours. Depressions in the constant-pressure surface correspond to areas of low pressure and raised areas to regions of high pressure.

contact cooling The cooling that occurs when warm air has contact with a surface at a lower temperature. This can reduce the temperature of the lower air to below the DEW POINT TEMPERATURE. Depending on the amount of moisture present and the thickness of the layer of air that is cooled, CONDENSATION produces DEW or FOG. If the air is chilled to below the FROST POINT, FROST forms.

contessa del vento A type of LENTICULAR CLOUD in which the base is rounded and the upper surface bulges. Sometimes several clouds of this type form one above the other, in a stack. The name is from the cloud that develops in a westerly airstream near Mount Etna.

continental air Air that is very dry and that forms AIR MASSES over all the continents (*see* SOURCE REGION). Continental air is hot in summer and cold in winter, except over the Arctic and Antarctic, where it is cold at all times of year.

continental arctic air *See* ARCTIC AIR.

continental climate A climate that is produced by CONTINENTAL AIR. It occurs in areas deep in the interior of continents, far from the ocean. Air loses its moisture as it crosses the continent, making continental climates dry. Because of the large difference in HEAT CAPACITY between water and land, the continental interior heats rapidly in summer and cools rapidly in winter. This produces a much wider annual temperature range than that of a MARITIME CLIMATE in the same latitude. Omaha, Nebraska, has a typical continental climate. Its annual temperature range is 73° F (41° C) and its average annual precipitation is 29 inches (737 mm). Eureka City, in Humboldt County, California, is in the same latitude as Omaha but lies on the western coast, where it enjoys a maritime climate. There the annual temperature range is 22° F (12° C) and the average annual precipitation is 38 inches (965 mm).

The climates of Omaha and Eureka are extreme examples of their types. In other places the maritime or continental influence is less extreme and its extent can be calculated as the CONTINENTALITY or OCEANICITY of the climate.

continental drift The theory that the continents have not always occupied the positions in which they are seen today and that they are still moving was first proposed in its modern form by the German meteorologist ALFRED WEGENER (1880–1930). In *Die Entstehung der Kontinente und Ozeane* (The origin of the continents and oceans), first published in 1915, Wegener described "continental displacement." The South African geologist Alexander Logie Du Toit (1878–1948) was the first person to use the term *continental drift.* Du Toit had found many similarities in the rock formations of South Africa and South America. This led him to support Wegener's theory. His book, *Our Wandering Continents: An Hypothesis of Continental Drift,* appeared in 1937.

It was not a new idea. Many people had noticed the apparent fit between the coastlines of the continents on each side of the Atlantic Ocean. The Dutch cartographer Abraham Ortelius (1527–98) suggested in 1596 that they had once been joined but had been torn apart by earthquakes and floods. In 1881, at a time when many scientists believed the Moon had been torn away from the Earth, the English geologist Osmond Fischer (1817–1914) suggested the Pacific Ocean filled the scar this had made. As Asia and America moved closer together to heal the scar, America was torn from Africa and Europe and the Atlantic opened between them.

Wegener noted that the Appalachian Mountains of North America matched the Scottish Highlands and that American coal deposits matched coal deposits in Europe. Similar fossils occurred on both continents, and there were fossils of tropical plants in Spitzbergen,

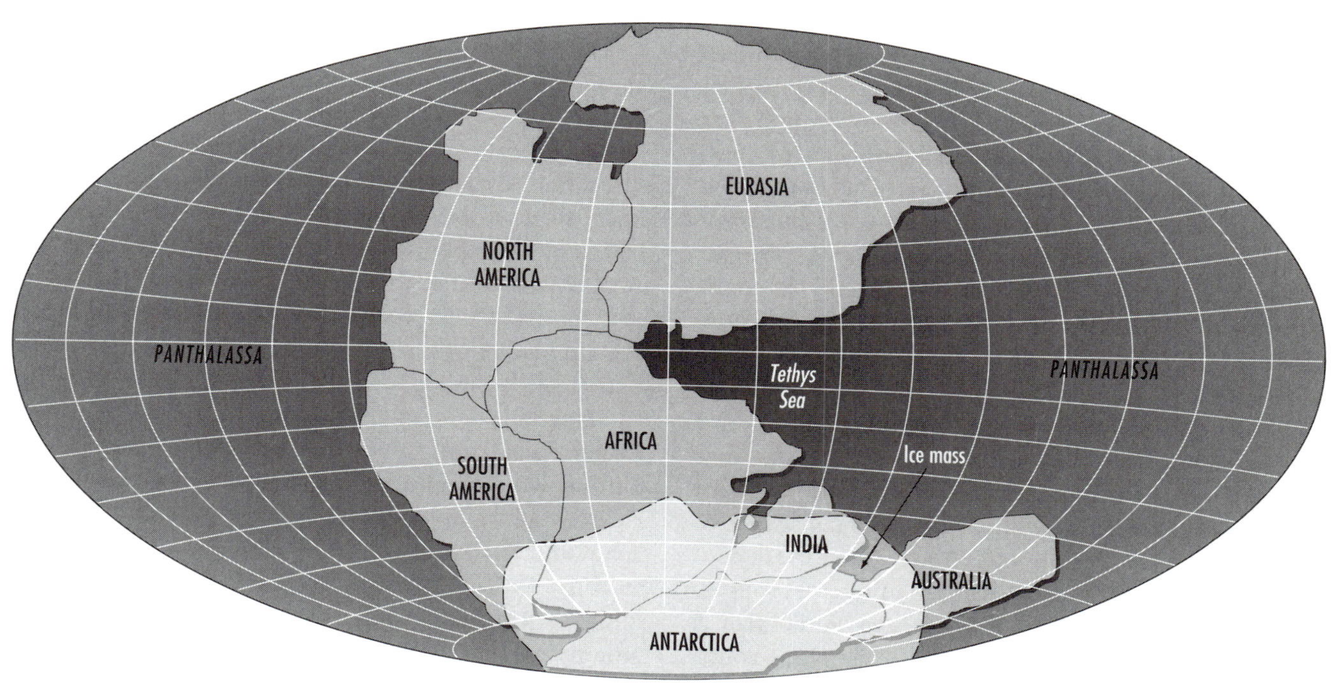

Continental drift. **Pangaea as it is believed to have been about 200 million years ago, when the supercontinent had begun to break apart and the Tethys Sea separated the northern and southern regions.**

suggesting that the climate in that part of the Arctic had once been tropical.

Measurements of the longitude of Greenland suggested to Wegener that over the course of a century it had moved away from Europe by about 1 mile (1.6 km). It also seemed that Washington, D.C., and Paris were moving apart by about 15 feet (4.6 m) a year and that San Diego, California, and Shanghai were moving toward each other by about 6 feet (1.8 m) a year. Unfortunately, the measurements were incorrect, and this helped to discredit his theory. Accurate measurements have since found the continents are moving at about one-tenth the speed he supposed.

Wegener assembled the evidence and then joined coastlines together where they seemed to fit. He concluded that about 300 million years ago all the continents had formed a single supercontinent, which he called *Pangaea,* a word meaning "all Earth" (from the Greek *pan,* "all," and *ge,* "Earth"). A single ocean, Panthalassa, which means "all sea," surrounded Pangaea. Then Pangaea started to break apart and its sections, forming the present continents, moved away from each other. South America and Africa began to separate around 150 million years ago and Australia and Antarctica separated about 40 million years ago.

In his book, Du Toit proposed that the southern continents had once been joined in a supercontinent he called *Gondwanaland,* and the northern continents had constituted Laurasia. As Pangaea began to split, an arm of the ocean separated them. This was called the *Tethys Sea.* The Mediterranean Sea is a remnant of Tethys.

It was not until the 1960s that the theory of continental drift won full scientific acceptance. It has now been incorporated in the larger theory of PLATE TECTONICS.

continental high The area of high AIR PRESSURE that covers the center of a continent. Siberia and northern Canada are covered by continental highs consisting of continental polar (cP) air (*see* CONTINENTAL AIR and POLAR AIR). These AIR MASSES are the principal sources of cold air in the Northern Hemisphere. Antarctica is also covered by a continental high consisting of continental arctic (cA) air (*see* ARCTIC AIR).

continental interior regime *See* RAINFALL REGIME.

continentality The extent to which the climate of a particular place resembles the most extreme type of CONTINENTAL CLIMATE. Although climates can be classified as continental or maritime (*see* MARITIME CLIMATES), these types grade from one into the other. Except on ocean islands and some, but not all, coasts, a place with a maritime climate also experiences more continental conditions for some of the time. Similarly, maritime influences extend a long way inland from the coasts of continents.

There is a need to refine the classification of continental and maritime climates in a way that reflects the gradations between them. The differences can be shown clearly on a HYTHERGRAPH, but only in a relative sense. The hythergraph compares the climates of two or three places (including more makes the graph too cluttered to be easily read) but gives no absolute value for the continentality of their climates. This is possible if the essential climatic features of an area, together with its latitude, can be used to calculate an index of continentality.

A simple way to achieve this is to note the extent to which the area lies beneath CONTINENTAL AIR. In 1940 the German climatologist H. Berg proposed a method based on this principle. If the frequency of continental AIR MASSES is related to the total number of air masses of all kinds in the course of a year, the result can be shown as a percentage:

$$K = (C/N) \times 100$$

where K is the index of continentality, C is the number of continental air masses that covered the area in the year, and N is the total number of air masses covering the area during the same period. Because it calculates percentages, an extreme maritime climate has a continentality index of 0 percent and an extreme continental climate has an index of 100 percent.

This index shows that continentality is only about 25 percent over most of Norway, all of Denmark, and Europe west of about 5° E longitude, a region that includes all of France west of Paris and the whole of Spain and Portugal. The index reaches only 50 percent at about 20° E. In Scandinavia the change from maritime to continental climates is abrupt, because of mountains that obstruct the eastward flow of maritime air. Elsewhere in western Europe, where there are no mountain ranges, the transition is much more gradual.

The index is easy to understand, but in order to calculate it, it is necessary to record the types of air mass. This means accurately checking the movements of air masses over several years in order to obtain an average. It is much more complicated than using the annual TEMPERATURE RANGE as the key datum. Accurate temperature records are kept for most places and they are readily available. The more continental the climate the greater the temperature range. Allowance must be made for latitude, because this affects the amount and seasonal distribution of solar radiation.

Several attempts have been made to devise a reliable method. The one that is now most widely used was introduced in 1946 by the American climatologist V. Conrad ("Usual formulas of continentality and their limits of validity" in *Transactions of the American Geophysical Union*, vol. 17, pages 663–64). According to the Conrad formula

$$K = 1.7\,A/\sin(\Phi + 10) - 14$$

where A is the average annual daytime temperature range and Φ is the latitude. The Conrad index should yield a value of 0 for a fully maritime climate and 100 for a fully continental climate. Thorshavn in the Faeroe Islands has an index of –2.5, and Verkhoyansk, in Siberia, has an index of 103. In North America, values for K range from 0 to about 10 along the western coast and are between 30 and 40 in New England. In the central United States, the values are generally between 60 and 65, but higher in some places. The value for Omaha, Nebraska, is 76.

continentality effect The production of extreme temperatures in summer and winter in water that is almost completely surrounded by a continental landmass. This produces a continental climate in places that would otherwise experience a maritime climate because of their proximity to an ocean. Hudson Bay, in northern Canada, is fully ice-covered in winter as a result of this effect, despite being farther south than parts of the Atlantic and Pacific Oceans that are free from ice. (Fort Severn, Ontario, on the shore of Hudson Bay, is at 56.58° N; Dundee, Scotland, is at 56.30° N.)

continental polar air *See* POLAR AIR.

continental subarctic climate In the STRAHLER CLIMATE CLASSIFICATION a climate in his Group 3, comprising climates controlled by polar and arctic AIR MASSES. The continental subarctic climate occurs in SOURCE REGIONS for continental polar air in latitudes 50°–70° N. In winter, the air is stable (*see* STABILITY OF AIR) and extremely cold. Summers are cool and short. There is a very large range of temperature through the year. The climate is moist, but this is because the rate of evaporation is low, not because there is heavy precipitation. Precipitation is light and falls during storms associated with FRONTAL SYSTEMS involving maritime polar air. This is a cold, snowy forest, or humid microthermal, climate. In the KÖPPEN CLASSIFICATION, if the climate is moist throughout the year, it is designated *Dfc* if summers are cool and *Dfd* if summers are cool but winters extremely cold. If the winter is dry it is designated *Dwc* if the summers are cool and *Dwd* if winters are very cold.

continental tropical air *See* TROPICAL AIR.

continental drought (**accidental drought**) One of the types into which DROUGHTS are formally classified. It is unpredictable and can occur anywhere, and its end is no more predictable than its start. If it is especially severe it is known as a *DEVASTATING DROUGHT.*

continuous-wave radar *See* RADAR.

contour line A line on a weather map that joins points where the value of some atmospheric feature is the same. Such lines have names beginning with *iso-*, from the Greek work *isos,* which means "equal." The lines resemble the contour lines joining places at the same elevation on a topographic map. ISOBARS join points where the AIR PRESSURE is the same, ISOHYETS join points where PRECIPITATION is equal, ISOTHERMS join places where the TEMPERATURE is the same, and there are several more.

contour microclimate A variation in a MICROCLIMATE that is directly due to a difference in elevation of the area experiencing it.

contrail (**condensation trail, vapor trail**) A long, narrow cloud that is produced by the exhaust from an aircraft engine. Exhausts from piston engines can produce contrails, but they are more often associated with jet aircraft. This is partly because jet engines burn more fuel than piston engines and consequently emit larger quantities of water vapor, and partly because they fly at higher altitudes, where contrail formation is more likely.

The water vapor is a product of the combustion of gasoline and kerosene. These are hydrocarbons (containing hydrogen and carbon), and combustion is the oxidation of carbon (C) and hydrogen (H) in an exothermic (energy-releasing) reaction.

$$C + O_2 \rightarrow CO_2 \quad \text{and} \quad 4H + O_2 \rightarrow 2H_2O$$

The hot, moist exhaust gas is pumped into cold air and immediately starts to cool. Cloud forms if sufficient CLOUD CONDENSATION NUCLEI (CCN) are present and the mixture of air and gas cools to below its DEW POINT TEMPERATURE. Engine exhausts release particles of unburned fuel and soot that act as CCN, so cloud forms provided the air reaches SATURATION. This is unlikely in the lower part of the TROPOSPHERE when the sky is clear, because the temperature is high enough for the water vapor in the exhaust to be absorbed without saturating the air. At high altitude, however, the air is very cold and therefore soon saturated. Most contrails form above 6,000 m (20,000 feet), but the precise altitude at which they form varies from day to day. They are usually composed of ice crystals. It takes a few seconds for the exhaust gases to cool to the dew point temperature, which is why contrails begin a short distance behind the airplane producing them.

Usually, high-level winds quickly disperse contrails by mixing the cloud with drier surrounding air until the RELATIVE HUMIDITY of the mixture falls below 100 percent and the ice crystals sublime (*see* SUBLIMATION). If the upper air is close to saturation, contrails survive much longer and spread into CIRRUS or CIRROSTRATUS cloud. Contrails that remain visible for an hour or longer indicate the presence of moist air aloft. This is probably air in the warm sector of an approaching frontal system, which indicates that the amount of cloud is likely to increase and precipitation will commence in a few hours. Where the air traffic is heavy, contrails increase cloudiness by an appreciable amount.

contrail-formation graph A graph on which atmospheric pressure, temperature, and RELATIVE HUMIDITY are plotted in order to determine the height at which CONTRAILS will form. This is of military importance, because contrails make high-altitude aircraft clearly visible.

contrastes Winds that blow along the shore, but from opposite directions on the two sides of a headland. They occur in winter along the Mediterranean coast of Europe.

control day A day on which folk custom holds that the weather will determine the weather over the coming weeks, months, or season. Many control days are associated with Christian saints or festivals. Candlemas Day (February 2), which is traditionally celebrated with lighted candles, is one example:

If Candlemas be fair and bright,
Winter will have another flight.
But if Candlemas Day be clouds and rain,
Winter is gone and will not come again.

The weather on Easter Day and that on Christmas Day are believed to be linked:

Easter in snow, Christmas in mud,
Christmas in snow, Easter in mud.

See also ST. BARTHOLOMEW'S DAY, ST. HILARY'S DAY, ST. MARY'S DAY, ST. MICHAEL AND ST. GALLUS, ST. PAUL'S DAY, ST. SIMON AND ST. JUDE, ST. SWITHIN'S DAY, ST. VITUS'S DAY.

control-tower visibility The VISIBILITY that is measured from an airfield control tower.

convection Transport of heat that occurs through vertical motion within a fluid. A convective circulation is driven by gravity. When a fluid is heated from below, the warmed fluid at the bottom expands and becomes less dense. Cooler, denser fluid descends from above and displaces it. Consequently, the warm fluid rises and its place at the bottom is taken by cool fluid. As the warm fluid rises, it moves away from the source of heat and cools. At the same time, the cool fluid that has sunk to the bottom is warmed. This now becomes warm water, the water that has risen becomes cool water, and they exchange places. This vertical circulation is maintained for as long as heat continues to be supplied at the bottom or, if the fluid is a liquid, the whole of the liquid has vaporized.

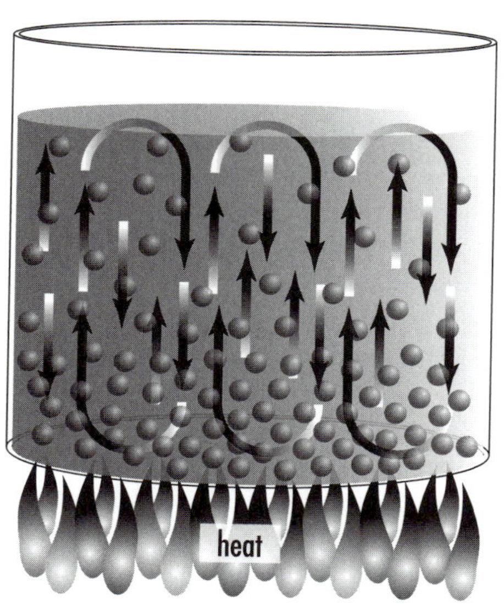

Convection. **As heat is applied at the base, water in the pot is heated at the bottom and expands. Cooler, denser water sinks to replace it. This motion establishes a vertical circulation.**

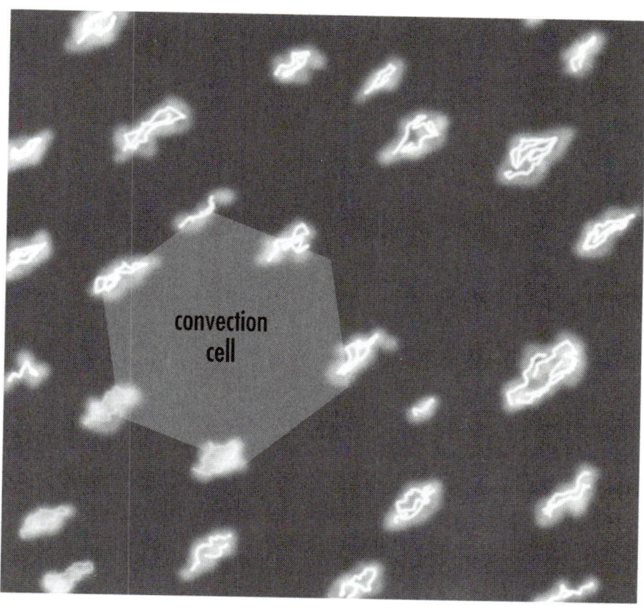

Convection cells. **Convection cells may be open or closed. These are open cells, with clear sky at the center surrounded by cumulus clouds.**

Convection is one of the three processes by which heat is transported from one place to another (the other two are CONDUCTION and radiation). Convection occurs only in fluids, because molecules must be free to move in relation to each other. It is the mechanism that drives the GENERAL CIRCULATION of the atmosphere and the formation of CUMULIFORM clouds.

convectional precipitation PRECIPITATION that results from thermal CONVECTION in moist air. The precipitation usually takes the form of rain, snow, or hail SHOWERS, which are sometimes heavy.

convection cell A form of circulation in which warm air rises, cools, and subsides again. Individual convection cells cover a fairly small horizontal area, from about 12 to 125 miles (20–200 km) in diameter, but they often occur in large groups, together occupying a large area. Columns of rising and descending air possess opposite VORTICITIES that may repel each other, keeping the columns separate.

Cells may be open or closed. An open cell appears as a patch of clear sky that is surrounded by an approximately hexagonal pattern of CUMULUS clouds. A closed cell consists of a hexagonal layer of STRATOCUMULUS that is surrounded by clear sky. Closed cells are smaller than open cells, both horizontally and vertically. Open cells are very common over the oceans and also form on the western sides of the SUBTROPICAL HIGHS, where air is moving eastward and toward the Pole and becoming unstable (*see* STABILITY OF AIR). Closed cells are most common on the eastern sides of the subtropical highs, where there is often a shallow INVERSION caused by subsiding air and UPWELLINGS lower the temperature of the sea surface.

convection street A type of CONVECTION CELL that is drawn out into long, parallel lines by a strong wind near the surface and WIND SHEAR at a high level. Convection streets can extend for 60 miles (100 km) or more over the ocean. Over land they are less regular and shorter and are usually aligned downwind from sloping ground that faces the Sun and is therefore a good source of THERMALS.

convection theory of cyclones A theory proposing that CONVECTION, especially of moist air, can affect a large enough area and be sustained long enough for air that is drawn in at low level to acquire a distinct CYCLONIC flow. This can lead to the development of a CYCLONE.

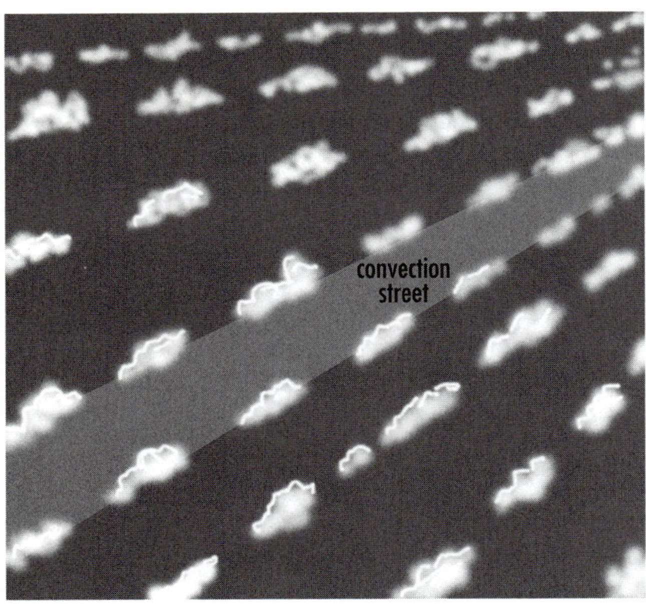

Convection street. **Convection can produce parallel lines of cumulus clouds that extend for many miles over the ocean, although they are shorter over land. The streets often exist even when the air is too dry for clouds to form.**

convective available potential energy *See* STABILITY INDICES.

convective cloud A CLOUD that develops vertically as a result of convection. CUMULIFORM clouds are of this type.

convective condensation level The height at which CONDENSATION occurs in a PARCEL OF AIR that is rising by CONVECTION, if the parcel becomes saturated while it is rising through air in which temperature falls at the DRY ADIABATIC LAPSE RATE and there is CONDITIONAL INSTABILITY above the height at which the parcel becomes saturated.

convective equilibrium (**adiabatic equilibrium**) The condition in a tall column of air that is being mixed predominantly by mechanical processes, such as air movements, and by CONVECTION. Because of mixing, any PARCEL OF AIR within the column is at the same temperature and pressure as the air around it. If it is displaced vertically, the parcel expands as it rises and is compressed as it sinks, so it remains at the same temperature and pressure as the surrounding air and the temperature decreases with height at very close to the DRY ADIABATIC LAPSE RATE. Convective equilibrium also ensures that despite their different weights, the molecules of atmospheric gases remain mixed and do not separate, and that solid particles present in the air remain evenly distributed. Most of the atmosphere in the TURBOSPHERE is in convective equilibrium. Above the TURBOPAUSE the air is in DIFFUSIVE EQUILIBRIUM.

convective inhibition The amount of energy a PARCEL OF AIR needs in order to reach its LEVEL OF FREE CONVECTION when its rise by CONVECTION is checked by an INVERSION; i.e. the energy needed to overcome the inhibition.

convective instability (**potential instability, static instability**) INSTABILITY that is caused by CONVECTION. It happens when the atmosphere is stratified, but the POTENTIAL TEMPERATURE decreases with height, and as a layer of moist air rises the lower part of the layer becomes saturated (*see* SATURATION) before the upper part. Rising air in the lowest layer, which is not yet saturated, is then cooling at the DRY ADIABATIC LAPSE RATE (DALR), saturated air in the layer above it is cooling at the shallower SATURATED ADIABATIC LAPSE RATE (SALR), and air in the uppermost layer is cooling at the DALR. If the lapse rate through the entire layer of rising air is greater than the SALR, saturated air is always rising into air that is cooler. It is always less dense, and therefore lighter, than the surrounding air; consequently its upward movement accelerates. The entire layer is then unstable and may overturn. As it continues to rise, however, it also cools and its DENSITY increases, so eventually the atmosphere becomes statically stable (*see* STATIC STABILITY).

convective region An area of the land or sea surface above which air is rising by CONVECTION, or where convection is especially common.

convective stability *See* STATIC STABILITY.

convergence A flow of air in which STREAMLINES approach an area from different directions. This happens when air flows into a region of low pressure. The effect is to accumulate air where the streamlines meet. This increases the quantity of air in that area and therefore increases the atmospheric pressure. The increased

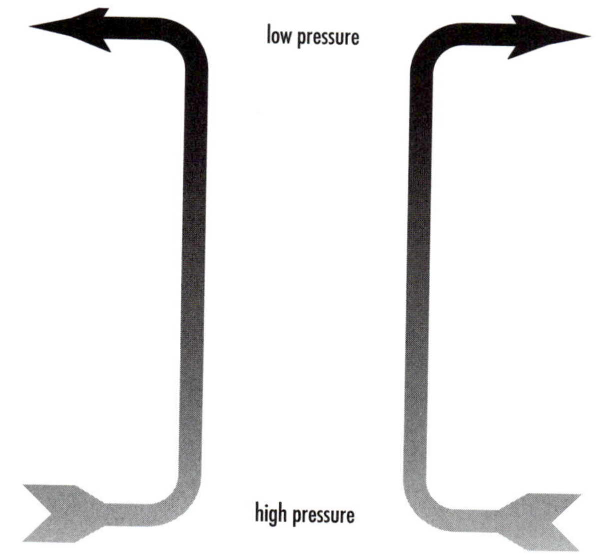

Convergence. **Air is approaching an area from several directions. This motion produces high pressure.**

pressure produced by convergence near the surface of land or sea causes air to rise, so a region of low-level convergence is also a region of rising air. The rising air eventually reaches an INVERSION level that constitutes a ceiling beyond which it can rise no farther. If the vertical motion is strong enough, air may rise all the way to the TROPOPAUSE. There air spreads out, moving away from an upper-level area that corresponds to the area of low-level convergence. This produces a region of DIVERGENCE and falling pressure above the area of low-level convergence and rising pressure.

Air that rises in a region of convergence cools in an ADIABATIC process as it does so. This favors the formation of CUMULIFORM clouds and PRECIPITATION.

convergence line A horizontal line along which CONVERGENCE is occurring.

conveyor belt A wind that blows up the slope of a front. The warm conveyor belt is a wind that blows at a height of about 3,000 feet (1,000 m) a little way ahead of a COLD FRONT at a speed of 55–65 mph (20–30 m s^{-1}). The wind carries air over the WARM FRONT and then turns eastward until it is approximately parallel to it. As it rises up the very shallow slope of the front, the moist air cools in an ADIABATIC process until it is saturated (*see* SATURATION) and STRATIFORM cloud forms. The air continues rising until it merges into the NIMBOSTRATUS layer associated with the cold front. The cold conveyor belt blows ahead of the warm front, carrying air from the cold sector. At first it blows approximately northwestward, parallel to the front but then turns in an ANTICYCLONIC manner. It often produces some MIDDLE CLOUD to the northwest of the DEPRESSION. (The directions refer to the Northern Hemisphere and should be reversed for the Southern Hemisphere.)

cooling degree days A measure that is used in calculating the amount of power that is needed to cool buildings. It assumes that 65° F (18° C) is a comfortable temperature and that cooling is required whenever the temperature rises above that value. Cooling degree days are registered in degrees Fahrenheit and are counted by subtracting 65° F from the mean temperature day by day. If the temperature rises to 75° F, for example, 10 cooling degree days are recorded on that day. At the end of the year the cooling degree days for each day are added to give an annual total. At Great Falls, Montana, there are an average of 350 cooling degree days each year; in Seattle, Washington, there are 200; in Detroit, Michigan, there are 700; and in Miami, Florida, there are 4,000.

Cooperative Holocene Mapping Project (COHMAP) A long-term study of the changes that have taken place in the climate over the last 18,000 years. It began by collecting data based on the distribution of fossil plankton in seabed sediments, POLLEN ANALYSIS, and the fossil remains of seeds, leaves, and other material. In the 1980s and early 1990s these data were used to construct a general outline of climatic changes, which was then used to test the changes indicated by a computer model simulation of the atmospheric changes over the same period. The results of this part of the study have been published on the Web. The continuation of COHMAP involves improving the model simulations by adding details of oceanic circulation and vegetation processes by using data that became available during the late 1990s. This phase of the project is called *Testing Earth System Models with Paleoclimatic Observations* (TEMPO).

(You can learn more about COHMAP from www.ncdc.noaa.gov/ogp/papers/kutzbac.html, and the

COHMAP results are published at www.ngdc.noaa.gov/paleo/paleo.html.)

cooperative observer In the United States, a voluntary weather observer who maintains a WEATHER STATION and supplies data to the National Weather Service without remuneration.

corange line *See* AMPHIDROMIC POINT.

coreless winter A winter in which the temperature falls to its minimum at the autumn EQUINOX, after which it falls no further. This occurs only in Antarctica. During the autumn months of February and March the temperature falls steadily by almost 1° F (0.5° C) a day, but in most years it decreases very little after the equinox. This unusual phenomenon is believed to occur because the amount of heat the continent loses by radiation during the six months of winter is balanced fairly precisely by the heat that is transported to the continent by winds carrying warm air from higher latitudes.

CorF *See* CORIOLIS EFFECT.

Coriolis effect (CorF) When a body that is not attached to the surface of the Earth moves with respect to the surface, its path is deflected to the right in the Northern Hemisphere and to the left in the Southern Hemisphere. The French physicist GASPARD GUSTAVE DE CORIOLIS (1792–1843) was the first to explain this phenomenon, which is known as the *Coriolis effect.* Sometimes called the *Coriolis force,* it is usually abbreviated as *CorF,* because it appears as though the moving body is being pushed to one side. In fact, though, no force is involved. The deflection is entirely due to the rotation of the Earth.

Every 24 hours the Earth completes one rotation on its axis. This means that a point anywhere on the surface travels through 360° at a rate of 15° every hour. The linear speed at which these points travel depends on their latitude, however. The circumference of the Earth at the equator is about 24,881 miles (40,034 km), so a place on the equator is traveling eastward at about (24,881 ÷ 24 =) 1,036.7 mph (1,668 km h^{-1}). A place at latitude 40° N, which is approximately the latitude of New York City and Madrid, Spain, has a shorter distance to cover, because the circumference of the Earth at 40° N is about 19,057 miles (30,662.7 km). People in these cities are traveling at about 794 mph (1,278 km h^{-1}).

Suppose an airplane set off to fly to New York from a point on the equator due south of New York, but the navigator knew nothing about the Coriolis effect and so the airplane headed due north. At takeoff, the airplane is traveling eastward at 1,037 mph (1,668 km h^{-1}). Once airborne and no longer attached to the surface, it continues to move to the east at the same speed. The flight to New York will take 6 hours, during which the airplane will travel (1,037 × 6 =) 6,222 miles (10,012 km) to the east. New York is also traveling eastward, but at only 794 mph (1,278 km h^{-1}), so during the 6 hours the airplane takes to make the journey the city will have moved 4,764 miles (7,665 km) to the east. By the time it arrives, the airplane will have traveled (6,222 – 4,764 =) 1,458 miles (2,346 km) farther to the east than New York and so instead of being over the city it will be far out over the Atlantic, not even in sight of land.

This is the Coriolis effect. It makes it seem that the airplane has been pushed far to the east (to the right) of the course its crew intended. For the same reason an airplane traveling in the opposite direction, south from New York, would experience an apparent drift to the west—also a deflection to the right. In reality, of course, navigators allow for CorF and adjust the airplane heading.

CorF also affects bodies traveling east and west. In this case there is no difference in speed between a moving body and the surface beneath it, but the orientation of the surface changes as the Earth rotates. This alters the angle at which a body moving in a straight line crosses the lines of longitude. Again, the result is an apparent deviation to the right in the Northern Hemisphere and to the left in the Southern Hemisphere.

The higher the latitude, the greater is the deflection. This is because the lines of longitude converge toward the North and South Poles and so the change in their orientation over any distance is greater than the change in a lower latitude. The magnitude of the CorF is zero at the equator and reaches a maximum at each pole.

The change in CorF magnitude with latitude is given by the Coriolis parameter. This is $2\Omega \sin \Phi$, where Ω is the ANGULAR VELOCITY of the Earth (7.29×10^{-5} rad s^{-1}) and Φ is the latitude. Note that when Φ =

0°, at the equator, sin $\Phi = 0$, and when $\Phi = 90°$, at the Pole, sin $\Phi = 1$.

CorF also varies according to the speed of the moving body. This is because the faster a body travels the greater the distance it covers in a given time. When this is taken into account, it is simple to calculate the magnitude of the CorF on a body at a given latitude moving at a given speed by CorF = $2\Omega \sin \Phi\, v$, where v is the speed. The result will be an ACCELERATION, because Ω, in units of radians per second, is multiplied by v, in units of meters (or feet) per second, to give a result measured in units of distance (feet or meters) per second per second.

(You can learn more about the Coriolis effect at www.windpower.dk/tour/wres/coriolis.htm and at zebu.uoregon.edu/~js/glossary/coriolis_effect.html.)

Coriolis, Gaspard Gustave de (1792–1843) French *Physicist, mathematician, engineer* The scientist who was the first to explain why moving bodies, such as winds and ocean currents, are deflected to the right in the Northern Hemisphere and to the left in the Southern Hemisphere was born in Paris on May 21, 1792. His family was from Provence, in the south of France. They were lawyers and were made aristocrats (hence the *de* in the family name) in the 17th century. The French Revolution stripped them of their privileges and wealth and Gaspard's father became an industrialist, living in the town of Nancy.

In 1808 he commenced his studies at the École Polytechnique, the school that trained government officials. He completed them at the École des Ponts et Chaussées. In the course of his studies there he spent several years in the Vosges mountains on active service with the corps of engineers.

He graduated in highway engineering and was determined to become an engineer, but his health was poor and his father's death left him with the responsibility of keeping the family. In 1816 he joined the staff of the École Polytechnique, first as a tutor and then as an assistant professor of analysis and mechanics. In 1829 he took up a position as professor of mechanics at the École Centrale des Arts et Manufactures, where he remained until 1836, when he became professor of mechanics at the École des Ponts et Chaussées. In 1838 he was made director of studies at the École Polytechnique. He was elected a member of the mechanics section of the Academy of Sciences in 1836.

Coriolis was a highly talented scientist but suffered from poor health, which prevented him from realizing his full potential. Nevertheless, he made several important contributions to science. He succeeded in establishing the term *work* as a technical term, which he defined as the displacement of a force through a distance. In dynamics he introduced a quantity, $1/2\ mv^2$, for which he coined the term *force vive,* now called *kinetic energy.*

It was in 1835 that he made the contribution to physics for which he is still remembered. In that year he published a paper, "Sur les équations du mouvement relatif des systèmes de corps" (On the equations of relative motion of a system of bodies), in volume 15 of *Journal de l'École Polytechnique.* This showed that when a body moves in a rotating frame of reference its motion relative to the frame of reference can be explained only if there is a force of inertia acting upon it. This inertial force causes the body to follow a path that curves to the right if the frame of reference is rotating counterclockwise and to the left if the rotation is clockwise. The inertial force came to be called the *Coriolis force* and is now known as the *CORIOLIS EFFECT.* It is of great importance in studies of the movements of air and of ocean currents. It is also relevant to ballistics, because missiles and projectiles traveling a long distance are subject to it.

Coriolis died in Paris on September 19, 1843.

(You can learn more about Coriolis at www.macmed.com/M%26C%20FILES/09maccs.html and in "Gustave Coriolis, and the Coriolis Effect" at camille-f.gsfc.nasa.gov/912/geerts/cwx/notes/chap11/gustave.html.)

Coriolis parameter *See* CORIOLIS EFFECT and GEOSTROPHIC EQUATION.

corner stream Air that is deflected around the sides of a tall building that is downwind of a lower building when a wind is blowing. The wind strikes the face of the building, producing a stagnation point about three-quarters of the way from its base. Air divides at the stagnation point, some rising over the building and some flowing downward. Some of this air contributes to EDDIES on the LEE side of the lower building, and the remainder travels around the sides of the tall building as corner streams that wrap around the back. As they pass the sides of the tall building, the corner streams

Corner stream. **Wind striking the face of the tall building divides at a stagnation point. Part of the air flows downward and accelerates as it goes around the sides of the building as corner streams.**

blow at a speed about 2.5 times greater than that of the wind as its approaches the buildings.

corona An optical phenomenon in which a whitish disk surrounds the Moon or less commonly the Sun. Colors can sometimes be seen, in which case the corona consists of two or more concentric rings that are reddish on the outside and bluish on the inside. The effect is caused by the DIFFRACTION of light through the water droplets forming a layer of cloud, commonly ALTOSTRATUS, between the Sun and the observer. Despite its superficial similarity, a corona is quite different from a HALO. If the corona grows larger over the space of an hour or two, it indicates the water droplets are becoming smaller and the cloud will disperse. If the corona becomes smaller, it means the water droplets are becoming bigger and before long are likely to fall as rain. A corona may also appear in fog. This indicates the fog is thinning and will soon clear.

coronazo de San Francisco A HURRICANE that forms over the Pacific Ocean, off the coast of Central America between Costa Rica and Point Eugenio in Baja California, Mexico. The tracks of these hurricanes usually carry them northward or northwestward and many strike the coast. They are less violent than hurricanes that form over the Atlantic and Caribbean and cover a smaller area.

corrected altitude (true altitude) The altitude measured by an ALTIMETER and adjusted to take account of the difference between the AMBIENT temperature and the temperature of the STANDARD ATMOSPHERE.

cosmic radiation A stream of particles possessing high energy that originates in space, some of it outside the solar system, and that falls onto the Earth. Primary cosmic rays consist of nuclei of the most common elements, predominantly of HYDROGEN; a hydrogen nucleus comprises a single proton. The radiation also includes electrons, positrons (particles identical to electrons, but carrying positive charge), neutrinos, and photons (electromagnetic radiation) at gamma-ray wavelengths of about 0.00001 µm. As they penetrate the atmosphere, the particles collide with atoms of nitrogen and oxygen. These collisions produce secondary radiation, consisting of more particles and gamma-ray photons. A single cosmic-ray particle is able to generate a large shower of secondary particles. Cosmic radiation contributes to the IONIZATION of gases that produces the IONOSPHERE.

cotidal line *See* AMPHIDROMIC POINT.

cotton-belt climate A climate with warm, wet summers and dry winters that occurs on the eastern sides of continents and that is characteristic of the cotton-growing regions of the southern United States and China.

coulomb (C) The derived SYSTÈME INTERNATIONAL D'UNITÉS (SI) UNIT of quantity of electricity, or electric charge, which is defined as the charge that is transferred by a current of one AMPERE per SECOND. The unit is named in honor of the French physicist Charles Augustin de Coulomb (1792–1806).

country breeze A light, cool wind that blows into a city from the surrounding countryside, especially on calm nights when the sky is clear. It is produced by the urban HEAT ISLAND effect, and it is on clear, calm nights that the effect is most pronounced. Warm air rising above the city produces an area in which the air pressure is lower than it is in the surrounding countryside. Cool air then flows toward the city center.

couple *See* TORQUE.

covalent bond An attractive force that holds two or more atoms together and that is based on the sharing of an electron between them; if they share two electrons they form a double covalent bond. An oxygen atom, for example, forms covalent bonds with two hydrogen atoms and forms a water molecule, and a carbon atom forms double covalent bonds with two oxygen atoms to form a molecule of carbon dioxide. The gases oxygen and nitrogen occur in the air as molecules comprising two atoms joined by a covalent bond.

cows According to weather folklore, the behavior of cows can predict a range of conditions. If a cow tries to scratch its ear there will soon be a shower. If it beats its flanks with its tail there will be a thunderstorm. If the cows gather at the top of a hill the weather will be fine, but if they move to lower ground it will be wet or stormy. If they stampede with their tails held high there will be rain and thunder. In cold weather, cows will lie down at the approach of rain or huddle together in a sheltered place with their tails to the wind. Some of these observations may be reliable, in particular the

Country breeze. **Warm air rises over the city, producing a region of low pressure near ground level. Cool air blows in from the surrounding countryside.**

stampeding reaction to an approaching storm. Cows behave like this because parasitic flies that lay their eggs on the skin of cows are at their most active when the air is warm and humid—the conditions that precede a storm. They make a high-pitched buzzing sound as they fly, and this sound makes cows raise their tails and run.

cowshee *See* KAUS.

crachin FOG, accompanied by low STRATUS cloud and DRIZZLE that is common between February and April along the coast of southern China and the Gulf of Tonkin. Sometimes it is caused by the mixing of AIR MASSES and sometimes by the ADVECTION of warm air across a cooler surface.

crassulacean acid metabolism *See* PHOTOSYNTHESIS.

crepuscular rays (Sun drawing water) An optical phenomenon in which what appear to be rays or bands of light radiate upward from the Sun when the Sun is low in the sky. They occur when the sky is partly covered by cloud, leaving gaps through which the Sun shines. If particles of dust or smoke are present in the sky above those gaps sunlight is reflected from them. In fact, light shining through the gaps illuminates the particles in its path. *Crepuscular* means "of the twilight" (from the Latin *crepusculum,* "twilight"), and crepuscular rays are seen only very early or late in the day. People used to believe the rays were caused by water being drawn from the sky toward the Sun and described the phenomenon as "the Sun drawing water," taking it as a sign of fine weather to come. This is not what is happening, of course, but crepuscular rays can be seen only when there are gaps in the cloud cover and this often means the cloud is breaking up and the weather turning fine.

crest cloud (cloud crest) A cloud that marks the crest of a LEE WAVE. It is seen above or slightly on the LEE side of a mountain peak or the top of a high hill and remains stationary in relation to the land below.

criador A westerly wind that produces rain in northern Spain.

critical point The temperature above which a gas cannot be liquefied, regardless of the pressure to which it is subjected. The gas becomes a supercritical fluid, which is a fluid that has the density of a liquid but the molecular freedom of a gas. The critical point for water is 705.9° F (374.4° C). At that temperature water vapor condenses into a liquid at a pressure of 220 ATMOSPHERES (32,000 lb in^{-2}).

crivetz A northeasterly wind that blows in spring and autumn through the lands bordering the lower Danube in Romania and southern Ukraine. It carries hot weather in late spring and early autumn and cold weather, sometimes with snow, during the colder months.

Croll, James (1821–1890) Scottish *Climatologist and geologist* A self-educated man, James Croll proposed in 1864 that the onset of GLACIAL PERIODS is triggered by changes in the ECCENTRICITY of the Earth's orbit and the PRECESSION OF THE EQUINOXES. This hypothesis developed and extended earlier ideas and aroused great interest. According to Croll, ice ages occur at intervals of 100,000 years, when the orbit reaches its maximum eccentricity. They are experienced alternately in the Northern and Southern Hemispheres, according to which hemisphere is having winter when the Earth is farthest from the Sun. Although interest in his astronomical theory waned at the end of his life, Croll's work did much to prepare the way for the theory of MILANKOVICH CYCLES, which is now widely accepted.

James Croll was born near Cargill, Perthshire, on January 2, 1821. The family was poor and James left school at 13. He read avidly but needed to earn money and so worked at a succession of jobs: as a millwright, then a carpenter, in a tea shop, and as a hotelkeeper. He made and sold electrical goods and sold insurance. Finally, in 1857, he published *The Philosophy of Theism,* which attracted some attention, and after working for a temperance newspaper, in 1859 he was made the keeper of the Andersonian Museum, in Glasgow. This gave him access to the library, where he continued his self-education. He published several papers on chemistry, physics, and geology, and in 1867 he was placed in charge of the Edinburgh office of the Geological Survey of Scotland, in a post he held until he retired, in 1880.

After his retirement he continued to work on his astronomical theory and other topics that interested him. He published several books, including *Climate and Time, in Their Geological Relations* (1875), *Discussions on Climate and Cosmology* (1885), *Stellar Evolution and Its Relations to Geological Time* (1889), and *The Philosophical Basis of Evolution* (1890).

Croll died of heart disease near Perth on December 15, 1890.

Cromerian interglacial An INTERGLACIAL period in Britain that preceded the ANGLIAN GLACIATION. Dating is uncertain, but it lasted from approximately 750,000 years ago until about 350,000 years ago.

Cromwell Current (Equatorial Undercurrent) An ocean current that flows from west to east beneath the surface of the Pacific Ocean between latitudes 1.5° N and 1.5° S and at a depth of 165–1,000 feet (50–300 m). The current is about 185 miles (300 km) wide and flows at up to 3.4 mph (5.5 km h^{-1}). The surface EQUATORIAL CURRENT is driven by the TRADE WINDS and carries warm surface water from east to west. The Cromwell Current counterbalances it by conveying water below the surface in the opposite direction.

crop yield model A climate MODEL that relates local weather conditions to crop yields. Farmers can use such models to help in planning farming operations. The models take account of temperature, water availability, carbon dioxide, and sunshine. They also make some allowance for crop pests and diseases, the seriousness of which is partly related to climate. Relationships between weather and yield for particular crops are often based on historical records.

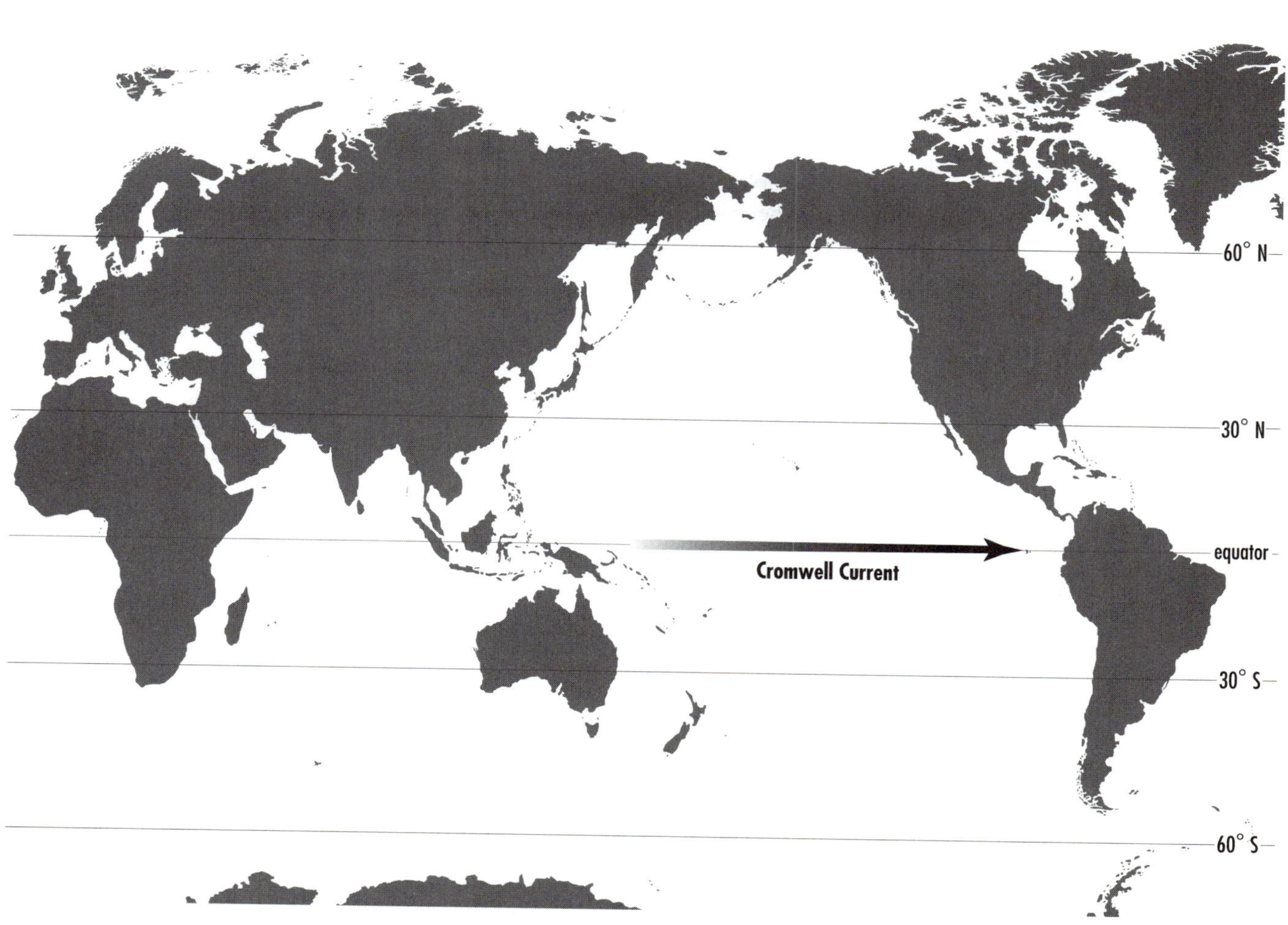

Cromwell Current. **A wide, strong current that flows below the surface of the tropical Pacific from west to east.**

crosswind A wind that is not blowing in the same direction as a moving object such as an airplane (a TAILWIND), nor in the opposite direction (a headwind). Consequently, there is a component of the wind that acts perpendicularly to the object's direction of motion.

Crutzen, Paul (born 1933) Dutch *Atmospheric chemist* Professor Paul Crutzen shared with F. SHERWOOD ROWLAND and MARIO J. MOLINA the 1995 Nobel Prize in Chemistry, which was awarded for their discovery of the processes that deplete the OZONE LAYER.

Paul Crutzen was born on December 3, 1933, in Amsterdam, the Netherlands. His father, Josef Crutzen, was a waiter; his mother, Anna Gurk, worked in a hospital kitchen. Paul commenced his education in September 1940. In May of that year the Netherlands had been invaded and occupied by the German army and Crutzen's primary education coincided with the period of occupation. His schooling was interrupted several times and conditions were especially severe from the fall of 1944 until the country was liberated in May 1945. Nevertheless he was able to complete his primary schooling, and in 1946 he entered the middle school, where he prepared to enter university. He specialized in mathematics, physics, and languages and was keen on sports, especially long-distance skating on the Dutch lakes and canals.

Unfortunately, illness meant he did not achieve the grades in his final examinations that would have won him a grant to pay for a university education. Instead of embarking on a four-year university course he enrolled at a technical school for a three-year course in civil engineering, from 1951 until 1954. The second year was spent working for a civil engineering company to gain practical work experience and he managed to live for two years on what he was paid.

On completing his course, Crutzen obtained a job with the Bridge Construction Bureau of the City of Amsterdam. He worked there from 1954 until 1958, with an interruption for his compulsory military service from 1956 to 1958. During his time there he met Terttu Soininen, a student of Finnish history at the University of Helsinki. They were married in February 1958. They made their home in Gävle, a town about 125 miles (200 km) north of Stockholm, Sweden, where Crutzen was then working for a construction company. Their two daughters, Ilona and Sylvia, were born in Gävle.

In 1958 Crutzen saw a newspaper advertisement for a computer programmer in the Department of Meteorology of the Stockholm Högskola (high school, and since 1961 Stockholm University). Despite knowing nothing at all about computer programming he applied for the post and won it. At the beginning of July 1959 the Crutzen family moved to Stockholm and Paul embarked on a second career. The Department of Meteorology (now the Meteorology Institute) and the International Meteorological Institute that was associated with it were at the forefront of research and housed some of the fastest computers in the world.

Until 1966, Crutzen spent much of his time building and running some of the first weather prediction MODELS, including one of a TROPICAL CYCLONE. High-level computer languages, such as Algol and Fortran, had not yet been developed, and all programs had to be written in machine code, using binary notation.

Because he worked at the university, Crutzen was able to attend some of the lectures. He had no opportunity to do laboratory work, however, and so he concentrated on mathematics, statistics, and meteorology. In 1963 he obtained his master of science degree in these subjects and in 1965 began to work for his doctorate. At that time he was assisting another scientist in a project to study the different forms (called *allotropes*) of oxygen in the upper atmosphere, and he chose to make stratospheric chemistry the subject for his doctoral thesis. He received his Ph.D. in 1968 and his D.Sc. (doctor of science, which is a higher degree than a Ph.D.) in 1973, for research into the photochemistry of ozone.

Crutzen left Stockholm in 1969 to work until 1971 as a European Space Research Organization fellow at the Clarendon Laboratory of the University of Oxford, England. From 1974 to 1980 he worked at Boulder, Colorado, on the Upper Atmosphere Project of the National Center for Atmospheric Research and as a consultant in the Aeronomy Laboratory of the Environmental Research Laboratories (NATIONAL OCEANIC AND ATMOSPHERIC ADMINISTRATION). He was an adjunct professor in the Atmospheric Sciences Department of the University of Colorado from 1976 to 1981. In 1980 he was appointed director of the Atmospheric Chemistry Division of the Max-Planck Institute for Chemistry, at Mainz, Germany, and from 1983 to 1985 he was executive director of the institute. He held a part-time professorship at the University of Chicago from 1987 until 1991 and since 1992 he has been a part-time professor at the Scripps Institution of Oceanography of the University of California.

Professor Crutzen has received many honors and holds honorary degrees from universities at York, Canada, Louvain, Belgium, East Anglia, England, and Thessaloniki, Greece.

(You can learn more about Paul Crutzen at www.nobel.se/chemistry/laureates/1995/crutzen-autobio.html and www.mpch-mainz.mpg.de/~air/crutzen/vita.html.)

Cryogenic Limb Array Etalon Spectrometer (CLAES) An instrument carried by the UPPER ATMOSPHERE RESEARCH SATELLITE that measures infrared radiation emitted from the atmosphere. Its ETALON is kept chilled (cryogenic). The instrument measures the temperature in the STRATOSPHERE and lower MESOSPHERE and the trace constituents of the atmosphere: ozone (O_3), nitric oxide (NO), nitrogen dioxide (NO_2), nitrous oxide (N_2O), nitric acid (HNO_3), dinitrogen pentoxide (N_2O_5), methane (CH_4), CFC-11 (CCl_3F), CFC-12 (CCl_2F_2), and $CIONO_2$.

cryosphere Snow and ice that lie on the surface of the continents and oceans. The perennial cryosphere is confined mainly to polar regions, where it covers 8 percent of the surface of the Earth. In addition, there is a seasonal cryosphere that covers 15 percent of the surface in January (Northern Hemisphere winter) and 9 percent in July (Southern Hemisphere winter). Over the Arctic Ocean the ice cover is not complete; it consists of large ice floes that move in relation to each other. The area covered by sea ice varies considerably from year to year. In 1968 the sea was covered with ice between Iceland and Greenland, and in the summer of 2000 an area at the North Pole was largely free from ice.

cryptoclimatology The CLIMATOLOGY of enclosed spaces. It is a branch of microclimatology, which is the study of MICROCLIMATES.

Cs *See* CIRROSTRATUS.

CTM *See* CHEMICAL TRANSPORT MODEL.

Cu *See* CUMULUS.

Cu_{fra} *See* FRACTOCUMULUS.

cumuliform An adjective that is used to describe the shape of a cloud that resembles a cloud belonging to the cloud genera CUMULUS and CUMULONIMBUS (*see* CLOUD CLASSIFICATION). A cumuliform cloud has a fleecy appearance, like cotton wool, or is heaped up with many protuberances, like a cauliflower.

cumulonimbus (Cb) A genus of dense cloud with a low base that often extends vertically to a great height, sometimes all the way to the TROPOPAUSE. It forms mountainous and towering shapes, but the uppermost part is usually smooth and the top flattened, marking the level beyond which air is unable to rise by CONVECTION. It may form the ACCESSORY CLOUD, INCUS. Cumulonimbus cloud carries precipitation, which is often heavy, and it is the cloud associated with THUNDERSTORMS, TROPICAL CYCLONES, and TORNADOES. The great depth of cloud in which light is scattered by water droplets makes the lower part of a cumulonimbus cloud very dark and often menacing.

Cumulonimbus is classified as a low cloud because of the height of its base (*see* CLOUD CLASSIFICATION). It occurs as the species CALVUS and CAPILLATUS, and sometimes with the ACCESSORY CLOUD, VIRGA.

cumulus (Cu) A genus of cloud (*see* CLOUD CLASSIFICATION) that develops vertically as warm air rises by CONVECTION. As it rises the air cools by an ADIABATIC mechanism and some of its WATER VAPOR condenses into droplets, forming the cloud, which is of a fleecy or billowing shape, sometimes called *cotton wool cloud*. Cumulus clouds are isolated from each other. Blue sky is often visible between them, so their boundaries are sharply defined. Small, scattered cumulus clouds seen on a fine day are known as *fair weather cumulus*. At other times cumulus may be immersed in clouds of other types, so it is more difficult to distinguish. The base of a large cumulus cloud is dark, because of the density of water droplets by which light is scattered on its way to the ground. The sunlit upper parts of the cloud are very bright.

Cumulus is classified as low cloud by the height of its base, although its upper parts may extend to middle altitudes. The cumulus species are CONGESTUS, FRAC-

TUS, HUMILIS, and MEDIOCRIS, and it also occurs as the variety RADIATUS.

cumulus fractus *See* FRACTOCUMULUS.

current A flow of fluid (gas or liquid) through a mass of a similar fluid that remains stationary. Wind is a current of air that moves horizontally through the surrounding air (if all the air were to move, it would leave behind a vacuum!). Air currents can be vertical as well as horizontal. CUMULIFORM clouds are produced by vertical air currents caused by CONVECTION. Ocean currents are streams of water that move through the ocean. Many form GYRES. The principal ocean currents are the AGULHAS CURRENT, ALASKA CURRENT, ALEUTIAN CURRENT, ANTARCTIC CIRCUMPOLAR CURRENT, ANTARCTIC POLAR CURRENT, ANTILLES CURRENT, BENGUELA CURRENT, BERING CURRENT, BRAZIL CURRENT, CALIFORNIA CURRENT, CANARY CURRENT, CARIBBEAN CURRENT, CROMWELL CURRENT, EAST AUSTRALIAN CURRENT, EAST GREENLAND CURRENT, EQUATORIAL COUNTERCURRENT, FALKLAND CURRENT, FLORIDA CURRENT, GUINEA CURRENT, GULF STREAM, IRMINGER CURRENT, KAMCHATKA CURRENT, KUROSHIO CURRENT, LABRADOR CURRENT, MOZAMBIQUE CURRENT, NORTH ATLANTIC DRIFT, NORTH EQUATORIAL CURRENT, NORWEGIAN CURRENT, OYASHIO CURRENT, PERU CURRENT, SOUTH EQUATORIAL CURRENT, WEST AUSTRALIAN CURRENT, and WEST GREENLAND CURRENT.

curvature effect *See* CLOUD CONDENSATION NUCLEI.

cutoff high A closed, middle latitude ANTICYCLONE that has moved into a higher latitude, out of the prevailing westerly air flow and detached from it. A cutoff high can cause BLOCKING.

cutoff low A closed, middle latitude CYCLONE that has become detached from the prevailing westerly air flow and has moved away from it into a lower latitude. A cutoff low can cause BLOCKING.

cutting-off The process by which a middle latitude ANTICYCLONE or CYCLONE is detached from the prevailing westerly air stream. This usually happens in the upper TROPOSPHERE, and it is often associated with BLOCKING, because it produces CUTOFF HIGHS and CUTOFF LOWS that are very slow-moving. Weather systems are deflected around them.

cyclogenesis The series of events by which a CYCLONE develops in the POLAR FRONT. When the POLAR FRONT THEORY was first proposed meteorologists had access only to data obtained from air near the surface. Cyclogenesis now takes account of conditions

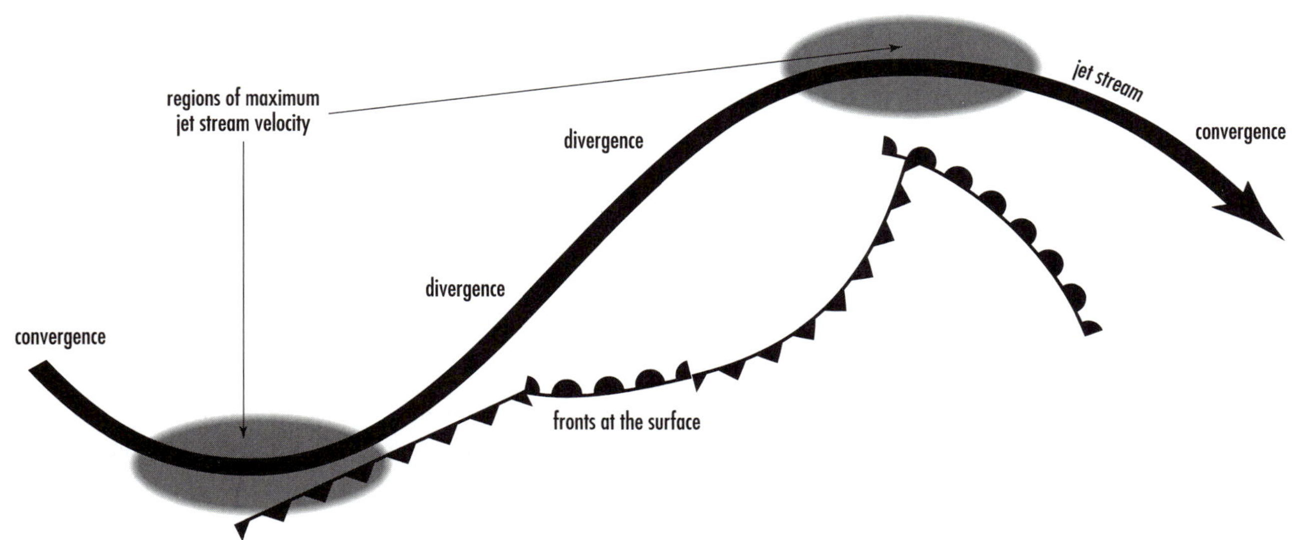

Cyclogenesis. **The formation of a low-level cyclone, associated with an area of convergence along the path of the jet stream.**

throughout the TROPOSPHERE and lower STRATOSPHERE and in particular of the JET STREAM and the INDEX CYCLE. These exert a strong influence on the formation of weather systems by imposing regions of high-level DIVERGENCE that remove air faster than low-level CONVERGENCE can replace it, thus producing an area of low pressure near the surface.

cyclolysis The weakening and disappearance of the cyclonic circulation of air that occur as a cyclone family dissipates and high pressure begins to dominate.

cyclone (depression) An area of low atmospheric pressure around which there is a clearly defined wind pattern, the winds flowing in a CYCLONIC pattern. In middle latitudes, cyclones develop in association with FRONTS between AIR MASSES. They last for only a limited time, forming (*see* CYCLOGENESIS) and finally dying away (*see* CYCLOLYSIS) as the low pressure fills. Often they occur in groups, known as CYCLONE FAMILIES, one following another.

The same word is used as the local name for a TROPICAL CYCLONE that forms in the Indian Ocean. They form in both Northern and Southern Hemispheres. Those in the Northern Hemisphere often move northward through the Bay of Bengal and cause severe damage in India and Bangladesh. Others form farther to the west and move northward into the Arabian Sea, sometimes reaching Oman and Pakistan, or westward to Madagascar. Those in the Southern Hemisphere may reach northern Australia.

Cyclones are often extremely severe (*see* GERALDA, for example). Of all the tropical cyclones that develop, 90 percent are either cyclones or TYPHOONS.

cyclone family Midlatitude FRONTAL WAVES (depressions) often occur as sequences of three or four, each sequence known as a family. The first frontal wave to form is called the *primary* and those following it are *secondaries* that develop along the trailing edge of a very extended COLD FRONT. Each secondary follows a track a little to the south of the one ahead of it. This is because polar air, to the poleward side of the POLAR FRONT, pushes farther south at the rear of each wave in the sequence. The sequence ends when the polar air has formed a large wedge extending a long way south of the primary wave and establishing a region of high pressure.

cyclonic The adjective that describes the direction in which the air flows around a CYCLONE. The direction is the same as that of the Earth's rotation as seen from

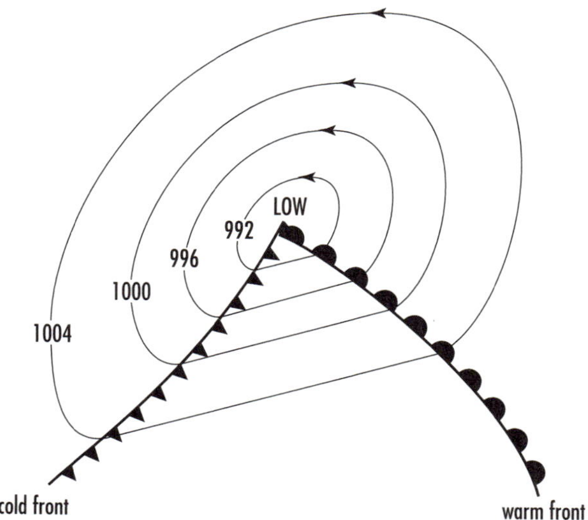

Cyclone. **An area of low pressure. In this case the cyclone has formed at the boundary between two air masses, at the point where a cold front is advancing toward a warm front. The entire system is traveling to the right. The thin lines are isobars, labeled in millibars with the pressure they indicate.**

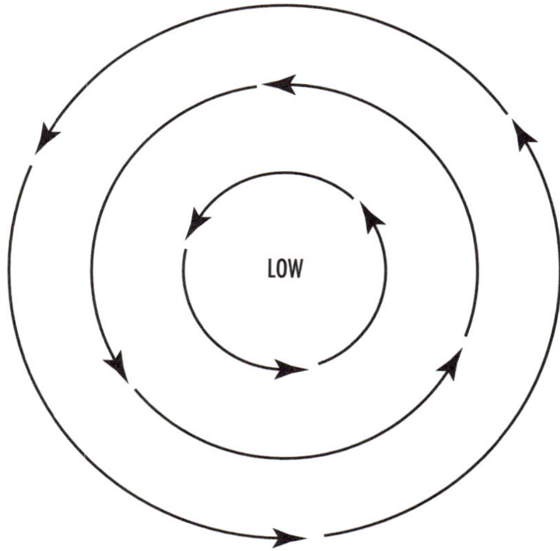

Cyclonic. **Well clear of the surface, the winds flow almost parallel to the isobars, around the center of low pressure. Cyclonic flow is counterclockwise in the Northern Hemisphere and clockwise in the Southern Hemisphere.**

directly above the North and South Poles: counterclockwise in the Northern Hemisphere and clockwise in the Southern Hemisphere. Air also flows cyclonically around a TROUGH. At a height of 33 feet (10 m), which is the standard height for observing surface winds, the winds cross the ISOBARS at an angle of 10° to 30°, depending on local topographical characteristics and the wind speed. Above the BOUNDARY LAYER the winds flow almost parallel to the isobars at speeds proportional to the PRESSURE GRADIENT.

cyclonic rain PRECIPITATION that is associated with a CYCLONE or DEPRESSION. Steady, persistent rain or snow falls from STRATIFORM clouds along the WARM FRONT, and showers, sometimes heavy, are produced by CUMULIFORM clouds along the COLD FRONT.

cyclonic scale *See* SYNOPTIC SCALE.

cyclonic shear WIND SHEAR that is associated with VORTICITY around CYCLONES. Looking downwind, the winds are stronger on the right in the Northern Hemisphere and on the left in the Southern Hemisphere. This tends to set the air rotating cyclonically along the line of the wind.

cyclostrophic wind A strong, low-level wind that follows a very tightly curved path, such as the wind that blows around the side of a hill or around a TORNADO. It is the type of wind that generates a DUST DEVIL when it blows around a very small but intense area of low surface pressure.

It occurs when the radius of the path followed by the wind is too small for the CORIOLIS EFFECT to be significant, or close to the equator, where the Coriolis effect is also small—winds around intense TROPICAL CYCLONES are often cyclostrophic. The very strong PRESSURE-GRADIENT FORCE exerts a CENTRIPETAL ACCELERATION that balances it, thus maintaining the flow along the curved path parallel to the ISOBARS. The equation for calculating the speed (V) of the cyclostrophic wind is

$$V = \sqrt{\{(r/\rho)(\delta p/\delta x)\}}$$

where r is the radius of the curved path, ρ is the air density, and $\delta p/\delta x$ is the pressure gradient.

dadur A northwesterly wind that blows from the Siwalik Hills, in the foothills of the Himalayas, down the valley of the Ganges River, in India.

daily forecast A weather forecast that is issued for the period from 12 to 48 hours ahead.

daily mean The average value of a meteorological factor over 24 hours counted from midnight to midnight. It is obtained by adding together the hourly readings for temperature, pressure, cloud cover, or the amount of precipitation that has fallen in the preceding hour, and dividing the total by 24. In the case of temperature and pressure the average can be calculated by adding the highest and lowest values recorded and dividing by 2.

DALR *See* DRY ADIABATIC LAPSE RATE.

Dalton, John (1766–1844) English *Meteorologist and Chemist* John Dalton was born at Eaglesfield, near Cockermouth, Cumberland (now Cumbria), on September 6, 1766, the son of a weaver and the third of six children. His father, Joseph, was a devout member of the Society of Friends; in accordance with Quaker practice at the time he did not register the date of his son's birth, so there is some uncertainty about it. John began his education at the Quakers' school in Eaglesfield. The teacher was John Fletcher; when he retired in 1778 John took his place.

Three years later, in 1781, Dalton moved to Kendal, where he continued to earn his living as a teacher and became a headmaster. John Gough, a wealthy Quaker who was also a classicist and mathematician, befriended Dalton. Under his influence Dalton began to write articles on scientific topics for two popular magazines, *Gentlemen's Diary* and *Ladies' Diary*. Gough also encouraged him to keep a diary of meteorological observations. He began this diary in 1787 and made entries in it regularly for 57 years until his death. It contained more than 200,000 observations.

In the 18th century only members of the Church of England were accepted as students at the universities of Oxford and Cambridge. As a Quaker, Dalton was excluded, and consequently he was very largely self-taught.

In 1793 Dalton moved to Manchester, where Gough had helped him obtain a position as a teacher of mathematics and natural philosophy at New College. He held this post until 1799, when the college was moved to York and Dalton remained behind in Manchester as a private teacher of mathematics and chemistry.

His first publication on weather, *Meteorological Observations and Essays*, appeared in 1793 but did not sell well. In 1794 Dalton was elected to the membership of the Manchester Literary and Philosophical Society and a few weeks later delivered his first paper to the society, "Extraordinary Facts Relating to the Vision of Colours." Dalton and his brother were both color blind and this was the first account of the way the world appears to someone with this condition. For a time color blindness was known as *Daltonism.* He also lectured to the society on the weather. Dalton became the honorary secretary and then president of the society, and after he resigned his teaching position in 1799 he lived for many years in a house the society bought for him, which he shared with the Reverend W. Johns.

In 1803 Dalton proposed the law of PARTIAL PRESSURES, known as *DALTON'S LAW.* He discovered that the density of water varies with its temperature, reaching a maximum at 42.5° F (6.1° C). In fact, water reaches its maximum density at 39.2° F (4° C), but Dalton's calculation was close. He studied what happens when substances dissolve in water and when gases mix and concluded that water and gases must consist of very small particles that intermingle, so the particles of a dissolved substance are located between water particles. In his book *New System of Chemical Philosophy,* published in 1808, he suggested that the particles of different elements have different weights. He compiled a list of these weights (relative atomic masses) and devised a system of symbols for the elements. These could be combined to represent compounds.

Dalton became very famous. He delivered two courses of lectures at the Royal Institution in London, the first in 1804 and the second in 1809–10. He was elected a fellow of the Royal Society in 1822. He became a corresponding member of the French Academy of Sciences and in 1830 was elected one of its eight foreign associates. In 1833 the government awarded him an annual pension of £150, raised to £300 in 1836. This was a substantial award that would have allowed Dalton to live comfortably at a higher standard than the one he chose for himself. It was also an unusual honor. British governments rarely recognized the worth of individuals in this way. Government support for science took the form then, as now, of providing funds for institutions and paying the salaries of scientists directly employed by government ministries and agencies.

Dalton spent almost all his time in Manchester, working in his laboratory and teaching. Each year he visited the Lake District, in his native Cumbria, and occasionally he visited London. He made a short visit to Paris in 1822. His only recreation was the game of bowls, which he played every Thursday afternoon.

John Dalton died in Manchester on July 27, 1844, and a statue was erected in his memory. The house in which he lived for so long contained many of his records and other relics; it was destroyed during a bombing raid in World War II.

Dalton's law A law that was proposed in 1803 by the English chemist JOHN DALTON (1766–1844). It states that in a mixture of gases, the total pressure is the sum of the pressures that each component of the mixture would exert if it alone occupied the same volume at the same temperature. This can be written as

$$p = \Sigma_i \, \rho_i \, R_i \, T$$

where p is the total pressure, Σ_i is the sum of the constituent gases, ρ_i is the density of each gas, R_i is the specific GAS CONSTANT for each gas, and T is the ABSOLUTE TEMPERATURE. The pressure exerted by each constituent gas is called the *PARTIAL PRESSURE* for that gas.

damp air Air in which the RELATIVE HUMIDITY (RH) is higher than is usual for the place and season in which it is experienced. In Minneapolis, Minnesota, for example, the average RH at noon in July is 52 percent. If the humidity reached 70 percent, people would think the air was damp. In Seattle, Washington, in contrast, the average RH at noon in July is 68 percent and a 70 percent RH would feel quite normal.

damp haze A very thin FOG that reduces VISIBILITY, but only by an amount that is typical of a HAZE. Visibility does not fall below 1.2 miles (2 km). The haze feels damp because it is caused by very small water droplets or by HYGROSCOPIC NUCLEI.

damping A decrease in the AMPLITUDE of an oscillation that occurs over time because resistance to the oscillation drains energy from it. A wind blowing into a forest loses energy by friction with the trees and is slowed. This is a form of damping. Damping mechanisms are built into certain instruments, such as magnetic compasses, to reduce the difficulty of taking a reading from an oscillating needle.

damping depth The depth within which the temperature of soil is affected by DIURNAL or annual temperature changes at the surface. Assuming the soil to be homogeneous throughout the layer, the damping depth (D) is given by

$$D = (\kappa P/\pi)^{1/2}$$

where κ is the THERMAL DIFFUSIVITY of the soil and P is the PERIOD of the surface temperature change. In addition to mineral particles, soils also contain variable amounts of air and water. These also have thermal diffusivities and damping depths that must be taken into account, depending on the proportion of air and water present in the soil. The diurnal damping depth for dry sand is 3 inches (7.9 cm) and for wet sand 5.5 inches (14 cm). This means that the change in surface temperature as the ground warms in the morning and cools at night is not felt below 3 inches in dry sand and 5.5 inches in wet sand. The annual damping depth, below which the yearly cycle of temperature is not felt, is 5 feet (1.5 m) in dry sand and 9 feet (2.7 m) in wet sand.

Dan A TYPHOON that struck the Philippines on October 10, 1989. It killed 43 people and rendered 80,000 homeless. Dan was also the name of a typhoon that struck the Philippines in October 1999 and then weakened to a TROPICAL STORM as it approached the outlying Taiwanese islands of Penghu and Kinmen. In the Philippines at least eight people were killed, thousands of homes were flooded, and the cost of damage to crops was estimated at $2 million.

dancing devil *See* DUST WHIRL.

dangerous semicircle The side of a TROPICAL CYCLONE where the winds are strongest and where they tend to push ships into the path of the approaching storm. The difference in wind speeds is due to the fact that the storm itself is moving. Consequently, on one side of the storm its own speed of motion must be added to the wind speed of the storm, and on the other side it must be deducted from it. Because the circulation around the storm is CYCLONIC and the storms move in a generally westerly direction, driven by the TRADE WINDS, then turn away from the equator, the dangerous semicircle is on the side of the storm farthest from the equator. This is the northern side in the Northern Hemisphere and the southern side in the Southern Hemisphere.

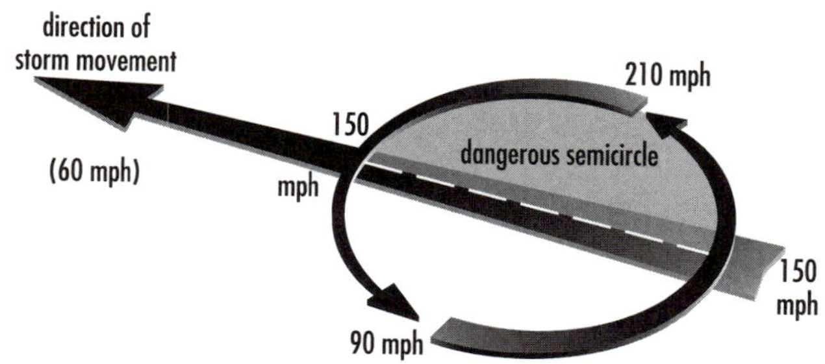

Dangerous semicircle. **The dangerous semicircle is on the northern side of a hurricane or typhoon in the Northern Hemisphere. If the storm is moving westward at 60 mph (96 km/h) and generating winds of 150 mph (241 km/h), in the dangerous semicircle the wind speed is 210 mph (338 km/h) and blowing toward the storm track, pushing ships into the path of the storm.**

Daniell, John Frederic (1790–1845) English *Meteorologist, chemist, and inventor* John Daniell was a prolific inventor and became one of the most eminent scientists of his day. He was born in London on March 12, 1790, the son of a lawyer. He was educated privately, mainly learning Latin and Greek, and either earned a degree from Oxford University or was awarded an honorary one. He then went to work in a sugar refinery and resin factory owned by a relative, where he was able to improve the technology being used.

In his spare time he attended lectures at the Royal Institution given by William T. Brande, the professor of chemistry. He met Brande and the two became close friends. Together they revived the fortunes of the Royal Institution, which were then at a low ebb. Daniell left the factory and became a scientist when, in 1813 at the age of 23, he was appointed professor of physics at the University of Edinburgh. He combined chemistry with the physics he researched and taught at the university, and for a time in 1817 he managed the Continental Gas Company, developing a new process for making gas by distilling resin dissolved in turpentine. This process was used for a time in New York. His interest in gases also extended to the atmosphere.

In 1820 he made his first major contribution to meteorology with his invention of a DEW POINT HYGROMETER that measured RELATIVE HUMIDITY. It comprised two bulbs made from thin glass that were connected by a tube. One bulb was filled with ether and held a thermometer. When the temperature of the air in the other bulb changed, the ether in the bulb was warmed or cooled and either evaporated or condensed. This caused water vapor to condense onto or evaporate from the bulb containing ether. The average temperature at which this happened was the dew point temperature. The Daniell hygrometer was in use for many years.

In 1823 he published a book describing his meteorological research, *Meteorological Essays*. In the same year he was elected a Fellow of the Royal Society.

He moved from Scotland to London in 1831, when he was appointed the first professor of chemistry and meteorology at King's College, which had recently been founded (it is now part of the University of London). Daniell remained in this post until his death.

Daniell was very active in the Royal Society. In 1830 he installed in the entrance hall a BAROMETER that used water to measure pressure. Over the following years he made many observations with it.

He investigated the climatic influence of solar radiation and the circulation of the atmosphere, and he pointed out the importance of maintaining a humid atmosphere in greenhouses. This practice revolutionized greenhouse horticulture, and in 1824 Daniell was awarded the silver medal of the Horticultural Society.

His interest in chemistry and physics had not diminished and in the 1830s Daniell became increasingly interested in electrochemistry. This led to his invention, in 1836, of the Daniell cell, the first reliable source of direct-current electricity.

On March 13, 1845, Daniell suffered a heart attack and died while attending a meeting of the Council of the Royal Society in London.

(You can learn more about John Daniell at www.cinemedia.net/SFCV-RMIT-Annex/rnaughton/DANIELL_BIO.html and www.bioanalytical.com/calendar/97/daniell.htm.)

Dansgaard–Oeschger event **(DO event)** One of the brief warm periods that occurred during the most recent ice age, the WISCONSINIAN GLACIAL (known in Britain as the *Devensian*, in northern Europe as the *Weichselian*, and in the Alps as the *Würm*). There were many of these events, lasting from hundreds to several thousand years. At the onset of a DO event the temperature rose in a few decades to a level not much cooler than that of today, and when it ended the temperature fell just as abruptly to its ice age level. The difference between glacial and DO event sea-surface temperatures in the North Pacific amounted to 5.4°–9° F (3°–5° C). DO events appear to be related to HEINRICH EVENTS and to rapid advances and retreats of GLACIERS, suggesting the events occurred throughout the world. There is also evidence to suggest DO events may also have occurred during the last INTERGLACIAL, known as the *SANGAMONIAN* in North America, the *Eemian* in northern Europe, *Riss–Würm* in the Alps, and the *Ipswichian* in Britain. During this interglacial average temperatures were similar to those of today. The existence of DO events was discovered in the early 1980s by the Danish climatologist Willi Dansgaard, the Swiss climatologist Hans Oeschger, and the American climatologist Chester C. Langway through their examination of ICE CORES taken from the GREENLAND ICE SHEET. These findings were later confirmed during examination of samples taken from the bed of Lake Gerzensee, near Bern.

Darcy's law A mathematical equation that expresses the relationships among the various factors that determine the rate at which GROUNDWATER moves through an AQUIFER. The law, proposed by the French engineer Henri Philibert Gaspard Darcy (1803–58), is stated as follows:

$$Q = kIA$$

where Q is the rate of groundwater flow, k is the PERMEABILITY of the aquifer, I is the gradient of the slope down which the water is moving, and A is the cross-sectional area of the aquifer through which the water is moving.

dark segment **(Earth's shadow)** A dark band that is sometimes seen above the horizon and just below the ANTITWILIGHT ARCH. It appears shortly before sunrise or after sunset under conditions of HAZE.

dart leader A small LIGHTNING stroke that constitutes the third stage in a lightning flash. The flash begins with a STEPPED LEADER. This forms a lightning channel, and before it reaches its destination it is met and neutralized by the RETURN STROKE. The stepped leader carries negative charge away from the cloud and the return stroke carries positive charge to the cloud. This is usually sufficient to neutralize only a part of the cloud and the lightning flash continues, this time starting from deeper inside the cloud. The new discharge begins with a dart leader, carrying negative charge—a stream of electrons. This travels along the lightning channel, ionizing it once more, and is followed by the second major flash. The process is repeated until the charge in the cloud has been fully neutralized.

data *See* DATUM.

data buoy An instrument package that is located in a fixed position offshore and transmits continuous measurements of the surface conditions at sea. These include wind speed and direction and sea-surface temperature. Data buoys have been sited along the East Coast and Gulf Coast of the United States since the early 1970s. They supply information that is used in the compilation of daily weather forecasts and form an important component of the hurricane warning system.

datum (*pl.* **data**) Something that is known or assumed to be true, a premise from which inferences may be drawn, or the fixed starting point of a scale.

datum level A point or level surface that is used as a base from which other elevations can be measured. The point or surface is therefore a DATUM. The most widely used datum level is the sea surface. Altitudes in the atmosphere, the elevation of cities, the heights of mountains, and the depth of the sea are all measured against a sea-level datum.

David A HURRICANE, rated category 5 on the SAFFIR/SIMPSON HURRICANE SCALE, that struck islands in the Caribbean and the eastern coast of the United States in late August and early September 1979. It had winds of up to 150 mph (240 kmh^{-1}), killed more than 1,000 people, and caused damage estimated at billions of dollars. The hurricane affected the Dominican Republic, Dominica, Puerto Rico, Haiti, Cuba, the Bahamas, Florida, Georgia, and New York State.

Dawn A TYPHOON that struck central Vietnam on November 19 to 23, 1998. It was the worst storm to affect the country in 30 years. More than 100 people were killed and about 200,000 were forced to leave their homes.

day The time that the Earth takes to complete one rotation about its axis. This is usually measured from noon on one day to noon on the following day. In SYSTÈME INTERNATIONAL D'UNITÉS (SI) units one day is 86,400 seconds. This varies slightly with changes in the rotation of the Earth and so the figure given is the mean calculated over several years.

The length of a day is not the same when measured in respect to the position of the Sun as it is when measured against the position of a fixed star. To distinguish the two, the day measured with reference to a fixed star is known as the SIDEREAL DAY. Its mean length is 86,164 seconds, which is 236 seconds shorter than the mean solar day. One solar day is therefore equal to 86,636 seconds of mean sidereal time.

Month N	S	0°	10°	20°	30°	40°	50°	60°	70°	80°	90°
Jan	Jul	12.07	11.35	11.02	10.24	9.37	8.30	6.38	0.00	0.00	0.00
Feb	Aug	12.07	11.49	11.21	11.10	10.42	10.07	9.11	7.20	0.00	0.00
Mar	Sep	12.07	12.04	12.00	11.57	11.53	11.48	11.41	11.28	10.52	0.00
Apr	Oct	12.07	12.21	12.36	12.53	13.14	13.44	14.31	16.06	24.00	24.00
May	Nov	12.07	12.34	13.04	13.38	14.22	15.22	17.04	22.13	24.00	24.00
Jun	Dec	12.07	12.42	13.20	14.04	15.00	16.21	18.49	24.00	24.00	24.00
Jul	Jan	12.07	12.40	13.16	13.56	14.49	15.38	17.31	24.00	24.00	24.00
Aug	Feb	12.07	12.28	12.50	13.16	13.48	14.33	15.46	18.26	24.00	24.00
Sep	Mar	12.07	12.12	12.17	12.23	12.31	12.42	13.00	13.34	15.16	24.00
Oct	Apr	12.07	11.55	11.42	11.28	11.10	10.47	10.11	9.03	5.10	0.00
Nov	May	12.07	11.40	11.12	10.40	10.01	9.06	7.37	3.06	0.00	0.00
Dec	Jun	12.07	11.32	10.56	10.14	9.20	8.05	5.54	0.00	0.00	0.00

The word *day* is also used to mean the length of daylight. In this sense *day* is contrasted with *night*. Hours of daylight are measured from the first appearance of the rim of the Sun above the eastern horizon to the disappearance of the upper rim of the Sun on the western horizon.

Daylength varies according to latitude and SEASON. Summer days are long and winter days are short, but the extent to which they are so differs greatly from place to place. At the March and September EQUINOXES the Sun is directly overhead at noon at the equator and on those two days there are 12 hours of daylight and 12 hours of darkness throughout the world. At the June and December SOLSTICES the Sun is directly overhead at noon at the TROPICS of Cancer and Capricorn, respectively. On those two days the difference in length between day and night is at its maximum.

Because the relative lengths of day and night are determined astronomically, they can be predicted with great accuracy for any place on Earth on any day of the year. Inside the ARCTIC CIRCLE and ANTARCTIC CIRCLE, the longest day of the year lasts a full 24 hours (86,400 s) and the shortest day of the year has no length at all, because the rim of the Sun does not appear above the horizon. At the equator, daylight on every day in the years lasts 12 hours 7 minutes (43,620 s). At New York City (40.72° N) the longest day of the year lasts 15 hours 6 minutes and the shortest 9 hours 15 minutes. At Los Angeles (34.05° N) the longest day lasts 14 hours 26 minutes and the shortest 9 hours 53 minutes.

The table shows the hours of daylight on the 15th of each month in hours and minutes for different latitudes. The months are listed twice, for the Northern (N) and Southern (S) Hemispheres.

(You can find the time of sunrise and sunset during the current year for any location in the world from the Astronomical Applications Department of the U.S. Naval Observatory at aa.usno.navy.mil/AA/data/docs/RS_OneYear.html and from the Custom Sunrise Sunset Calendar at www.sunrisesunset.com/custom_srss_calendar.asp.)

day degree (DD) A value that is calculated by multiplying together the number of days on which the mean temperature is above or below a particular DATUM LEVEL by the number of degrees by which it deviates from the datum level. Plants and many animals, especially invertebrate animals, are able to grow and reproduce only when the temperature is above a certain threshold. Consequently, their development is directly related to the length of time during which the temperature exceeds that threshold. Calculating the number of day degrees allows scientists to predict the date when a crop plant will be ready to harvest and the date when particular insect pests will emerge. The DD concept is similar to that of ACCUMULATED TEMPERATURE.

dayglow Very weak light that is emitted in the MESOSPHERE and that contributes to daylight. It is caused by the bombardment of oxygen molecules by sunlight in the far ULTRAVIOLET part of the spectrum, at WAVELENGTHS below 200 nm. Dayglow becomes weaker as the Sun sets.

day-neutral plants *See* PHOTOPERIOD.

DCI *See* STABILITY INDICES.

DD *See* DAY DEGREE.

DDA value *See* DEPTH–DURATION–AREA VALUE.

DDT (***pp*'-dichlorodiphenyltrichloroethane**) An insecticide that came into widespread use in the 1940s, first to control insect vectors of human diseases and later to control agricultural pests. Restrictions imposed on its use in most countries started in the late 1960s because, although it posed very little danger to human health, traces of it accumulated along food chains and caused harm to wildlife, especially to birds of prey. DDT evaporates from the soil and also adheres to dust particles. Its HALF-LIFE of about three years allowed time for airborne DDT to become distributed throughout the world at very low concentrations. In the 1960s it was detected in air by means of the ELECTRON CAPTURE DETECTOR invented by JAMES LOVELOCK. It was this discovery that gave rise to campaigns to have it banned that led to the emergence of the environmentalist movement. DDT was synthesized by the Swiss chemist Paul Hermann Müller (1899–1965), although it was first synthesized in 1873 by an Austrian chemist, Othmar Zeidler. Müller recognized its insecticidal properties, and in 1943 the compound was patented in Switzerland and Britain. For this achievement Müller was awarded the 1948 Nobel Prize for Chemistry.

death assemblage See POLLEN ANALYSIS.

débâcle The breakup of river ice that occurs in spring in Eurasia and North America. It begins in March in the south and progressively later farther north.

decay constant *See* HALF-LIFE.

declination The latitude of an astronomical body, such as the Sun, measured in relation to the celestial equator. This is a projection of the Earth's equator to the outermost limit of the universe (the celestial sphere). Declination is the angle north (designated +) or south (–) of the celestial equator. MAGNETIC DECLINATION is the difference between the direction of true (geographic) north and magnetic north.

Deep Convective Index *See* STABILITY INDICES.

deepening A fall in AIR PRESSURE that occurs at the center of a CYCLONE.

Defense Meteorological Satellite Program (DMSP) A program that is run by the Air Force Space and Missile Systems Center to collect meteorological and oceanographic data and to monitor the space environment through which Earth moves. The program involves the design, building, launching, and maintenance of a number of satellites. These are in near SUN-SYNCHRONOUS ORBITS, with an orbital PERIOD of about 101 minutes, that carry them close to the poles at a height of about 516 miles (830 km). They cross every point on the Earth up to twice every day. Their instruments monitor a swath 1,860 miles (3,000 km) wide, recording images in visible and INFRARED light, and their scanning RADIOMETERS gather information that is used to determine cloud height and type, land and water surface temperatures, ocean currents, and ice and snow. The data are transmitted to ground-based terminals and eventually used in planning U.S. military operations and in compiling civilian weather forecasts. Every day, data are also sent to the Solar Terrestrial Physics Division of the National Geophysical Data Center, where it is added to an accumulating archive.

(You can learn more about the DMSP at www.ngdc.noaa.gov/dmsp/descriptions/dmsp_desc.html and from the Aerospace Corporation at www.laafb.af.mil/SMC/CI/overview/index.html.)

deflation The removal of material from the land surface by the action of the wind. This can produce a depression in the surface, called a *deflation hollow.* Dry sand grains and small soil particles are readily lifted and transported by the wind, so deflation is common on beaches, on dry lakebeds and river beds, and in deserts.

deflation hollow *See* DEFLATION.

deforestation The permanent removal, by clear felling, of an area of forest. On steep slopes, the removal of tree cover can leave the ground unprotected from heavy rain and RUNOFF. This can lead to serious EROSION. Deforestation can also cause local climatic changes, principally by altering the movement of water. TRANSPIRATION is reduced and wind speed increases as the sheltering effect of the trees is lost. Together, the effect is to lower the HUMIDITY of the air. This reduces both PRECIPITATION and EVAPORATION. A change from dark-colored trees to paler grasses or crops also increases the surface ALBEDO. This offsets the increased intensity of sunlight once the shading effect of the trees is lost, and, combined with the increased wind, this means that the surface temperature changes little. There is no evidence that deforestation can have a large direct effect on climates far from the region in which it occurs, but it may have an indirect one. As the root systems of the forest decompose the carbon they contain is oxidized to carbon dioxide, most of which escapes into the air. Consequently, it is possible that widespread deforestation may be followed some time later by a major release of CO_2.

de Geer, Gerhard Jacob *See* VARVES.

deglaciation The melting of an ICE SHEET or GLACIER that exposes the land beneath.

degree-days The number of degrees by which the mean daily temperature is above the minimum temperature needed for growth—called the *zero temperature*—for a particular crop plant. This indicates the time it will take for a crop plant to mature. The zero temperature for corn (maize), for example, is about 55°

F (12.8° C) and in northern Utah corn needs 1,900–2,600 degree days. The sum of all the individual degree days is called the *total degree-days*. *See also* HEATING DEGREE-DAY.

degrees of frost An informal way to describe the number of degrees Fahrenheit by which the temperature is below freezing. For example, a temperature of 26° F might be described as "six degrees of frost." The expression was once widely used in Britain, but its use declined with the adoption of the CELSIUS TEMPERATURE SCALE in radio and television weather forecasts.

dehumidifier A device that is used to dry air that is too humid for comfort. Dehumidifiers are incorporated in many air-conditioning systems, especially those used in large buildings. There are two ways air can be dried. It can be passed through a spray of very cold water, which chills it to below its DEW POINT TEMPERATURE, causing water vapor to condense. Alternatively, the air can be passed across a bed of crystals of a hygroscopic substance such as common salt. The crystals absorb water directly from the air. (*See also* HUMIDIFIER.)

delayed oscillator theory An explanation for the development of EL NIÑO events that proposes these begin with an accumulation of warm water in the western equatorial South Pacific. This is due to ROSSBY WAVES. These are reflected from the western boundary of the atmospheric system, causing KELVIN WAVES to propagate eastward, driving the cold water that gives rise to an El Niño.

delta region A part of the atmosphere in which DIFFLUENCE is occurring. The diverging STREAMLINES make a triangular shape, reminiscent of the Greek letter Δ (delta).

Deluc, Jean André (1727–1817) Swiss *Geologist, meteorologist, and physicist* Jean Deluc is principally remembered for having invented the dry pile, a type of electric battery, but his contributions to geology and meteorology were no less important. He was born in Geneva on February 8, 1727. Until he was 46 he worked in commerce and politics and made scientific excursions among the Swiss mountains. It was then, in 1773, that he moved to England. He was elected a Fellow of the Royal Society and appointed a reader to Queen Charlotte, a post that gave him an income but did not make excessive demands on his time and freed him to pursue his scientific interests.

In 1778 and 1779 he published a six-volume work on geology, *Lettres physiques et morales sur les montagnes et sur l'histoire de la terre et de l'homme* (Physical and moral letters on mountains and on the history of the Earth and of man). In this he suggested that each of the six days of the Biblical creation was an epoch.

Deluc also discovered that water is more dense at 40° F (4° C) than it is at either higher or lower temperatures. In 1761 he also found that the heat required to melt ice or vaporize liquid water does not raise its temperature. This was the concept of LATENT HEAT discovered independently at about the same time by JOSEPH BLACK. Jean Deluc was also the first scientist to propose that the amount of water vapor that can be contained in a given space is independent of any other gases that may be present in that space. He invented a HYGROMETER, though not a very successful one, and he was the first to devise a way of measuring height by means of a BAROMETER. He showed that an increase in elevation is proportional to a decrease in the logarithm of the air pressure, and that a change in elevation is also inversely proportional to the air temperature.

Jean Deluc died at Windsor, England, on November 7, 1817.

dendrite A sharp spike, hexagonal in cross section, that projects from an ICE CRYSTAL. Crystals accumulate dendrites by diffusing (*see* DIFFUSION) through a cloud where the temperature is warmer than about 14° F (–10° C). Water droplets solidify by SUBLIMATION, thereby becoming attached to the growing crystal. Where adjacent dendrites interlock, their crystals are held together, and once crystals have begun to aggregate in this way, they grow into SNOWFLAKES, which can become large.

dendrochronology The dating of wood by means of counting the TREE RINGS it contains. The possibility of dating material in this way was discovered in 1901 by Andrew Ellicott Douglass, an astronomer working at the Lowell Observatory, in Flagstaff, Arizona, who was interested in SUNSPOTS and who thought it might be possible to correlate tree rings with climate. Later,

Douglass used the technique to date buildings in the prehistoric settlement of Pueblo Bonito, New Mexico, by matching tree rings from timber in the buildings with samples taken from trees of a known age.

The technique involves obtaining several samples that are first compared with each other to make certain no rings have been missed and there are no inconsistencies. If trees are being dated, the samples are also checked against specimens taken from other trees in the immediate vicinity and in the region. Finally, the tree rings are matched with an accepted reference standard. Using BRISTLECONE PINE rings, dendrochronology can date material that is up to about 9,000 years old.

(You can learn more about dendrochronology at dizzy.library.arizona.edu/library/teams/set/earthsci/treering/html, www. personal.umich.edu/~dushanem/whatis.html, and www.sonic.net/bristlecone/dendro.html.)

dendroclimatology The interpretation of TREE RINGS to estimate the growing conditions experienced by the trees during the past and from that to reconstruct details of the climate in which they grew.

Denekamp interstade (**Zelzate interstade**) An INTERSTADE that occurred in the Netherlands about 30,000 years ago. During the Denekamp, the average July temperature was about 50° F (10° C).

denoxification *See* POLAR STRATOSPHERIC CLOUDS.

density The mass of a unit volume of a substance. It is measured in pounds per cubic foot (lb ft^{-3}) or pounds per cubic inch (lb in^{-3}), and in SYSTÈME INTERNATIONAL D'UNITÉS (SI) units in kilograms per cubic meter (kg m^{-3}). Under a STANDARD ATMOSPHERE, pure water has a density of 10 kg m^{-3}. The density of air varies with altitude. Under a standard atmosphere its density is 1.23 kg m^{-3}, assuming a RELATIVE HUMIDITY of 50 percent and a content of 0.04 percent carbon dioxide by volume. If this were the density of the air throughout the whole of the atmosphere, the thickness of the atmosphere would be 8.4 km (5.2 miles). This is called the *scale height.* In fact, the air density at that height is 0.5 kg m^{-3}. At a height of 30 km (18.6 miles) the air density is 0.02 kg m^{-3}, which is about 1.6 percent of its sea-level density.

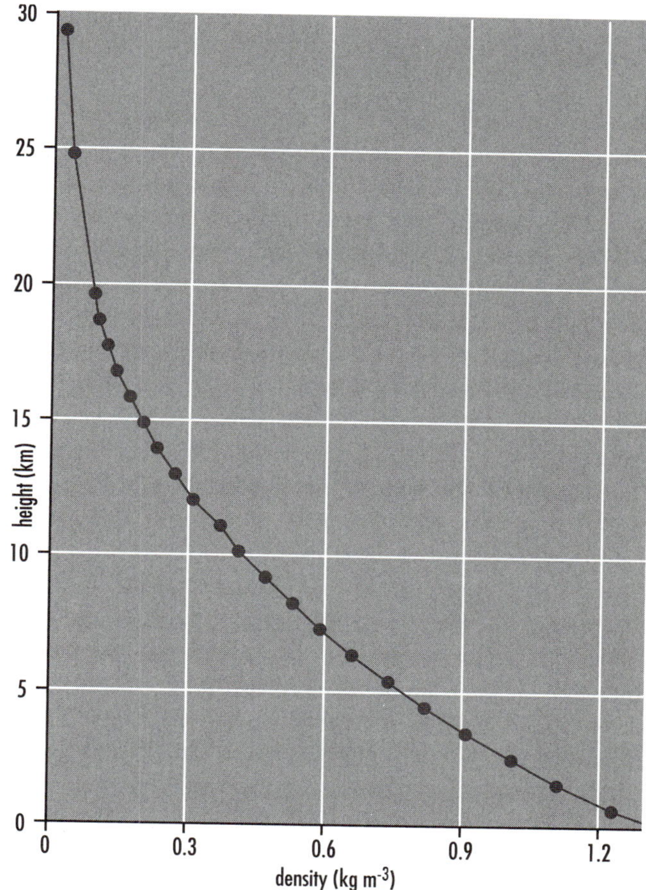

Density. **The decrease in air density with altitude.**

density altitude The height above the surface at which the DENSITY of the air has a specified value. Air density determines the amount of lift the wing and tailplane surfaces of an airplane produce, the amount of drag the airplane experiences, and the power that its engines yield. An instrument in the cockpit shows the density of the outside air. By finding this figure in a table that shows the temperature, pressure, and density of air at various altitudes in a STANDARD ATMOSPHERE, the pilot can read the density altitude. This is the altitude at which air has the measured density in a standard atmosphere, regardless of the actual height of the airplane above the surface. When approaching an airfield after a long flight, the pilot needs to know the density altitude of the runway in order to calculate the way the airplane will respond to its controls.

(You can learn more about density altitude at usatoday.com/weather/wdenalt.html.)

density current A flow of air that is caused by a difference between the DENSITY of the moving air and that of adjacent air. Density currents occur when cold air undercuts warmer air at a COLD FRONT. KATABATIC WINDS are also density currents (*see* MOUNTAIN BREEZE). These occur when air high on a mountainside is chilled by contact with a snow-covered surface. The density of the air increases as its temperature falls, and it begins to move down the slope, displacing the warmer, less dense air at a lower level.

density ratio The ratio of the DENSITY of the air at a specified altitude to the density of air at the same altitude in a STANDARD ATMOSPHERE.

denudation The stripping away of the material that covers the surface of the land, leaving bare rock exposed. The word is derived from the Latin verb *denudare,* which means "to strip completely naked." Denudation includes the processes of WEATHERING, EROSION, and the transport of material that has been detached from the surface.

departure The extent to which the value of a meteorological factor, such as TEMPERATURE, PRECIPITATION, AIR PRESSURE, or HUMIDITY, differs from the mean value.

deposition The formation of ice on a solid surface by the direct conversion of water vapor into ice without passing through the liquid PHASE. This is the mechanism by which HOAR FROST forms. Many scientists prefer to use the word *deposition* to describe this process rather than *sublimation,* which they reserve for the direct change of ice into water vapor.

depression The name that is commonly used to describe a midlatitude FRONTAL CYCLONE; it refers to a well-defined area of low atmospheric pressure: it is the AIR PRESSURE that is "depressed." *Cyclone* refers to the circulation of air around the low-pressure center; this is CYCLONIC. The two words both describe the same phenomenon, but from different points of view. *Depression* is used informally rather than *cyclone,* because *cyclone* is popularly associated with TROPICAL CYCLONES. Mention of an approaching cyclone in a TV weather forecast might distress viewers, when all they need fear is a period of rain.

depression family *See* WAVE DEPRESSION.

depth–duration–area value (DDA value) The average depth of the PRECIPITATION that falls during a specified time over a specified area.

depth hoar (sugar snow) A layer of FROST that forms by DEPOSITION just beneath the surface of a layer of snow. In Antarctica, depth hoar forms a layer up to 1 inch (2.5 cm) deep in autumn, beneath a layer of hard snow that is only about 0.2 inch (5 mm) thick. These alternate layers of depth hoar and snow are preserved, and by cutting a trench down through the snow it is possible to use them to measure the amount of snow that fell each year.

descriptive climatology CLIMATOLOGY that is presented as descriptions of climates, using verbal accounts, graphs, tables, and other illustrations, but that omits discussions of the causes of climatological phenomena or of climatological theory.

descriptive meteorology (aerography) METEOROLOGY in which the composition and structure of the atmosphere and atmospheric phenomena are described through verbal accounts, graphs, tables, and other illustrations, but with no discussion of the causes of those phenomena or of meteorological theory.

desert climate In the CLIMATE CLASSIFICATION devised by MIKHAIL I. BUDYKO, a climate in which the RADIATIONAL INDEX OF DRYNESS has a value of more than 3.0. *See also* MIDDLE-LATITUDE DESERT AND STEPPE CLIMATE.

desert devil *See* DUST WHIRL.

desertification The deterioration of land until it has the characteristics of a dry desert. The fact that this happens leads to the supposition that some or all deserts may be expanding as they encroach onto adjacent land.

Modern fears of this process began in the early 1970s. Prolonged DROUGHT in the SAHEL region of

Africa forced people to migrate and led to the deaths of more than 100,000 people and up to 4 million livestock. In 1973, drought in Ethiopia is believed to have claimed up to 250,000 lives and the drought returned in 1984–85.

Aerial and satellite photographs revealed that the vegetation was being damaged over large areas where farm livestock was being allowed to graze with no control. The pictures also showed fenced areas where grazing was controlled and the vegetation was in much better condition. This led some people to conclude that human behavior was responsible for the spread of the desert.

The United Nations (UN) held the Conference on Desertification in Nairobi in 1977. Officials of the United Nations claimed to have coined the term *desertification* to describe the process, but the word was first used, in French, in 1949, in the book *Climats, Fôrets et Desertification de l'Afrique Tropicale* by A. Aubreville. In fact, fears of the southward advance of the Sahara were being expressed in the 1930s. The UN Environment Program defines desertification as "land degradation in arid, semi-arid, and dry sub-humid areas resulting mainly from adverse human impacts."

It is not confined to Africa. According to UN statistics, the area of degraded land worldwide increases by more than 23,000 square miles (59,570 km^2) every year. There are almost 400,000 square miles (1,000,000 km^2) of "desertified" land in Brazil alone. In Asia there are almost 5,400,000 square miles (14,000,000 km^2).

Scientists now believe that droughts in the semiarid lands bordering deserts are entirely natural phenomena in which human activity plays no part. They occur at irregular intervals, sometimes last for several years, and end with the return of the rains. Both the Sahel and Ethiopia recovered from the droughts that devastated them.

During the drought, however, land management becomes critically important. As the pasture dries, seminomadic peoples whose livelihoods depend on their cattle, sheep, goats, and camels are likely to be crowded into the areas where the pasture survives. This leads to overexploitation of the vegetation. Similarly, where once-nomadic peoples are encouraged to settle in permanent villages, taking their livestock with them, the pasture around the village is often destroyed. Trees and shrubs are also removed, for use as fuel.

In this situation, desert soils dry to dust, and, blown by the wind, they bury the plants on land nearby. Dust coats plant leaves, inhibiting PHOTOSYNTHESIS, and more vegetation dies. Most scientists now agree that this deterioration can be described as desertification only if it affects vegetation as well as the soil, if it is caused partly by human activity, and if the deterioration continues for at least 10 years. If rain returns, all or most of the affected land can be made to recover.

Desertification was debated at the Rio Summit (the United Nations Conference on Environment and Development) held in June 1992 in Rio de Janeiro, where it was agreed that a treaty should be prepared as a guide to nations in halting land degradation. This led to the UN Convention to Combat Desertification (CCD), which is based on local schemes that are coordinated internationally.

(You can learn more about desertification in Michael Allaby's *Ecosystem: Deserts* [New York: Facts On File, 2001] and at www.rona.unep.org/action/15f.htm. Information about the UN Convention to Combat Desertification is at www.unccd.de/main2.html.)

desert wind A wind that blows off the desert. It is hot in summer, cold in winter, and very dry.

desiccation A long-term decrease in the amount of surface water and/or GROUNDWATER in a region as a consequence of climatic change.

desorption The release of a gas that had previously been held in or on the surface of another substance. It is the opposite of both ABSORPTION and ADSORPTION.

determinism The idea that comprehensible natural laws govern the transition of a system from its present to a future state. In other words, laws acting upon the present condition determine the future condition. It follows that if it were possible to know the present condition in sufficient detail and to understand the laws completely, the entire future could be predicted. The weather, for example, could be forecast accurately for years or even centuries in advance. If the weather system is deterministic, it means its state results entirely from forcing (*see* CLIMATIC FORCING). If a system is not deterministic, it may be STOCHASTIC or chaotic (*see* CHAOS).

detrainment An outflow from a body of moving air into the surrounding air. The process occurs when a CUMULIFORM cloud dissipates. Its CLOUD DROPLETS evaporate, the smallest ones first. The cloud grows darker in color, because the larger droplets scatter less light than the small ones. Finally all the droplets have gone and the water vapor and air with which it is mixed disperse into the surrounding air. Detrainment from the top of a towering CUMULONIMBUS cloud can release water vapor that forms CIRROSTRATUS at a higher level.

devastating drought One of the types into which DROUGHTS are formally classified. It is a particularly severe CONTINGENT DROUGHT that occurs in summer and causes plants to wilt and die.

development The generation of vertical motion by BUOYANCY forces and the rise of warm air. This leads to a DIRECT CIRCULATION.

Devensian glacial The most recent GLACIAL PERIOD in Britain, which began about 70,000 years ago, or possibly a little earlier, and ended about 10,000 years ago. The LAST GLACIAL MAXIMUM occurred about 21,000 years ago. It is known as the *Weichselian glacial* in Northern Europe and *Würm glacial* in the Alps, and it is approximately equivalent in date to the WISCONSINIAN glacial of North America. The Devensian was preceded by the IPSWICHIAN INTERGLACIAL and followed by the present FLANDRIAN INTERGLACIAL.

dew Moisture that condenses from the air onto the surface, most commonly onto plant leaves. It forms during cool nights when there is little wind. Heat absorbed during the day is radiated from the ground surface. This produces a layer of cool air adjacent to the surface. If the air is chilled sufficiently for its RELATIVE HUMIDITY to exceed 100 percent, water condenses onto surfaces. This is dew. If the air is completely still, the water that has condensed cannot be replaced from the moister air above, and so only a small amount of dew forms. If the wind is too strong, warmer air replaces the cool surface air and condensation does not occur. The wind speed at which dew forms varies according to the roughness of the surface.

dewbow A very faint RAINBOW that is sometimes seen in DEW drops on the ground.

dew cell (dew point hygrometer) A HYGROMETER that measures the DEW POINT TEMPERATURE directly. It has a mirrored surface that is cooled electrically. As soon as water starts condensing onto it the reflectance of the surface changes. This change is detected by a photoelectric sensor, which trips a switch, causing the mirrored surface to be heated. As soon as the water evaporates from it the change in reflectance is once again detected, and the cooling circuit is activated. This alternation between heating and cooling continues until the two stabilize. The temperature at which this happens is the dew point temperature, and its value appears on the display.

dew gauge (surface wetness gauge) An instrument that is used to measure DEW. It comprises a polystyrene plastic (Styrofoam) ball of a standard size. The weight of the ball changes as dew condenses onto it and the instrument records this change. The ball is held at the end of a vertical arm that is connected to a system of balances. A pen at the end of one of the balance arms makes a continuous record on a chart fixed to a rotating drum. The Styrofoam ball is exposed, but the remainder of the device is enclosed by a weatherproof case. Siting of the gauge is important. If it is to give a true reading for only condensation it

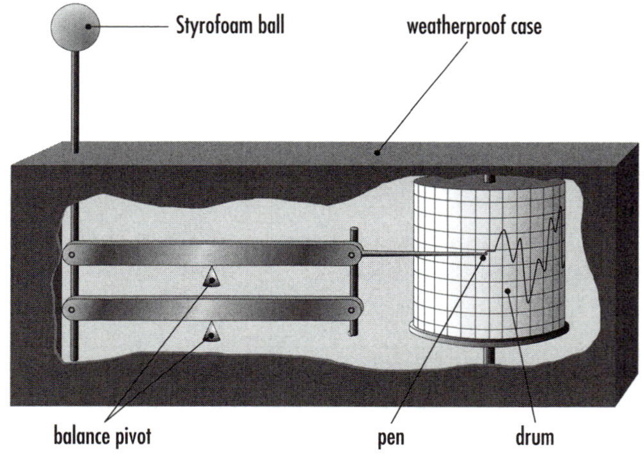

Dew gauge. **The weight of moisture that condenses as dew onto the plastic (Styrofoam) ball is weighed by a system of balances and a pen records the changing weight on a rotating drum.**

must be protected from rain and from water dripping from foliage.

dew in the night Weather folklore that predicts a fine day in summer. The saying is

Dew in the night,
Next day will be bright.

This is often true.

dew point depression The difference between the AMBIENT temperature and the DEW POINT TEMPERATURE. This is approximately double the WET-BULB DEPRESSION. Dew point depression is one of the ways in which meteorologists specify the amount of water vapor present in the air. For example, at an ambient temperature of 68° F (20° C) and a RELATIVE HUMIDITY of 75 percent, the dew point depression is 7.9° F (4.4° C) and the wet-bulb depression is 15.7° F (8.7° C).

dew point front *See* DRY LINE.

dew point hygrometer See DEW CELL.

dew point lapse rate *See* LAPSE RATES.

dew point temperature The temperature at which a PARCEL OF AIR would become saturated if it were cooled with no change in the amount of moisture it contained or in the atmospheric pressure. As the air is chilled to below its dew point temperature, water condenses onto surfaces as dew. The dew point temperature can be read from a table in which it is calculated from the dry-bulb temperature and WET-BULB DEPRESSION measured by the PSYCHROMETER. It can be read directly from a DEW CELL. If the air is chilled to below freezing, it passes the FROST POINT.

dew pond A shallow pond that is excavated by a farmer on pasture land to provide water for livestock. DEW collects in the bottom of the pond, hence the name, but most of the water arrives as rain.

diabatic temperature change A change in the temperature of air that is due to contact with the surroundings; the change is not ADIABATIC. Diabatic temperature changes occur in the layer of air adjacent to the surface of the Earth. This air is in contact with the surface, by which it is warmed or cooled, and the horizontal movement of air is strongly affected by the roughness of the surface. This produces eddies that have the effect of mixing the air thoroughly.

diamond dust *See* ICE PRISMS.

Diane A HURRICANE that struck the United States in August 1957. It killed more than 190 people and caused $1.6 billion of damage.

diapause A temporary cessation of growth and development that an insect enters during a period of adverse conditions, usually cold or dry seasonal weather. Development resumes as soon as conditions improve. Insects can enter diapause as eggs, larvae, pupae, or adults.

diathermancy (*adj.* **diathermanous**) The property of being transparent to radiant heat.

diathermanous *See* DIATHERMANCY.

differential heating The warming of surfaces at varying rates when they are all equally exposed to sunshine. It occurs for a number of reasons. A dark surface warms faster than a pale surface, because its ALBEDO is lower. Sand warms faster than peat, because it has a higher THERMAL CONDUCTIVITY and a lower HEAT CAPACITY. Wet material warms more slowly than dry material, because of the high heat capacity of water and also because of cooling of wet surfaces by EVAPORATION. Differential heating causes ANABATIC WINDS and LAND AND SEA BREEZES, and in hot, dry climates it produces DUST DEVILS and desert WHIRLWINDS.

diffluence A flow of air in which two or more STREAMLINES move away from each other. This slows the rate of flow, because there is a widening stream to carry the flow, but unlike DIVERGENCE, it does not result in an outflow of air from the area or any vertical movement.

diffraction The bending of light as it passes close to the sharply defined edge of an object by an amount that is proportional to the wavelength of the light (red light is diffracted more than blue light because its wavelength is greater). This causes the edge to appear

DRY-BULB TEMPERATURE (°C)	WET-BULB DEPRESSION (°C)														
	0.5	1.0	1.5	2.0	2.5	3.0	3.5	4.0	4.5	5.0	7.5	10.0	12.5	15.0	17.5
-10.0	-12.1	-14.5	-17.5	-21.3	-26.6	-36.3									
-7.5	- 9.3	-11.4	-13.8	-16.7	-20.4	-25.5	-34.4								
-5.0	-6.6	-8.4	-10.4	-12.8	-15.6	-19.0	-23.7	-31.3	-78.6						
-2.5	-3.9	-5.5	-7.3	-9.2	-11.4	-14.1	-17.3	-21.5	-27.7	-41.3					
0.0	-1.3	-2.7	-4.2	-5.9	-7.7	-9.8	-12.3	-15.2	-18.9	-23.9					
2.5	1.3	0.1	-1.3	-2.7	-4.3	-6.1	-8.0	-10.3	-12.9	-16.1					
5.0	3.9	2.8	1.6	0.3	-1.1	-2.6	-4.2	-6.1	-8.1	-10.4	-47.7				
7.5	6.5	5.5	4.4	3.2	2.0	0.7	-0.8	-2.3	-4.0	-5.8	-21.6				
10.0	9.1	8.1	7.1	6.0	4.9	3.8	2.5	1.2	-0.2	-1.8	-12.8				
12.5	11.6	10.7	9.8	8.8	7.8	6.7	5.6	4.5	3.2	1.9	-6.8	-28.2			
15.0	14.2	13.3	12.5	11.6	10.6	9.6	8.6	7.6	6.5	5.3	-1.9	-14.5			
17.5	16.7	15.9	15.1	14.3	13.4	12.5	11.5	10.6	9.6	8.5	2.3	-7.0	-35.1		
20.0	19.3	18.5	17.7	16.9	16.1	15.3	14.4	13.5	12.6	11.6	6.1	-1.4	-14.9		
22.5	21.8	21.1	20.3	19.6	18.8	18.0	17.2	16.3	15.5	14.6	9.6	3.2	-6.3	-37.5	
25.0	24.3	23.6	22.9	22.2	21.4	20.7	19.9	19.1	18.3	17.5	12.9	7.3	-0.2	-13.7	
27.5	26.8	26.2	25.5	24.8	24.1	23.3	22.6	21.9	21.1	20.3	16.1	11.1	4.7	-4.7	-31.7
30.0	29.4	28.7	28.0	27.4	26.7	26.0	25.3	24.6	23.8	23.1	19.1	14.5	9.0	1.6	-11.1
32.5	31.9	31.2	30.6	29.9	29.3	28.6	27.9	27.2	26.5	25.8	22.1	17.8	12.8	6.6	-2.4
35.0	34.4	33.8	33.1	32.5	31.9	31.2	30.6	29.9	29.2	28.5	24.9	21.0	16.4	11.0	3.9
37.5	36.9	36.3	35.7	35.1	34.4	33.8	33.2	32.5	31.9	31.2	27.7	24.0	19.8	14.9	8.9
40.0	39.4	38.8	38.2	37.6	37.0	36.4	35.8	35.1	34.5	33.9	30.5	26.9	23.0	18.5	13.3

To calculate the dew point temperature:

1. determine the wet-bulb depression;
2. find the dry-bulb temperature in the column on the left;
3. find the wet-bulb depression along the top row;
4. look down the column below the wet-bulb depression and along the row next to the dry-bulb temperature. The number where the column and row intersect is the dew point temperature.

For example, suppose the dry-bulb temperature is 7.5° C and the wet-bulb depression is 1.5° C. Look down the column below 1.5 and along the row next to 7.5. The column and row intersect at 4.4, so the dewpoint temperature is 4.4° C.

Note that dewpoint temperature tables use temperatures in degrees Celsius (°C). To convert from Fahrenheit (°F), remember to distinguish between measurements of temperature and heat (the magnitude of degrees on the two scales). For the dry-bulb temperature: °C = (° F - 32) x 5 ÷ 9
For the wet-bulb depression: °C = °F x 5 ÷ 9

Dew point temperature. **To calculate the dew point temperature: (1) Find the dry-bulb temperature in the column on the left; (2) find the wet-bulb temperature along the row at the top; (3) follow the column below the wet-bulb temperature and the row next to the dry-bulb temperature; the two intersect at the dew point temperature.**

blurred, but it also gives rise to INTERFERENCE, as different parts of the spectrum meet. This produces certain optical phenomena.

diffusion Mixing that occurs when one fluid is added to another, but the two are not stirred or otherwise agitated. It is due to the random movement of

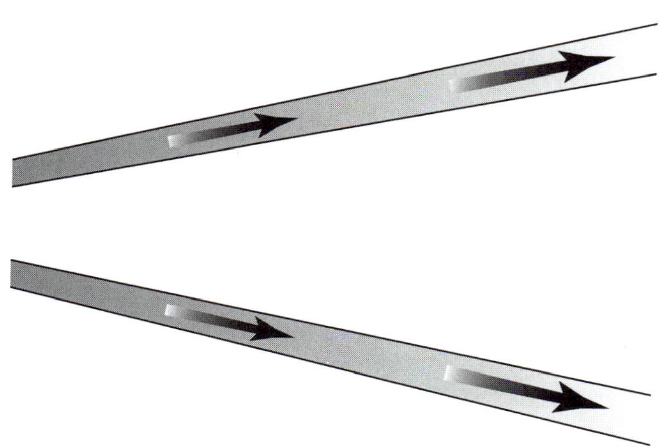

***Diffluence.* Airstreams flow away from each other and slow.**

molecules. Molecules of one component mingle with those of the other until the two types are distributed evenly throughout the fluid, which is then a mixture. This can be seen, for example, when a few drops of a food dye are added to clean water. Almost at once, the color begins to spread outward and eventually spreads throughout the water so it is no longer possible to identify the place where the colorant was added. This is a slow process, however, and it might take months for the color to spread itself evenly. Gas molecules move faster and with much more freedom than molecules in a liquid, so diffusion is much faster in gases than in liquids. A gas added to air diffuses through it fairly quickly, provided the volume of air does not extend very far vertically. If a column of air is more than about 3,000 feet (1 km) tall, diffusion is affected by gravity and molecules tend to sort themselves by weight.

diffusion diagram A graphic representation of several of the processes that are involved in DIFFUSION, such as the MEAN FREE PATH or MIXING LENGTH of molecules together with their VELOCITY.

diffusive equilibrium A condition in which the downward movement of air molecules under the influence of gravity is balanced by the upward movement of molecules drifting into a region containing fewer molecules per unit volume. This occurs above the TURBOPAUSE, at heights greater than about 60 miles (100 km). At lower levels, collisions between molecules contribute to the mixing of the atmospheric constituents, but the air in the upper atmosphere is so rare that molecules seldom collide. The overall effect is that the mean density of the atmosphere decreases with height and the heavier molecules are concentrated at lower levels, so the atmospheric constituents tend to separate by weight. The region in which diffusive equilibrium determines the atmospheric structure is known as the *HETEROSPHERE*.

diffusograph An instrument that measures diffuse radiation from the sky. It consists of a PYRANOMETER that is surrounded by a circular strip set at such an angle that as the Sun crosses the sky the sensor on the pyranometer remains always in shade. Consequently, the sensor is exposed only to diffuse light from above.

digital image A picture that is made from a continuously varying stream of DATA received from an orbiting satellite or other remote source. The continuous variation is converted into discrete variation, in which the data change in small steps. Each discrete change is then given a numerical value as a picture element (conventionally abbreviated to *pixel*) and assigned a location defined by coordinates.

dimensionless number A quantity that is completely abstract, in that it is not a quantity of anything. The most familiar dimensionless number is π (pi), which defines the ratio of the diameter of a circle to its circumference. Meteorologists use several dimensionless numbers and they are extremely useful. The REYNOLDS NUMBER, used in calculations of fluid flow, is a dimensionless number. Others are the magnitude of the CORIOLIS EFFECT, which acts on bodies moving in respect to the surface of the Earth, and the ROSSBY NUMBER, which is used in calculations of the ACCELERATION of bodies that is caused by the rotation of the Earth.

dimethyl sulfide (DMS) A chemical compound, $(CH_3)_2S$, that is produced by the decomposition of dimethylsulfoniopropionate, a substance present in the cells of many single-celled marine algae, where it prevents the salt concentration from rising to a harmful level. When the organisms die the dimethylsulfoniopropionate decomposes and DMS enters the water, where its decomposition continues. A proportion of the DMS escapes into the air, however, and is rapidly oxidized. One product of DMS oxidation is sulfuric acid (H_2SO_4). Small droplets of H_2SO_4 form ideal CLOUD CONDENSATION NUCLEI, and DMS produced by marine algae is believed to be the principal source of such nuclei over the open ocean, where the air is almost completely free of dust. DMS is also the most important compound involved in the transport of sulfur from the sea to the land as part of the SULFUR CYCLE.

direct cell One of the cells that form part of the GENERAL CIRCULATION of the atmosphere and that are driven by CONVECTION. The HADLEY CELLS and POLAR CELLS are direct cells.

direct circulation The situation in which warm air is rising and cold air subsiding within a FRONTAL ZONE. This is a stage in CYCLOGENESIS, which is the intensification of the front. At a weakening front the circulation is in the opposite direction, with cold air rising and warm air sinking.

discomfort index *See* TEMPERATURE–HUMIDITY INDEX.

dishpan experiments Laboratory experiments that simulate the GENERAL CIRCULATION of the atmosphere and demonstrate the importance of the horizontal transport of heat and MOMENTUM. Descriptions of the general circulation began in the late 17th century, when EDMUND HALLEY offered an explanation for the TRADE WINDS. Some years later, GEORGE HADLEY improved on Halley's explanation, and in 1856 WILLIAM FERREL completed the explanation. The result came to be known as the *three-cell model* of the way heat is transported by the atmosphere from the equator to the Poles, and by the end of the 19th century the model was assumed to be correct.

Climatologists still accept the three-cell model, but by the 1920s some scientists were beginning to suspect that it does not tell the whole story. They thought it possible that, in addition to the vertical transport of the three cells, heat and momentum traveled horizontally. At the time there seemed no way to test the idea. The PREVAILING WINDS that blow horizontally near the surface were well known, but to test the idea the scientists needed to know how the air was moving in the upper TROPOSPHERE, an inaccessible region about which little was known.

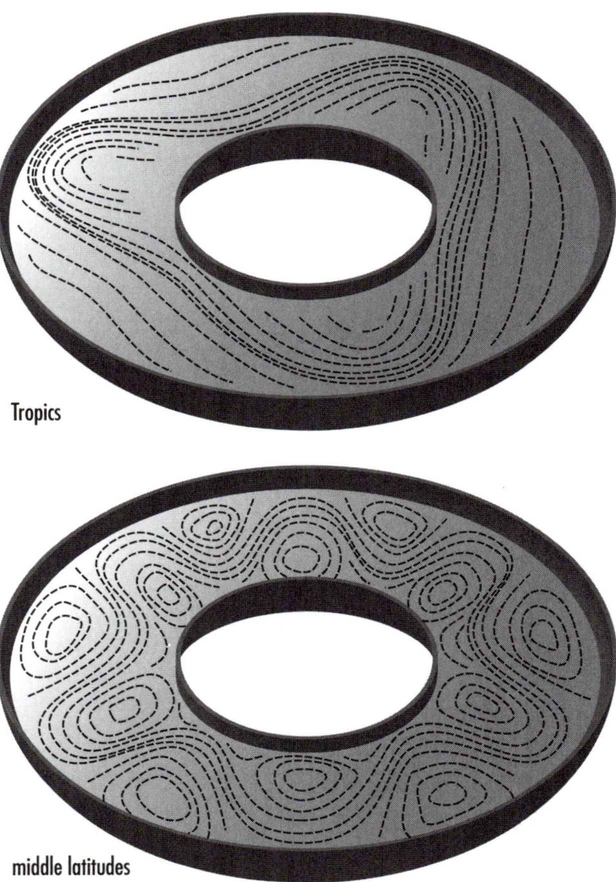

Dishpan experiments. **The dishpan is an open-topped, doughnut-shaped container that is cooled at the center and warmed at the outer edge. When it is filled with water and made to rotate, currents develop in the water with a pattern that is identical to the pattern made by the horizontal movement of the atmosphere.**

The three-cell model was based on what happens when a pan of water, or "dishpan," is heated over a stove. Warming at the base and cooling at the top establish one or more CONVECTION CELLS. The atmosphere is not really like a dishpan on a stove, however, because it is subject to forces that arise from the rotation of the Earth and also because the equator is not a simple source of heat.

That is how the laboratory dishpan was invented. It is shallow and shaped like a doughnut. The center is kept cold while heat is applied around the outer edge, and the dishpan rotates. It is filled with water to which drops of dye are added. As currents develop in the water their directions are traced by the dye they carry.

Water in a rotating dishpan represents the atmosphere as seen from above the North or South Pole (depending on the direction of rotation). The outer rim is the "equator" and the cold center is the pole.

A simple refinement of the device makes it possible to simulate the atmosphere within particular zonal (latitudinal) belts. All this requires is an adjustment of the rate of rotation. Slow rotation produces a "low-latitude atmosphere," and faster rotation produces a simulation of atmospheric motion in higher latitudes.

Dishpan experiments showed that in addition to movement through the vertical convection cells, air develops horizontal waves with a very long wavelength. These are now called *ROSSBY WAVES*. In the Tropics there are about three waves at any time, with a WAVELENGTH of about 8,300 miles (13,340 km). In middle latitudes there are about six and the wavelength is about 5,860 miles (9,435 km). The middle latitude circulation also generates many eddies. All the waves move slowly eastward, and those in middle latitudes periodically change in AMPLITUDE over an INDEX CYCLE.

These large waves carry warm air away from the equator and cold air toward the equator and they also confirm that the atmospheric circulation in middle latitudes differs substantially from that of the Tropics. They also show that the surface wind belts—the low-latitude easterlies, middle-latitude westerlies, and high-latitude easterlies—are produced by traveling waves, not by the meridional (north–south) circulation.

Results from dishpan experiments confirm predictions of atmospheric behavior that have been made mathematically and by computer simulations.

dispersion The separation of the component wavelengths as white light passes through a prism. Each wavelength is then seen as a band of a distinct color.

dissipation trail *See* DISTRAIL.

distorted water Water that has a molecular structure different from that of the main body of water of which it forms part. This water forms a layer, several molecules thick, at the surface of a body of water and adjacent to the BOUNDARY LAYER.

distrail (dissipation trail) A line of clear air that appears in thin cloud behind an aircraft. It is caused by vortices in the wake of the aircraft. These draw dry air down into the cloud, causing the CLOUD PARTICLES to evaporate. The distrail may form a straight line or a line of holes in the cloud. What appears to be a distrail is sometimes a shadow cast onto the cloud by a CONTRAIL above it.

disturbance Any small-scale variation that occurs in the general state of the atmosphere at a particular time and place.

disturbance line A weather system that occurs in spring and fall in West Africa. Moist air flowing from the southwest as part of the developing or fading MONSOON circulation is overrun by dry air from the Sahara. This produces a SQUALL LINE several hundred miles long that travels westward at about 30 mph (50 km h^{-1}). It produces SQUALLS and THUNDERSTORMS but dissipates once it has crossed the coast and encounters the cold water of the Atlantic. Disturbance lines are a major cause of rainfall in the periods immediately preceding and following the main monsoon seasons, between April and June and September and November.

diurnal Of the day, from the Latin word *dies,* which means "day." This can be interpreted in two ways. A diurnal animal, for example, is active by day, in contrast to a nocturnal animal, which is active only by night. In METEOROLOGY, however, *diurnal* refers to a phenomenon that occupies and is completed within a full 24 hours, counted from midnight to midnight, or that returns at intervals of 24 hours.

diurnal range The difference between the daytime and nighttime temperature for a particular place. This may be calculated as the mean range, in which case the difference is between average temperatures, or the absolute range, in which case it is counted between the highest and lowest temperatures that have been recorded. In Atlanta, Georgia, for example, the mean diurnal range is 18° F (10° C) and the absolute range is 66° F (37° C).

divergence A flow of air in which STREAMLINES move outward from an area. This happens when air flows away from a region of high pressure. The effect is to disperse air from the region where the streamlines separate. This decreases the quantity of air in

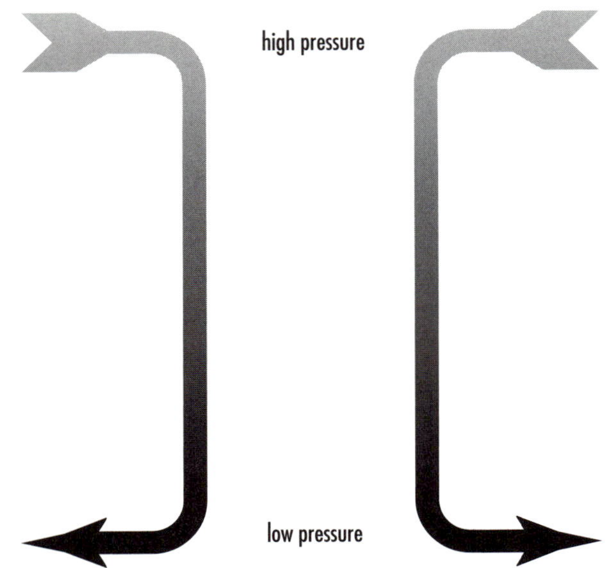

Divergence. **Air flows outward, producing low pressure.**

that area and therefore decreases the atmospheric pressure. The decreased pressure produced by divergence near the surface of land or sea causes air at a higher level to sink and fill the low pressure, so a region of low-level divergence is also a region of sinking air. High-level air flows inward to an area that corresponds to the area of low-level divergence. This produces a region of CONVERGENCE and rising pressure above the area of low-level divergence and falling pressure. Air that sinks in a region of divergence warms ADIABATICALLY as it does so. This raises the RELATIVE HUMIDITY of the air, favoring the evaporation of clouds and clear skies.

DMS *See* DIMETHYL SULFIDE.

DMSP *See* DEFENSE METEOROLOGICAL SATELLITE PROGRAM.

Dobson spectrophotometer An instrument that was invented in 1924 by the British physicist G.M.B. Dobson (1889–1976). It measures the intensity of different wavelengths of ULTRAVIOLET (UV) RADIATION, from which the concentration of OZONE present in the atmosphere can be inferred. The spectrophotometer contains a photoelectric cell and a filtering device that allows UV radiation at four wavelengths to fall on the cell one at a time in a sequence. Of these wavelengths, two are absorbed by ozone and two are not. Those that are absorbed by ozone give a lower reading that those that are not absorbed. The readings from the stronger wavelengths are gradually reduced until they equal those from the weaker wavelengths. From this, the ratio of the stronger to the weaker sets of readings can be measured and the concentration of ozone is inferred from the amount of radiation the ozone absorbs. A spectrophotometer in Switzerland has produced continuous data on the ozone layer since the 1920s. The instrument can use UV emitted by the Moon and stars as well as solar UV, but its measurements can be distorted by AEROSOLS and pollutant gases.

Dobson unit (DU) A unit of measurement that is used to report the concentration of a gas that is present in the atmosphere or in a particular part of the atmosphere. It refers to the thickness of the layer that gas would form if all the other atmospheric gases were removed and the gas in question were taken to sea level and subjected to standard sea-level pressure. The amount of OZONE present in the stratospheric OZONE LAYER is usually reported in Dobson units. In the case of ozone, 1 Dobson unit corresponds to a thickness of 0.01 mm (0.0004 inch) and the amount of ozone in the ozone layer is typically 220–460 DU, corresponding to a layer 2.2–4.6 mm (0.09–0.18 inch) thick. The unit is named for Gordon Miller Bourne Dobson (1889–1976), the British physicist who studied stratospheric ozone in the 1920s and who invented the DOBSON SPECTROPHOTOMETER.

doctor *See* HARMATTAN.

DO events *See* Dansgaard–Oeschger events.

dog days July and the first half of August, which is the hottest part of the summer in the Northern Hemisphere. At this time Sirius, the "dog star" and the brightest star, rises in conjunction with the Sun. The expression is from the Latin *caniculares dies* and arises from the Roman belief that the hot weather is due to the fact that heat from Sirius adds to the heat from the Sun. This is not so. The amount of energy Earth receives from Sirius is infinitesimal and has no effect on the weather.

doldrums A sea area in which the winds are light and variable. The extent of the doldrums varies considerably with the seasons, but they are located in the INTERTROPICAL CONVERGENCE ZONE, on the side nearer the equator of the region in which the TRADE WINDS originate. The calm weather of the doldrums is interrupted at intervals by fierce storms.

There are three principal doldrum zones. In the Pacific, the doldrums are located in the east, but from July to September they extend westward as a tongue reaching to about longitude 110° W. A second zone is located in the western Pacific, north of Australia and in the vicinity of Indonesia, and in the Indian Ocean. Its area increases from October to December, but it reaches its maximum extent in March and April, when it reaches from the coast of East Africa to longitude 180° E, a distance of about 10,000 miles (16,000 km). The third doldrum zone is in the eastern Atlantic. For most of the year it extends only a short distance from the African coast, but from July to September it reaches all the way across the ocean to Brazil.

Sailing ships could be becalmed in the doldrums. The variability in their location and the fact that at times they extended from one side of the Atlantic to the other made it difficult for captains to avoid them. Lack of sufficient wind to shift the vessel was a very real hazard, because stores of food and drinking water could run low. Until modern times, sailors had no means of making sea water drinkable. The plight of sailors in this condition was described by the English poet Samuel Taylor Coleridge (1772–1834) in *The Rime of the Ancient Mariner:*

All in a hot and copper sky,
The bloody Sun, at noon,
Right up above the mast did stand,
No bigger than the Moon.

Day after day, day after day,
We stuck, nor breath nor motion;
As idle as a painted ship
Upon a painted ocean.

Water, water, everywhere,
And all the boards did shrink;
Water, water, everywhere,
Nor any drop to drink.

The origin of the word *doldrums* is obscure, but it probably is from the Old English word *dol,* which meant "dull" or "stupid." By the early 19th century a *doldrum* was a dull or stupid person; from it arose the use of "the doldrums" to refer to low spirits and "in the doldrums" to mean "down in the dumps." Coleridge never used *doldrums* to describe the weather his Ancient Mariner experienced, because it was not until the middle of the century, after his death, that "in the doldrums" came to be associated with a geographical locality. Despite its association with the days of sailing ships, giving an impression of antiquity, it is a fairly recent word.

Dolly A TROPICAL STORM that struck islands in the Caribbean on August 20, 1996, strengthening to hurricane force as it reached Punta Herrero, Mexico. It then weakened but strengthened to hurricane force again as it moved across the sea toward northeastern Mexico and Texas.

Dolores A HURRICANE, with extremely heavy rain, that struck an area housing poor people on the outskirts of Acapulco, Mexico, on June 17, 1974. At least 13 people died and 35 were injured.

Domoina A CYCLONE that struck Mozambique, South Africa, and Swaziland from January 31 to February 2, 1984. It caused severe flooding in which at least 124 people died and thousands were rendered homeless.

Donau Glacial A GLACIAL PERIOD that occurred in Europe very early in the PLEISTOCENE EPOCH. It probably ended about 1 million years ago. Few traces of it remain. It was preceded by the TIGLIAN INTERGLACIAL and followed by the WAALIAN INTERGLACIAL. *Donau* is the German name for the River Danube.

Donau–Günz interglacial *See* WAALIAN INTERGLACIAL.

Donora A town (pop. 7,500) 28 miles (45 km) to the south of Pittsburgh, Pennsylvania, where one of the most serious of all air pollution incidents took place in October 1948. For a week, fog was trapped beneath an INVERSION and smoke and fumes from a zinc works and an iron works both owned by the American Steel and Wire Company mixed with the water droplets. On Friday, October 28, sulfur dioxide from the factories had formed a sulfuric acid mist, and by Saturday people started arriving at hospitals complaining of breathing difficulties, headaches, nausea, and abdominal pains and the first deaths occurred. On Sunday, the town was closed to traffic, including ambulances, partly as a result of poor visibility. Firefighters visited homes with oxygen to help people with respiratory problems. On Sunday evening, the zinc works closed, but rain washed the acid from the air and it opened again on Monday. About 6,000 people—half the population at the time—were made ill and 17 died. Two more people died later of the effects of the pollution.

Doppler, Johann Christian (1803–1853) Austrian *Physicist* An Austrian physicist who discovered the DOPPLER EFFECT. This was first tested in 1845 at Utrecht, in the Netherlands. A train carried a group of trumpeters in an open carriage past a group of musicians located beside the rail track. They reported a change in pitch as the trumpeters approached and receded. Doppler was born in Salzburg and educated in Vienna as a physicist and mathematician. In 1835, he was appointed professor of mathematics at the State Secondary School in Prague and subsequently held professorships at the State Technical Academy, Prague, and the Mining Academy, Schemnitz. In 1850, he returned to Vienna as director of the Physical Institute and professor of experimental physics at the Royal Imperial University of Vienna.

He died in Venice, Italy, on March 17, 1853.

Doppler effect The rise in pitch of a sound that is approaching rapidly and the fall in pitch of a sound that is retreating rapidly. It was discovered by JOHANN CHRISTIAN DOPPLER and was first tested by using sound but occurs with any form of wave radiation.

As the source of emission approaches, the distance waves must travel becomes progressively shorter. Their speed remains constant—at the speed of sound or the speed of light—but the number of waves reaching the observer each second increases. The distance between each wave crest and the next decreases, decreasing the wavelength and increasing the frequency. An increase in the frequency of a sound is detected as a rise in pitch. An increase in the frequency of a light beam is detected as a shift in the color toward the blue end of the spectrum. If the source of emission is retreating, the opposite occurs. The wavelength increases, frequency decreases, the pitch of a sound becomes lower, and light is shifted toward the red end of the spectrum. A Doppler effect is conventionally measured as an amount of blue shift or red shift and a moving object can be said to be red-shifted or blue-shifted.

Doppler radar A technique that uses two radar devices to measure the DOPPLER EFFECT on water droplets and, from that, to determine the speed at which the droplets are rotating about a vertical axis. Doppler radar is employed to study air movements and measure wind speed inside TORNADOES.

WEATHER RADAR is widely used in the study of clouds, precipitation, and storms. A tornado funnel is visible because water vapor condenses in the low atmospheric pressure prevailing inside it, producing water droplets that are detectable by radar. The funnel also extends upward, through the WALL CLOUD and into the MESOCYCLONE. These regions cannot be seen, because they are obscured by the cloud surrounding them, but they are visible to radar, allowing the entire structure of a tornadic storm to be examined.

If two radar transmitters are set some distance apart horizontally, the two beams they transmit enter the cloud at different points. If water droplets at these points are moving horizontally toward or away from the transmitter, there is a measurable Doppler effect on the reflected beams. This reveals whether the air inside the cloud is rotating, because if it is, the reflection from one side of the center of rotation is red-shifted and that from the other side is blue-shifted. If the air is rotating, the extent of the red and blue shifts indicates the speed of rotation.

The Doppler radar devices used to study tornadoes and other severe storms produce real-time color displays, allowing storms to be studied and measured while they are still active. This has made it possible to issue increasingly accurate severe storm and tornado warnings.

The use of Doppler radar for meteorological research was developed during the 1970s, principally at the National Severe Storms Laboratory, at Norman, Oklahoma. Its first success was in 1973, with the study of a SUPERCELL storm at Union City, Oklahoma. It was by means of Doppler radar that meteorologists discovered the way a mesocyclone forms high in the storm cloud and then extends downward shortly before a tornado emerges below the cloud.

Until Doppler radar became available there was no way to measure accurately the wind speed around the center of a tornado. People guessed at the speeds, and some scientists suggested the winds might occasionally reach or even exceed the speed of sound. At 68° F (20° C) this is about 770 mph (1,239 km h^{-1}). The radar detected winds approaching 300 mph (483 km h^{-1}) and meteorologists now believe this is the highest speed they ever reach.

These successes led to the establishment of the Joint Doppler Operational Project (JDOP), based at Norman, which ran from 1976 until 1979. This led in turn to the NEXT GENERATION WEATHER RADAR PROGRAM, which began in 1980.

Portable Doppler radar are sometimes used to study tornadoes. These can be set up within a few miles of a tornado. They transmit at a relatively short wavelength. This allows them to emit a narrow beam with a resolution high enough to measure conditions at different parts of the storm, and therefore wind speeds, but their reflections can sometimes be difficult to interpret. Bigger radars study storms from a distance of 100 miles (160 km) or more, but at long distances the curvature of the Earth places the lowest part of a storm below the horizon and out of the reach of a radar beam. Two sets of Doppler radars are used to study storms forming along SQUALL LINES, where the storms themselves are moving across the field of view.

(More information about Doppler radar can be found at http://www.nssl.noaa.gov/~trapp/burgess-abstract.html.)

Dot A TYPHOON that struck Luzon, Philippines, on October 19, 1985. It destroyed 90 percent of the buildings in the city of Cabanatuan, killed at least 63 people, and caused damage estimated at $1 billion.

downburst A DOWNDRAFT from a CONVECTION CELL in a CUMULONIMBUS cloud that reaches the surface and spreads to the sides. It produces strong gusts of wind and WIND SHEAR. In extreme cases a downburst from a SUPERCELL cloud can produce gusts of more than 75 mph (121 km h^{-1}). If the downburst affects a surface area no larger than 2.5 miles (4 km) in diameter it is known as a *MICROBURST.*

downdraft A current of sinking air that is produced inside a CUMULONIMBUS cloud. Downdrafts usually travel at less than about 11 mph (18 km h^{-1}) but can be much stronger in a SUPERCELL cloud. A typical multicell cumulonimbus cloud comprises a number of CONVECTION CELLS in each of which air is both rising and sinking, so the cloud contains several downdrafts.

downrush The strong DOWNDRAFT that occurs during the dissipation of a large CUMULONIMBUS cloud. It is caused by the failure of the convective upcurrents that sustained the cloud. Snow and cold rain falling through the cloud chill the air and this process cools the warm air that is rising by CONVECTION, so it ceases to rise. There is then no mechanism for supporting the RAINDROPS and SNOWFLAKES. They all fall, as the cloud loses its moisture, producing a CLOUDBURST. As they fall they drag cold air with them, and it is this cold air that constitutes the downrush.

downwash The transport of air to the surface when it becomes caught in an EDDY on the LEE side of a hill or building. If the caught air is polluted, the pollutant is carried to ground level. Pollutants do not disperse well from a chimney that is situated on the roof of a tall building surrounded by lower buildings and the top of the chimney is only a short distance from the roof. The CHIMNEY PLUME is carried downward by the eddy on the lee side of the building. This does not happen if the chimney is tall enough to rise above the eddy. The plume from a chimney on the roof of a low building that is located beside a taller building is also caught in the eddy from the tall building and carried to the ground.

downwind *See* UPWIND.

drag The retarding effect that is caused by FRICTION when air crosses a rough surface (*see* AERODYNAMIC ROUGHNESS). Surface winds and moving vehicles are slowed by drag, which acts as a force applied in the opposite direction to the motion. Frictional drag also moves objects, literally dragging them (*see* SURFACE SHEARING STRESS). Surface currents in the ocean are driven by the wind: the wind drags the water with it. Drag also causes CUMULONIMBUS clouds to dissipate. As RAINDROPS and SNOWFLAKES fall through the cloud they drag with them the small pockets of air that surround them. This air is very cold. It chills the warm air that is rising by CONVECTION to feed moisture into the cloud. The loss of moisture causes the cloud to dissipate.

drainage wind *See* KATABATIC WIND.

dreikanter A VENTIFACT that has three edges. Ventifacts are formed by the abrasive action of wind-blown sand. In the case of dreikanters, this is believed to loosen and remove mineral grains from the surface directly exposed to the wind until the pebble falls over, exposing a second side. This is abraded in turn until the pebble falls again, exposing a third side.

driven snow Snow that has been transported by the wind and deposited in SNOWDRIFTS. Wind may drive snow that is falling, and it may also lift dry, powdery snow from the surface. If the wind speed is great enough the driven snow may be classed as a BLIZZARD.

drizzle Liquid precipitation in which the droplets are smaller than 0.02 inch (0.5 mm) in diameter, are all of approximately similar size, and are very close together. They form by the COALESCENCE of smaller droplets near the base of STRATUS cloud. As soon as enough have coalesced to reach a size that is just heavy enough to fall, droplets start to sink slowly. There are no upcurrents in stratus, so they have no opportunity to coalesce to a larger size before they leave the cloud base, and the base is low enough for them to reach the surface before evaporating.

dropsonde An instrument package attached to a parachute that is dropped from an airplane. As it falls, its radio transmits data on pressure, temperature, and humidity recorded by its instruments at various altitudes. A similar package that can be tracked to provide information on the wind speed and direction at different altitudes is called a *dropwindsonde*. Modern dropwindsondes are tracked by means of the Global Positioning System and are especially useful in studying TROPICAL CYCLONES.

dropwindsonde *See* DROPSONDE.

drought A prolonged period during which the amount of precipitation falling over a particular area is markedly less than the usual amount that falls over that period in that place. A drought is longer than a DRY SPELL, but the length of time without rain that is needed to define a drought obviously varies greatly from place to place. In Britain, a drought is declared after a period of 15 days without rain. A drought is declared in the United States after a period of 21 days during which the precipitation does not exceed 30 percent of the average for that place and that time of year. In the northern Sahara Desert a drought is considered to exist if no rain has fallen for at least 2 years and parts of the Atacama Desert in Chile have gone for 20 years without rain without this period's being considered a drought. Antofagasta, a city in the Atacama located on the Chilean coast and almost exactly on the tropic of Capricorn, has an average 0.5 inch (13 mm) of rain a year.

It is impossible, therefore, to define a drought in terms of absolute precipitation. The concept is more economic and social than meteorological. A dry spell becomes a drought when the lack of water threatens to restrict human activities. The authorities may ban the use of hose pipes for watering gardens and lawns or for washing cars, for example, so ornamental plants wilt and cars become dusty (but not mud-spattered, of course!). As the drought continues, the domestic water supply may be rationed, and to prevent wastage the supply to private houses is sometimes shut down and standpipes installed in the street so people have to collect their water in containers.

During a drought, plant litter lying naturally on the ground becomes very dry. In this condition a lightning strike, the focusing of sunshine by a piece of glass such as the bottom of a bottle, or a carelessly dropped match or cigarette is enough to ignite the material. The resulting grass, heath, or forest fire can spread rapidly. There were severe wild fires in California in 1991, and between June and December 1997 forest fires that swept Indonesia spread to farms and destroyed crops and produced a pall of smoke that seriously polluted the air over a large part of southeastern Asia. Although many of these fires were lit deliberately by farmers wishing to clear land, the vegetation was so dry after the prolonged drought caused by EL NIÑO that the fires raged out of control. The destruction of crops then contributed to a famine in which hundreds of people died.

The PALMER DROUGHT SEVERITY INDEX provides a means for measuring the severity of a drought and it is frequently used in the United States. Droughts can also be classified according to their types as

PERMANENT DROUGHT, SEASONAL DROUGHT, CONTINGENT DROUGHT, DEVASTATING DROUGHT, and INVISIBLE DROUGHT. They can also be classified according to their effects as AGRICULTURAL DROUGHT and HYDROLOGICAL DROUGHT.

(You can learn more about droughts from Michael Allaby, *Dangerous Weather: Droughts* [New York: Facts On File, 1998] and *Ecosystem: Deserts* [New York: Facts On File, 2001].)

dry adiabat The continuous rate at which the temperature of dry air changes in an ADIABATIC manner with height. It is a sequence of states, each of which is defined by temperature and pressure. Dry adiabats are shown on THERMODYNAMIC DIAGRAMS. On a TEPHIGRAM, they appear as straight lines that are also ISOTHERMS of POTENTIAL TEMPERATURE.

dry adiabatic lapse rate (DALR) The rate at which rising unsaturated air cools through an ADIABATIC process and descending air warms. Cooling is due to the energy expended by the rising air as it expands on entering a region of lower density. Warming is due to the energy that is absorbed by sinking air as it is compressed on entering a region of higher density. The DALR is 5.4° F per 1,000 feet (9.8° C km^{-1}). This lapse rate is maintained only while the air remains unsaturated. As soon as it reaches the CONDENSATION LEVEL and its water vapor begins to condense, the rate of cooling changes to the SATURATED ADIABATIC LAPSE RATE.

dry air stream A flow of air that develops in the middle and upper TROPOSPHERE behind a midlatitude CYCLONE. It descends once the cyclone has passed.

dry-bulb temperature The temperature that is registered by a THERMOMETER the bulb of which is dry and directly exposed to the air. A dry-bulb thermometer is used to measure the air temperature. This reading is compared with the WET-BULB TEMPERATURE to determine the RELATIVE HUMIDITY and DEW POINT TEMPERATURE.

dry-bulb thermometer *See* DRY-BULB TEMPERATURE.

dry climate A climate in which the average annual PRECIPITATION is less than the POTENTIAL EVAPOTRANSPIRATION and plant growth is restricted by the lack of moisture (*see* CLIMATE CLASSIFICATION). In the KÖPPEN CLIMATE CLASSIFICATION, dry climates are designated category B and include the semiarid steppe (categories BSh and BSk) and desert climates (categories BWh and BWk). In the THORNTHWAITE CLIMATE CLASSIFICATION, dry climates are those with a MOISTURE INDEX lower than zero. These include the dry subhumid (category C_1, moisture index –20 to 0), semiarid (category D, moisture index –40 to –20), and arid (category E, moisture index –60 to –40) climates.

dry deposition The transfer of particles from dry air to a surface onto which they are adsorbed. It occurs as the turbulent flow of air places the particles in contact with the surface, and the rate at which this takes place can be calculated to give the deposition velocity. Particles or molecules that are adsorbed onto leaves may then be absorbed into plant cells through the stomata (*see* PHOTOSYNTHESIS). The report of a study in the Los Angeles area, published in 1980, found that dry deposition of acid pollutants was 15 times more important than wet deposition, or ACID RAIN, in the harm it caused to vegetation.

(For more information about the Los Angeles study, see J. J. Morgan and H. M. Liljestrand, *Final Report, Measurement and Interpretation of Acid Rainfall in the Los Angeles Basin* [Sacramento: California Air Resources Board, 1980].)

dry haze A HAZE that contains no water droplets, in contrast to a DAMP HAZE.

dry ice Solid carbon dioxide (CO_2), which, at standard sea-level pressure, must be kept at a temperature below –109.3° F (–78.5° C, 194.7 K) because at this temperature CO_2 sublimes (*see* SUBLIMATION). When dry ice is released into the air it has an immediate cooling effect. As the temperature of air that is almost saturated (*see* SATURATION) falls below –40° F (–40° C) around the dry ice particles, water vapor instantly forms ICE CRYSTALS in the absence of FREEZING NUCLEI. This is because air that is close to saturation with respect to liquid water is supersaturated (*see* SUPERSATURATION) with respect to ice, and at below 32° F (0° C) air is always supersaturated with respect to ice. One gram (0.035 ounce) of dry ice falling through air with a temperature of 14° F (–10° C) and a RELATIVE HUMIDITY close to 100 percent triggers the formation of about

100 billion (10^{11}) ice crystals before it sublimes. Once ice crystals have formed, water vapor condenses onto them. This removes water vapor from the air, leaving only ice crystals. These are fairly large and fall quite rapidly, gathering CLOUD DROPLETS as they pass through the lower, warmer parts of the cloud. This is the method used to seed clouds with dry ice (*see* CLOUD SEEDING).

Dry ice is also used to disperse cold FOG, in which the droplets are below freezing temperature. The dry ice particles initiate the formation of ice crystals, which grow at the expense of the supercooled water droplets (*see* SUPERCOOLING). In most cases the ice crystals quickly grow big and heavy enough to fall to the ground, clearing the fog, but even if they remain suspended in the air, the VISIBILITY improves markedly. This is because less light is scattered by a few large ice crystals than by many tiny water droplets.

Dry ice is also used to create theatrical effects. Released into the warm, moist air of a theater it immediately causes formation of a cloud.

dry impingement A technique that is used to remove particulate matter from a stream of waste gases. The gases are blown against a surface to which the particles adhere.

dry line (dew point front) A boundary that often forms over the Great Plains in spring and summer between hot, dry air to the west and warm, moist air to the east. The dry line develops over western Texas or eastern New Mexico, extending north into Oklahoma, Kansas, and Nebraska, and then moves east. Dry lines rarely occur east of the Mississippi River. Differences in the surface terrain between eastern Texas and the mountains of New Mexico provide the conditions in which dry lines form. Over New Mexico the barren ground is heated strongly. Over eastern Texas, clouds develop in moist air from the Gulf.

Advancing dry air lifts the moist air ahead of it, often producing huge CUMULONIMBUS clouds. Sometimes dry air about 1,000 feet (330 m) above the surface moves eastward faster than air at the surface. It overruns the lower air, causing a CAPPING INVERSION. This inhibits the development of storm clouds, but if moist air penetrates the inversion, storms develop very quickly. The location of the dry line is commonly measured by comparing the DEW POINT TEMPERATURES to each side. Ahead of the dry line these are commonly between about 60° F and 70° F (15°–20° C) and behind it between 20° F and 30° F (–7° to –1° C). The air temperature ahead of the dry line is usually between 70° F and 80° F (20°–27° C) and behind it between about 85° F and 95° F (29°–35° C).

Storms associated with dry lines frequently become tornadic.

(There is more information about dry lines at ww2010.atmos.uiuc.edu/(Gh)/guides/mtr/as/frnts/dfdef.rxml and www.usatoday.com/weather/wdryline.htm.)

dry season The time of year when PRECIPITATION is much lower than it is at other times. The dry season may occur in winter or in summer. Dry summers occur around the Mediterranean Sea and on the western sides of continents in latitudes 30°–45° N and S. San Francisco, California, receives an average 22 inches (561 mm) of rain a year, but only 1.1 inches (28 mm) falls between May and September. Gibraltar, at the southern tip of Spain, has a very similar climate. Its average annual rainfall is 30 inches (770 mm), of which only 2.2 inches (56 mm) falls between May and September.

In regions with a MONSOON climate it is the winter that is dry. Bombay, India, receives an average 71 inches (1,811 mm) of rain a year. Only 4 inches (104 mm) falls between October and May.

A dry season has an effect on plants that is similar to that of a cold winter. In both cases, plant growth ceases for lack of moisture. In regions with a dry season the aridity is caused by lack of precipitation. In regions with a cold season it is caused by temperatures below freezing that turn liquid water to ice, rendering it unavailable to plant roots.

dry snow *See* SNOW.

dry spell A period during which no rain falls that is of shorter duration than a DROUGHT. In the United States a dry spell is said to occur if no measurable precipitation falls during a period of not less than two weeks.

dry subhumid climate In the THORNTHWAITE CLIMATE CLASSIFICATION, a climate in which the MOISTURE INDEX is between –33 and 0 and the POTENTIAL EVAPOTRANSPIRATION is 5.6–11.2 inches (14.2–28.5 cm). It is designated C_2. In terms of THERMAL EFFICIENCY, this is a tundra climate (D').

dry tongue A tongue-shaped extension of dry air that protrudes into a region of moister air.

DU *See* DOBSON UNIT.

duplicatus A variety of cloud (*see* CLOUD CLASSIFICATION) that comprises layers, sheets, or patches of clouds that are at different heights but merge or overlap each other as seen from the ground. Duplicatus is seen in association with the cloud genera CIRRUS, CIRROSTRATUS, ALTOCUMULUS, ALTOSTRATUS, and STRATOCUMULUS. *Duplicatus* is a Latin word that means "duplicated."

düsenwind A strong northeasterly wind that blows through the Dardanelles, which is the strait in Turkey that links the Sea of Marmara with the Aegean Sea. It blows when there is an area of high atmospheric pressure over the Black Sea and lower pressure over the Aegean.

dust Solid particles that are lifted into the air by the wind or ejected as ash during volcanic eruptions. Certain industrial processes also generate dust that is removed in the FLUE GAS. Wind-raised dust consists of soil particles and is made from a variety of minerals, of which silica (silicon dioxide, SiO_2) is the most abundant. Soil particles range in size from those of clay (less than 2 µm across) and silt (2–60 µm) to fine sand (60–200 µm). In order for soil particles to be blown from the surface, the ground must be bare and dry. Deserts are the main source of dust, but farming also contributes a large amount. The process of lifting is called *DEFLATION*. The particles that become airborne most readily are about 40 µm in diameter. Larger particles are too heavy to be lifted easily and when they are lifted soon fall back to the surface. Very small particles are also difficult to lift because their large surface area in relation to their volume makes them tend to adhere to adjacent particles, and small particles are often sheltered by stones that are much too big for the wind to move. At higher wind speeds, however, both bigger and smaller particles become airborne. Dust can be carried for long distances. Eventually it settles. Sand dunes are accumulations of wind-blown sand and fine soil, made from silt-sized particles, is deposited as a type of soil called *loess,* which is very fertile. There are large loess deposits in China and loess covers much of the central United States. A belt of loess, very thick in places, lies along the eastern side of the Mississippi River in Mississippi, Tennessee, and Kentucky. It originated during the WISCONSINIAN GLACIAL, when the brief summer thaw caused VALLEY GLACIERS to retreat, releasing water that flooded the flat-bottomed valleys. As the water drained away the valley bottoms were turned into mudflats, and as these dried the wind carried away the dust.

Dust Bowl A region of the Great Plains in the United States, covering about 150,000 square miles (388,500 km²) in southwestern Kansas, southeastern Colorado, northeastern and southeastern New Mexico, and the panhandles of Oklahoma and Texas, that experienced a severe drought from 1933 until the winter of 1940–41. During the drought, soil blew away as clouds of dust that blanketed much of the eastern United States and extended far into the Atlantic.

Droughts in this region occur at intervals of 20–23 years and they returned in the 1950s, 1970s, and 1990s. They vary in severity, but "megadroughts," such as that of the 1930s, are known to have occurred in the 13th and 16th centuries and more recently in the 1750s, 1820s, and 1890s. Some of the earlier megadroughts lasted much longer than the drought of the 1930s. The 16th-century drought lasted 20 years. Despite their recurrence, which is somewhat irregular, too little is known about their causes to allow them to be predicted.

The drought that gave the region the name *Dust Bowl* caused much more human suffering than any other drought, because of the large agricultural population of the area. From the middle of the 19th century, settlers were moving into the region and plowing up the natural grasses to grow wheat. Periodic droughts destroyed crops and ruined some farmers, but the return of the rains allowed farming to continue. Farming methods intensified during the early 20th century and the cultivated area increased. Cereal prices fell during the Great Depression that began in 1929, forcing farmers to cultivate the land even more intensively in order to maintain their incomes. They were able to do so because the annual rainfall from 1927 to 1933 was above average.

During this period the climate was becoming warmer. One consequence of the warming in middle latitudes was an increase in the number of days each year when the wind blew from the west. Westerly winds in North America lose their moisture crossing the Rocky Mountains and carry very dry air to the Great Plains. It was in 1933 that the effect of this first became evident, when the annual rainfall was 7 inches (178 mm) below the average of 23 inches (584 mm). Crops started to fail, the natural grasses died back, and the land was left bare.

Natural prairie grasses bind the soil into large lumps that become very hard during drought. Over many years, cultivation had broken up these lumps, producing the fine soil texture that is needed for sowing, but when the land dried, instead of hard lumps the soil turned to dust. The loss of soil to the westerly winds was almost continuous, but the most severe dust storms were in May 1934 and October 1935. As well as soil, the wind carried away seed.

The Soil Conservation Service was established in 1935 in response to the tragedy of the Dust Bowl years, and the federal government began to teach and encourage practices to protect vulnerable soils. It was also recognized that on some prairie soils the natural grasses should be reestablished and the land should not be farmed.

dust burden The weight of DUST that is suspended in a volume of gas, such as air or FLUE GAS. It is measured in grams of dust per cubic meter at STANDARD TEMPERATURE AND PRESSURE (S.T.P.).

dust collector A device that removes DUST from industrial waste gases. ELECTROSTATIC PRECIPITATORS are used for this purpose, and dust may also be removed by filtration, IMPACTION, or IMPINGEMENT.

dust devil A twisting wind that resembles a tornado but is much smaller, a great deal less violent, and not associated with a violent storm. Dust devils rarely exceed about 300 feet (100 m) in height and last for only a few minutes. Whirlwinds, which form under similar conditions, are much bigger and more destructive.

Dust devils are common in dry tropical and subtropical regions during the hottest part of the day, when the sky is clear. Unlike tornadoes, they rise upward from the ground, rather than descending from a cloud.

A dust devil develops over a patch of ground that has absorbed more heat than its surroundings. The layer of air in contact with the hot ground is heated until its temperature is several degrees higher than that of the air above the surface layer. The lower air is then unstable and begins to rise rapidly. This produces a very local area of low pressure into which air flows. Vorticity causes the converging air to turn, and the conservation of its angular momentum accelerates it as it approaches the center of low pressure and its radius of curvature is reduced. Close to the center it is drawn into the upward flow, which then begins to spiral. The flow is strong enough to carry dust and any loose, light material into the upward spiral. Dust devils may pass unnoticed if they occur over vegetation, where there is nothing they are able to lift.

They die quickly because the upward movement of air carries away the excess surface heat that triggered them. Once the patch of ground cools to the temperature of its surroundings the dust devil has no source of energy to sustain it.

dust dome The increased concentration of DUST particles that occurs beneath an URBAN DOME. CONVECTION due to the higher temperature of the URBAN CLIMATE generates surface winds blowing toward the center of the urban area. These winds carry dust from adjacent rural areas and also raise dust within the urban area, and the dust is kept airborne by the convection currents.

dust horizon The upper surface of air that is held beneath a temperature INVERSION and that is made visible from a distance by the large amount of DUST it holds. Seen partly silhouetted against the sky the top of the dust-laden air resembles a horizon.

dust storm A strong wind that blows across bare ground in an arid region, where it lifts dust and keeps it aloft. The frequent removal of fine particles explains why most desert surfaces consist either of bare rock or boulders or of sand. A wind that blows over a sandy surface may cause a SAND STORM, which differs from a dust storm only in the size of the particles it transports. The THRESHOLD VELOCITY at which the wind begins to raise particles is proportional to the size of the particles, and a moderate breeze of 13–18 mph (21–29 km h^{-1}) is sufficient to raise dust. Although it is moving, the dust rises no more than a few hundred feet above ground level unless there are also vertical air currents to carry it to a greater height. Consequently, major dust storms occur in unstable air (*see* STABILITY OF AIR). Then the dust can rise much farther, occasionally to 15,000 feet (4,600 m) or even higher, and it advances like a wall. A storm covering an area of 5,000 square miles (13,000 km^2) to a height of 10,000 feet (3,050 m) may carry 8 million tons (7.7 million tonnes) of dust. The FALL SPEED of dust particles is only about 0.4 inch per second (1 cm s^{-1}). This means that once it is aloft it can be carried a long distance and when the vertical currents die it falls slowly and over a wide area. Dust raised by haboob storms in the Sudan is responsible for the DRY HAZE over much of Central and West Africa.

Winds that are associated with an area of low pressure and that blow over a long distance often cause dust storms. The KHAMSIN, GHIBLI, and SHAMAL are winds of this type. Dust storms can also be produced by the winds such as the HABOOB that are generated by large THUNDERSTORMS.

(For more information about dust and sand storms see Michael Allaby, *Ecosystem: Deserts* [New York: Facts On File, 2000].)

dust veil index *See* LAMB'S DUST VEIL INDEX.

dust whirl (dancing devil, desert devil, sand auger, sand devil) A small column of rapidly rotating air that carries DUST, sand grains, leaves, scraps of paper, and other light material. It is caused by CONVECTION above a patch of ground that has been heated more strongly than the surrounding area. A dust whirl is a small version of a DUST DEVIL.

DVI *See* LAMB'S DUST VEIL INDEX.

dynamic climatology The scientific study of the movements of air and the thermodynamic processes that cause them. The term was coined in 1929 by TOR HAROLD PERCIVAL BERGERON to describe the way the concept of AIR MASSES and fronts (*see* FRONTOGENESIS) might be developed.

dynamic meteorology The scientific study of atmospheric motion that predicts the future state of the atmosphere in terms of the physical variables of temperature, pressure, and VELOCITY. The laws of fluid mechanics and thermodynamics are expressed in the form of complex mathematical equations (partial differential equations).

dynamic soaring A flying technique that is employed by some large seabirds, most notably the wandering albatross (*Diomedea exulans*), which has an average wingspan of 10 feet (3 m). The bird glides downwind from a height of about 50 feet (15 m) until it is very close to the sea surface. Then it turns into the wind to fly across the wind gradients behind wave crests or into the FRICTION LAYER. In doing so it enters air that is moving more slowly than the air it leaves, and this increases its speed in relation to the air (its airspeed). Increasing its airspeed increases the amount of lift generated by its wings, allowing it to climb back to its original height. In this way the bird can remain airborne for long periods with very little need to beat its wings.

dynamic trough *See* LEE TROUGH.

Earth Observing System (EOS) Part of the Earth Science Enterprise, a program launched by the National Aeronautics and Space Administration (NASA) in 1991 with the aim of studying the Earth as an environmental system. The first part of the program involved collecting and studying data from a number of satellites. EOS forms the second part. It uses satellites dedicated to the task, the first of which was launched into POLAR ORBIT in 1999 and the second into a low-inclination orbit (*see* INCLINED ORBIT) in 2000. The EOS will study the solid Earth, oceans, and atmosphere as an integrated system.

(More information about the EOS can be obtained from NASA, starting at the EOS home page: http://eospo.gsfc.nasa.gov/eos_homepage/description.html.)

Earth Radiation Budget Experiment (ERBE) A study that uses both scanning and nonscanning RADIOMETERS carried on three satellites to measure all the solar radiation reaching the Earth and all the long-wave radiation leaving the surface. The aims of the ERBE are to determine, for a minimum of one year, the monthly average radiation budget on regional, zonal, and global scales; to determine the equator-to-Pole energy transport gradient; and to determine the average diurnal variation of the radiation budget on a regional and a monthly scale. In the course of a month, the instruments provide measurements throughout almost the whole daily cycle for most regions of the Earth. The scanning radiometers transmit high-resolution measurements for each 2.5° latitude × 2.5° longitude region, and the nonscanning radiometers have a wide field of view (5° × 5° and 10° × 10°) that allows long-term monitoring on the scale of continents. The instruments are carried on the satellites *NOAA 9,* launched in December 1984, and *NOAA 10,* launched in September 1986, into SUN-SYNCHRONOUS ORBIT and the EARTH RADIATION BUDGET SATELLITE (ERBS).

(More detailed information about the ERBE can be found at http://eosdis.larc.nasa.gov:12000/campaign_documents/erbe_project .html.)

Earth Radiation Budget Satellite (ERBS) A satellite that is dedicated to the EARTH RADIATION BUDGET EXPERIMENT. (ERBE). It was launched by the space shuttle *Challenger* in October 1984 and is operated by the National Aeronautics and Space Administration (NASA) at the Goddard Space Flight Center. It travels in an INCLINED ORBIT, with an inclination of 57°. In addition to its ERBE instruments, the ERBS carries the STRATOSPHERIC AEROSOLS AND GAS EXPERIMENT II.

Earth Resources Technology Satellite *See* LANDSAT.

Earth's shadow *See* DARK SEGMENT.

Earth tides *See* TIDES.

Earthwatch Program A program to monitor changes in the environment that was established in

1973 under the terms of the Action Plan that was agreed on at the United Nations Conference on the Human Environment, held in Stockholm, Sweden, in June 1972. The Earthwatch Program is managed by the United Nations Environment Program (UNEP) and based in Geneva, Switzerland.

(You can learn more about Earthwatch at www.unep.ch/earthw.html.)

East Australian Current An ocean current that flows southward carrying warm water parallel to the eastern coast of Australia. The current is narrow, only 330–660 feet (100–200 m) wide, and slow-moving, flowing at 0.6–1.2 mph (1–2 km h^{-1}).

Easter Day According to WEATHER LORE, a day that is believed to indicate what the weather will be like in the following months. According to one tradition,

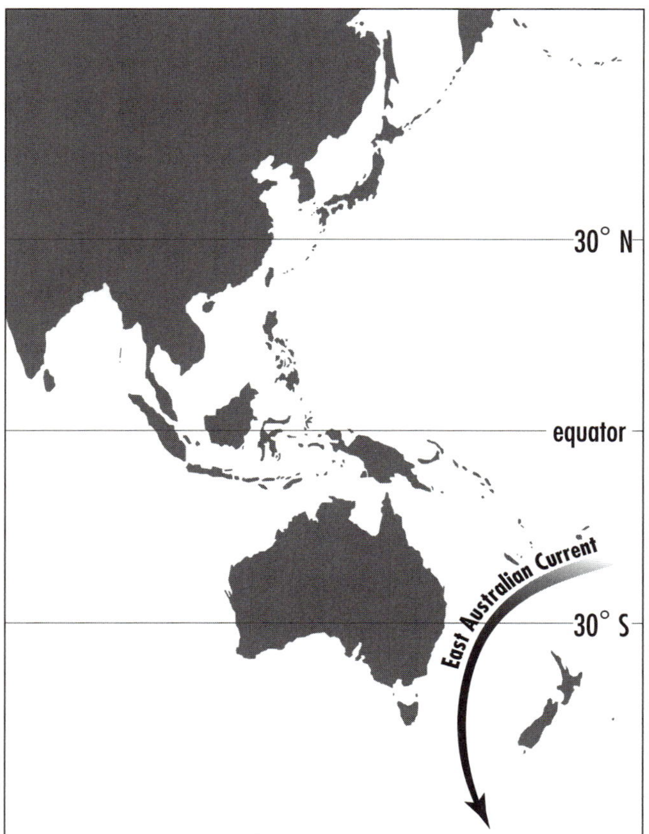

East Australian Current. **A warm ocean current that flows parallel to the eastern coast of Australia.**

Easter in snow, Christmas in mud
Christmas in snow, Easter in mud.

This is self-explanatory. Another saying is that rain on Easter Day means June will be a wet month:

If it rains on Easter Day,
There shall be good grass but very bad hay.

Rain will make the grass grow well, but after mowing the grass must dry thoroughly if it is to make good hay.

easterly jet A JET STREAM that blows from east to west over India and Africa. The jet forms in summer in the upper TROPOSPHERE at a height of about 9 miles (15 km) and extends from the South China Sea to the southeastern Sahara. In the summer the EQUATORIAL TROUGH that is associated with the INTERTROPICAL CONVERGENCE ZONE (ITCZ) moves north to about latitude 25° N. The THERMAL EQUATOR lies along the equatorial trough, and to the south of it, on the side nearer the equator, warm air is rising vigorously. There is cooler air throughout the troposphere between the thermal equator and the geographic equator. This situation produces a strong north–south pressure and tem-

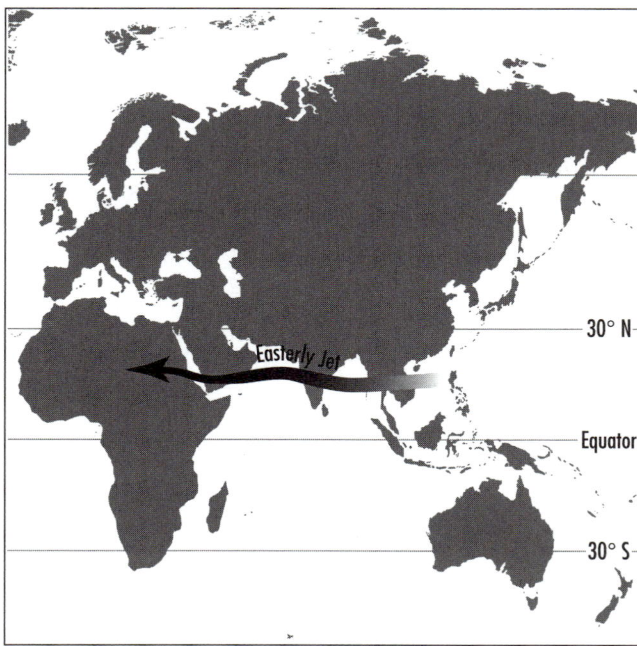

Easterly jet. **The location of the easterly jet in summer, at about 15° N. It extends from the South China Sea, across southern Asia, southern India, and the Arabian Sea, and into the southeastern Sahara.**

perature gradient, reversing the normal baroclinicity (*see* BAROCLINIC). The gradient is most marked at high level, where it generates a THERMAL WIND at about 15° N. Thermal winds blow with the cooler air on their left in the Northern Hemisphere. Consequently, this high-level thermal wind blows from east to west. The easterly jet reaches its maximum force in about July. In September, as the southwesterly MONSOON retreats from India, the thermal gradient weakens, the midlatitude westerly winds move farther south, and a branch of the subtropical jet stream, blowing to the south of the Himalayas, replaces the easterly jet.

easterly wave (African wave, tropical wave) A long, weak, low-pressure TROUGH that moves from east to west across the tropical oceans. It deflects the easterly TRADE WINDS, producing a wave pattern in the surface STREAMLINE. The WAVELENGTH is usually 1,200–2,500 miles (1,900–4,020 km) and the waves travel about 6°–7° of longitude per day. They last for one to two weeks before disappearing. Easterly waves are especially marked in the Caribbean region. The troughs producing easterly waves usually slope toward the east with increasing height above the surface, so the weather associated with them occurs behind the line at which the trough lies on the surface. Ahead of the trough there is a RIDGE of high pressure, with generally fine weather, scattered CUMULUS cloud, and some HAZE. Close to the line of the trough the cumulus clouds are bigger, giving some showers, and improved visibility as the rain washes away haze. Behind the trough, the wind veers (*see* VEERING), the temperature is lower, and the cloud thickens, with some CUMULONIMBUS. Showers are heavy, with some thunder. Easterly waves sometimes intensify to become TROPICAL DISTURBANCES.

East Greenland Current An ocean current that flows southward from the Arctic Ocean into the North Atlantic Ocean, parallel to the northeastern coast of Greenland. In about the latitude of Iceland it merges with a branch of the NORTH ATLANTIC DRIFT. The East Greenland Current carries cold water with low salinity.

East Greenland Current. **A cold ocean current that flows parallel to the eastern coast of Greenland.**

EBM *See* ENERGY BALANCE MODEL.

eccentricity The extent to which the orbital path of a planet or satellite deviates from a circle. Planetary orbits are elliptical; an ellipse is a geometrical figure with two foci, so the bodies move about one focus of the ellipse. This is not located at the center of the figure, so the orbit is eccentric—not about the geometrical center. Unlike a circle, an ellipse can vary in shape, becoming more circular or more elongated, and variations alter the distance between the orbiting body and the focus when the orbiting body is at its closest and farthest. The eccentricity of the Earth's orbit affects the amount of solar energy received at the surface at aphelion and perihelion, and changes in eccentricity are linked to major changes in climate (*see* MILANKOVICH CYCLES).

The extent of eccentricity can be measured. If the geometrical center of the ellipse is C, and the focus about which the body orbits is F_1, the distance between C and F_1 is the linear eccentricity, *le*. The location of the second focus, F_2, is at the center of the major axis, α, from C to the point at which the orbiting body is

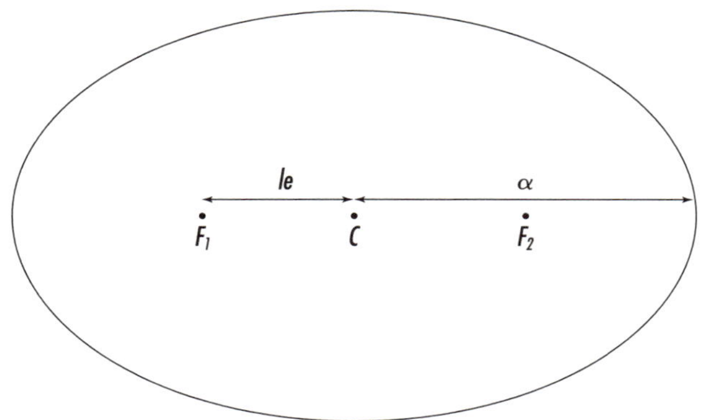

Eccentricity. **The extent to which the orbit of a satellite or planet deviates from a circle.**

farthest from F_1. Eccentricity, e, is then given by $e = le/\alpha$.

At present the eccentricity of Earth is 0.017. It varies over a cycle of about 100,000 years from 0.001, which is almost circular, to 0.054.

eclipse The temporary disappearance from the sky of either the Moon or the Sun as it is completely shaded by the passage of another body. When the Earth passes directly between the Sun and Moon, its shadow hides the Moon, producing a lunar eclipse. This does not occur at every full Moon, when the Sun, Earth, and Moon are aligned with the Earth in the center, because the lunar orbit is inclined at 5° to the Earth's orbit. Lunar eclipses occur when both the Earth and Moon arrive at one of the two points where their two orbits intersect when the Moon is full.

If they arrive at one of these points when the Moon is new, so the Earth, Moon, and Sun are aligned with the Moon at the center, a solar eclipse occurs. This is possible because, although the Moon is tiny when compared to the Sun, it is also much closer to the Earth. In fact, the diameter of the Sun is 400 times greater than that of the Moon, but the Sun is also 400 times farther from the Earth.

The eclipse is not visible everywhere on Earth. The shadow of the Moon crosses the Earth along a track inside which the Sun appears fully obscured for a length of time that decreases with distance from the center to the edges of the track. Outside the track, the Sun appears partially eclipsed over a much larger belt. If the track of the full eclipse misses the Earth completely, a partial eclipse may nevertheless be visible from some places.

Both the Earth and Moon follow ECCENTRIC orbits. Consequently, the distances between them and between both of them and the Sun vary. This alters the apparent diameter of the Moon and Sun as these are seen from Earth, by about 2 percent in the case of the Sun and 8 percent in the case of the Moon. If the three bodies are aligned at a time when the Moon is more distant from the Earth than the average 239,000 miles (384,000 km), its shadow is too small to obscure the Sun completely. The center of the Sun disappears, but a bright ring of sunlight remains. This is known as an annular eclipse. If the alignment occurs when the Moon is closer than average to the Earth, the Sun is obscured fully, producing a total eclipse.

During a solar eclipse, the shadow of the Moon traveling from west to east causes complete darkness and also a sharp drop in temperature. This produces a wind that may be felt either as itself or as a weakening or strengthening of the wind that was blowing before the shadow approached.

eclipse year *See* ORBIT PERIOD.

ecliptic *See* PLANE OF THE ECLIPTIC.

eddy Air that moves turbulently, its speed and direction changing rapidly and irregularly. Its movement is vertical as well as horizontal. It is also hierarchical, in that large eddies produce smaller ones in a cascade that ends only when the viscosity of the air is sufficient to damp them out. Moving air is subject to many disturbances, all of them producing eddies. As it flows across the surface of land or sea, but especially that of land because the land surface is irregular, the air nearest the surface is slowed by friction and air immediately above it tends to roll forward. This produces a ceaseless jostling among small pockets of air, with air descending on one side of each roll and rising on the other. Air flowing around a building also rolls, but about a vertical axis. On a much larger scale, the GENERAL CIRCULATION of the atmosphere produces eddies and the ROSSBY WAVES that disrupt the ZONAL FLOW of air (*see* ZONAL INDEX) also produce eddy patterns. Large-scale midlatitude eddies are of major importance in the transport of heat from the Tropics.

eddy correlation A technique that is used to study the effect of the sea surface on the air immediately above it. It involves measuring the mean speed and direction of air movement and comparing them with the fluctuations that occur in the vertical and horizontal components of that movement. Measurements are usually made at intervals of 30 to 60 minutes and at heights from 1.6 to 6.5 feet (0.5–2 m) above the water surface. The correlation is then calculated by applying mathematical equations.

eddy diffusion The transport of heat energy and momentum by DIFFUSION in air that is flowing turbulently. The rate at which diffusion occurs is known as the *eddy diffusivity* and is measured in units of area per second.

eddy diffusivity *See* EDDY DIFFUSION.

eddy viscosity The DIFFUSION of momentum that takes place in air that is moving as a turbulent EDDY. It is caused by the mingling of air at the boundaries between eddies. Intermingling transfers air that is moving along the path and at the speed of one eddy to another eddy that is moving at a different speed and in a different direction. This quickly diffuses the energy of the air, causing it to lose momentum. Eddy viscosity is many thousands of times more effective than molecular VISCOSITY as a mechanism for the transport of energy in air. Measurements of the eddy viscosity are important in determining the rate at which water evaporates into moving air and the extent to which wind cools surfaces exposed to it.

Eemian interglacial An INTERGLACIAL period in Northern Europe that lasted from about 130,000 years ago until about 72,000 years ago and that is equivalent to the IPSWICHIAN INTERGLACIAL of Britain and partly coincides with the SANGAMONIAN INTERGLACIAL of North America. It followed the Saalian glacial and preceded the Weichselian glacial. Sea levels rose during the Eemian, forming a large inland sea, called the *Eem Sea,* in what are now the Netherlands and northern Germany. Average summer temperatures were about 3.6°–5.4° F (2°–3° C) warmer than those of today.

effective gust velocity The upward vertical component of a GUST of wind that would produce a given ACCELERATION on an aircraft flying straight and level through air of a constant density at the recommended cruising speed of that aircraft.

effective precipitable water The proportion of the PRECIPITABLE WATER that can fall as PRECIPITATION.

effective precipitation A value for the aridity of a climate that is important in many schemes for CLIMATE CLASSIFICATION. The aridity of a climate determines its suitability for agriculture and the type of natural vegetation it supports. Aridity is the amount of water available to plants and is equal to the difference between precipitation and evaporation, or in other words to the effective precipitation. Effective precipitation is calculated as r/t, where r is the mean annual precipitation in millimeters and t is the mean annual temperature in degrees Celsius (° C). Precipitation that falls as snow must be converted to its rainfall equivalent. If r/t is less than 40 the climate is considered to be arid and if r/t is greater than 160 the climate is perhumid (extremely wet).

Between these extremes, the boundary between steppe grassland and desert, and the boundary between forest and steppe can be determined by values for r/t, depending on how the precipitation is distributed. If precipitation falls mainly in winter, the two boundaries lie where $r/t = 1$ and $r/t = 2$, respectively. If precipitation is distributed evenly through the year they fall where $r/(t + 7) = 1$ and $r/(t + 7) = 2$. If precipitation falls mainly in summer, they fall where $r/(t + 14) = 1$ and $r/(t + 14) = 2$.

effective stack height The height at which a CHIMNEY PLUME begins to move downwind after it has emerged from the top of a smokestack. Except in very strong winds, it does not move downwind immediately upon leaving the stack, because it is traveling vertically upward and is warmer than the air into which it discharges and consequently possesses BUOYANCY. The effective stack height is equal to the sum of the height of the smokestack and the extent to which the plume rises above it, known as the *PLUME RISE.* The greater the effective stack height, the less pollution the plume causes, especially close to the stack, because it remains well clear of the ground longer, giving more time for it to be diluted by mixing with the surrounding air.

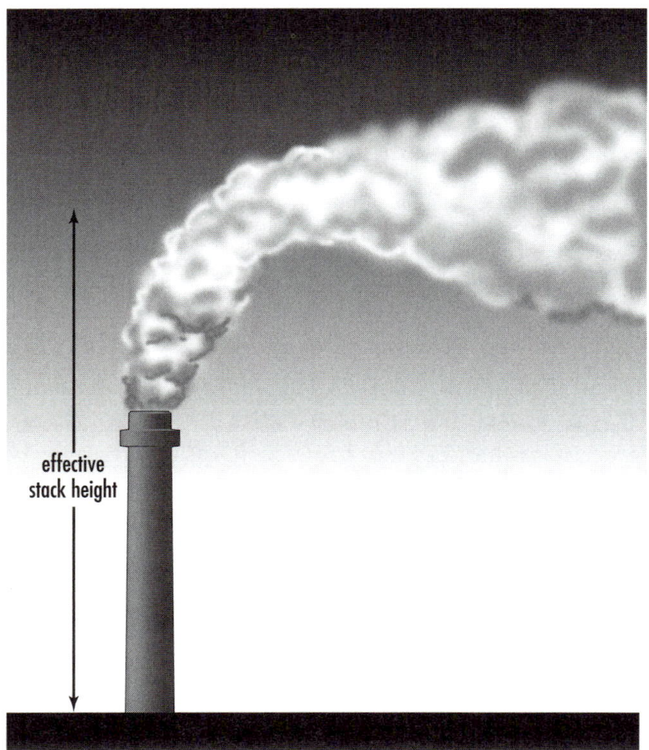

Effective stack height. **As it emerges from the stack, a chimney plume rises before bending over and heading downwind.**

effective temperature The temperature of saturated air with an average WIND SPEED of no more than 0.2 m s^{-1} (0.45 mph) that would produce the same sensation of comfort in a sedentary person wearing ordinary indoor clothes as air with the actual movement, humidity, and temperature to which that person is exposed. The concept is used to define COMFORT ZONES. Effective temperature is approximately equivalent to the TEMPERATURE–HUMIDITY INDEX. The term is also used to describe the temperature of the surface of a planet in the absence of an atmosphere. The effective temperature on Earth is lower than the actual temperature, largely because of the GREENHOUSE EFFECT.

einkanter A VENTIFACT that has only one edge. This indicates it was formed by the action of wind-blown sand arriving from predominantly one direction.

Ekman layer Part of the PLANETARY BOUNDARY LAYER of the atmosphere in which the wind blows at an angle across the ISOBARS. Within this layer the movement of air is balanced among the PRESSURE GRADIENT FORCE, CORIOLIS EFFECT, and FRICTION. The balance is maintained by the air's flowing toward a center of low pressure and outward from a center of high pressure. Wind accelerates as it moves toward a center of low pressure as a result of the conservation of its ANGULAR MOMENTUM as it spirals inward. For the same reason it slows as it spirals outward from a center of high pressure. This is why winds are always stronger around a CYCLONE than they are around an ANTICYCLONE. *See also* UPWELLING.

Ekman spiral The spiral pattern that emerges if the changing direction of the wind with increasing height is plotted on a two-dimensional surface. It is named after VAGN WALFRID EKMAN, who discovered the similar spiral that appears when the changing direction with depth of wind-driven ocean currents is plotted.

Because of the CORIOLIS EFFECT, air does not flow directly from a region of high pressure to a region of low pressure. Instead, a balance is achieved between the Coriolis effect and the PRESSURE GRADIENT FORCE that causes the air to flow parallel to the isobars. This is the GEOSTROPHIC WIND, but it blows only above a height of about 1,650 feet (500 m). Below this height, friction with the surface slows the wind. This reduces the magnitude of the Coriolis effect, thus increasing the pressure gradient force component and causing the wind to flow obliquely across the isobars. The angle at which it does so is proportional to the frictional drag and consequently is greatest at the surface, decreasing to zero above the BOUNDARY LAYER where the wind is geostrophic. At the surface the wind blows across the isobars at an angle of 10°–20° over the sea and 25°–35° over the land, where friction is greater. With increasing height, therefore, the wind direction changes in a spiraling fashion.

Ekman, Vagn Walfrid (1874–1954) Swedish *Oceanographer and physicist* Vagn Ekman was born in Stockholm on May 3, 1874. He was educated at the University of Uppsala, where he received his doctoral degree in 1902. His doctoral thesis described his research into the cause of a phenomenon first reported in the 1890s by the Norwegian arctic explorer Fridtjof Nansen (1861–1930).

Nansen had noticed that ice drifting on the surface of the sea did not move in the direction of the wind, but at about 45° to the right of it. Ekman was able t o relate the movement of wind-driven ocean currents to friction between different layers of water and to the CORIOLIS EFFECT. Together these produced a change in the direction of currents with depth that became known as the *EKMAN SPIRAL*. A similar effect occurs in the atmosphere. In 1905 he explained the spiral more fully in the paper "On the Influence of the Earth's Rotation on Ocean-Currents," published in the Swedish journal *Arkiv för Matematik, Astronomi och Fysik*.

Vagn Walfrid Ekman. **The Swedish oceanographer and physicist who discovered the effect of friction on the flow of air and water.** ***(University of Lund, Sweden)***

After receiving his doctorate, in 1902 Ekman moved to Norway (which was still part of Sweden) to take up a post as an assistant at the International Laboratory for Oceanographic Research in Oslo (which was then called *Christiania*). He remained there until 1908 and while there he came to know Nansen.

Ekman returned to Sweden in 1910, as professor of mechanics and mathematical physics at the University of Lund, a post he held until he retired in 1939. Vagn Ekman died on March 9, 1954.

(You can read a description of an experiment demonstrating the Ekman spiral at "Ekman Pumping Experiment,"taylor.math.ualberta.ca/~eifl/teaching/ekman/index.html.)

elastic modulus *See* SPEED OF SOUND.

El Chichón A volcano in Mexico that became active in 1982, after having remained dormant for several centuries. It began to erupt on March 26 and continued erupting until the middle of May. The final death toll from the eruption was estimated at about 2,000. There was an especially violent eruption on March 29 that killed 100 people living in nearby villages, and on April 4 a huge explosion released a cloud of dust and gas. The gas included up to 3.6 million tons (3.3 million tonnes) of SULFUR DIOXIDE, all of which was converted into sulfuric acid. By May 1 this cloud had circled the Earth. In late June the densest part of the cloud was detected at a height of 17.4 miles (28 km), and by the end of July the top of the cloud reached about 22 miles (36 km). A year later the cloud had spread to cover almost the whole of the Northern Hemisphere and a large part of the Southern Hemisphere. It was the first cloud of volcanic material to be tracked by instruments on satellites and it produced a marked warming of the lower STRATOSPHERE. In June the cloud reduced the average global surface temperature by about 0.4° F (0.2° C).

Electra *See* BENEDICT.

Electrically Scanning Microwave Radiometer (ESMR) A RADIOMETER transmitting in the MICROWAVE waveband that was launched on the *NIMBUS 5* satellite in December 1972 and continued to function until the end of 1976. Its purpose was to monitor sea ice. It operated on a single channel, collecting data at a WAVELENGTH of 1.55 cm. At this wavelength water has an EMISSIVITY of about 0.44, but the emissivity of ice is between 0.80 and 0.97, so the rate of emission is much greater for sea ice than for water.

electric field A region in which a force is exerted on any electrically charged body that enters it. The strength of the field is measured in VOLTS per unit distance. In the atmosphere, there is a natural electric field with a strength of about 60 volts per foot (200 volts per meter), although this varies greatly in the vicinity of THUNDERSTORMS and is reduced by CONVECTION and RAIN.

electrification ice nucleus A fragment of ice that is produced when a DENDRITE shatters in an ELECTRIC FIELD with a strength of several hundred volts per inch. Electric fields of this strength occur in CUMULONIMBUS clouds.

electrodialysis *See* OSMOSIS.

electrogram A record that shows changes over time in the atmospheric ELECTRIC FIELD.

electrometer An instrument that is used to measure the atmospheric ELECTRIC FIELD. Electrometers are raised a measured distance above the surface and then used to measure the change over time in the electric potential between the instrument and the surface. They are also used to measure the electric field between two balloons at different heights. The first electrometer is believed to have been the one used in 1766 by the Swiss naturalist HORACE BÉNÉDICT DE SAUSSURE (1740–99).

electron capture detector An instrument, invented by JAMES LOVELOCK, that can measure extremely small concentrations of substances in the air. It contains a radioactive source that emits a stream of electrons in an electromagnetic potential field. As a flow of air passes through the field, certain atoms present in the air capture electrons from the electron stream. This reduces the electric current by a measurable amount. The instrument is most sensitive to the halogens bromine, chlorine, fluorine, and iodine. This makes it very suitable for detecting the presence of organochlorine pesticides, such as *pp*'-dichlorodiphenyltrichloroethane (DDT), and also chlorofluorocarbons (CFCs).

Electronic Numerical Integrator and Calculator (ENIAC) The first fully electronic computer, which was built at the University of Pennsylvania by J. P. Eckert and J. W. Mauchly and was completed in 1946. It comprised 20 electronic adding machines and contained 18,000 thermionic valves and consumed 100 kW of power when it was at its maximum output. Programming it involved setting switches and plugging in connections manually.

electrostatic charge A positive or negative electric charge that is at rest (it does not flow).

electrostatic coalescence The merging of CLOUD DROPLETS that carry opposite ELECTROSTATIC CHARGES on their surfaces.

electrostatic filter A filter to which an ELECTROSTATIC CHARGE is applied. Particles bearing a charge are attracted to the opposite charge on the filter. This increases the efficiency with which the filter removes small particles from the gas or air passing through it.

electrostatic precipitator A device that is used to remove solid and liquid particles from a gas. The gas is made to flow between two electrodes across which a high voltage is applied. This produces an ELECTRIC FIELD. As the particles pass through the electric field they acquire a charge and move to the electrode bearing the opposite charge, where they are held. Electrostatic precipitators are extremely efficient at collecting small particles, and they are used widely to clean industrial emissions.

***Elektro* satellite (Geostationary Operational Meteorological Satellite)** A Russian meteorological system that was launched on October 31, 1994, into a GEOSTATIONARY ORBIT above 76.83° E. It provides continuous monitoring of atmospheric processes, detects natural hazards, monitors SEA-SURFACE TEMPERATURE and WIND VELOCITY at various altitudes, and measures variations in the Earth's magnetic field and solar ULTRAVIOLET and X-RAY radiation. It also transmits pictures taken in visible and infrared light.

(You can learn more about this system at sputnik.infospace.ru/goms/engl.goms_1.htm.)

Eline A cyclone (*see* TROPICAL CYCLONE) that struck Mozambique on February 22, 2000. It produced winds of up to 160 mph (257 kh^{-1}) and torrential rain. There were reports of coastal villages swept away by the STORM SURGE.

El Niño A change in the prevailing winds over the equatorial South Pacific that occurs at intervals of two to seven years. Ordinarily, the winds blow from the southeast and drive a surface current from east to west. The current carries warm water away from the South American coast and toward Indonesia, producing a relatively thin layer of warm surface water in the east and a deep pool of warm water in the west.

During an El Niño, the southeasterly winds weaken or may even cease to blow or reverse direction. The wind-driven surface ocean current ceases or reverses and warm water accumulates off the coast of South America. The meteorological effects of this change usually appear at about Christmas, the source of the name, which means "male child" (i.e., the Christ child). *See* ENSO.

Eloise A HURRICANE, rated category 4 on the SAFFIR/SIMPSON SCALE, that struck several Caribbean islands and the eastern United States in September 1975. It reached Puerto Rico on September 16, bearing winds of up to 140 mph (225 km h^{-1}) and heavy rain and killing 34 people. It then moved to Hispaniola, where 25 people died. It crossed Haiti and the Dominican Republic and from there reached Florida, where 12 people were killed. Eloise then moved northward along the coast as far as the northeastern United States, where a state of emergency was declared.

ELR *See* ENVIRONMENTAL LAPSE RATE.

Elsasser's radiation chart A chart that is used to determine the components that together make up the outgoing radiation from the Earth's surface. There are several types; the one most widely used in the United States was devised in 1942 by WALTER MAURICE ELSASSER and improved in 1960 by him and M. F. Culbertson. The components of outgoing radiation are the total energy radiated from the surface, which is determined by the EMISSIVITY and temperature of the surface, and the downward counterradiation from the atmosphere, which is determined by the air temperature, PRECIPITABLE WATER vapor, and cloud cover. RADIOSONDE data for the temperature and water-vapor content of the air at different heights are plotted onto the chart. This allows the effective infrared radiation from the surface, the net amount of radiation at a CLOUD BASE or CLOUD TOP, and the rate at which the surface is cooling by radiation to be determined.

Elsasser, Walter Maurice (1904–1991) German-American *Physicist* Walter Elsasser is the scientist who developed the theory that the Earth's core acts as a dynamo, generating the magnetic field. He also pioneered the analysis of the magnetic field recorded in the orientation of rock particles as a tool for studying the history of crustal rocks. This has been of great importance in tracing the movements of continents (*see* PLATE TECTONICS) and the history of climate.

Elsasser was born at Mannheim, Germany, and was educated at the University of Göttingen, where he was awarded his doctorate in 1927. He then taught at the University of Frankfurt, but after the Nazis came to power in 1933 he left Germany. He spent three years in Paris and then moved to the United States and joined the staff of the California Institute of Technology. In 1940 he became an American citizen. In the course of his career Elsasser was professor of physics at the University of Pennsylvania, professor of geophysics at Princeton University, and a research professor at the University of Maryland.

Elsie A TYPHOON that struck the Philippines on October 19, 1989; it killed 30 people and left 332,000 homeless.

elvegust (sno) A cold, descending SQUALL that occurs in the upper parts of Norwegian fjords.

embata An onshore wind from the southwest that blows across the Canary Islands. It is caused by a reversal of the TRADE WINDS in the LEE of the islands.

emergency A situation in which people are at risk of injury or death and must move to a place of safety. Extreme weather conditions can cause emergencies. A medical emergency can occur when a person remains outdoors in hot weather.

Prolonged exposure to heat causes heat exhaustion. The patient has an accelerated pulse and cold and sweaty skin, may feel nauseous or vomit, and feels drowsy, although the body temperature remains normal. Drinking regularly to compensate for fluid lost by sweating prevents the condition. Once it occurs, the patient recovers by resting somewhere cool. If heat exhaustion is prolonged the patient may become apathetic, hysterical, or aggressive. The condition is then known as *heat neurasthenia.*

Overheating can increase the blood flow to the skin while decreasing the flow to the brain. This can cause the patient to lose consciousness. The condition is known as *heat collapse.*

Extreme overexposure to heat can cause heat stroke, which can be fatal. The body temperature rises; the skin is dry, hot, and usually flushed. The patient seems confused, lacks full control of the limbs, and may lose consciousness. Medical help must be sought immediately and the patient cooled gently, by loosening clothing, fanning, and sponging with cool (not cold) water.

Prolonged exposure to cold can cause hypothermia. The patient has a weak pulse, feels cold to the touch, speaks or behaves irrationally, and has difficulty with speech and vision. If untreated, hypothermia can be fatal. Treatment consists of gently warming the patient, but not by applying strong external heat, which can be dangerous.

Frostbite is physical damage to tissues that results from the loss of blood supply to a part of the body exposed to a very low temperature. The bodily extremities—ears, nose, cheeks, fingers, and toes—are the parts of the body most likely to be affected. The condition is painless, because there is no blood supply to the affected nerves, but the skin is locally very pale. This paleness allows the early visual detection of frostbite, and people outdoors alone in very cold weather are advised to carry a pocket mirror and regularly examine their faces and ears. In its early stages, frostbite can be treated by warming the affected parts in tepid water. Massaging is dangerous, because it may increase the tissue damage. If it remains untreated for too long, the tissue damage may be irreversible and necessitate amputation.

More widespread general emergencies are caused by FLASH FLOODS, violent STORMS, BLIZZARDS, TROPICAL CYCLONES (such as HURRICANES), and TORNADOES. Warnings that are issued in advance of these events must be heeded.

emissary sky A sky that is covered by patchy CIRRUS cloud. This type of cloud often forms in humid air on a WARM FRONT. Its appearance indicated the approach of weather associated with the lower part of the front, so the cloud acts as an emissary of the rain and winds that are to come.

emissivity The amount of radiation a body emits, expressed as a proportion of the radiation that would be emitted at the same wavelength by a BLACKBODY at the same temperature. The emissivity of gases varies considerably with wavelength, but that of solid objects remains fairly constant. Taking account of emissivity (conventionally symbolized by ϵ) requires a modification of the STEFAN–BOLTZMANN LAW, so the equation becomes

$$E = \epsilon\sigma T^4$$

where σ is the Stefan–Boltzmann constant and T is the temperature in kelvins. The modification means the Earth does not behave precisely as a blackbody; it is therefore described as a gray body. For most surfaces, the emissivity is greater than 0.90, where blackbody emissivity is 1.0 (or 90 percent if blackbody emissivity is given a value of 100).

Typical emissivities

Water	0.92–0.96
Fresh snow	0.82–0.995
Desert	0.90–0.91
Tall, dry glass	0.90
Oak woodland	0.90
Pine forest	0.90
Dry concrete	0.71–0.88
Plowed field	0.90

emittance *See* EXITANCE.

endemic windthrow *See* BLOWDOWN.

energy balance model (EBM) A model of the atmosphere that describes the change in sea-level temperature with latitude and the balance of energy input and output in each latitudinal belt. Such a model is one-dimensional (it deals only with temperature changes with latitude in a horizontal plane), but its simplicity means it can be installed and run on small computers. The model is used as an instructional tool and as a check of the accuracy of more complex models. The relationship it describes is summarized by

$$\rho C\Delta T(\theta) \div \Delta T = R{\downarrow}(\theta) - R{\uparrow}(\theta) + \text{transport into belt } \theta$$

where ρC is the HEAT CAPACITY of the area being studied, ΔT is the change in temperature, θ is the latitudinal belt, $R{\downarrow}$ is the radiant energy entering the system, and $R{\uparrow}$ is the INFRARED RADIATION leaving the system.

energy exchange The warming of a body that comes into contact with another body at a higher temperature. Energy is exchanged between an AIR MASS and the land or sea surface beneath it, and it is this exchange that gives the air mass its defining characteristics and that generates weather phenomena. As the air mass moves away from its SOURCE REGION this exchange alters until eventually the character of the air mass changes. Energy is exchanged between every surface and the air above it, except when the surface and the air are at the same temperature.

ENIAC *See* ELECTRONIC NUMERICAL INTEGRATOR AND CALCULATOR.

ENSO **(El Niño–Southern Oscillation event)** A change in the distribution of surface atmospheric pressure and the wind direction that affects the equatorial South Pacific at intervals of two to seven years. In normal (non-ENSO) years the Hadley cell circulation produces prevailing southeasterly surface winds (the TRADE WINDS). These drive the warm South Equatorial Current. This flows from east to west, carrying warm water that accumulates as a relatively deep pool in the region of Indonesia. Off the South American coast the layer of warm surface water is much thinner.

Near Indonesia, the high sea-surface temperature warms the air. The rate of evaporation is high and rainfall is heavy. In eastern South America, the prevailing winds blow from land to sea, crossing the Andes, and the coastal climate is extremely dry. Off the coast, the layer of warm surface water is thin enough to allow cooler water to rise close to the surface as a series of upwellings in the cold Peru Current, flowing northward from the Southern Ocean. The Peru Current is

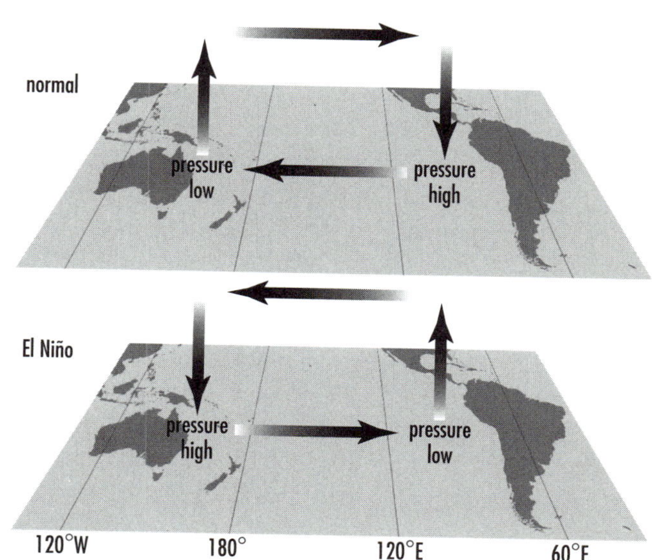

ENSO. **In the normal situation, surface pressure is high in the eastern Pacific and low in the western Pacific. This condition drives the air circulation and the wind drives a surface sea current flowing from east to west. During El Niño the situation is reversed.**

rich in nutrients, gathered as it moves across the ocean floor. These nourish a large planktonic population, which in turn large numbers of fish and seabirds that feed on the fish.

At high level, the WALKER CIRCULATION carries rising air from the west to the east, where it sinks to the surface.

During the ENSO event, the pressure pattern changes. Surface pressure rises over Indonesia. This reduces the PRESSURE GRADIENT across the ocean and the winds weaken. As the winds weaken, so does the South Equatorial Current, allowing the layer of warm water in the east to deepen. This is the southern oscillation and, in a strong ENSO, the pressures actually reverse, so there are high pressure in the west and low pressure in the east, and the flow of air at both high and low levels reverses direction.

This is the El Niño phase of the ENSO. Warm water off South America suppresses the upwellings of the Peru Current. The plankton disappear, followed by the fish and seabirds. Warm, moist air now flows toward the South American coast, taking heavy rain to the coastal strip. There are also wider effects, including drought in northeastern South America, Indonesia, parts of Africa, and Australia, and heavy rain and storms in the western and southern United States. The extreme weather can also affect human health, through outbreaks of such diseases as malaria, Rift Valley fever, and possibly dengue or breakbone fever, which is an acute, infectious disease transmitted by mosquitoes.

With the passing of the El Niño, the situation returns to normal but may then swing into the opposite pattern, called *LA NIÑA;* the full southern oscillation, El Niño, La Niña, and the final return to normal constitute the full ENSO event. El Niño is not always followed by La Niña, and La Niña sometimes occurs independently of an El Niño.

Especially strong ENSO events occurred in 1982–83 and 1997–98. Other ENSOs, some continuing for longer than one season, were recorded in 1396, 1685–88, 1789–93 (associated with the European crop failures that preceded the French Revolution), 1877–79, 1891–92, 1925–26, 1957–58, 1972–73 (associated with severe drought in the Sahel region of Africa), 1976–77, and 1986–87.

(There is more information about ENSO and El Niño at http://nic.fb4.noaa.gov/products/analysis_monitoring/ensostuff/ensoyears.html and at http://www.elnino.noaa.gov/lanina_new_faq.html.)

enthalpy The heat that can be felt (sensible heat) and that is transferred between bodies at different temperatures. It is defined as

$$H = U + pV$$

where H is the enthalpy, U is the internal energy of the system, p is the pressure, and V is the volume. If p remains constant, U is equal to the HEAT CAPACITY at that pressure (c_p), and enthalpy is then

$$H = c_pT$$

where T is the temperature.

If energy is added to the system either its internal energy increases or the system does some work according to the first law of thermodynamics (*see* THERMODYNAMICS, LAWS OF), which states that energy can neither be created nor destroyed. Provided no energy is added to air, its enthalpy remains constant, and where changes are ADIABATIC the enthalpy of the air is conserved and can be known from its temperature.

If MOISTURE is added, the heat capacity of water and the absorption and release of LATENT HEAT associated with phase changes modify the enthalpy of the air. The moist enthalpy for a specified mass of air is also conserved, but calculating it is more complicated. It is given by

$$H = (c_p + m_tc_{pl})T + Lm_v$$

where m_t is the MIXING RATIO for the total amount of water in the air, c_{pl} is the heat capacity of liquid water, L is the latent heat of vaporization, and m_v is the mixing ratio of WATER VAPOR.

entrainment Mixing that takes place between a body of air and the air surrounding it. This happens when air rises by CONVECTION and the process can be inferred from the way a CUMULIFORM cloud expands and dissipates at its edges. It expands because the air in which it forms is expanding and it does so partly by incorporating surrounding air. That air is drier than the air in which CONDENSATION is occurring and consequently cloud droplets EVAPORATE at the edges of the cloud.

Entrainment occurs whenever air moves vertically, and as well as limiting the lateral expansion of cumuliform cloud it contributes to the final dissipation of CUMULONIMBUS cloud. Once PRECIPITATION commences, the falling ICE CRYSTALS, SNOWFLAKES, and RAINDROPS chill the air around them and drag cool air downward, forming downdrafts. The downdrafts are joined by cold, dry air that is drawn into the cloud from its surroundings. Because it is dry, the entrained air causes some of the precipitation to evaporate, drawing LATENT HEAT from the air inside the cloud and further cooling the downdraft. The air from the downdraft leaves the base of the cloud and spreads to the sides when it meets the surface below. This causes more air to be entrained, the downdrafts to increase, and when the downdrafts dominate air movement inside the cloud, the cloud dissipates.

entrance region The area in which air is being drawn toward the core of a JET STREAM. Air entering from the side nearer the pole (the cold side) is subjected to CONVERGENCE and subsides. Some of this air is drawn down from the lower STRATOSPHERE and the jet stream contains more OZONE (of stratospheric origin) than does the air adjacent to the jet stream but outside it. Air entering the core from the side nearer the equa-

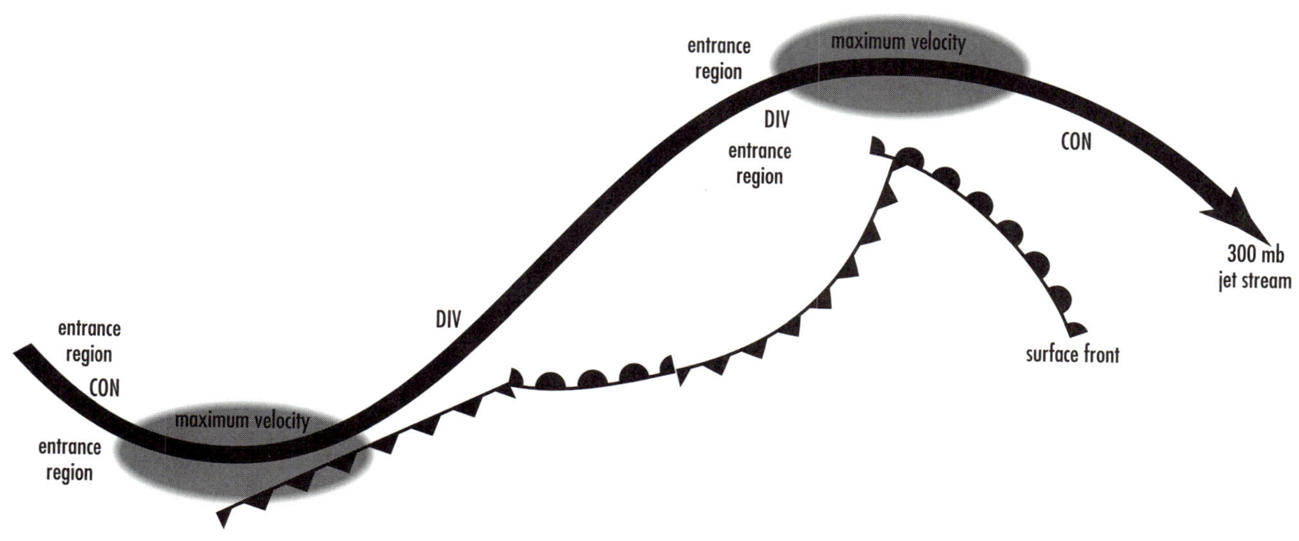

Entrance region. **The entrance region lies on the upstream side of the core, where air is being drawn into the jet stream.**

tor (the warm side) is subjected to DIVERGENCE and tends to rise.

entropy A measure of the amount of disorder that is present in a system. As the amount of disorder increases, so does its entropy, and disorder increases all the time and has throughout the history of the universe. This is because there is a much greater statistical probability that the random motion of atoms and molecules will lead them to form chance arrangements than highly organized structures. If a vase falls onto a stone floor it will probably shatter into many pieces. This is a less ordered state than the one that existed when the fragments were joined seamlessly together and the vase was complete. The shattering of a vase is a common event and one that increases disorder. History records no instance when the fragments of a shattered vase moved back together again to reconstruct the vase as it was before it fell.

Order can be described as a difference in energy level between an object and its surroundings. As entropy increases, this difference decreases. Consider what happens when a cup of hot coffee is left to stand on a table. It cools until it is at the same temperature as the air around it. The energy difference between the coffee and the air has disappeared and both are at the same energy level. It never happens that the hot coffee absorbs energy from the air and its temperature increases.

In exactly the same way, when solar radiation warms the surface of the Earth that heat is dissipated. Some is lost immediately as INFRARED RADIATION (*see* BLACKBODY) into space and some warms the air in contact with the surface. This warming produces all our weather phenomena, but the energy driving the weather also dissipates, because the atmospheric gases also radiate energy into space. The weather continues because there is a constant supply of solar energy to drive it. Without that energy, the Earth would cool until it was at the same temperature as the space surrounding it.

Entropy means that most everyday events are irreversible. For this reason it is sometimes known as the "arrow of time." It is described by the second law of thermodynamics (*see* THERMODYNAMICS, LAWS OF).

envelope orography A technique that is sometimes used in the mathematical models used in weather forecasting. It assumes the valleys and passes in mountain ranges are filled mainly with stagnant air. This allows them to be ignored, effectively increasing the average height of the mountains. The disadvantage of this simplification is that increasing the average height of the mountains also increases the extent to which they block the passage of air to a value greater than the real value.

environmental lapse rate (**ELR**) The rate at which the air temperature decreases with height as this is measured at a particular time and place. This temperature change is not the result of ADIABATIC cooling, but simply the actual temperature that is observed, and it is very variable. Strong daytime heating of the ground may produce a layer of very warm air with much cooler air above it and a steep ELR. During the evening, the ground cools and the ELR becomes less steep. It may even reverse, with warmer air lying above cooler air producing a temperature INVERSION. If the ELR is greater than the DRY ADIABATIC LAPSE RATE, it is said to be *superadiabatic.*

environmental quality standards The maximum limits or concentrations of polluting substances that are permitted in air, water, or soil. In the United States, there are primary and secondary standards. With an allowance to provide a safety margin, primary standards represent those maxima that present no threat to human health. Secondary standards are those maxima that present no threat to public welfare.

Eocene The epoch of geological time (*see* GEOLOGICAL TIME SCALE) that began about 56.5 million years ago and ended about 35.4 million years ago. It constitutes the central part of the PALEOGENE EPOCH.

eolian (**aeolian**) Transported by the wind. The word is derived from the name of Aeolus, the Greek god of winds. An eolian deposit consists of dust or soil that has been transported by the wind to the place where it is found. An eolian sand sheet is an area of sand that has been deposited by the wind. Such areas occur around the edges of areas covered with sand dunes. Eolian EROSION is another name for wind erosion.

EOS *See* EARTH OBSERVING SYSTEM.

ephemerides *See* EPHEMERIS.

ephemeris (**pl.** ***ephemerides***) A table that sets out the height above the horizon (altitude), DECLINATION, and other data for astronomical bodies. An ephemeris showing the declination of the Sun for every day of the year is used in calculating the changing intensity of solar radiation at a particular location. The word *ephemeris* also designates a book that contains a collection of these and other astronomical tables.

equal-angle projection A map projection in which all the angles between lines on the map are equal to the corresponding angles on the surface of the spherical Earth. A map is a two-dimensional representation of a three-dimensional solid object (the Earth or part of it) and is drawn by projecting the location of surface features onto a flat sheet. In the equal-area projection, this sheet is imagined to be at right angles to the plane of the equator. The equator appears as a straight horizontal line and one meridian as a straight vertical line at right angles to the equator. All other lines of latitude and longitude appear as arcs of circles, those of latitude (known as *small circles*) decreasing in size with distance from the equator. The equal-angle projection represents all the angles correctly, but areas, and therefore distances, are distorted.

equation of mass conservation *See* NUMERICAL FORECASTING.

equation of motion The mathematical equation that describes the movement of air according to Newton's second LAW OF MOTION. It is most often written as

$$a = F/M$$

where a is ACCELERATION and F is the force acting on a body of air with a mass M. The equation assumes that the mass of a PARCEL OF AIR remains constant, and therefore it cannot be applied to the development of a CUMULIFORM cloud in which the mass changes as a consequence of ENTRAINMENT.

This equation can be broken down into its constituent elements of acceleration and force. Relative acceleration (a_r) is essentially linear and occurs as air moves in relation to the surface beneath. Acceleration due to the CORIOLIS EFFECT (a_c) takes account of the rotation of the Earth. The forces causing or affecting the movement of air consist of the PRESSURE GRADIENT FORCE (f_p), the gravitational force (f_g), and FRICTION (f_f). It is the pressure gradient force that causes the air to move, the gravitational force acts vertically downward, and friction acts in the opposite direction to the movement of the air. Substituting these terms for those in the first equation, the equation becomes

$$a_r + a_c = f_p + f_g + f_f$$

equation of state (ideal gas law) An equation or law that relates the temperature, pressure, and volume of an ideal gas. The basic equation is $pV = R{*}T$, where p is the pressure, V the volume, T the temperature, and $R{*}$ is the universal GAS CONSTANT (8.31434 J K^{-1} mol^{-1}). In the case of air, which is unconfined and so has no precise volume, the equation substitutes the density for the volume. In meteorology, therefore, the equation is $p = \rho RT$ where ρ is the density and R is the specific gas constant for air (= $10^3 R{*} \div M$, where M is the relative molecular mass). Since air is a mixture of gases, each with its own specific mass and molecular weight, M has to be calculated from the individual molecular weights of its constituents. This shows that for dry air in which the component gases are well mixed M = 29.0 g and the specific gas constant is 287 J K^{-1} kg^{-1}.

equations of continuity *See* NUMERICAL FORECASTING.

equatorial air Warm, humid air that forms an AIR MASS covering the equatorial belt in both hemispheres (*see* SOURCE REGION). The air is rising on the equatorward side of the HADLEY CELLS and consequently equatorial air is usually cooler than TROPICAL AIR. Most of the equatorial region is covered by ocean, so the air mass over it is classified as MARITIME equatorial (mE), and CONTINENTAL equatorial air does not occur.

equatorial climate The climate of the region approximately bounded by latitudes 10° N and 10° S. It is warm and humid throughout the year, with little seasonal variation. This is the climate of tropical rain forests, known as the *WET EQUATORIAL CLIMATE* in the STRAHLER CLIMATE CLASSIFICATION.

Equatorial Countercurrent A narrow ocean current that flows from west to east between the NORTH and SOUTH EQUATORIAL CURRENTS.

equatorial easterlies The northeasterly and southeasterly TRADE WINDS during the summer, when they extend almost to the TROPOPAUSE and the westerly winds blowing above them are either nonexistent or too weak to affect the lower TROPOSPHERE.

equatorial regime *See* RAINFALL REGIME.

equatorial trough A wide belt of low surface pressure that encircles the Earth where the TRADE WINDS of the Northern and Southern Hemispheres converge (*see* CONVERGENCE) in the INTERTROPICAL CONVERGENCE ZONE, causing air to rise. This is part of the HADLEY CELL circulation. Winds in the trough are generally light and easterly, but this is also the location of the DOLDRUMS. Rising air causes water vapor to condense, and the trough is the region where most equatorial precipitation occurs, although the amount of cloud is very variable and there are large areas of clear skies.

Equatorial Undercurrent *See* CROMWELL CURRENT.

equatorial upwelling *See* UPWELLING.

equatorial vortex The VORTEX at the center of the CYCLONIC circulation of air that is produced by an EQUATORIAL WAVE over the Pacific Ocean. The vortex travels westward toward the Philippines but rarely intensifies into a TROPICAL CYCLONE.

equatorial wave A wave disturbance that sometimes develops in the EQUATORIAL TROUGH over the Pacific Ocean when the trough is far enough from the equator for the CORIOLIS EFFECT to generate CYCLONIC motion. Equatorial waves are more common in the Northern Hemisphere than the Southern Hemisphere and they seldom generate winds that intensify into TROPICAL CYCLONES.

equatorial westerlies Westerly winds that occur in summer between the northeasterly and southeasterly TRADE WINDS. These westerlies are most strongly evident over continents, especially over Africa and southern Asia, where heating of the ground surface produces a pressure distribution that shifts the EQUATORIAL TROUGH northward. The westerlies extend to a height of about 1–2 miles (2–3 km) over Africa and 3–4 miles (5–6 km) over the Indian Ocean. They are associated with the summer MONSOON over Asia, but their cause is quite different from the pressure pattern

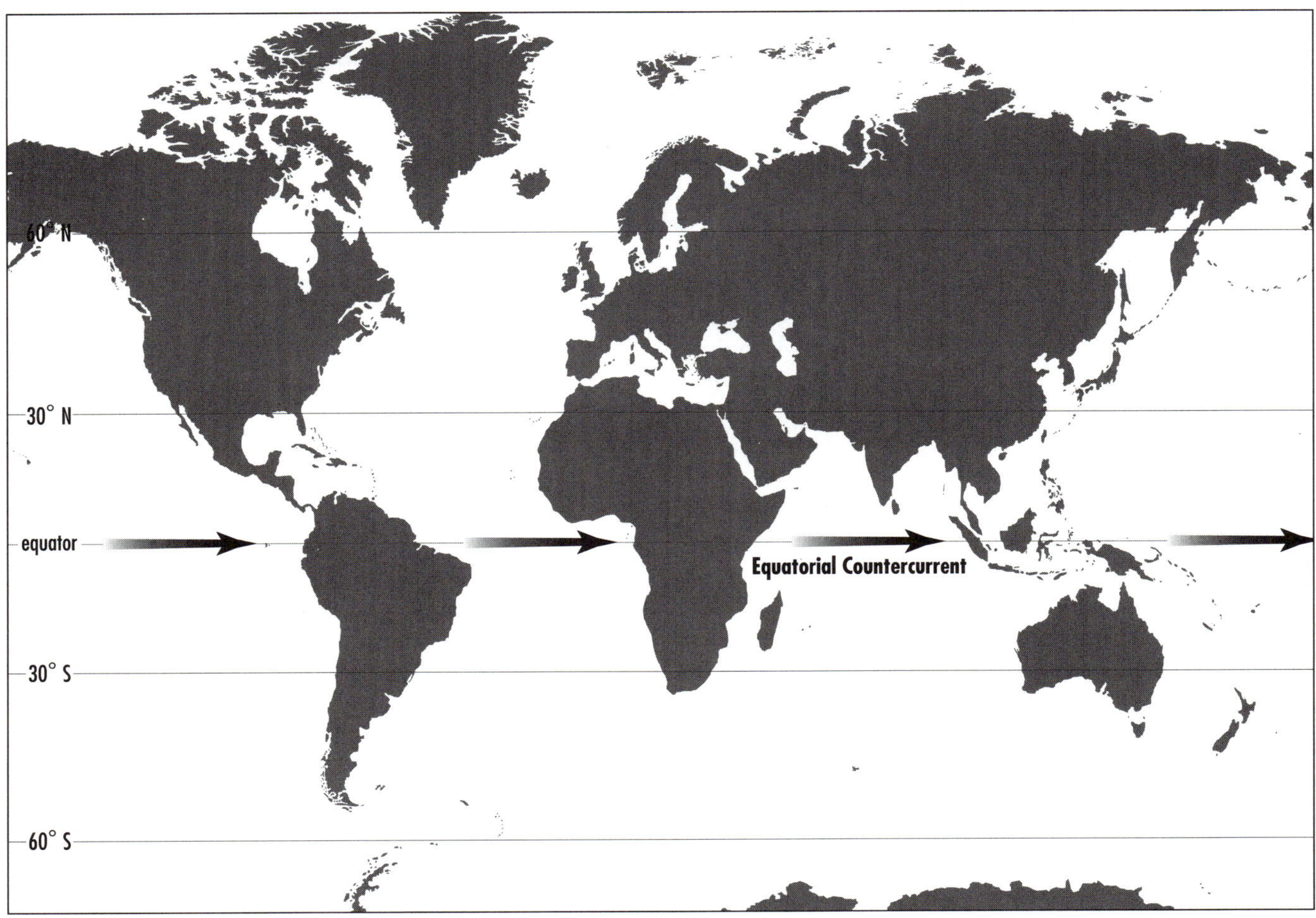

Equatorial Countercurrent. **An ocean current that flows close to the equator and from west to east in all oceans.**

that produces the monsoon winds. The equatorial trough does not move far enough from the equator to produce westerly winds over the Pacific or Atlantic Ocean.

equilibrium level (**level of zero buoyancy**) The height at which the VIRTUAL TEMPERATURE of a PARCEL OF AIR that has risen above the LEVEL OF FREE CONVECTION is equal to the temperature of the surrounding air. At this height the air is neutrally buoyant and rises no farther.

equilibrium vapor The WATER VAPOR in the very shallow layer of air that is in contact with an exposed surface of liquid water. The layer of air is about 0.04 inch (1 mm) thick, and the water vapor in it is in equilibrium because its amount does not change. Water molecules are constantly leaving the liquid surface by EVAPORATION, but molecules are returning to the liquid by CONDENSATION at the same rate. If the number of molecules leaving the liquid increases, the number returning to the liquid also increases by the same amount. If molecules leave the layer by moving into the air above the layer, they are replaced immediately by molecules escaping from the liquid, but in this case the volume of the liquid is decreased. If the loss continues the water diminishes by evaporation.

equinoctial gale A wind storm that occurs around the time of an EQUINOX. According to a folk belief in both the United States and Britain, gales are more common at the equinoxes. There is no basis for this belief.

equinoctial rains Rainy seasons that occur in some parts of the Tropics at around the time of the equinoxes (*see* EQUINOX).

equinox One of the two dates in each year when day and night are of equal length, an interval of 12 hours passing between dawn (measured as the Sun's appearing above the horizon) and sunset (measured as the Sun's disappearing below the horizon), and 12 hours between sunset and dawn. At each equinox, the Sun is directly overhead at noon over the equator. The equinoxes fall on March 20–21 and September 22–23. The varying length of the day is due to the fact that the rotational axis of the Earth is tilted with respect to the PLANE OF THE ECLIPTIC, and it is this tilt that produces the SEASONS.

equivalent-barotropic model A model of the atmosphere in which it is assumed that air movements are not affected by friction and are ADIABATIC (that is, warmed and cooled as a consequence of the vertical movement of the air), and the vertical WIND SHEAR of the horizontal wind is proportional to the horizontal wind itself. The atmosphere is in HYDROSTATIC EQUILIBRIUM and QUASIGEOSTROPHIC BALANCE. In this atmosphere the wind direction does not change with height, all the contours on any ISOBARIC SURFACE are parallel, and vertical movements are presumed to be equivalent to those at an intermediate level, the equivalent-barotropic level.

equivalent potential temperature The temperature a PARCEL OF AIR would have if it were decompressed at the SATURATED ADIABATIC LAPSE RATE almost to zero pressure and then recompressed at the DRY ADIABATIC LAPSE RATE to 1,000 mb (sea-level pressure). As the saturated air is decompressed its temperature falls and its water vapor condenses. This releases LATENT HEAT of condensation, warming the air, and the warmth is retained, so the equivalent potential temperature is higher than the actual temperature of saturated air at the 1,000-mb level. The difference can be estimated from:

$$\theta_e - \theta_w = 2.5q_s$$

where θ_e is the equivalent potential temperature, θ_w is the temperature of saturated air at the 1000-mb level, and q_s is the SPECIFIC HUMIDITY of the saturated air at the 1,000-mb level expressed in grams of water vapor per kilogram of air including the water vapor.

equivalent temperature (isobaric equivalent temperature) The temperature a PARCEL OF AIR would have if all the water vapor it contained were condensed from it, the LATENT HEAT of condensation were allowed to warm the air, and the pressure remained constant. *Equivalent temperature* is also sometimes used as synonym of EQUIVALENT POTENTIAL TEMPERATURE.

ERBE *See* EARTH RADIATION BUDGET EXPERIMENT.

ERBS *See* EARTH RADIATION BUDGET SATELLITE.

Eric One of two CYCLONES that struck Viti Levu, Fiji, on January 22, 1985. The other was Nigel. Together they caused the deaths of 23 people. It is most unusual for two cyclones to occur so close together.

erosion The wearing away of exposed rock surfaces and the removal of soil by the action of wind and flowing water. Although the term is often used rather loosely, erosion involves only the physical process of wearing away. It is not the same as either WEATHERING or DENUDATION.

Provided they are dry and not adhering to one another, mineral particles up to the size of sand grains move readily under pressure from wind and water. A wind of more than 15 mph (24 km h^{-1}) is strong enough to produce a SAND STORM. Once they are in motion, the particles are hurled against rocks and detach more particles from them. This is how rocks are eroded. It is a gradual process in which soft rocks erode first, leaving harder, more resistant rocks exposed. This can produce desert landforms with curious shapes. Continued over millions of years, this kind of erosion has produced the sands of the deserts and, because water carries particles to the sea, the sands, gravels, and pebbles of river beds, beaches, and the ocean floor.

Extreme weather can cause erosion that is more rapid. Heavy rain and melting snow can wash exposed soil from hillsides. In this type of erosion the water cuts temporary channels that are left as barren, irregular gullies when dry conditions return but that grow wider and deeper with every subsequent storm.

If all the vegetation is removed and finely textured soil is allowed to become very dry, the slightest wind lifts it. DUST STORMS are fairly common, but the worst examples in modern times occurred in the 1930s on lands that became known as the *DUST BOWL*.

Erosion is a natural phenomenon. Poor farming practices can accelerate soil erosion and modern techniques aim to minimize it. Plowing is a major cause of erosion. Land is plowed in order to destroy weeds. The growing of crops that have been genetically modified to tolerate herbicides allows farmers to kill weeds chemically without risk to the crop and without a need to plow. This has been found to reduce soil erosion considerably. Soil erosion is considered serious only if the rate at which the soil is being eroded exceeds the rate at which new soil is forming.

Rates of erosion are measured in bubnoff units, named after the German geologist S. von Bubnoff (1888–1957). One bubnoff unit (B) is equal to the erosion of one micrometer (1 μm = 0.00004 inch) per year, or one millimeter (1 mm = 0.04 inch) per thousand years. GLACIERS scour away the soil at a rate of about 1,000 B. In a temperate, MARITIME CLIMATE, soil erodes naturally at 1–5 B; poor farming can cause soil to erode at 2,000 B or more.

erosion of thermals The mechanism by which a rising THERMAL dissipates. As the warm air rises, cooler air from its surroundings is incorporated around its edges by ENTRAINMENT. The air at the edges then reaches its own EQUILIBRIUM LEVEL, leaving a smaller mass of air that is still rising. Entrainment continues, steadily eroding the warm air until all of it is neutrally buoyant (*see* BUOYANCY), at which stage the thermal has ceased to exist.

ERTS *See* LANDSAT.

ESMR *See* ELECTRICALLY SCANNING MICROWAVE RADIOMETER.

estivation (aestivation) A state of sluggishness or dormancy that occurs in some animals, including snails and lungfishes, when the weather is hot and dry. Estivation helps the animals to survive adverse conditions. It is analogous to HIBERNATION, by which some animals survive cold weather.

etalon An INTERFEROMETER that comprises half-silvered plane-parallel glass or quartz plates a fixed distance apart with a film of air enclosed between them.

etesian The name given to cool, dry winds that blow from the north over the eastern Mediterranean. They are most frequent from May to October and commonly attain speeds of 40 mph (64 km h^{-1}) but can be so strong it is impossible for sailing vessels to travel against them. In Turkey, where they are called *meltemi*, they blow from the northwest and often yield relief from the intense summer heat. The winds are associated with low-pressure systems and are very dependable. The name *etesian* is from the Greek *etesiai*, meaning "annual."

Eumetsat The European Organization for the Exploitation of Meteorological Satellites, an intergovernmental organization, founded in 1986, that establishes and maintains operational meteorological satellites on behalf of 17 nations (Austria, Belgium, Denmark, Germany, Finland, France, Greece, Ireland, Italy, the Netherlands, Norway, Portugal, Spain, Sweden, Switzerland, Turkey, and the United Kingdom). The satellites are operated from a control center in Darmstadt, Germany, which is linked to stations in Rome and Fucino, Italy; Bracknell, United Kingdom; and Toulouse and Lannion, France.

Eurasian high An area of high atmospheric pressure that develops across Eurasia in winter. It collapses fairly rapidly in April.

EUROCORE Project A European project that aims to collect data held at universities and other institutions within Europe that are obtained from sediment cores and samples dredged from the seabed and from ICE CORES. These are assembled into a directory published on the Internet and eventually will allow direct Internet access to the data.

(Information about the project is at www.eu-seased.net/eurocore/content.htm.)

eustasy The worldwide change in sea level that is due either to tectonic movements of the Earth's crust (*see* PLATE TECTONICS) or to the expansion or melting of GLACIERS.

evaporation The change from the liquid to the gaseous phase. In the case of water, evaporation is the change from liquid into WATER VAPOR. Water is unusual among everyday chemical compounds in that it can exist in all three phases at temperatures commonly found at the Earth's surface.

Liquid water molecules are linked in small groups by HYDROGEN BONDS. The groups are constantly breaking and reforming, they can move freely, and they slide past one another. Molecules at the surface are held there by SURFACE TENSION. If the water is heated, its molecules absorb the heat energy and move faster. This increases the VAPOR PRESSURE the molecules exert at the surface. When they have absorbed enough energy, molecules begin to break away from their groups and enter the air. The energy that is used to break the bonds between molecules, called *LATENT HEAT*, does not raise the temperature.

Immediately above the surface the molecules add to the vapor pressure due to the water molecules already present. This pressure returns molecules to the liquid surface. Evaporation occurs if more molecules enter the air than leave it, so the amount of liquid decreases and the amount of vapor increases.

If the water surface is at a higher temperature than the air immediately above it, water molecules always escape and evaporation occurs. The high HEAT CAPACITY of water means it warms and cools more slowly than land. Air moving from the land is usually cooler than the sea in late fall and winter, so the evaporation rate is high. In spring and summer the air is warmer than the water, so less evaporation occurs.

Wind reduces the vapor pressure in the air above the surface. This is because it generates a TURBULENT FLOW of air that removes moist air from above the liquid surface and replaces it with dry air. The rate of evaporation (E) can be calculated by using DALTON'S LAW of PARTIAL PRESSURES:

$$E = Ku(e_w - e_a)$$

where K is a constant, u is the mean wind speed, e_w is the vapor pressure at the water surface, and e_a is the vapor pressure in the air above the surface.

Salt water evaporates more slowly than freshwater, by about 5 percent.

evaporation pan A device for measuring the rate of evaporation that consists of a container of a standard size holding water that is exposed to the air. The pan is set in the open, filled with water, and left for a convenient length of time. The exposure period, of one day, one week, or longer, must be sufficient for a reliably measurable amount of water to evaporate. This is determined by the temperature and RELATIVE HUMIDITY of the air, and so the period varies according to local conditions. The depth of water is measured at the start and end of the period, and the change in depth is converted to the volume of water lost.

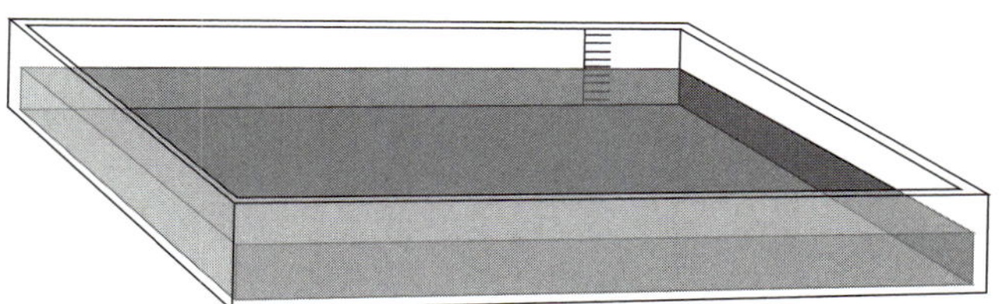

Evaporation pan. **The pan consists of a rectangular container of standard size. The volume of water that evaporates over a given period is measured.**

evaporation pond A body of seawater that is enclosed and allowed to evaporate in order to obtain the salts that crystallize from it. At one time this was the method by which almost all common salt (sodium chloride, NaCl) was obtained, and it is still used to produce sea salt in coastal areas that enjoy a hot climate.

evaporative capacity *See* EVAPORATIVE POWER.

evaporative power **(evaporative capacity, evaporativity, potential evaporation)** The ability of the climate of a region, or the weather at a particular time, to evaporate water. It is measured as the rate of evaporation from the surface of chemically pure water at the temperature of the air in contact with it. Since in practice it is impossible to separate the rate of evaporation from an exposed water surface from the water leaving plants by TRANSPIRATION, evaporation and transpiration are usually considered together as EVAPOTRANSPIRATION.

evaporimeter A simple instrument that is used to measure the rate of evaporation. There are several designs, all of which measure the volume of water that evaporates from a known surface area over a measured

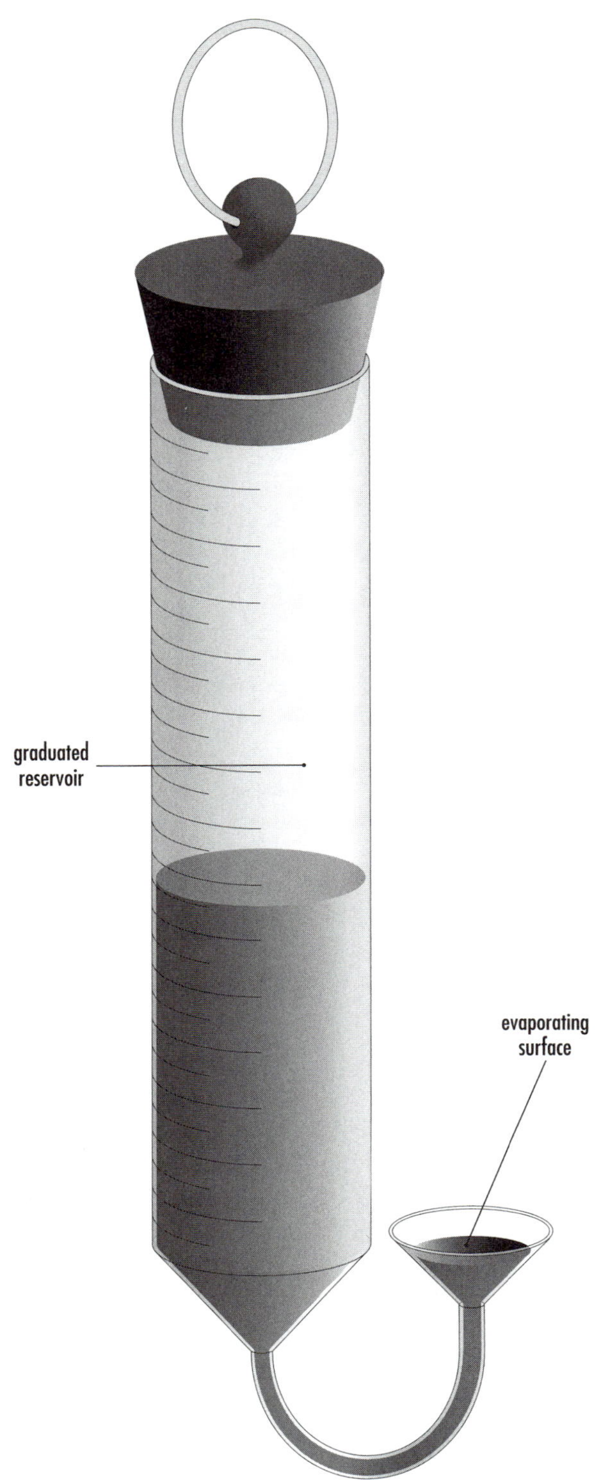

Evaporimeter. **Water is held in a graduated reservoir opening at the bottom into a U tube that expands into an evaporating surface of known area. The surface is covered by filter paper. There are several types of evaporimeter. This is the Piché evaporimeter.**

period. An evaporimeter may comprise nothing more elaborate than an open dish of known surface area that is weighed periodically. It can be placed in the branches of a tree to measure evaporation above ground level. An alternative design consists of a graduated reservoir, sealed with a cork and with a ring from which it can be hung. At the bottom of the reservoir a U tube expands to an open surface of known area that is covered with filter paper. As water evaporates from the filter paper the level falls in the reservoir. For more precise measurements of evaporation scientists use a LYSIMETER.

evaporite A mineral salt, most commonly ordinary salt, NaCl, that is precipitated by the EVAPORATION of water in which it was dissolved. Later geological processes may convert an evaporite deposit into sedimentary rock. There are large evaporite beds in Asia, Europe, and North America. The Bonneville salt flats in Utah are an evaporite deposit that was formed by the evaporation of LAKE BONNEVILLE.

Evaporites are precipitated when, over a prolonged period, the amount of water evaporating from a body of water exceeds the amount flowing into it and entering it from precipitation. In-flowing water carries dissolved salts, which are concentrated by the loss of freshwater through evaporation until the water becomes saturated. The salts then accumulate on the bed as the body of water gradually disappears.

The presence of evaporite deposits indicates that the area once lay beneath salt water and that a climatic or geological change caused that water to evaporate. This can happen if the climate becomes warmer or the air over the water becomes drier. It can also happen if rock movements or a fall in sea level closes a narrow, shallow strait linking the water to a sea or ocean from which it is replenished.

evapotranspiration The loss of water from the surface due to the combined effects of EVAPORATION and TRANSPIRATION. Evaporation is measured fairly simply, using an EVAPORIMETER or EVAPORATION PAN. It is more difficult to measure transpiration, but it can be done for an individual plant, although not for a stand of plants growing in the open. Measuring each of these separately introduces an element of artificiality, however, and for climatological or meteorological purposes it is unnecessary. Except in the most barren deserts, water vaporizing from the ground surface cannot be distinguished from water vaporizing from plants. The two processes are entirely distinct, but they both produce the same result, that of adding water vapor to the air, and they both operate at the same time. Consequently, it is usual to consider them together, combining evaporation and transpiration as evapotranspiration.

evapotranspirometer An instrument that is used to measure POTENTIAL EVAPOTRANSPIRATION, which is an important value in calculations connected with the THORNTHWAITE CLIMATE CLASSIFICATION. Potential evapotranspiration is equal to the difference between the amount of water reaching the surface as precipitation and the amount that percolates downward through the soil. This can be measured, assuming the amount of moisture that is retained by the upper soil remains constant. Two devices are installed, because the average of the two readings provides a more reliable measure than the reading from one device alone.

For each installation a 6-inch (15-cm) layer of gravel is placed beneath an enclosed block of sandy loam soil 18 inches (45 cm) deep and 22 inches (56 cm) wide and covered with vegetation 2 inches (5 cm) tall. Galvanized iron piping, its end protected by a fine gauze filter, leads down an incline from the base of the gravel to collecting can placed at the bottom of a pit closed by a lid. The collecting cans contain the water that has percolated through the soil over a convenient period (a day, week, or month, depending on the amount of precipitation). If this value is subtracted from the amount of precipitation over the same period the remainder is equal to the amount of evapotranspiration. This is then converted into a standard unit of millimeters per day.

evapotron An instrument that measures the extent and direction of vertical EDDIES that are involved in the vertical transfer of water vapor. This makes it possible to measure directly the rate of EVAPORATION over a very short period.

Eve A TYPHOON that struck the southern tip of Kyushu, Japan, on July 18, 1996, carrying winds of 199 mph (191 km h^{-1}).

event deposit *See* STORM BED.

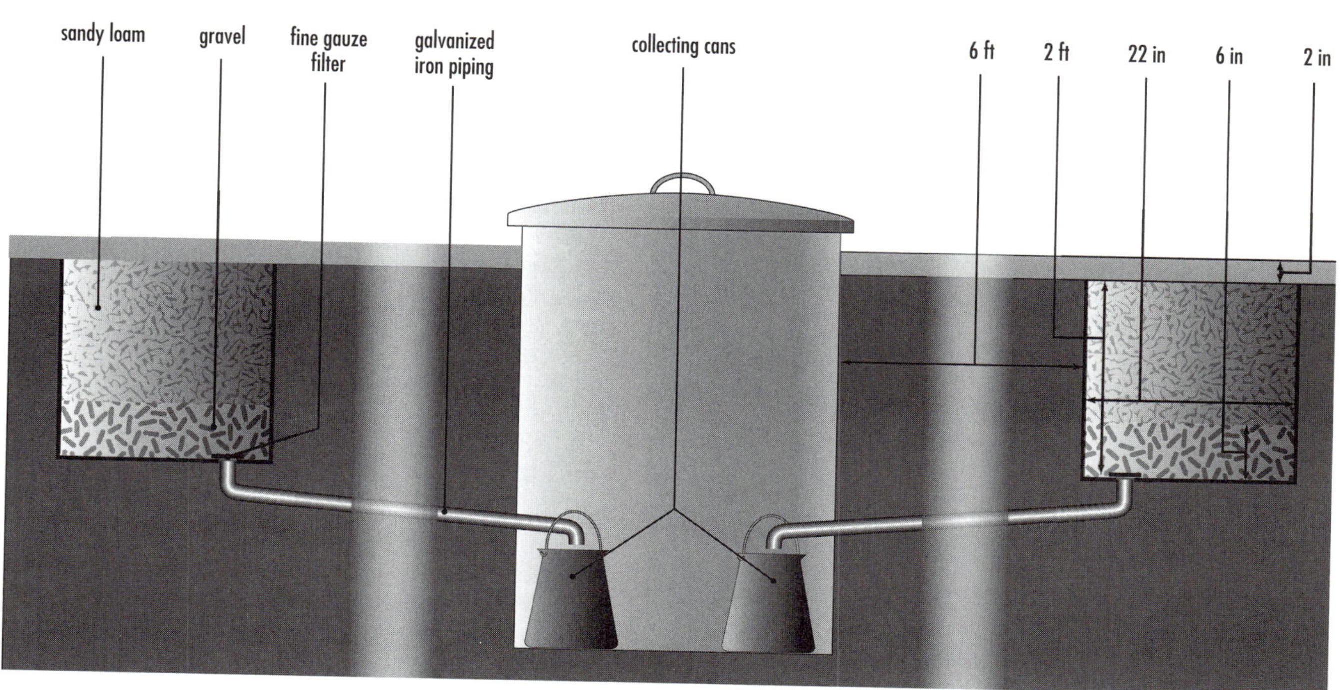

Evapotranspirometer. **A device that measures the loss of moisture from potential evapotranspiration. It consists of two installations. The final value is taken as the average of the two results.**

exine The hard external layer of a pollen grain. Because the layer resists decay, the exteriors of pollen grains can be preserved for many thousands of years (the pollen ceases to be viable very soon after it leaves the plant). Grooves, called *colpi* (sing. *colpus*), form patterns on the exine that are characteristic for particular plant families and in some cases for genera and even species. This allows the source of the pollen to be identified and the past vegetation to be reconstructed, indicating the climatic conditions at that time.

exitance (emittance) A measure of the amount of electromagnetic radiation that is released from a unit area of a surface. The exitance of the PHOTOSPHERE of the Sun is about 70 MW m^{-2}. Multiplied by the area of the photosphere, this figure indicates that the total solar output is about 4.2×10^{20} MW.

exit region The area in which air is being expelled from the core of a JET STREAM. Air leaving on the side nearer the Pole (the cold side) is subjected to DIVERGENCE and tends to rise. Air leaving the core on the side nearer the equator (the warm side) is subjected to CONVERGENCE and tends to subside.

exosphere The layer of the upper atmosphere that lies between about 300 and 450 miles (480–725 km) above the surface. The predominant gases are atomic oxygen, helium, and hydrogen, about 1 percent of which are ionized (*see* ION). The GAS LAWS do not apply. Some hydrogen escapes constantly into space and is replaced by the breakdown of water vapor (H_2O) and methane (CH_4) near the MESOPAUSE. Helium also escapes and is replenished by the action of COSMIC RADIATION on atomic nitrogen and from the decay of radioactive elements in the Earth's crust.

expansional cooling The mechanism by which an expanding fluid cools. Expansion means that the molecules composing the fluid move farther apart. In order to do so they must push other molecules out of the way. When molecules collide, KINETIC ENERGY is transferred from one to another. If the colliding molecules remain in the same body of fluid and at a constant temperature the amount of kinetic energy remains unchanged, because no energy is lost in the transfer of energy between molecules. In an expanding fluid, however, the molecules in one body of fluid (the expanding body) transfer kinetic energy to outside molecules (in the surrounding fluid). Consequently, energy leaves the expanding body. The loss of kinetic energy means the molecules move more slowly. When they collide with a surface, such as that of the bulb of a thermometer or of heat sensors in human skin, the

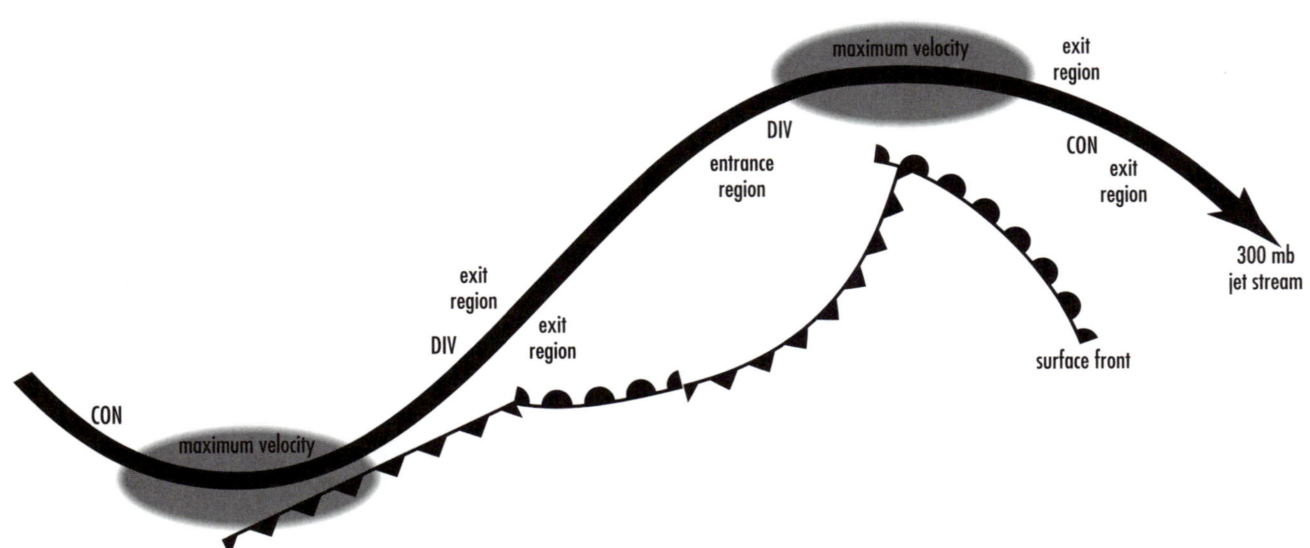

Exit region. **The exit region lies on the downstream side of the core, where air is being expelled from the jet stream.**

impact is less violent than it was formerly. This is measured by the thermometer and felt by the skin as a drop in temperature. The opposite happens when a fluid is compressed. This is known as COMPRESSIONAL WARMING.

exponential An adjective describing a function that varies as the power of a particular quantity. If $x = y^a$, x is said to vary exponentially with a. An exponential change is one in which a quantity changes by a constant proportion that is calculated in each period on the accumulated total of the changes in previous periods. Suppose, for example, that an initial value of 1,000 increases by 20 percent in each period. At the end of the first period the 1,000 is equal to 1,000 + 200 (20 percent of 1,000) = 1,200. At the end of the second period the 20 percent is calculated from the new total of 1,200, so 1,200 + 240 = 1,440. Over a number of periods, therefore, the sequence is 1,000, 1,200, 1,440, 1,728, 2,073.6, and so on. Although the rate of change (20 percent in this example) remains constant the accumulating total grows increasingly rapidly. At a 20 percent rate of increase, a quantity doubles in four periods. Plotted as a graph, an exponential curve has a characteristic J shape. It grows slowly at first, but then more rapidly until it rises almost vertically. An exponential decrease produces a curve of similar shape, but that falls rather than rises. The decay of radioactive substances proceeds as an exponential decrease.

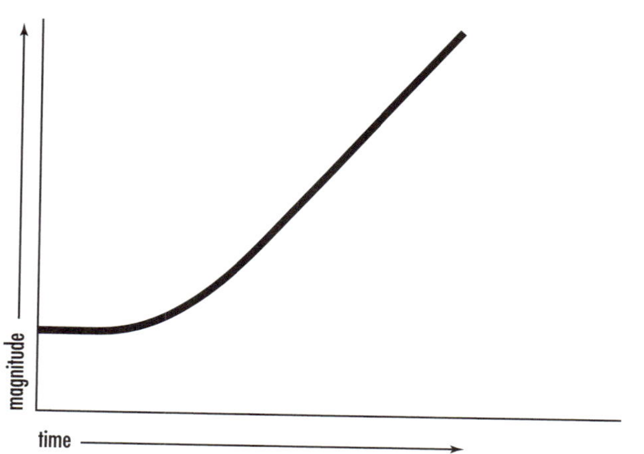

Exponential. **The characteristic J shape of an exponential curve begins slowly but is soon rising (or falling) almost vertically.**

exposure The extent to which a site experiences the full effect of such meteorological events as wind, sunshine, frost, and precipitation. It is a measure of the lack of protection against the weather.

extinction coefficient *See* VISIBILITY.

extratropical cyclone A region of low pressure (*see* DEPRESSION) that occurs in middle latitudes (outside the Tropics) and around which the air circulates CYCLONICALLY. It is generated by a FRONTAL SYSTEM that occurs where different AIR MASSES meet. It is a CYCLONE, because air circulates it cyclonically, but the process causing it is quite different from that which causes a TROPICAL CYCLONE.

extratropical hurricane A severe storm that occurs in the Arctic or Antarctic. The term was first used in 1954 by TOR HAROLD PERCIVAL BERGERON to describe Arctic storms. The CYCLONES that produce extratropical hurricanes are seldom as much as 620 miles (1,000 km) in diameter, and some are no more than about 185 miles (300 km) across. Pressure at the center falls to around 970 mb, and they produce winds of about 45 mph (72 km h^{-1}) gusting to 70 mph (113 km h^{-1}). Although these are much less severe than winds of hurricane force (more than 75 mph [121 km h^{-1}] on the BEAUFORT WIND SCALE) such storms resemble TROPICAL CYCLONES in certain respects. Their sustained winds follow a circular path with a calm eye at the storm center, they produce clearly defined spiral bands of CUMULONIMBUS cloud, and air flows outward from them at high level, producing CIRRUS cloud. They extend to the TROPOPAUSE and dissipate rapidly if they cross over land.

Extratropical hurricanes differ from tropical cyclones in other respects. They develop much more quickly, reaching their full strength in 24 hours or less; travel much faster, at up to 35 mph (56 km h^{-1}); and survive for only 48 hours or less before crossing a coast and dissipating. Tropical cyclones can spend a week or longer crossing the ocean. Extratropical hurricanes form where there is a low-pressure disturbance close to the edge of the sea ice and where streams of air with very different characteristics merge along an ARCTIC FRONT. In both cases the effect is to draw cold air from above the ice and air from above the sea that may be 72° F (40° C) warmer.

extreme The highest or lowest temperature that is observed at a particular place over a specified period. Contrasted with the average daily temperature for the hottest or coldest month the extreme represents the range of temperatures that are possible. In New York City, for example, July is the hottest month and January the coldest. The average maximum temperature in July, measured over 40 years, is 85° F (29° C) and the highest temperature recorded (the extreme) is 106° F (41° C). The lowest average temperature in January, over the same 40-year period, is 27° F (–2.8° C) and the coldest extreme is –15° F (–26° C).

eye of storm The region of calm air that lies at the center of a deep CYCLONE. TROPICAL CYCLONES and EXTRATROPICAL HURRICANES have clearly defined eyes surrounded by towering CUMULONIMBUS clouds that form an EYEWALL in which winds and precipitation are more intense than they are elsewhere in the cyclone. Inside the eyewall, in the eye, winds are light and variable in direction and the sky is often clear.

eye of wind The direction from which the wind is blowing or the point on the horizon from which it appears to blow. A person facing that point is said to be facing into the eye of the wind.

eyewall The circle of towering CUMULONIMBUS clouds that surround the calm center of a TROPICAL CYCLONE or EXTRATROPICAL HURRICANE. CONVECTION in the eyewall is more intense than anywhere else in the storm, and it is in the eyewall that the strongest winds are generated and PRECIPITATION is heaviest. Further bands of cumulonimbus surround the eyewall and spiral outward, away from the center.

F *See* FARAD.

Fabry, Marie Paul Auguste Charles (1867–1945) French *Physicist* Charles Fabry was one of the most distinguished, and famous, physicists of his generation. He was born in Marseilles on June 11, 1867, and after commencing his education in Marseilles in 1885 enrolled at the École Polytechnique in Paris and graduated in 1889. He studied physics and mathematics but became increasingly drawn toward astronomy and optics. After his graduation, Fabry moved to the University of Paris, where he obtained his doctorate in physics in 1892.

Fabry then spent two years teaching physics at lycées (high schools) in several cities before joining the staff at the University of Marseilles in 1894; there he devoted himself to teaching and pursuing his own research. In 1904 he was appointed professor of industrial physics. In 1914 the French government called him to Paris to investigate interference in sound and light waves, and in 1921 Fabry moved to Paris permanently as professor of physics at the Sorbonne. Later he combined this post with that of professor of physics at the École Polytechnique and director of the Institute of Optics. In 1935 he became a member of the International Committee on Weights and Measures. He retired two years later, in 1937.

It is not uncommon for research scientists to find that further advances are impossible because the available instruments are inadequate. They cannot penetrate the phenomena being investigated deeply enough or provide sufficiently accurate measurements. In this situation the scientist, who is the only person in a position to know precisely what is needed, often modifies an existing device or invents a new one. That is what Charles Fabry and his colleague the French physicist Albert Pérot (1863–1925) did in 1896. The instruments they invented were based on two perfectly parallel half-silvered plates. If the distance between the plates is fixed the instrument is known as the *Fabry-Pérot* INTERFEROMETER and as the *Fabry-Pérot* ETALON if the distance can be varied. These instruments break light into its constituent wavelengths with a much greater resolution than was possible with other devices. Fabry and Pérot spent 10 years designing, improving, and using them. They were able to confirm the DOPPLER EFFECT for light in the laboratory and applied the instruments to a variety of astronomical questions.

A question of particular interest concerned the absorption of solar ULTRAVIOLET RADIATION in the atmosphere. Clearly some atmospheric gas was filtering it, and in 1913, Fabry used the interferometer to discover the presence of abundant OZONE in the upper atmosphere.

Fabry died in Paris on December 11, 1945.

facsimile chart A weather chart that is distributed as a fax from a central meteorological office. The device that transmits the chart is known as a *facsimile recorder*. In the United States, facsimile charts are sent daily from a center in Washington, D.C., to stations throughout the country.

facsimile recorder *See* FACSIMILE CHART.

Fahrenheit, Daniel Gabriel (or Gabriel Daniel) (1686–1736) Polish–Dutch *Physicist* The scientist whose name is still used every day because of the temperature scale he devised was born on May 14, 1686, in Danzig (now called Gdansk), an ancient city at the mouth of the Vistula River on the coast of the Baltic Sea. Culturally the city is Polish, and it now lies deep inside Poland, but at various times in the past changes in frontiers have meant it was in Prussia or, more recently, in Germany.

Daniel Fahrenheit began his education in Danzig, but in 1701 he moved to Amsterdam in order to learn a business. He became interested in the making of scientific instruments, and in about 1707 he left the Netherlands to tour Europe, meeting scientists and other instrument makers and learning the craft he had chosen to follow. In the course of his travels, in 1708 he met the Danish physicist and instrument maker Olans (Ole) Christensen Römer (1644–1710). Fahrenheit returned to Amsterdam in 1717, established his own business making instruments, and remained there for the rest of his life.

At that time there was intense scientific interest in studying the atmosphere and the weather it produced, but meteorologists were greatly hindered by the lack of a reliable THERMOMETER. GALILEO GALILEI had made thermometers and GUILLAUME AMONTONS had improved on them, but both of their instruments relied on the expansion and contraction of air to raise and lower a column of liquid and they were very inaccurate. Fahrenheit turned his attention to the problem.

His first thermometer used alcohol. Unlike earlier designers, however, Fahrenheit filled the bulb with liquid, so changes in temperature were indicated by the expansion and contraction of the column of liquid, not of a pocket of air. This was a great improvement, but an alcohol thermometer cannot measure very high temperatures, because of the low boiling point of alcohol (pure ethanol boils at 172.94° F [78.3° C]). He tried mixing alcohol and water. This raised the boiling point, but the volume of the mixture did not change at a constant rate as the temperature increased or decreased, making the thermometer very difficult to calibrate. Finally, in 1714, he tried mercury. Mercury boils at a much higher temperature than water and freezes at a much lower temperature, so a mercury thermometer can be used over a much wider temperature range than an alcohol thermometer. First, though, he had to devise a way to purify the metal, because impurities caused it to stick to glass surfaces. Once he had achieved this, he found mercury changed its volume at a fairly constant rate with changing temperature, although it changed by a smaller amount than alcohol.

He then had to calibrate his thermometer. Ole Römer had invented a thermometer in about 1701, and during his visit to Copenhagen, Fahrenheit had watched him calibrate one. He based his calibration on the RÖMER TEMPERATURE SCALE, using two FIDUCIAL POINTS. After some later adjustments, he produced the FAHRENHEIT TEMPERATURE SCALE that is still in use today, on which ice melts at 32°, water boils at 212°, and average body temperature is 98.6°.

The thermometer was far more reliable and accurate than any that had existed before, and the mercury thermometers in use today are made in the way Fahrenheit devised. Amontons had earlier suggested that water always boils at the same temperature. Fahrenheit set out to check this assertion and found it to be true, but only if the air pressure remained constant. He also examined many other liquids and found all of them had characteristic boiling and freezing temperatures.

In 1724 Fahrenheit described his method for making thermometers in a paper he submitted for publication in the *Philosophical Transactions of the Royal Society.* He was elected to the Royal Society in the same year.

Fahrenheit died at the Hague on September 16, 1736.

Fahrenheit temperature scale The temperature scale that was devised by DANIEL GABRIEL FAHRENHEIT in about 1714 and that remains in use in Britain, the United States, and other English-speaking countries, although it is being replaced by the CELSIUS TEMPERATURE SCALE. Scientific publications always use the Celsius scale.

Fahrenheit derived his scale from the RÖMER TEMPERATURE SCALE, in which two FIDUCIAL POINTS are used, the freezing point of water and average body temperature (or blood heat). Fahrenheit modified the Römer scale by using the freezing point of a mixture of ice and salt for his lower fiducial point. This was then believed to be the lowest temperature possible, so calling it 0° allowed all temperature values to be positive. He marked this point on his thermometer and then measured body temperature. In order to be able to measure small temperature differences, Fahrenheit divided the distance between the upper and lower points into 96 degrees (eight times more than Römer used). On this scale the freezing point of pure water was 30° and body temperature was 90°.

Later he adjusted the scale. He substituted the boiling point of pure water for the upper fiducial point and divided the distance between the new upper point and the freezing point of pure water into 180 degrees, while still retaining the lower fiducial point. On the revised scale pure water freezes at 32° and boils at 212° and average body temperature is 98.6°.

1° F = 0.56° C; 1° C = 1.8° F. To convert Fahrenheit temperatures into Celsius, °F = (°C × 1.8) + 32; °C = (°F – 32) × 5 ÷ 9.

fair An adjective that is used to describe weather that is pleasant for a particular place at a particular time of year. The term is subjective and has no precise meaning, but it generally implies light winds, no precipitation, and a sky less than half covered by cloud.

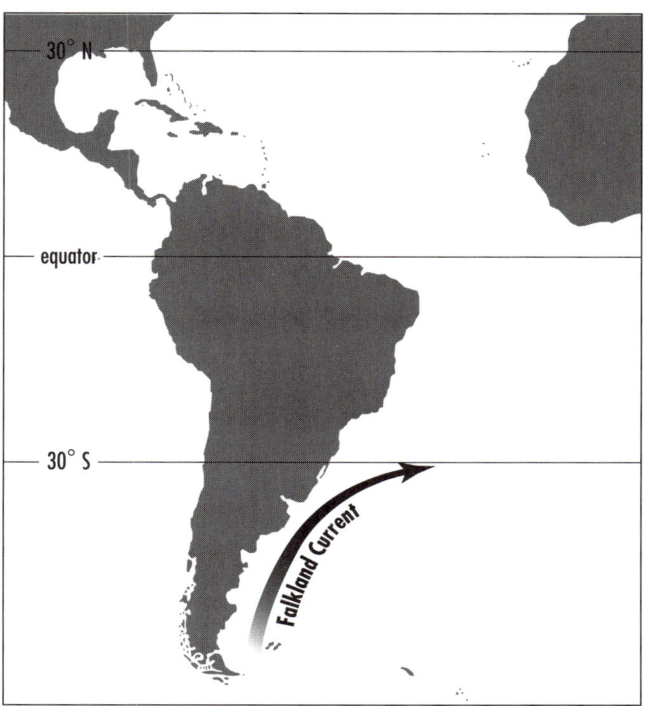

Falkland Current. **The current flows northward, parallel to the coast of Argentina, carrying cold water**

fair weather cumulus Small, white fleecy clouds that appear in fine weather and that deliver no precipitation. Their scientific name is *CUMULUS HUMILIS*. The clouds are all at the same height and have a somewhat flattened appearance. This is due to the presence of a temperature INVERSION immediately above them. The CLOUD BASE is determined by the CONDENSATION LEVEL, but there is warmer air above the clouds. CLOUD DROPLETS that rise by CONVECTION enter air in which they immediately evaporate.

fair weather electric field The electric field that exists in the atmosphere and at the ground and sea surface. It has an average value of 120 volts per meter (37 volts per foot) and is directed downward, carrying positive charge from the ionosphere to the ground.

Faith One of two TROPICAL STORMS that struck central Vietnam in December 1998. Together Faith and Gil generated rain that caused floods in which at least 22 people died and thousands had to leave their homes.

Falkland Current An ocean BOUNDARY CURRENT that carries cold water northward past the Falkland Islands (Malvinas) and parallel to the coast of Argentina as far as about latitude 30° S and rather farther north in winter. Its influence greatly reduces the effect ordinarily produced by boundary currents outside the Tropics, of transporting warm water to western coasts.

fallout The removal from the air of solid particles that descend to the surface by gravity. The rate at which particles fall varies inversely with their mass but is also influenced by the wind and EDDY DIFFUSIVITY. Particles are also removed from the air by IMPACTION, RAINOUT, and WASHOUT.

fallout front The lower boundary of an area of FALLOUT. The fallout front marks where solid particles that are descending by gravity reach the surface.

fallout wind A wind that carries particles that are at the same time descending to the surface by gravity.

Fanning. **The plume of gas and particles moves downwind in a fairly straight line, without dispersing.**

fall speed The speed with which a body, such as a RAINDROP, SNOWFLAKE, or HAILSTONE, falls through the air. It is equal to the TERMINAL VELOCITY of the body minus the velocity of any upward air current to which it is exposed.

fallstreak hole A hole that sometimes develops in clouds that are composed of supercooled water droplets (*see* SUPERCOOLING). Droplets in part of the cloud freeze and then grow into RAINDROPS by the BERGERON–FINDEISEN MECHANISM. The raindrops fall from the cloud, often in the form of VIRGA, or fallstreaks, leaving behind clear air from which the cloud droplets have been removed.

fallstreaks *See* VIRGA.

fall wind *See* KATABATIC WIND.

false cirrus Cloud that resembles CIRRUS but that is formed from the upper part of a CUMULONIMBUS cloud that has dissipated or from which it has become detached.

false color Colors that are used in satellite images but that differ from the actual colors of the surfaces they represent. Instruments carried on satellites are able to detect radiation at any WAVELENGTH, including wavelengths outside the wave band to which the human eye is sensitive. The different wavelengths all convey useful information, and the allocation of fairly arbitrary colors to them makes the areas emitting them clearly visible. In many false-color images INFRARED RADIATION appears as visible red. Vegetation is highly reflective to infrared wavelengths, and so it often appears red in false color pictures.

fanning One of the patterns a CHIMNEY PLUME may make as it moves away downwind. The gases and particles travel smoothly and horizontally without dispersing. Fanning occurs when there is a strong INVERSION in the layer of air extending from the surface to beyond the top of the plume. In this situation the ENVIRONMENTAL LAPSE RATE marks an increase of temperature with height. Fanning is sometimes seen early on winter mornings.

farad (F) The derived SYSTÈME INTERNATIONAL D'UNITÉS (SI) UNIT of CAPACITANCE, which is defined as the capacitance of a capacitor that has a potential difference of 1 VOLT between its plates when it is charged with 1 COULOMB. This is a very large unit and so the unit most commonly used is the microfarad (10^{-6} F). The unit is named in honor of the English chemist and physicist Michael Faraday (1791–1867).

far-infrared radiation INFRARED RADIATION that has a WAVELENGTH greater than 15 μm.

far UV *See* ULTRAVIOLET RADIATION.

fastest mile The greatest WIND SPEED, measured in miles per hour, that is recorded over a specified period (usually 24 hours) for one mile of wind (*see* RUN OF WIND.)

Fata Morgana A MIRAGE that occurs when atmospheric pressure increases with height within a thin layer of air lying above a cold surface, most commonly of the sea. Light is refracted to produce a greatly magnified superior image of distant buildings or cliffs that can sometimes resemble great castles partly in the sky and partly beneath the sea. The phenomenon is especially common in the Strait of Messina, between the Italian mainland and Sicily. This is the site associated with the mythical submarine palace of Morgan le Fay, or Fata Morgana, the fairy sister of King Arthur, which the mirage is imagined to reflect.

Faye A TYPHOON that struck South Korea in July 1995; it killed at least 16 people.

February fill dyke An English country expression that refers to the fact that in Britain, February is usually a cold, wet month. There is a saying "February fill dyke, black or white." A dyke is a ditch, and the saying means that in February the ditches will be full either of water (black) or of snow (white).

feedback The behavior of a system that is regulated by a part of the system itself, so that a signal generated by a component affects the whole. The concept is derived from engineering. When steam engines were first employed to drive factory machines, it was very important that they ran at a constant speed. The speed of a steam engine is regulated by the opening and closing of a valve that allows steam into the cylinders. To maintain a constant speed, engines were fitted with devices called *governors*. These were weights mounted on the ends of arms. The arms were pivoted and attached to a rotating vertical axis. They were linked to the valve by levers. If the engine speed increased, the weights spun faster and were thrown outward. This action moved the levers in such a way that the valve descended, reducing the supply of steam. If the engine ran more slowly, the weights fell inward, raising the valve and increasing the steam supply.

Feedback can be positive or negative and the governor of a steam engine worked by negative feedback. It is called *negative* because the response of the system (in this case the engine) is to reduce the activity of the component that generated the signal. If the engine ran faster, the governor slowed it, and if it ran slower, the governor accelerated it.

Like all natural systems, the climate is also subject to both positive and negative feedback. The idea of a SNOW BLITZ is based on positive feedback. A fall in temperature leads to an increase in the area covered by snow and ice. This increases the ALBEDO of the planet. More solar radiation is reflected rather than being absorbed, and so the temperature falls further.

If the temperature increases, on the other hand, it may be checked by negative feedback. In this case the increase in temperature leads to an increase in the rate of evaporation. The atmosphere contains more water, so condensation also increases and clouds form. Clouds reflect solar radiation, and this process cools the surface.

Although feedback mechanisms are extremely important, in reality their effects are not simple. A rise in

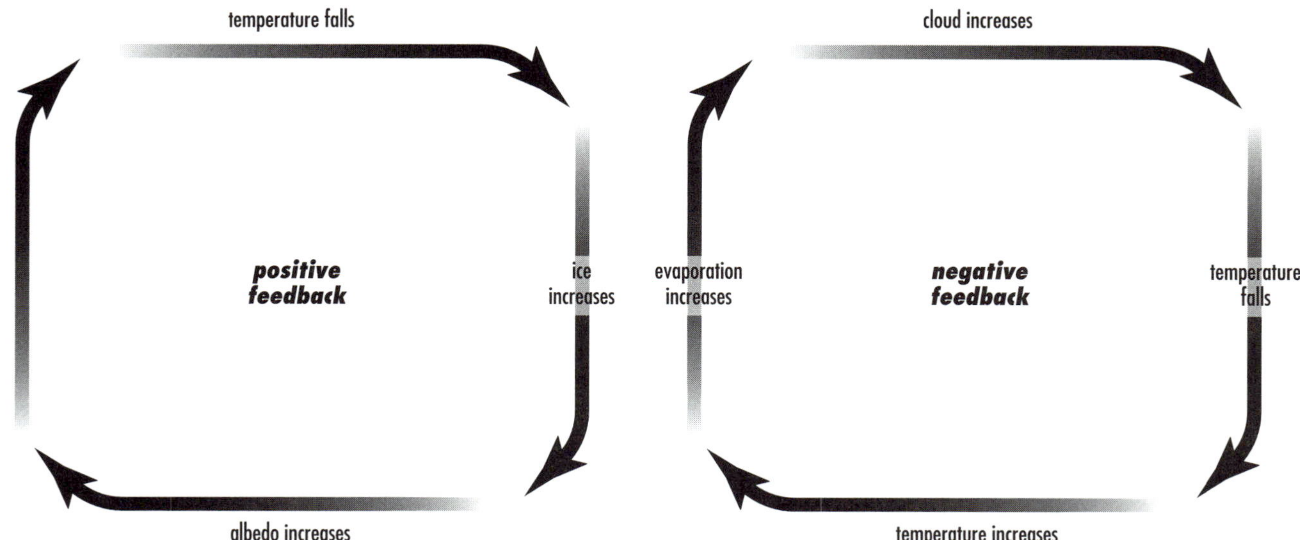

Feedback. **In positive feedback, falling temperature causes the temperature to fall further. In negative feedback a rising temperature is checked.**

temperature increases evaporation, for example, but WATER VAPOR is the most powerful of all GREENHOUSE GASES and so it may cause the atmosphere to warm and the feedback to be positive. Whether the feedback effect of clouds is positive or negative depends on the type of cloud. Clouds reflect sunlight, which is negative, but are composed of water, which absorbs radiation, and so the effect is positive. CIRRIFORM clouds are too thin to have a major effect on the albedo, but they are able to absorb radiation, and so they tend to have a warming effect. Low clouds are usually thicker and more reflective, and so they have a cooling effect.

In the snow blitz example, the positive feedback may also be increased by the reduction in evaporation as the temperature falls. Cold air is drier than warm air and so the greenhouse warming effect decreases as the temperature falls.

Fennoscandian ice sheet The ICE SHEET that covered Northern Europe during the most recent PLEISTOCENE Ice Age, known as the *DEVENSIAN GLACIAL* in Britain, the *Weichselian* elsewhere in Northern Europe, and the *Würm* in the Alps. At its greatest extent the ice sheet covered all but the southernmost part of Britain; all of Scandinavia, Belgium, and the Netherlands; and about half of Germany and Poland. Everywhere from the North Cape, in the far north of Norway, to Kiev, Ukraine, on the banks of the River Dnieper, lay beneath ice.

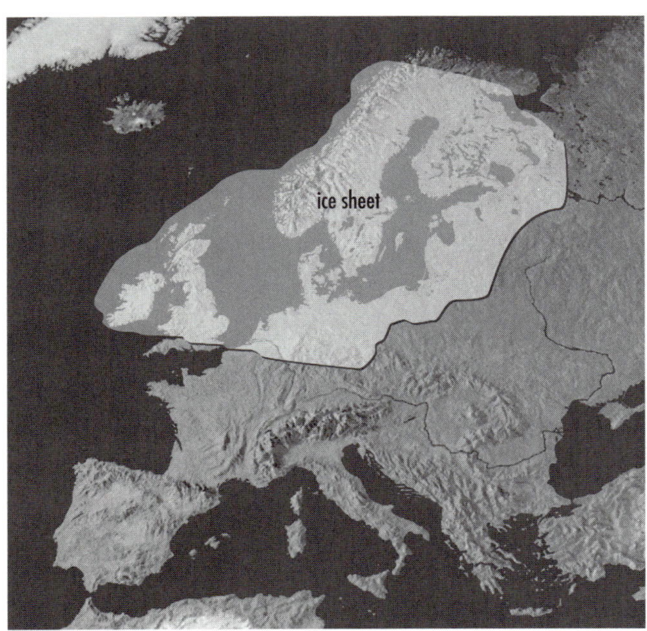

Fennoscandian ice sheet. **The ice sheet that covered northern Europe from the North Cape to the Dnieper during the last ice age.**

Ferdinand II. **The Italian physicist who designed a thermometer and invented the hygrometer.** ***(John Frederick Lewis Collection, Print and Picture Collection, The Free Library of Philadelphia)***

Ferdinand II (1610–1670) Italian *Physicist* A member of the powerful Medici family, Ferdinand was born on July 14, 1610, the son of Cosimo II, grand duke of Tuscany. His father died in 1620. Ferdinand was 10 years old when he became ruler.

He was not a strong ruler and was unable to protect GALILEO GALILEI from his trial by the Inquisition in 1633, but he took a keen interest in science, and especially in atmospheric science. One of the scientific challenges of the time was to find a way to measure temperature. Galileo had attempted this with his AIR THERMOSCOPE, and in 1641 Ferdinand improved it by inventing a thermometer consisting of a tube that contained liquid and was sealed at one end. He improved the design further in 1654, and his thermometer provided the basis for the design that would be made by DANIEL GABRIEL FAHRENHEIT about 60 years later.

Ferdinand also designed one of the earliest accurate HYGROMETERS. It consisted of a tapering cylinder that was filled with ice. Water vapor condensed on the outside of the cylinder and ran down it into a collecting funnel and from there to a flask. The amount of water that accumulated in the flask indicated the humidity of the air.

In 1657, the year he produced his hygrometer, Ferdinand and his brother Leopold founded the Accademia del Cimento (Academy of Experiments) in Florence. Members of the accademia were especially interested in studying the atmosphere. CARLO RENALDINI was one of those who worked on developing the thermometer. The accademia itself was the forerunner of other scientific academies, including the Royal Society of London (founded in 1665) and the Royal French Academy of Sciences (founded in 1666). The accademia ceased to function in 1667.

Ferdinand died in May 24, 1670, and was succeeded by his son, Cosimo III. Ferdinand's grandson, Gian-Gastone, had no male heir, and the Medici family ended with his death in 1737.

(You can learn more about the Medici family from es.rice.edu/ES/humsoc/Galileo/People/medici.html.)

fern frost FROST that forms in patterns that resemble fern fronds. It is most often seen early in the morning on windows and is a variety of frozen DEW. In the early part of the night WATER VAPOR condenses onto the window from the warm, moist air indoors. As the outside temperature falls, the temperature of the window drops below freezing. The water droplets on the window are also chilled to below freezing but remain liquid (*see* SUPERCOOLING). Eventually, ICE CRYSTALS start to form between the water droplets. The supercooled water freezes onto them, and a chain reaction occurs in which the ice grows rapidly to cover a large area with the fern pattern. Large supercooled droplets do not produce fern frost. Being larger, they freeze more slowly and form a layer of clear, unpatterned ice.

Ferrel cell In the THREE-CELL MODEL of atmospheric circulation, the INDIRECT CELL that lies between the HADLEY CELL and the POLAR CELL in each hemisphere. Air moves through the Ferrel cell in the direction opposite to its movement through the two DIRECT CELLS, rising at the POLAR FRONT and subsiding in the subtropics, where it contributes to the SUBTROPICAL HIGH. The cell was discovered in 1856 by the American climatologist WILLIAM FERREL. Air movement is weak in the Ferrel cell, with the result that the predominant movement in middle latitudes is controlled by the weather systems that are carried by the generally westerly airflow.

Ferrel, William (1817–1891) American *Climatologist* William Ferrel was born on January 29, 1817, in Bedford County, Pennsylvania. When he grew up he earned his living as a school mathematics teacher but combined this with an intense interest in the tides and weather. His study of the circulation of the atmosphere led him to publish in 1856 a mathematical MODEL of the circulation of the atmosphere. He revised his model in 1860 and again in 1889. The model proposed the existence of a midlatitude cell. In this cell the vertical movement of the air is driven by the HADLEY CELL on the side nearer the equator and by the POLAR CELL on the side nearer the Pole. This third cell is known as the *FERREL CELL*.

In 1857 his particular interest in and understanding of tides led to an invitation to join the staff of *The American Ephemeris and Nautical Almanac,* which was published in Cambridge, Massachusetts. It was while working for this publication that Ferrel calculated that the combined effect of the PRESSURE GRADIENT FORCE and the CORIOLIS EFFECT must be to cause winds generated by a PRESSURE GRADIENT to flow at 90° to it, parallel to the ISOBARS rather than across them. A few months after he reached this conclusion, the same phenomenon was announced by the Dutch meteorologist CHRISTOPH HENDRICK DIDERIK BUYS BALLOT, and it became known as *BUYS BALLOT'S LAW.*

On July 1, 1867, Ferrel was appointed to a position at the United States Coast and Geodetic Survey. His task there was to develop the general theory of the tides, to which he had already devoted a considerable amount of work and which he had advanced further than any of his contemporaries.

Tides are affected by winds and atmospheric pressure, and so Ferrel widened his study of tides to include relevant meteorological phenomena. This led him to a more general investigation of meteorology, and for some time he alternated between studying tides and studying weather.

He wrote extensively on the subject of meteorology. His works include *Meteorological Researches,* published in three volumes between 1877 and 1882; *Popular Essays on the Movements of the Atmosphere* (1882); *Temperature of the Atmosphere and the Earth's Surface* (1884); *Recent Advances in Meteorology* (1886); and *A Popular Treatise on the Winds* (1889).

On August 9, 1862, Ferrel tendered his resignation from the Coast Survey in order to accept a position in the Army Signal Service. The superintendent accepted his resignation and Ferrel continued to work for the Signal Service until his retirement in 1886. The Signal Service already had an interest in meteorology and in November 1870 its newly established Division of Telegrams and Reports for the Benefit of Commerce became a national weather service.

The acceptance of his resignation from the Coast Survey was conditional. Ferrel was asked to complete the investigations on which he was engaged at the time and to continue supervising the tide-predicting machine he had invented. This was a mechanical device, worked by levers and pulleys, that took account of 19 constituents of the forces affecting tides and gave readings, on five dials, of the predicted times and heights of high and low water. Ferrel submitted plans and an explanation of the machine to the Coast Survey in the spring of 1880, and in August of that year he described it in Boston at the annual meeting of the American Association for the Advancement of Science. The idea was accepted, but it proved difficult to find a machinist with the adequate skills. Work on constructing the device did not commence until the late summer of 1881, and the machine was not completed until the autumn of 1882. The tide-predicting machine, first used to predict the tides for 1885, remained in use until 1991. Computers are now used to predict tides.

After his retirement, William Ferrel moved to Maywood, Kansas, where he died on September 18, 1891.

fetch The distance over which air moves across the surface of the sea or ocean. A long fetch modifies the characteristics of an AIR MASS that was formerly over a continent by moderating its temperature and increasing the amount of WATER VAPOR it carries. Together with its speed and duration, the fetch of a wind determines the height of waves (*see* WAVE CHARACTERISTICS).

fetch effect *See* ADVECTION.

fib *See* FIBRATUS.

fibratus (fib) A species of clouds (*see* CLOUD CLASSIFICATION) that consist of long filaments that are almost straight or irregularly curved and that do not end in hooks. Fibratus may occur as a detached cloud or form a thin veil across part of the sky. The species is most often seen in clouds of the genera CIRRUS and CIRROSTRATUS.

The name of the species is derived from the Latin word *fibra,* which means "fiber."

fibril A trail of cloud that is sometimes seen extending from a CUMULONIMBUS cloud. It consists of droplets the size of DRIZZLE droplets. These are large enough to have a TERMINAL VELOCITY that exceeds the force of the air currents within the cloud, so they are able to escape from it.

fibrous ice *See* ACICULAR ICE.

FIDO *See* FOG INVESTIGATION DISPERSAL OPERATIONS.

fiducial point A position that is fixed and from which other positions can be measured. The word *fiducial* is from the Latin verb *fidere,* which means "to trust." The fiducial point of a BAROMETER, also called the *standard temperature,* is the TEMPERATURE at which a particular barometer at latitude 45° gives a correct reading. At any other temperature or latitude the barometer reading must be adjusted. The fixed point that marks zero on the scale of a FORTIN BAROMETER is also called the *fiducial point.*

field capacity The amount of water that a particular soil will retain under conditions that allow water to drain freely from it. It is measured by thoroughly soaking a measured weight of oven-dried soil, then leaving it to drain for a day or two before weighing it again. Field capacity is usually reported as a percentage of the oven-dried weight of the soil.

field changes The rapid fluctuations in the vertical component of the electrical field that occur near the surface during a THUNDERSTORM.

Fifi A HURRICANE, rated category 3 on the SAFFIR/SIMPSON SCALE, that struck Honduras on

September 20, 1974. It generated winds of 130 mph (209 km h^{-1}) and heavy rain. An estimated 5,000 people were killed and tens of thousands were rendered homeless.

filling An increase in AIR PRESSURE that occurs at the center of a CYCLONE.

fire The burning of vegetable matter releases gases, including CARBON MONOXIDE, CARBON DIOXIDE, SULFUR DIOXIDE, NITROGEN OXIDES, and a range of organic compounds. Some of these are harmful to human health. Fires also release PARTICULATE MATTER. Particles acts as CLOUD CONDENSATION NUCLEI, but those released by fires include very fine particles with diameters smaller than 2.5 µm (0.0001 inch) that are believed to cause respiratory damage. Most fires in remote areas are caused by LIGHTNING, but in populated areas most are started by people, accidentally or deliberately. Fires occur every one to three years in forests that grow in MONSOON climates, and in North America and Eurasia 12–50 million acres (5–20 million ha) of forest burns out of control every year. During the 1982–83 EL NIÑO about 35 million acres (14 million ha) of forest and cropland burned in Borneo and Indonesia. During the hot, dry summer of 2000, wildfires swept across the western United States, consuming almost 1 million acres (600,000 ha). In the world as a whole, vegetation covering an average of 81,269 square miles (210,282 km^2) is burned every year.

(You can learn more about fires and their health effects from the World Health Organization at www.who.int/.)

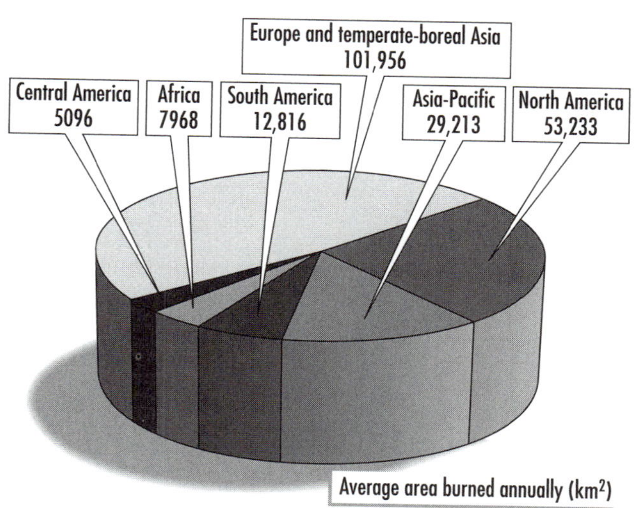

Fire. **The average area of vegetation that is burned each year, by continent. (1 km^2 = 0.386 mi^2)**

fire storm A wind storm that is caused by an intensely hot fire. During World II, bombing raids in which very large numbers of incendiary bombs were dropped with the intention of starting fires caused fire storms in Hamburg and Dresden, Germany, and in Tokyo, Japan. The atomic bomb that was dropped on Hiroshima also caused fire storms, and the one dropped on Nagasaki may have done so.

Hot air rising by CONVECTION from the fire creates local areas of low AIR PRESSURE. The hotter the fire, the more vigorous are the convectional upcurrents and the lower is the pressure at the base. Air is drawn in at the base to fill the low pressure, and if the fire is hot enough the air enters as a GALE force or even HURRICANE force wind. The in-flowing air delivers oxygen to the flames, fanning them to still higher temperatures. As the air approaches the low-pressure center, CONVERGENCE starts it rotating CYCLONICALLY. The wind picks up loose material lying in its path and carries it into the fire. Much of this is likely to be flammable, and so the wind also feeds fuel into the fire. The fire storm is then self-sustaining until it exhausts its supply of fuel.

The heat of a fire storm is great enough to ignite other fires some distance away. In this case ignition is due not to sparks or burning fragments falling onto objects or structures, but to radiant heat. Just as a piece of paper held close to a hot fire bursts into flames before the fire touches it, so objects some distance from the storm start to burn.

fire weather Weather conditions that favor the ignition and spread of forest fires in a specified area. When there is a risk of fire a fire weather forecaster at the NATIONAL WEATHER SERVICE issues a fire weather watch. There is a serious fire risk between fall and early spring, when deciduous trees are leafless, if the sustained average wind speed is 15 mph (24 km h^{-1}) or more, the RELATIVE HUMIDITY is 25 percent or less, and the temperature is greater than 75° F (24° C).

Firinga A CYCLONE, rated category 4 on the SAFFIR/SIMPSON SCALE, that struck the island of Réunion, in the Indian Ocean, on January 28 and 29, 1989. It generated wind speeds of more than 125 mph (200 km h^{-1}). At least 10 people died and 6,000 were rendered homeless.

firn SNOW that is lying on the ground during the second winter since it fell, having failed to melt during the intervening summer. As further falls of snow cover it during succeeding winters, firn is compressed until its DENSITY reaches 50–52 pounds per cubic foot (800–840 kg m^{-3}). The snow is then impermeable to both air and water and is known as *glacier ice*.

firn limit See FIRN LINE.

firn line (annual snow line, firn limit) A line on a GLACIER that marks the highest elevation at which snow that falls during the winter melts during the summer. Snow above the firn line becomes FIRN. The firn line is often clearly visible, because below it there is clear, blue ice and above it there is snow.

firn wind (glacier wind) A KATABATIC WIND that blows from a GLACIER during the day, especially in summer. Air in contact with the glacier is chilled and becomes denser, and its increased DENSITY causes it to flow down the slope.

first gust A sudden sharp increase in the wind speed that is felt as a CUMULONIMBUS cloud enters its mature stage. The gust is caused by the arrival at the surface of the cold downdraft.

first-order station Any weather station in the United States that is staffed partly or wholly by personnel employed by the NATIONAL WEATHER SERVICE.

fitness figure (fitness number) A value, used in Britain, that is calculated for the suitability (fitness) of weather conditions for the safe landing of aircraft at an airport. The value takes account of VISIBILITY, CLOUD BASE, and crosswinds.

fitness number *See* FITNESS FIGURE.

FitzRoy, Robert (1805–1865) English *Naval officer, hydrographer, and meteorologist* In 1860 *The Times* of London became the first newspaper in the world to publish a daily weather forecast. It was prepared by Admiral FitzRoy, the head of meteorology at the Board of Trade. This was the department that was to become the British METEOROLOGICAL OFFICE.

Robert FitzRoy was born on July 5, 1805, at Ampton Hall, in Suffolk, an English county to the northeast of London. He was an aristocrat, the grandson on his father's side of the duke of Grafton and on his mother's side of the marquis of Londonderry and directly descended from King Charles II. FitzRoy was educated at the Royal Naval College, Portsmouth. After graduating, on October 19, 1819, he entered the Royal Navy, and he received his commission as an officer on September 7, 1824.

He served in the Mediterranean and was then sent to South America on HMS *Beagle,* which was conducting a surveying mission. When the captain of the *Beagle* died FitzRoy assumed command, completing the survey and returning to England. He applied to lead a second survey and in 1831 the naval hydrographer, Admiral Sir FRANCIS BEAUFORT, granted the request. FitzRoy sailed once more as captain of the *Beagle,* this time accompanied by Charles Darwin.

The *Beagle* was well equipped for its scientific mission, and the equipment included several barometers. FitzRoy used these to prepare short-term weather forecasts, and this was the first voyage in which wind observations were based on the BEAUFORT WIND SCALE. A well-trained and experienced sailor, FitzRoy knew how important it was to predict weather accurately.

With the survey complete, the *Beagle* arrived back in Portsmouth on October 2, 1836, and FitzRoy settled down to write his accounts of the voyage. These were published in 1839 (as two volumes of *Narrative of the Surveying Voyages of His Majesty's Ships* Adventure *and* Beagle *Between the Years 1826 and 1836, Describing Their Examination of the Southern Shores of South America, and the* Beagle's *Circumnavigation of the Globe*). He was elected a fellow of the Royal Society for his surveying work.

By then an admiral, in 1841 he became a member of Parliament for Durham and in 1843 he was made governor general of New Zealand. At the insistence of the British settlers he was recalled from New Zealand in 1845, mainly because he believed the Maori claims to land were as legitimate as theirs. Admiral FitzRoy retired from active service in 1850. In 1854 he took up

Robert FitzRoy. **The British naval officer and meteorologist who encouraged the collection of weather observations and who designed an efficient barometer.** ***(John Frederick Lewis Collection, Print and Picture Collection, The Free Library of Philadelphia)***

his post at the Meteorological Office and devoted himself wholly to meteorology.

At the Meteorological Office he encouraged the collection of weather observations, established barometer stations, and used TELEGRAPHY to gather data. These allowed the Meteorological Office to issue weather forecasts and, in 1861, the first storm warnings. In 1863 he published *The Weather Book,* in which he set out principles to guide sailors in forecasting the weather.

These included 47 "instructions for the use of the barometer to foretell weather." FitzRoy believed a barometer should be installed at every port. Sailors could examine the instrument, use FitzRoy's instructions to interpret the reading, and then decide whether to set sail. "FitzRoy barometers" became very popular and FitzRoy himself invented some versions. There were domestic versions that included thermometers, STORM GLASSES, and various other devices, as well as a set of the admiral's instructions. FitzRoy barometers were still being manufactured in the early 20th century and reproductions are still made.

His work saved many lives, but criticisms of his forecasting methods by the eminent meteorologist MATTHEW MAURY and of his humanitarian political beliefs by newspapers and politicians greatly troubled him. He also experienced conflict between his strongly held religious views and Darwin's theory of evolution by natural selection, which he had helped to develop. Unable to resolve these difficulties, on April 30, 1865, he committed suicide at his home at Upper Norwood, near London, by cutting his throat.

(You can learn more about Robert FitzRoy from "FitzRoy of *The Beagle,*" at www.intranet.ca/~jedr/fitzroy.htm, and "Heavy Weather: Admiral FitzRoy and the FitzRoy Barometer" at www.sciencemuseum.org.uk/collections.exhiblets/weather/fitzroy.)

Flandrian interglacial (Holocene interglacial) The present INTERGLACIAL period, which began when the most recent GLACIAL PERIOD came to an end, approximately 10,000 years ago. The warmest part of the Flandrian, known in Europe as the *ATLANTIC PERIOD* and in North America as the *Hypsithermal,* lasted from about 7,500 to 5,000 years ago. The climate then was warmer than that of the present day. Soon after the postglacial warming began it was interrupted by a return to near–ice age conditions (*see* YOUNGER DRYAS). The Flandrian has already lasted for about the average duration of an interglacial, and scientists agree that eventually it will give way to a new glacial period. There is no agreement, however, about when the present interglacial will end, or how abruptly, nor about how far the ICE SHEETS are likely to extend during the next glacial period.

flaschenblitz Lightning that flashes upward from the top of a CUMULONIMBUS cloud.

flash flood A severe flood that occurs very suddenly when extremely heavy rain, usually a CLOUDBURST, falls on high ground, delivering water much faster than it can soak into the soil and be dispersed safely.

Prolonged rain that fell earlier may have left the ground saturated, so additional water cannot drain through it, but flash floods can be caused by rain falling on dry ground if the water falls faster than it can soak away. In that case the water may fill previously dry gullies, turning them into torrential streams. In either case, water flows across the ground surface, moving downhill with sufficient energy to dislodge and transport soil, rocks, and sometimes trees. The flood may then advance as a front, resembling a wall of water, up to 20 feet (6 m) high, moving at 25 mph (40 km h^{-1}) or more. In the summer of 1996, light rain that suddenly grew into a huge cloudburst sent water, rocks, and mud through a campsite and trailer park at Biescas in northern Spain. Trees were uprooted and carried by the water; tents, campers, and trailers were swept away; and more than 70 people died.

flash flood warning A notification that is issued by the NATIONAL WEATHER SERVICE to inform people that serious flash flooding is imminent or has already begun somewhere close to them. It may mean, for example, that extremely heavy rain has fallen in nearby hills and that flood water is already flowing across the ground surface. A flood warning is similar but warns of more gradual flooding than that experienced in a FLASH FLOOD. Both warnings are broadcast to WEATHER RADIOS. Persons receiving such a warning should move immediately to a place of safety. The warning may be accompanied by instructions to evacuate. Such an instruction should be obeyed immediately.

flash flood watch A notification that is issued by the NATIONAL WEATHER SERVICE to inform people that there is a risk of a FLASH FLOOD. A similar warning, known as a *flood watch,* is issued to warn of impending floods. Both are broadcast to WEATHER RADIOS. Receipt of a watch notification means flooding is possible within the next few hours. Persons receiving such a notification should check their emergency supplies of food, fresh, water, first aid equipment, and gasoline.

flash frost A FROST that appears on roads very suddenly soon after dawn. As the Sun begins to rise, frost on roadside vegetation begins to melt and evaporate. This causes a rapid and large increase in the RELATIVE HUMIDITY close to ground level. Air over the road surface is chilled and ice forms on the road by DEPOSITION. An ice-free road can be covered by frost within 15 minutes. The flash frost soon disappears as the day advances and the temperature rises, but while it lasts it constitutes a driving hazard.

flat plate solar collector *See* SOLAR ENERGY.

flaw An old English name for a sudden SQUALL of wind.

flight forecast A weather forecast that is prepared for a specific air journey.

flight visibility The forward VISIBILITY from an aircraft that is in flight.

Flo A TYPHOON that struck Honshu, Japan, on September 16 and 17, 1990; it killed 32 people.

flo *See* FLOCCUS.

floccus (flo) A species of cloud (*see* CLOUD CLASSIFICATION) that consists of elements, each of which has a CUMULIFORM appearance. The base is ragged to a greater or lesser extent. Floccus often occurs with VIRGA. The species is most often seen with the cloud genera CIRRUS, CIRROCUMULUS, and ALTOCUMULUS.

Floccus is a Latin word meaning a "tuft of filaments" or "woolly hairs."

Flohn classification A CLIMATE CLASSIFICATION that was proposed in 1950 by the German climatologist Professor Hermann Flohn ("Neue Anshcauungen über die allgemeine Zirkulation der Atmosphäre und ihre klimatische Bedeutung" (New views of the general circulation of the atmosphere and its climatic significance) in *Erdkunde* (Earth science), 3, pp. 141–62). It is a genetic classification: that is, it is based on the influence of AIR MASSES and PREVAILING WINDS. It is considered one of the best schemes of this type and especially useful as an introductory outline of climate classification.

Flohn divides the climates of the world into eight groups. These are based mainly on the global wind belts and the type and amount of precipitation each receives.

1. Equatorial western zone, where the climate is constantly wet.

2. Tropical zone, dominated by the winter TRADE WINDS, where rain falls mainly in summer.
3. Subtropical dry zone, dominated either by the trade winds or by the SUBTROPICAL HIGH, where the climate is arid.
4. Subtropical winter-rain zone, where the climate is of the MEDITERRANEAN type and rain falls mainly in winter.
5. Extratropical westerly zone, where there is precipitation throughout the year.
6. Subpolar zone, where there is little precipitation, but it is distributed throughout the year.

6a. Boreal, continental subtype, where rain falls in summer and a limited amount of snow falls in winter.

7. High polar zone, where precipitation is very low, falling as rain in summer and snow in winter.

flood warning *See* FLASH FLOOD WARNING.

flood watch *See* FLASH FLOOD WATCH.

Florida Current An ocean current that flows northward parallel to the coast of Florida and that forms part of the GULF STREAM. It extends from the southern tip of Florida to Cape Hatteras, North Carolina. It is narrow, 30–47 miles (50–75 km) wide, and fast, flowing at 2.2–6.7 mph (3.6–11 km h^{-1}).

Floyd A HURRICANE that formed in September 1999 and at its peak generated winds of 155 mph (249 km h^{-1}), making it category 4 bordering category 5 on the SAFFIR/SIMPSON SCALE. It was also unusually large, with a diameter of about 600 miles (965 km). Floyd struck the Bahamas on September 14 and then moved northward along the eastern coast of the United States, reaching New York and New Jersey on September 16, by which time it had been downgraded to a TROPICAL STORM, with winds of 65 mph (105 km h^{-1}). It had heavy rain, causing flooding, and STORM SURGES of up to 7 feet (2.1 m). More than 2.3 million people were evacuated from their homes in Florida, Georgia, and the Carolinas. Crop damage in North Carolina was estimated to cost more than $1 billion. A total of 49 people lost their lives, 1 of them in the Bahamas and the remainder in the United States, 23 of them in North Carolina.

flue A passage that is designed to remove the hot waste gases and other by-products of COMBUSTION from an incinerator or other combustion site. It must be built from materials that are capable of withstanding the temperatures to which the combustion products will subject them and arranged in such a way as to produce a steady flow of air to carry the products.

flue gas Any gas that is produced by a COMBUSTION process and that travels through a FLUE.

fluorosis A disease affecting ruminant animals, such as cattle and sheep, that consume excessive amounts of fluorine compounds. It is usually caused by AIR POLLUTION that deposits fluorides on grass, which is then grazed. Fluorosis causes weakening and mottling of the teeth and thickening of the bones.

flurry A sudden, brief shower of SNOW that is accompanied by a GUST of wind. A mild wind SQUALL is sometimes called a *flurry,* even if it carries no snow.

flux The rate at which a fluid or radiation flows across an area. In the case of a VECTOR QUANTITY, the flux is calculated as the rate of flow past a line that crosses the area at right angles to the direction of flow, multiplied by the area. In the case of radiation, flux is measured as the number of particles per unit volume and their average velocity.

fly ash Very fine particles of ash that are produced by COMBUSTION and carried in FLUE GASES. Fly ash may contain unburned hydrocarbons and other pollutants, and it may be acidic and contribute to ACID DEPOSITION. The inhalation of small particles can cause damage to the lungs and respiratory passages.

fog Precipitation in the form of STRATUS cloud that extends to the ground or sea surface and reduces horizontal visibility to less than 1,094 yards (1 km). Cloud that reduces visibility less than this is often called *MIST.* Fog consists of water droplets smaller than about 0.004 inch (100 μm) in diameter. If pollutants, such as soot particles, are mixed with the water droplets the fog is called *SMOG*. Fog forms either when air is cooled to below its DEW POINT TEMPERATURE or when evaporation adds water vapor to the air, increasing its RELATIVE HUMIDITY to SATURATION. Fog formed by cooling is classified according to the mechanism that causes the cooling as ADVECTION FOG, RADIATION FOG, and HILL

FOG (or upslope fog). Fog caused by evaporation is classified as FRONTAL FOG and STEAM FOG. ARCTIC SEA SMOKE is a type of steam fog.

fog bank A well defined mass of FOG that is seen from a distance, especially at sea.

fogbow A RAINBOW that is seen in FOG. The water droplets that form fog are much smaller than the raindrops that act like prisms to separate white light into its constituent spectral colors, so although they refract light in the same way the rays remain so close to each other that they merge, and the fogbow is white.

fog dispersal The deliberate clearance of FOG in order to improve the horizontal visibility sufficiently to permit aircraft and vehicles to maneuver on the ground and vessels to move safely in coastal waters and harbors. The first attempts to disperse fog were made during World War II and were code-named *FOG INVESTIGATION DISPERSAL OPERATIONS*. Since then several other methods have been developed. Which of these is appropriate in a particular situation depends on the type of fog that is to be dispersed, and for this purpose fogs are classified not by the manner of their formation, but by their temperature. Warm fogs consist of water droplets all of which are liquid and warmer than 32° F (0° C). Supercooled fogs contain a mixture of supercooled liquid droplets and ice crystals and have a temperature between 32° F and –22° F (0° C and -30° C). ICE FOG consists only of ice crystals; its temperature is below –22° F (–30° C).

Warm fog can be dispersed by heating, mechanical mixing, or seeding with particles. Jet engines sited beside an airfield runway and running at full power disperse fog over the runway by heating it directly. This raises the DEW POINT TEMPERATURE, causing water droplets to evaporate. The method works, but it is costly.

The RELATIVE HUMIDITY may also be reduced if dry, warm air can be mixed with the saturated air of the fog. There is usually suitable air above the fog, and all that is needed to achieve thorough mixing is a large fan located above the fog and directed downward. The downwash from a helicopter hovering above the fog works well. Its disadvantage is that one helicopter can clear fog from only a very local area.

Hygroscopic particles—particles such as salt crystals that have a strong affinity for water—also disperse fog. The particles must be of the right size. If they are too small, they remain suspended in the air and reduce visibility even more. If they are too big, they simply fall to the ground without having any useful effect. Particles of the right size are scattered into the fog upwind of the area that is to be cleared. Water condenses onto them, forming droplets that are heavy enough to fall to the ground. This removes moisture from the air, allowing many more droplets to evaporate. Within about 10 minutes, a fog-free area is drifting downwind.

Supercooled fog is cleared by a technique that exploits the difference in SATURATION VAPOR PRESSURE over ice and over water. The pressure is slightly lower over an ice surface; therefore, if ice crystals and supercooled water droplets occur together water evaporates from the droplets and freezes onto the ice crystals, so the ice crystals grow at the expense of the droplets. Fog dispersal therefore encourages this process so the crystals turn into snowflakes and fall to the ground.

It is achieved by seeding the cloud with particles that act as FREEZING NUCLEI. Dry ice (solid carbon dioxide) and liquid propane are commonly used. Dry ice is released from an aircraft above the fog and propane is sprayed from the ground. Dry ice particles readily act as freezing nuclei. Propane droplets rapidly expand and vaporize. This chills the air around them, causing water to freeze and form ice crystals onto which more water will freeze.

At present there is no practicable way to disperse ice fog.

fog drip Water that is deposited on trees and other tall structures by FOG and that drips to the ground. This can deliver as much water to the ground as a light shower. Fog drip is common in the redwood groves of northern California.

fog droplet An individual particle of FOG, consisting of a droplet of water that is 0.00004–0.0008 inch (1–20 μm) in diameter. In a typical fog there is less than 0.001 ounce of water in every cubic foot (1 g m^{-3}).

fog horizon The boundary between the sky and the upper surface of a layer of FOG that is trapped beneath an INVERSION. The boundary is seen from a position above the fog, where it resembles the true horizon. The fog conceals the true horizon. A fog horizon can be misleading, because its flat, featureless surface may not be horizontal.

Fog Investigation Dispersal Operation (FIDO) A method that was devised during World War II to clear fog from military airfields. In the days before pilots could be "talked down" by a ground controller observing their glide paths by means of radar, this was necessary to allow aircraft to land in fog. Aircraft could take off in fog, using their own instruments for guidance, provided their crews could see well enough to taxi to the runway.

The FIDO employed perforated metal pipes that were placed on each side of each runway. Gasoline vapor was pumped through the pipe and ignited at the perforations (by an airman with a torch!). The result was a fierce heat that raised the DEW POINT TEMPERATURE, causing the fog to evaporate. The system remained in operation for about 2.5 years, during which it allowed 2,500 aircraft to land and consumed about 112,000 tons (102,000 tonnes) of gasoline to do so—almost 45 tons (41 tonnes) of fuel for each landing.

fog shower A type of PRECIPITATION that can occur on mountains at elevations that are higher than the LIFTING CONDENSATION LEVEL when the mountain is engulfed by a passing CUMULONIMBUS or large CUMULUS cloud that contains supercooled (*see* SUPERCOOLING) water droplets. These freeze on contact with small objects, producing a coating of RIME FROST or GLAZE. The cloud appears as FOG to an observer on the mountainside and the impact of the very cold droplets feels like a shower of rain.

fog streamer A wisp of FOG that forms near the surface when cold air blows across a lake. Several fog streamers sometimes merge to form a STEAM DEVIL.

fog wind *See* PUELCHE.

föhn air Warm air that is carried down a mountainside by a FÖHN WIND. The air is warm because it has been heated in an ADIABATIC process by its descent.

föhn cloud A cloud, usually of the LENTICULARIS type, that forms on the LEE side of a mountain range. Such clouds are often associated with a FÖHN WIND. They are produced by LEE WAVES in air that has crossed the mountains; air flowing down the lee side of the mountains produces the föhn wind.

föhn cyclone A CYCLONE on the LEE side of a mountain that draws air down the mountainside, creating a FÖHN WIND.

föhn island A low-lying area on the LEE side of a mountain that is affected by a FÖHN WIND. Adjacent areas remain under the influence of the cold air, so the area that lies beneath FÖHN AIR is like an island of warm air in a sea of cold air.

föhn nose The characteristic shape of the ISOBARS that indicates a fully developed FÖHN WIND on a SYNOPTIC CHART. There are a RIDGE on the WINDWARD side of the mountains and a FÖHN TROUGH on the LEE side, producing a pattern of isobars that are reminiscent of a nose.

föhn pause The boundary that exists between FÖHN AIR and the cold air adjacent to it. There can also be a temporary cessation in exposure to föhn air that occurs when cold air intrudes and lifts the warm air clear of the ground. This is also called a *föhn pause* and its effect can be dramatic, especially when it is repeated at fairly frequent intervals. Rain, falling from clouds at a higher level, is warmed as it passes through the layer of föhn air and reaches the ground as warm rain. As the cold air intrudes, the temperature drops sharply. The ground surface temperature falls to below freezing, so the layer of water left by the warm rain freezes. Repeated changes from warm to cold air allow a thick layer of ice to accumulate.

föhn period The length of time during which a specified place lies beneath FÖHN AIR.

föhn phase The stage that has been reached in the development of a FÖHN WIND when this is the result of BLOCKING by an INVERSION at the level of the mountain summit on the WINDWARD side. There are three phases. In the first, a subsidence inversion (*see* ANTICYCLONIC GLOOM) separates cold air at the surface from warmer air aloft. In the second, SUBSIDENCE increases the surface AIR PRESSURE on the LEE side of the mountain and cold air is pushed away. In the third, the FÖHN WALL forms and the föhn wind blows down the mountainside and across the plain.

föhn trough *See* FÖHN NOSE.

föhn wall The upper surface of the CAP CLOUD blanketing a mountain peak, as seen from the LEE side of the mountain. The cloud extends for some distance down the mountainside, carried by the airflow that produces the FÖHN WIND, and its upper side appears as a solid wall of cloud.

föhn winds Warm, dry winds that occur on the northern side of the European Alps, most commonly in spring. They are of the same type as the North American CHINOOK wind, and they also occur on the eastern side of the New Zealand Alps and on the leeward side of the mountains of the Caucasus and in Central Asia. At Tashkent, Uzbekistan, this wind can rapidly raise the temperature from about freezing to 70° F (21° C).

There are two ways that a wind of the föhn type may develop. Both require that the mountains are high and that the range extends a long way across the path of the moving air and also in depth, so the air must travel a considerable distance at high level.

The first and simplest mechanism involves the forced ascent and descent of air as it crosses a mountain range. As the air rises on the windward side of the range it cools, first at the DRY ADIABATIC LAPSE RATE (DALR) and above the LIFTING CONDENSATION LEVEL at the SATURATED ADIABATIC LAPSE RATE (SALR). Cloud forms, precipitation falls, and by the time the air reaches the top of the mountains it is fairly dry. The air then flows down the leeward side. As it does so it warms at the DALR and at the same time its RELATIVE HUMIDITY and ABSOLUTE HUMIDITY decrease. Because the SALR is lower than the DALR, this mechanism warms the air by about 2° F for every 1,000 feet (4° C per kilometer) the air is made to climb and descend. If the mountains rise 8,000 feet above the plain, for example, then air that starts its ascent at, say, 30° F (–1° C) should be at about 46° F (8° C) by the time it reaches the plain on the far side.

This is certainly what happens in some cases, but it is not the whole story, because sometimes there is no precipitation on the windward side. In this case there is a temperature INVERSION at the level of the mountain tops. Air approaching the mountain below the level of the inversion is unable to rise because of the inversion. This prevents the formation of cloud and precipitation. Air approaching the mountain above the inversion is not barred in this way. It crosses the mountains, then slides over the top and down the lee side, warming adiabatically as it does so. The result is a warm, dry föhn wind, but with dry air on the windward side of the mountains.

This explains the temperature, but it does not explain the speed of these winds, which can reach gale force. The wind is an effect of the GRAVITY WAVES that develop as air is forced to rise, then sinks by gravity, overshoots, and rises again. Where the moving air passes through a constricted vertical space, close to the mountain tops, and again as it sinks close to the ground on the lee side, wind speed increases. In the intervening regions, where the air is not constricted in this way, wind speed slows.

following wind A TAILWIND, or a wind that blows in the same direction as that it which waves are moving over the surface of the sea.

forced convection CONVECTION that occurs when air that is at the same temperature as the surrounding air, and therefore has neutral BUOYANCY, is made to rise or sink. This vertical movement can be due to OROGRAPHIC LIFTING, FRONTAL LIFTING, or CONVERGENCE or DIVERGENCE. The TURBULENT FLOW of air across an uneven surface can produce EDDIES that also cause forced convection. Once the air is moving vertically, it enters regions where the surrounding air is at a different temperature. The air may return to its former level, but it is also possible for CONVECTIVE INSTABILITY to develop.

forecast period The length of time that is covered by a weather forecast. This may range from less than 12 hours to several months or a season. Common sense would suggest that the shorter the forecast period the more accurate the forecast is likely to be, but this is not necessarily so, because more detail is usually expected in forecasts for very short periods than in those for longer periods. Individual showers and storms are short-lived and affect small areas, for example. Their likelihood can be reliably predicted over a particular area, but they cannot be predicted to strike a specific neighborhood more than one hour ahead. A forecast for a season, on the other hand, would be expected to state no more than whether it would be warmer, colder, wetter, or drier than usual over an entire region or even continent. Most forecasts cover the period up to 24 hours ahead and add an "outlook" for two or three days beyond that. These forecasts achieve a fair degree of accuracy in predicting the track and behavior of middle latitude weather systems and in anticipating large-scale events such as the approach of CYCLONES and ANTICYCLONES over the outlook period.

forecast-reversal test A test that is used to measure the usefulness of a method for FORECAST VERIFICATION. The same verification method is applied simultaneously to two weather forecasts. One is an actual forecast and the other is a fabricated forecast that predicts the opposite conditions. If the real forecast predicts rain, for example, the fabricated forecast predicts dry weather, and if the real forecast predicts wind, the fabricated one predicts calm. Each forecast is given an accuracy score on the basis of the test and the two scores are compared. The comparison amounts to an evaluation of the verification test, because the real forecast should achieve a markedly higher score for accuracy than the fabricated one.

forecast skill A measure of the accuracy of a weather forecast on a scale that ranges from 0 (completely wrong) to 1 (completely correct). The skill measures the predictive power of the forecast or of a forecasting method. For example, a forecast that it will not rain tomorrow in Death Valley is very likely to be correct, but making it demands little of the forecaster beyond a knowledge of the climate in Death Valley. Consequently the forecast has very little predictive power. Forecast skill is usually measured by comparing the accuracy of a forecast with that of a CLIMATOLOGICAL FORECAST or a PERSISTENCE FORECAST.

Forecast Systems Laboratory (FSL) A research laboratory located in Boulder, Colorado, that is part of the NATIONAL OCEANIC AND ATMOSPHERIC ADMINISTRATION (NOAA). The FSL develops weather forecasting systems and the hardware and software associated with them. It then transfers new technologies to the organizations that will use them. FSL comprises six divisions: Science, Facility, Demonstration, Systems Development, Aviation, and Modernization. It employs scientists from a number of institutions, and scientists doing postdoctoral research are contracted through the National Research Council and the National Center for Atmospheric Research. The FSL is open to the public and its annual publication *FSL in Review* describes its activities.

(You can learn more about the FSL at www.fsl.noaa/gov.)

forecast verification Any technique that is used to measure the accuracy of a weather forecast. Verification is based on comparisons that are made between the conditions that were predicted and those that actually occurred. The accuracy of the forecast is then given a numerical score.

forensic meteorology The branch of METEOROLOGY that is concerned with the relevance of atmospheric conditions to legal problems. A forensic meteorologist might be asked to comment on such matters as whether a particular flood should have been anticipated in the design of the building it damaged or whether the reduced VISIBILITY that caused an accident was due to natural FOG or industrial pollution. The issue may relate to an insurance claim or to the possibility of individual or corporate liability. Forensic meteorologists are consulted on LIGHTNING strikes, TORNADO and TROPICAL CYCLONE damage, and ICE STORMS, as well as on damage or injury that might have resulted from AIR POLLUTION. Several universities now offer postgraduate courses in forensic meteorology and many companies have been established to offer forensic meteorological services.

Forest A TYPHOON that struck the Japanese islands of Honshu, Kyushu, and Shikoku on September 19, 1983. It delivered up to 19 inches (483 mm) of rain, causing flooding that inundated 30,000 homes and resulted in the deaths of 16 people.

forest climate In the CLIMATE CLASSIFICATION devised by MIKHAIL I. BUDYKO, a climate in which the RADIATIONAL INDEX OF DRYNESS has a value of 0.33–1.0.

forked lightning LIGHTNING flashes that are seen as brightly luminous, jagged lines between a cloud and the ground or between two clouds. It is the most dramatic form of lightning. Humans and animals directly struck by forked lightning are severely injured or killed, trees are often split apart, and lightning strikes are the cause of most naturally occurring fires.

Fortin barometer A portable mercurial BAROMETER that was invented in 1800 by the French instrument maker Jean Nicholas Fortin (1750–1831). It overcomes the difficulty with mercury barometers that the AIR PRESSURE must be calculated from the difference between the level of the mercury in the barometer tube and the level in the reservoir. Fortin achieved this by making the bottom of the reservoir from flexible leather that can be raised or lowered by means of an adjusting screw. Before a reading is taken, the height of the mercury in the reservoir is raised or lowered until it just makes contact with a point fixed to the top of the reservoir. This places the level of the mercury in the reservoir in predetermined position, so no further account need be taken of it. In the tube there is a vernier slide that can be moved by means of a screw on the side of the tube. The slide is lowered until its base just touches the top of the mercury and a reading is taken against the fixed scale on the tube. To prepare the barometer for a journey, the adjusting screw is used to push the base of the reservoir upward until both it and the tube are filled with mercury.

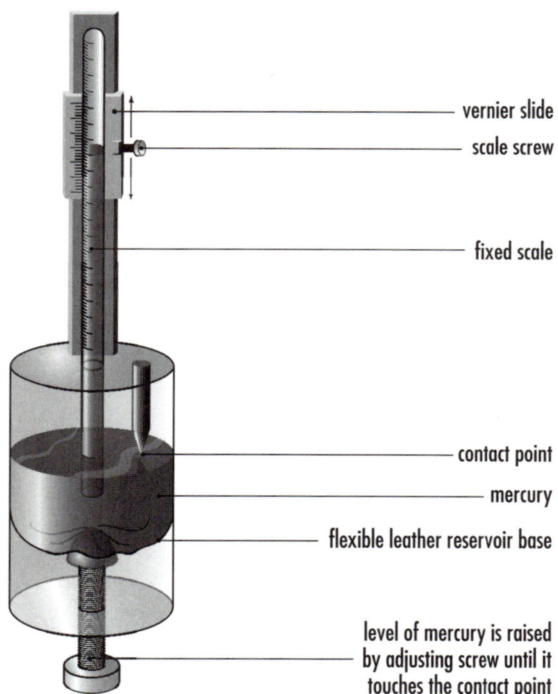

Fortin barometer. **The upper part of the Fortin barometer comprises the tube, fixed scale, vernier slide, and the screw for adjusting the slide. The lower part comprises the reservoir, the point against which the mercury is set, and the adjusting screw.**

The length of the fixed scale varies from one instrument to another, depending on the locations in which it is to be used. Its upper limit must be extended if the barometer is to be used, for example, in a mine, where the pressure is greater than it is at sea level.

The design makes no allowance for changes in temperature, which affect the volume of the mercury. It is also subject to errors due to CAPILLARITY, to the fact that the vacuum above the mercury is not complete, to small errors in the scales on the tube and vernier slide, and to errors in setting the instrument and in taking the reading. Despite these drawbacks, the Fortin barometer is sufficiently accurate for most purposes and its ease of use and portability make it a very useful and popular type of barometer.

fossil fuel *See* CARBON CYCLE.

fossil turbulence Local variations in temperature and humidity that are produced by TURBULENT FLOW and that persist in the air after the movement that caused them has ceased and the DENSITY of the air has become uniform. Fossil turbulence scatters radio waves and can cause small clouds to form where air is made to rise.

Fourier, Jean Baptiste Joseph (1768–1830) French *Mathematician and physicist* Fourier was born on March 21, 1768, at Auxerre, France, a town southeast of Paris. His father was a tailor. Jean was orphaned when he was eight, but the bishop exerted his influence to have him admitted to the Auxerre military academy, where boys were educated to become artillery officers. It was there that Jean first encountered mathematics and showed he had an enthusiasm and great aptitude for the subject.

His humble origin meant he was unable to become an artillery officer, so when he left the academy he went to a Benedictine school in St. Bênoit-sur-Loire. He returned to Auxerre in 1784 and taught mathematics at the military academy. When the French Revolution began in 1787 Fourier took an active part locally, but he did not support the Terror that occurred later. Despite his initial support, Fourier was arrested in 1794, but he was released after a few months, after the execution of Robespierre.

The École Normale opened in Paris in 1795 and Fourier taught there, gaining such a reputation that it was not long before he was made professor of analysis at the École Polytechnique. In 1798 he was one of the *savants* chosen to accompany Napoleon on his campaign in Egypt. He was made governor of lower Egypt and remained there until 1801. On his return to France he was made prefect of Isère and lived at Grenoble, in the south of the country. In 1808 Napoleon conferred the title of baron on him and later made him a count. Fourier rejoined Napoleon in 1815, and after Napoleon's final defeat the following year he settled in Paris. Fourier was elected to the Academy of Sciences in 1817 and in 1822 became its joint secretary, sharing the position with the zoologist Georges Cuvier (1769–1832). He was also elected to the Académie Française and to foreign membership of the Royal Society of London.

Fourier had resumed his mathematical studies during the years he lived in Grenoble. He was particularly interested in the conduction of heat and sought to describe it in purely mathematical terms. He explained the theory he had devised to do so in *Théorie analytique de la chaleur.* Published in 1822 and translated into English (*Analytical Theory of Heat*) in 1872, this proved to be one of the most influential scientific books of the 19th century.

The rate of conduction varies with the temperature gradient, as well as with the composition of the material and the shape of the conducting body. In order to comprehend this, Fourier developed what came to be known as *Fourier's theorem.* This is a technique that allows the overall description to be broken down into a series of simpler, trigonometric equations, known as the *Fourier series,* the sum of which is equal to the original description. The Fourier series can be applied to any complex function that repeats, and so it is of value in many branches of physics. It is widely used by meteorologists. He also developed the use of linear partial differential equations for solving boundary-value problems. This, too, is relevant to NUMERICAL FORECASTING.

His theorem and series were only part of his contribution to mathematics. He also investigated probability theory and the theory of errors.

Fourier had contracted an illness while he was in Egypt that may explain his conviction that the heat of the desert was beneficial. He lived wrapped up in thick, warm clothes in overheated rooms. He died in Paris on May 16, 1830.

fra *See* FRACTUS.

fractocumulus Fragments of broken or ragged cloud that have been torn from CUMULUS or that are the remains of cumulus that has dissipated. Fractocumulus is the cloud species (*see* CLOUD CLASSIFICATION) cumulus FRACTUS (Cu_{fra}).

fractostratus FRACTUS that is the cloud species St_{fra} (*see* CLOUD CLASSIFICATION), which remains as STRATUS dissipates.

fractus (fra) A species of clouds (*see* CLOUD CLASSIFICATION) that is seen only with the cloud genera CUMULUS and STRATUS. It consists of fragments of cloud that look as though they have been torn from the parent cloud or are all that remains after the parent cloud has dispersed, and this is how fractus is formed.

Fractus is a Latin word that means "broken."

fragmentation nucleus A tiny splinter of ice that is snapped off from a larger ICE CRYSTAL during a collision between crystals. The fragment then serves as an ICE NUCLEUS onto which a new ice crystal will grow.

Framework Convention on Climate Change A United Nations agreement that was reached at the United Nations Conference on Environment and Development held in Rio de Janeiro, Brazil, in June 1992 (and sometimes called the *Rio Summit* or the *Earth Summit*). The Framework Convention aims to address the issue of GLOBAL WARMING by seeking the agreement of national governments to promote relevant research and to reduce emissions of GREENHOUSE GASES. Its most direct achievement is the KYOTO PROTOCOL, which sets targets for reduced emissions.

(You can learn more about the convention at www.unfcc.de.)

Fran A TYPHOON, rated category 3 on the SAFFIR/SIMPSON SCALE, that struck southern Japan from September 8 to 13, 1976. It generated winds of 100 mph (160 km h^{-1}) and produced 60 inches (1,524 mm) of rain. It caused the deaths of 104 people and made an estimated 325,000 homeless. HURRICANE Fran, rated as category 3, struck the eastern United States in September 1996. It triggered TORNADOES and produced a STORM SURGE with waves of up to 12 feet (3.6 m) and in some places 16 feet (4.8 m). It passed Cape Fear, North Carolina, on September 6 and the following day crossed North and South Carolina, Virginia, and West Virginia. More than 500,000 people were evacuated from the coastal areas of South Carolina, and parts of North Carolina were also evacuated. It killed 34 people.

Frankie A TROPICAL STORM that crossed from the Gulf of Tonkin to the Red River delta, Vietnam, on July 23, 1996. It killed 41 people.

Franklin, Benjamin (1706–1790) American *Statesman, physicist, inventor, author, publisher* One of the best known and most admired men in the world during the second half of the 18th century, Franklin had achievements that were summarized by the French economist Anne Robert Jacques Turgot (1727–81) in the words "He snatched the lightning from the skies and the scepter from tyrants." To Americans he is known as one of the founding fathers, but he is no less famous in Europe. There, in his own day, he was famous mainly as a natural philosopher—a person who would nowadays be called a scientist.

He was a complex man, with many sides to his personality; his attitude to science was expressed in a letter he wrote in 1780 to the English chemist Joseph Priestley (1733–1804): "The rapid progress true science now makes occasions my regretting sometimes that I was born too soon. It is impossible to imagine the height to which may be carried, in a thousand years, the power of man over matter. . . . O that moral science were in as fair a way of improvement, that men would cease to be wolves to one another, and that human beings would at length learn what they now improperly call humanity!"

Benjamin Franklin was born on January 17, 1706, in Boston, Massachusetts. His father, Josiah, a soap and candle maker, had emigrated from Banbury, Oxfordshire, in England. It was a large family—Benjamin was the 15th of 17 children. The family could not afford to send him to college, and he spent only one year at a grammar school. He was educated privately but was mainly self-taught.

His greatest scientific achievements arose from his study of electricity. In 1745, the German physicist E. G. von Kleist had discovered a way to condense electric charges. His device was first investigated thoroughly by the Dutch physicist Pieter van Musschenbroek (1692–1761) at the University of Leiden (or Leyden) in the Netherlands and so it became known as a *Leyden jar.* It consists of a glass jar lined on the inside with metal and sealed by a cork through which a metal rod is inserted. There were already machines that used friction to produce static electrical charges, and the Leyden jar was very good at storing these charges. The stored charge would be discharged as a spark if a hand was brought close to the metal rod, and if enough charge had accumulated, anyone approaching too close would receive a strong electric shock. If a metal object was held close to the rod a spark would leap across the gap and there would be a loud crackling noise.

Like many other scientists, Franklin experimented with a Leyden jar. His observations led him to wonder whether the spark and crackle might not be a tiny demonstration of LIGHTNING and THUNDER and, therefore, whether during a THUNDERSTORM the sky and Earth might act as a giant Leyden jar.

It was an idea that needed testing, and to do so Franklin performed the most famous of all his experiments—and by far the most dangerous. In 1752, he flew a kite in a thunderstorm. He had attached a pointed piece of wire to the kite and tied a long silken thread to the wire. At the bottom of the thread he tied a metal key. As the kite flew near the base of the storm cloud and lightning began flashing nearby, Franklin held his hand close to the key and a spark jumped to it. Then he held a Leyden jar to the key and it accumulated an electric charge. He had proved that lightning is a discharge of electricity and thunder is the sound of the spark. In the same year, the French scientist Thomas-François d'Alibard (1703–99) also proved, independently of Franklin, that lightning is an electrical phenomenon.

He was extremely lucky. The next two people to repeat his experiment were killed, and on no account should anyone try to perform it.

Benjamin Franklin.* The American statesman and physicist who proved that lightning is an electric spark. *(John Frederick Lewis Collection, Print and Picture Collection, The Free Library of Philadelphia)

Franklin had also noticed that electricity sparks more readily and over a greater distance if there is a pointed surface to which it can travel. This led him to suggest that buildings could be protected from damage by lightning if pointed metal rods were fixed to their roofs and connected to the ground by metal wires. He had invented the lightning conductor and these were soon being fitted throughout America. Many lives were saved by them and much damage to property was prevented.

Franklin did make one mistake. It was known that there are two types of electricity and that they repel or attract each other, apparently on the principle of like repelling like and opposites attracting each other. This might be explained, Franklin thought, if electricity is some kind of fluid that can be present either in excess or in deficiency. Then, two bodies each of which contained either an excess or a deficiency of it would repel each other. If one body had an excess and the other a deficiency, on the other hand, they would attract each other and electricity would flow from the excess to the deficiency. This is wrong, of course, because electricity is not a fluid. We do retain a little of the terminology Franklin introduced, however. He proposed that an excess of the fluid be called *positive* electricity and a deficiency *negative* electricity.

At the end of a very long and very distinguished political, diplomatic, and scientific career, Benjamin Franklin at the age of 79 retired to Philadelphia. Already ill, he was bedridden for the last year of his life. He died on April 17, 1790. He was given the most impressive funeral there had ever been in Philadelphia; in France, many eulogies were composed to the man the French regarded as the embodiment of freedom and enlightenment.

frazil ice Ice crystals that form on the surface of the sea as the temperature falls below freezing. They dampen down small wave movements, thus making the water look oily, and as freezing continues and more of them form the sea becomes covered with slush.

Fred A TYPHOON that struck Zhejiang Province, China, on August 20 and 21, 1994. It killed about 1,000 people and caused damage estimated at more than $1.1 billion.

Frederic A HURRICANE that struck the eastern coast of Florida, Alabama, and Mississippi in September 1979. At least eight people died and about 500,000 were evacuated.

free atmosphere That part of the atmosphere that is not directly influenced by the land or sea surface because it lies above the PLANETARY BOUNDARY LAYER. The free atmosphere constitutes about 95 percent of the total mass of the atmosphere.

free convection CONVECTION that is caused by the warming of the air from below. When the TEMPERATURE of the air is higher than that of the air above it the warm air has positive BUOYANCY and rises.

free radical *See* COMBUSTION.

freezing The change of phase from liquid to solid and in the case of WATER the change from liquid to ice.

Change in the opposite direction, from solid to liquid, is called *melting*.

Like all molecules, water molecules move. Their freedom to move depends on the amount of energy they possess. It is the vigor with which they move that defines the quality we measure as TEMPERATURE. The less energy they possess the less vigorously they move and the less vigorously they move the lower is their temperature. As molecules cool, therefore, they move with less and less vigor. In liquid water the molecules form small groups that slide over and past each other quite freely, but as the temperature falls, the water molecules lose energy and move more slowly. Molecules pack together more closely and the water contracts to occupy a smaller volume.

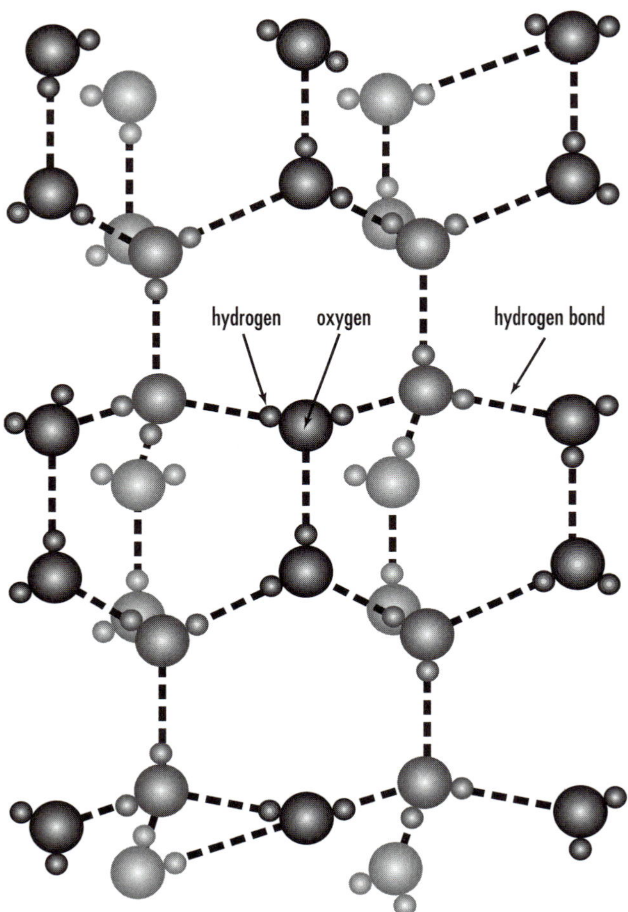

Freezing. **The arrangement of water molecules in ice. Each molecule is linked by hydrogen bonds to four neighboring molecules (not all the bonds are shown in this two-dimensional representation of a three-dimensional structure). The result is an ordered, but very open structure.**

Then, as the temperature approaches freezing, the molecules rearrange themselves. The molecules are linked to one another by HYDROGEN BONDS. In the liquid phase these are constantly breaking and reforming, because the individual molecules have sufficient energy to break free of them. As water turns into ice, the molecules no longer have enough energy to break free and the hydrogen bonds hold them together. The molecules lock together. This requires less energy than they possessed when they moved freely, and the surplus energy is released as LATENT HEAT. Ice molecules are not motionless, but they are no longer able to move freely. Their movement is restricted to vibrating about a fixed point.

As they form ice, each molecule forms hydrogen bonds with three closely neighboring molecules. This produces a very open structure. Consequently, as water freezes it also expands and it does so with sufficient force to fracture metal pipes.

Ice consists of water molecules that are arranged in a regular, repeating pattern called a *lattice*. At the ice surface, however, there are fewer hydrogen bonds to hold the molecules in place. The lattice is disordered where surface molecules project into the air. These molecules vibrate more vigorously than those in the solid lattice of the interior, and at a temperature well below freezing the surface layer of the ice, just a few molecules thick, is able to move as though it were a liquid. This is called *surface melting* and scientists believe it explains why ice is slippery—the characteristic of ice that makes skating possible.

(You can learn more about surface melting in John S. Wettlaufer and J. Greg Dash, "Melting Below Zero," *Scientific American*, February 2000, pp. 34–37.)

freezing drizzle DRIZZLE comprising supercooled droplets that freeze on contact with the ground. Inside the STRATUS cloud from which the precipitation falls, the temperature is a little above freezing. Droplets falling through the cloud do not freeze, but when they leave the base of the cloud their temperature is close to freezing. If the air temperature between the cloud base and the ground is below freezing, the droplets are chilled further as they fall, but because the cloud base is low they do not remain airborne long enough to freeze. Instead, they freeze immediately on contact with the ground, the temperature of which is also below freezing.

freezing fog FOG that forms when the air temperature is below freezing. The low temperature chills surfaces to below freezing and fog droplets freeze onto them. The droplets themselves may be supercooled (*see* SUPERCOOLING). Freezing fog coats surfaces with ice and makes driving conditions hazardous by icing the windshields of moving vehicles.

freezing index The cumulative number of DEGREE DAYS when the AIR TEMPERATURE is below freezing (32° F [0° C]). Freezing indices are used to predict the distribution of PERMAFROST; to estimate the thickness of ice on lakes, rivers, and the sea; to estimate the depth to which frost penetrates below ground; and to classify types of SNOW. The indices are available for the entire world, broken into areas measuring 0.5° longitude by 0.5° latitude.

(You can learn more about freezing indices at nsidc.org/NSIDC/CATALOG/ENTRIES/nsi-0063.html.)

freezing level The lowest height above sea level at which the air temperature is 32° F (0° C). This varies from place to place and from time to time, but if all the points where the temperature is at freezing are joined to form an imaginary surface, the spatial variations appear on this constant-temperature surface, like the hills and valleys of a landscape. The freezing level is determined by the ENVIRONMENTAL LAPSE RATE. Suppose the temperature at sea level is 75° F (24° C), the TROPOPAUSE is at a height of 46,000 feet (15 km), and the temperature at the tropopause is –76° F (–60° C). The temperature decreases from 75° F (24° C) to –76° F (–60° C) over a vertical distance of 46,000 feet (15 km), which is a decrease, or lapse rate, of 3.3° F per 1,000 feet (5.6° C per kilometer). Assuming this remains constant through the TROPOSPHERE (in reality it may not) and the air temperature will be 32° F (0° C) at a little over 13,000 feet (about 4 km).

freezing-level chart A SYNOPTIC CHART that uses contour lines to show the height of the constant-temperature surface of the FREEZING LEVEL. It shows the fronts between masses of warm and cold air and gives an indication of the likely availability of ice crystals that may accelerate CONDENSATION, leading to precipitation.

freezing nuclei Small particles onto which SUPERCOOLED water droplets freeze. They are crystals with shapes similar to that of an ice crystal, and actual ice crystals formed by deposition onto SUBLIMATION NUCLEI become freezing nuclei for further ice formation. Splinters of ice that break away from aggregations of crystals as these are moved violently by vertical air currents also act as freezing nuclei. Most of the mineral particles onto which water freezes are believed to be fine soil particles, and a small proportion may be meteoric (entering from space) or injected by volcanic eruptions. They are much less numerous than CLOUD CONDENSATION NUCLEI: there are seldom more than about three freezing nuclei per cubic foot (100 m^{-3}) of air.

Depending on their crystal structure, different freezing nuclei trigger the formation of ice at different temperatures. Ice starts to form on kaolinite particles, for example, at 16° F (–9° C), and once ice is forming it continues to do so on particles at temperatures up to 25° F (4° C). Kaolinite is a widespread clay mineral that occurs in many soils. Most freezing nuclei do not become active until the temperature falls to about 14° F (–10° C), however, so that clouds above this temperature consist predominantly of supercooled droplets. Liquid droplets and ice crystals occur in approximately equal numbers at temperatures between 14° F (–10° C) and –4° F (–20° C). Below –4° F (–20° C) ice crystals predominate.

freezing rain RAIN that freezes immediately on contact with the ground. The raindrops form in a cloud where the temperature in the lower part is a little higher than freezing. As they fall through this region liquid droplets remain liquid and ice crystals melt. Below the cloud they enter air at below freezing temperature. They fall quickly and are not exposed to the cold air long enough to freeze, but they are chilled to slightly below freezing temperature. Supercooled, they strike the ground, which is also at below freezing temperature, and freeze on contact.

frequency The rate at which a regularly repeating event recurs. In the case of waves, the frequency is the number of vibrations or oscillations that occur in a given time, usually one SECOND. The frequency of a wave (f) is calculated by

$$f = c/\lambda$$

where c is the speed at which the wave is moving and λ is the WAVELENGTH. The unit of frequency is the HERTZ.

frequency-modulated radar *See* RADAR.

fresh breeze In the BEAUFORT WIND SCALE, force 5, which is a wind that blows at 19–24 mph (31–37 km h^{-1}). In the original scale, devised for use at sea, a force 5 wind was defined as "or that to which a well-conditioned man-of-war could just carry royals, etc. in chase, full and by." On land, a fresh breeze makes small trees that are in full leaf wave about.

fresh gale In the BEAUFORT WIND SCALE, force 8, which is a wind that blows at 39–46 mph (63–74 km h^{-1}). In the original scale, devised for use at sea, a force 8 wind was defined as "or that to which a well-conditioned man-of-war could just carry treble-reefed topsails, etc. in chase, full and by." On land, a fresh gale tears small twigs from trees.

freshwater Water that contains very little salt, usually defined as less than 0.03 percent by volume. When water evaporates or freezes it is only water molecules that enter the air or form ice crystals. Molecules of any substances that were mixed with or dissolved in the water remain behind. Consequently, water vapor that condenses to form clouds is fresh, and PRECIPITATION consists of only freshwater (although substances present in the air may dissolve in atmospheric water droplets). Similarly, ice, including sea ice and ICEBERGS, is made from freshwater, although small amounts of salt water may be held between ice crystals.

friagem A spell of cold, cloudy weather, with occasional rain, that occurs in winter in the middle and upper Amazon basin and lasts four or five days. It is caused by the incursion of polar air from beyond the southern tip of the continent, and possibly all the way from Antarctica. This air forms the western flank of a large AIR MASS and sweeps as a COLD FRONT through northern Argentina and the lowlands of eastern Bolivia into equatorial Brazil. The Andes prevents air behind the front from moving farther westward. With the arrival of the friagem the temperature may fall by 7° F (4° C), and it has been known to drop to 34° F (1° C) at Cuyabá, at 15° S.

friction The force that resists the motion of a solid body or of a fluid that is in contact with a surface that is stationary or moving at a different speed or in a different direction. Wind that blows over a surface is slowed by friction with the surface, because the surface exerts a force acting in the opposite direction to that of the wind. The extent to which it is retarded depends on the AERODYNAMIC ROUGHNESS of the surface and the square of the speed of the wind. This type of frictional effect is called *DRAG*.

According to Newton's third LAW OF MOTION, the force with which a surface retards the wind must be equal to a force exerted by the wind in the opposite direction. Since this force is proportional to the square of the wind speed, a wind blowing at 120 mph (193 km h^{-1}) exerts 100 times more pressure on objects in its path than a wind blowing at 12 mph (19 km h^{-1}), not 10 times more. The force the wind exerts on a surface is called the *SURFACE SHEARING STRESS*.

Friction transfers momentum from the wind to the surfaces on which it exerts pressure. Mountain ranges do not bend with the wind, but westerly winds, blowing from west to east, accelerate the rotation of the Earth. Fortunately, they are balanced by winds blowing in the opposite direction. Nevertheless, the rotational speed of the Earth does vary by a very small amount with the changing winds. Wind over the ocean causes waves. This transfers energy from the wind to the water. Friction between layers of water produces an EKMAN SPIRAL, which in turn causes UPWELLING. Friction with a smooth surface also produces an Ekman spiral in the air.

Surface friction affects air in the PLANETARY BOUNDARY LAYER. Beyond the boundary layer friction also occurs between layers of air moving in different directions. This arises mainly from TURBULENT FLOW and occurs in EDDIES that turbulence produces. Its magnitude is proportional to the VISCOSITY of the air (*see* EDDY VISCOSITY).

friction layer **(surface boundary layer)** The lowest part of the PLANETARY BOUNDARY LAYER, constituting about 10 percent of the depth of the boundary layer.

TURBULENT FLOW within the friction layer ensures thorough mixing of the air. Consequently, the characteristics of the air are fairly constant throughout the layer.

friction velocity (**shear velocity**) The VELOCITY of air in the PLANETARY BOUNDARY LAYER above the LAMINAR BOUNDARY LAYER. It is symbolized by $u*$ and is equal to $\tau/\rho)^{1/2}$, where τ is the tangential stress on the horizontal surface and ρ is the DENSITY of the air. The friction velocity is used in the LOGARITHMIC WIND PROFILE equation.

Frida *See* BENEDICT.

Frigid Zone *See* MATHEMATICAL CLIMATE.

fringe region (**spray region**) The uppermost part of the EXOSPHERE, where the atmosphere is so rare that individual atoms seldom collide. Theoretically, atoms in the fringe region can escape into space without experiencing a collision. Those that do not escape move in free orbits about the Earth at speeds determined by their most recent collision.

frog storm (**whippoorwill storm**) In North America, the first bad weather to occur in spring after a spell of warm weather.

frontal analysis The study of data recorded on weather charts in order to identify the boundaries between adjacent AIR MASSES and mark the fronts separating them.

frontal contour A line that marks the intersection between a front and a surface. Weather maps show frontal contours with respect to the Earth's surface and the contours mark the location of the fronts with respect to the surface. Frontal contours can also be drawn for atmospheric CONSTANT-PRESSURE SURFACES.

frontal cyclone A region of low atmospheric pressure, around which the air circulation is CYCLONIC, that is associated with a FRONTAL SYSTEM. The term is synonymous with *FRONTAL WAVE* but is sometimes used to distinguish a cyclone of this type from a TROPICAL CYCLONE.

frontal decay *See* FRONTOLYSIS.

frontal depression *See* FRONTAL WAVE.

frontal fog (**precipitation fog**) FOG that is associated with a FRONT, where warm air is being lifted above colder air. The warm air cools as it rises, and when its temperature falls below the DEW POINT TEMPERATURE, its water vapor begins to condense. Cloud forms and may produce rain. The rain, falling from warm air, crosses the front into the colder air below. A BOUNDARY LAYER of air around each raindrop is warmed by contact with the water, and once its temperature rises it is able to contain more water vapor than the colder air around it. Raindrops evaporate rapidly, and the air containing them mixes with the surrounding air. Mixing raises the relative humidity of the cold air, and if it exceeds 100 percent the water vapor starts to condense again, but this time to form cloud beneath the front. The effect is to produce cloud that extends from ground level to the cloud above the front.

frontal inversion A temperature INVERSION that occurs at a front, where warm air lies above cold air. Air temperature decreases with height from the surface to the FRONTAL ZONE. There it increases with height as a result of the transition from one AIR MASS to the other. The front therefore forms a barrier to air that is rising convectively in the cold air mass beneath it.

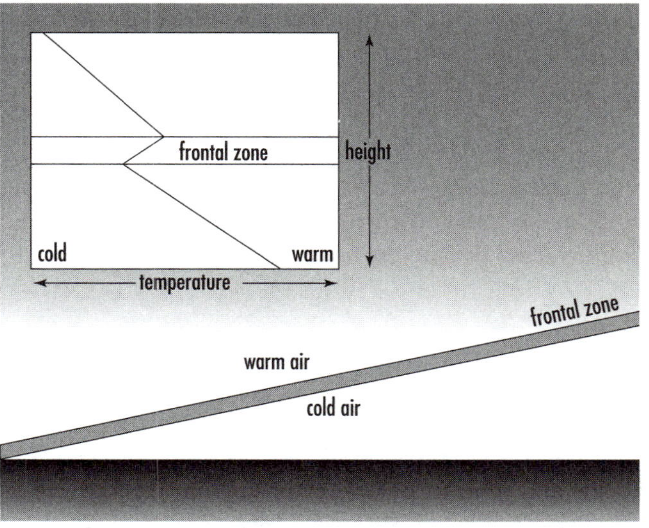

Frontal inversion. **Warm air is held above the front, with cold air below it. This produces a temperature inversion in the frontal zone, where temperature increases with height.**

frontal lifting The forced ascent of warm air as it rises over an adjacent mass of cold air at a WARM FRONT, or as it is undercut by advancing cold air at a COLD FRONT. It occurs because warm air is less dense than cold air and AIR MASSES (or water masses) of different densities do not mix readily.

frontal passage The movement of a front over a point on the surface. This is not an instantaneous event. A front is defined by the character of the air behind it. If its passage represents a change from cold to warm air it is a WARM FRONT, and if the transition is from warm to cold air it is a COLD FRONT. A warm front travels at an average 15 mph (24 km h^{-1}) and a cold front at an average 22 mph (35 km h^{-1}). Fronts are 60–120 miles (100–200 km) wide, so a warm front may take 4–8 hours to pass and a cold front 2.75–5.5 hours to pass.

frontal precipitation Precipitation that falls from clouds associated with a weather front, rather than AIR MASS PRECIPITATION or precipitation from clouds produced by OROGRAPHIC LIFTING.

frontal profile A diagram that shows a vertical cross section through a front, sometimes with the clouds that are associated with the front at different heights.

frontal slope The gradient of a warm or COLD FRONT, which is measured either as the angle between the front and the surface or as the ratio of vertical to horizontal distance. A WARM FRONT has an average slope of between 0.5° and 1°, or between about 1:115 and 1:57. CIRRUS is often the cloud that forms at the top of a warm front, close to the TROPOPAUSE. When cirrus associated with a warm front is overhead, the point where the front is at the surface is about

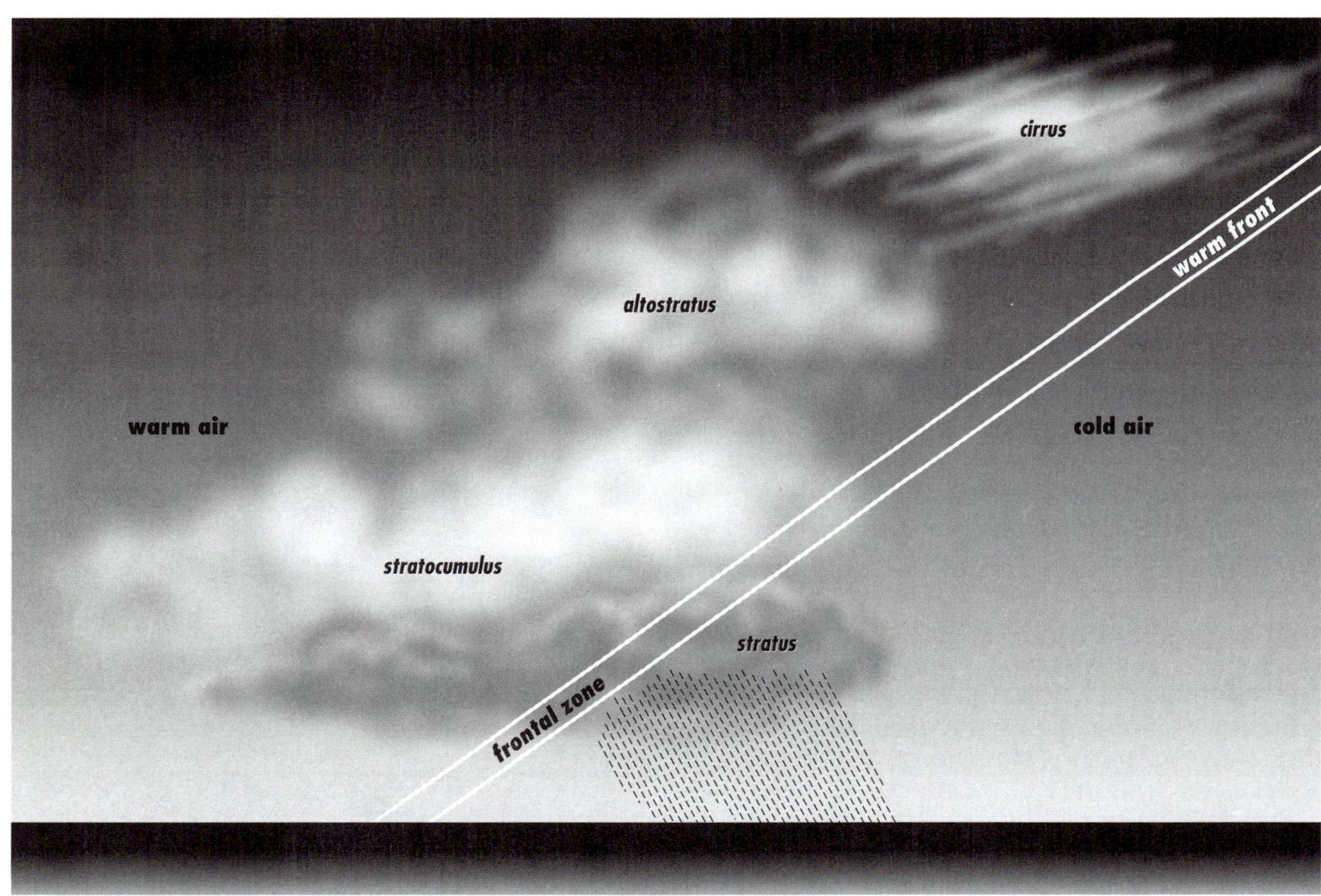

Frontal profile. **A cross-sectional view of a front, together with the cloud and precipitation associated with it.**

350–715 miles (570–1,150 km) away. A cold front slopes much more steeply, at about 2°, or about 1:30. When the cold front reaches a point on the surface, the upper edge of the front is about 185 miles (300 km) away.

frontal strip The representation of a front on a weather map as two parallel lines, rather than as a single line. This illustrates the fact that the front is a band of transition (*see* FRONTOGENESIS) between two air masses, rather than the abrupt change suggested by a single line. Despite its ability to show the location of the boundaries of the FRONTAL ZONE, a front is rarely shown as a frontal strip.

frontal structure The way air moves, cloud forms, and precipitation develops in a FRONTAL SYSTEM. At first, before a FRONTAL WAVE starts to appear, air is rising throughout the TROPOSPHERE. This generates STRATIFORM cloud along the front. Air converges to replace the rising air and planetary VORTICITY causes it to start rotating CYCLONICALLY. This increases the temperature gradients, because warm air is being carried toward the Poles on one side of the circulation and cold air is being carried toward the equator on the other side. As the frontal wave begins to develop, precipitation starts falling over a large area. There is then much middle and high cloud, including some CIRROSTRATUS. Air at high level rotates in an ANTICYCLONIC direction because of high-level DIVERGENCE. Divergence then starts removing air at high level faster than it can be replaced by CONVERGENCE at low level and the surface air pressure falls sharply. NIMBOSTRATUS covers a large part of the sky in the warm and cold FRONTAL ZONES, producing heavy precipitation, and there are ALTOSTRATUS and cirrostratus above it. As the fronts occlude the center of low pressure moves farther toward the Pole. Then, as the fronts dissipate, the temperature gradient slackens, the clouds start to clear, and precipitation ceases.

frontal system A complete system of warm, cold, and occluded fronts as these are shown on a weather map.

frontal thunderstorm A THUNDERSTORM that develops in warm, moist air that has been made unstable by frontal lifting. Such storms can be violent, and where the COLD FRONT is advancing very fast overriding the WARM FRONT, the rapid lifting along a substantial frontal length can produce a SQUALL LINE.

frontal wave (frontal depression) The wave that develops in the course of FRONTOGENESIS. It is produced by the CYCLONIC circulation caused by WIND SHEAR along the front, and the resulting wave forms a wedge of warm air protruding into the cold air. Once the wave has formed, the weather system enters its open stage. Cyclonic circulation surrounds the center of an area of low pressure at the crest of the wave. This constitutes the DEPRESSION. The system travels with a WARM FRONT at the leading edge of the wedge of warm air and a COLD FRONT at the trailing edge. The cold air to the rear of the system travels faster than the warm air, causing the cold front to override the warm front. The tip of the wedge of warm air is lifted clear of the surface and the fronts appear on a weather map as an OCCLUSION. This is the occluding stage in the life cycle of the frontal wave. When the whole of the wedge of warm air has been lifted above the surface the system dissolves, and the occluded fronts disappear as the warm and cold air mix.

frontal zone The region of a front where the temperature gradient is strongest. Because fronts slope at a very shallow gradient (*see* FRONTAL SLOPE) the position of the frontal zone in the upper TROPOSPHERE is above a surface position that is a long distance from its position at the surface. As a FRONTAL WAVE develops, the frontal zones become more sharply defined and can reach a gradient of about 14.5° F per 100 miles (5° C per 100 km).

frontogenesis The formation and subsequent development of a boundary between two AIR MASSES. Such a boundary is known as a *weather front* and it exists because air in one air mass does not mix readily with air in an adjacent mass that has substantially different characteristics. The name *front* was introduced during World War 1 by the team of meteorologists at the BERGEN GEOPHYSICAL INSTITUTE led by VILHELM BJERKNES.

Where two air masses meet there is a boundary between them, about 60–120 miles (100–200 km) wide, across which the temperature, pressure, wind, and humidity change sharply. Although this makes the boundary seem wide, it is thin enough to be shown on a weather map as a thick line. Air on one side of the

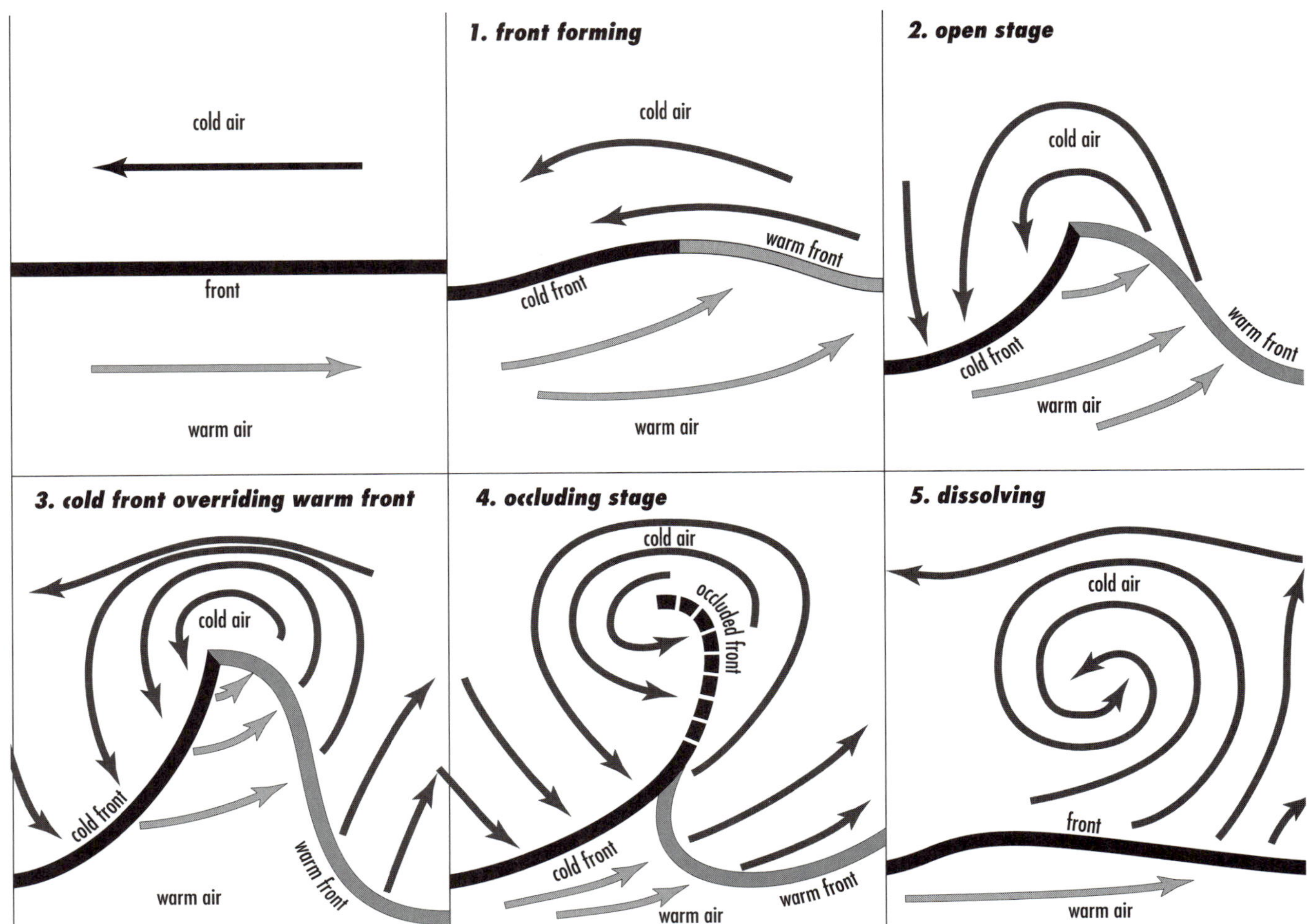

Frontal wave. **The sequence of events by which an undulation that appears in a front between two air masses develops into a wave, or depression, that occludes and finally dissolves.**

boundary is warmer than air on the other side. The warmer air is less dense and at a lower pressure than the cooler air. CONVERGENCE into the warmer air and DIVERGENCE from the cooler air establish winds blowing in opposite directions. In the Northern Hemisphere the direction of the wind is CYCLONIC (counterclockwise) around the center of the low-pressure, warmer air and ANTICYCLONIC (clockwise) around the center of the high-pressure, cooler air.

Usually both air masses are moving, but not at the same speed. As one advances against the other, the warmer air, which is less dense, rises over the cooler, denser air. The boundary, or front, does not rise vertically, therefore, but along a shallow FRONTAL SLOPE. Warm air cools in an ADIABATIC manner as it rises over the cooler air. If it is moist its water vapor condenses to form cloud that may produce precipitation. Sometimes cold air advances so fast against warm air that it triggers enough instability to produce a SQUALL LINE. A front at which air is rising is called an *ANAFRONT,* but it can also happen that air subsides down a front. One where this is happening is called a *KATAFRONT.* The two types of front produce quite different clouds and weather. An anafront produces large amounts of cloud, mainly of a STRATIFORM type but sometimes including CUMULONIMBUS near the cold front, and extensive, continuous, and often heavy precipitation. A katafront often produces a complete cloud cover, with showers at the cold front and a belt of light rain or DRIZZLE at the warm front.

WIND SHEAR on each side of the front causes a wave to develop with air circulating cyclonically around it and warm air projecting into the cold air mass. The center of the cyclonic flow is at the crest of the wave and is also the region of lowest atmospheric pressure. At this stage it has become a FRONTAL WAVE, or frontal depression, and there are two fronts, with a wedge of warm air between them.

If the cold air is traveling faster than the warm air, it starts to lift the wedge of warm air clear of the surface. On a weather map, which illustrates the situation at the surface, the cold front appears to override the warm front. The fronts are then said to be *occluded.* The cold front continues to advance against the warm front, causing the occluded section of the fronts to grow longer. Eventually all of the wedge of warm air has been lifted clear of the surface. A section of occluded front survives for a time, but it quickly shortens and then disappears. With the frontal wave dissipated, the original front reappears as a fairly straight boundary between the two air masses and the stage is set for the cycle to repeat itself.

frontolysis (frontal decay) The dissolution and disappearance of a weather front that occur when there is no longer any difference in the characteristics of two adjacent AIR MASSES. The disappearance of a front in what had been an active frontal system is often marked by OCCLUSION, as warm air is lifted clear of the surface and gradually absorbed into the surrounding air. It can also happen when two air masses remain stationary for a long time over a similar surface and both become modified by contact with that surface until they are both at the same temperature, pressure, and humidity. If the air masses are both moving they can also acquire similar characteristics if they spend a long time moving side by side and at the same speed over a similar surface, or if one travels behind the other at the same speed and along the same track.

frost A coating of ICE CRYSTALS that forms on solid surfaces. It is most often seen on objects close to ground level, such as plants, parked cars, and windows. Frost also forms on the surface of fallen snow and in air pockets in the snow. HOAR FROST is the most common type of frost. It is caused by the direct DEPOSITION of water vapor as ice when the DEW POINT TEMPERATURE is below freezing. DEW may also freeze if the temperature falls after it has been deposited. Frozen or white dew is usually hard and transparent and has rounded surfaces produced by the shape of the original dewdrops. Under different circumstances frozen dew may form FERN FROST. When the air is very dry its temperature may fall low enough to freeze the fluid inside plant tissues but without the formation of ice on external surfaces. This type of frost is known as BLACK FROST. Frost formation reduces further cooling of the bodies it covers. The layer of ice greatly restricts the radiation of heat from the underlying surface, and the deposition of ice or freezing of dew releases LATENT HEAT, warming the air immediately adjacent to the surface. *See also* AIR FROST and GROUND FROST.

frostbite *See* EMERGENCY.

frost climate *See* SEMIARID CLIMATE.

frost day A day on which frost occurs.

frost-freeze warning A warning that is issued by the NATIONAL WEATHER SERVICE and broadcast on radio, television, and WEATHER RADIO to alert people in an area where cold weather is not expected that the temperature is expected to fall below freezing. Some plants may be at risk and should be protected. People whose homes lack central heating should check that their heaters are working and that they have adequate supplies of blankets and warm clothes.

frost hazard The risk that growing plants will be damaged by frost. This can be expressed in several ways. It can be given as the probability (*see* PROBABILITY FORECAST) that a KILLING FROST will occur at a specified place on a particular date during the growing season. Alternatively, it can be expressed as the frequency with which killing frosts have occurred during the growing season in previous years. Or it can be a series of the dates on which the last spring and first autumn frosts have occurred over a number of years.

frost hollow (frost pocket) A sheltered, low-lying area, usually small in extent, that experiences freezing temperatures more frequently than the surrounding area. Frost hollows are found in hilly regions. At night, gentle KATABATIC WINDS carry cold air down the hillsides to the hollows, where the air accumulates. The process is called ponding. The upper slopes are kept relatively warm, because the cold air that flows away from them is constantly replaced by warmer air. By dawn there is often a sharp temperature difference, sometimes amounting to tens of degrees, between air in the hollow and air near the tops of the surrounding hills. The effect is most severe in hollows that are shaded from the late afternoon Sun, so they begin to cool during the afternoon.

frostless zone The part of a hillside that remains free of frost on nights when frost forms in the valley. Cold, dense air flows downhill by gravity as a KATABATIC WIND and accumulates (ponds) in the hollow at the base of the hill. The cold air is replaced at a higher level by air that is warmer as a result of mixing with air above ground level, and it is this that keeps the hillside relatively warm. This produces a valley INVERSION with a surface that is contoured in the same way as the ground surface. The air flowing down the slope is not always colder than the air at the bottom, especially if TURBULENT FLOW has caused mixing with air from above the surface or there has been ADIABATIC warming. In that case the warmer air rests above the cold air, reinforcing the inversion. The frostless zone lies between the pool of cold air at the base of the hill and the exposed top of the hill.

frost pocket *See* FROST HOLLOW.

frost point The temperature at which water vapor turns directly into ice. As a result, SATURATION does not occur until the air is cooled to below freezing.

frost smoke A type of STEAM FOG that forms when the temperature is well below freezing. Its particles are ICE CRYSTALS rather than water droplets.

frost table *See* TJAELE.

frost tolerance *See* BLACK FROST.

frozen fog FOG that is a low CLOUD consisting of ICE CRYSTALS. It forms when supercooled (*see* SUPERCOOLING) water droplets freeze. The crystals then grow by the BERGERON–FINDEISEN MECHANISM and soon become heavy enough to fall. Consequently, frozen fog usually clears quickly. Supercooled fog over airfields is sometimes cleared by seeding it with DRY ICE to accelerate this process.

frozen precipitation PRECIPITATION of any kind that reaches the ground in the form of ice, including FREEZING DRIZZLE, FREEZING FOG, FREEZING RAIN, FROZEN FOG, FROST SMOKE, GRAUPEL, HAIL, SLEET, and SNOW.

Fujita Tornado Intensity Scale In 1945, the atomic bomb dropped on Hiroshima, Japan, caused fierce FIRE STORMS that generated several TORNADOES. Tornadoes also developed in the fire storms that followed heavy bombing in several other cities—not only in Japan—and, in 1923, tornadoes in Tokyo were associated with the fire storms among the mainly wooden buildings that followed an earthquake. The Hiroshima tornadoes caught the attention of a young student, Tetsuya Fujita (1920–98), who embarked on what became a lifelong study of them.

Fujita later moved to the United States, and, in 1968, he took the middle name *Theodore.* For many years he was professor of meteorology at the University of Chicago and he was recognized as one of the leading world authorities on tornadoes.

He and his colleagues found there was a need to classify tornadoes. Ordinary winds were classified by the BEAUFORT WIND SCALE, in which the strongest wind was of hurricane force. The SAFFIR/SIMPSON SCALE expanded this to include hurricanes, but tornadoes were even stronger. In collaboration with Allen Pearson, formerly the chief tornado forecaster for the National Weather Service, in 1971 Fujita devised a six-point tornado scale, from F-0 to F-5. He also allowed the possibility of an F-6 tornado but believed its effects would be indistinguishable from the total destruction caused by an F-5 tornado. Even in a hurricane, it is usually possible to measure the wind speed, but, in 1971, there was no instrument capable of withstanding the wind inside a major tornado and so the wind speed could not be measured. Also, the short lifetime of most tornadoes made measurement difficult. Instead, experiments revealed the type of damage winds of different speeds would cause, and the speed of the wind in a tornado was then calculated from the type and extent of the damage it had caused.

The scale groups tornadoes as weak, strong, and violent, with two categories in each group. A weak tornado rated as F-0 may rip branches from trees and loose tiles from roofs. An F-1 tornado can knock down trees and break windows.

Strong tornadoes cause more serious damage. At F-2, full-grown trees may be torn from the ground and mobile homes demolished. An F-3 tornado can flatten entire stands of trees, demolish some walls, and overturn cars.

The most extreme tornadoes are classed as violent. An F-4 tornado can reduce a building to a pile of rubble. One rated at F-5 reduces buildings to rubble and then scatters the rubble over a wide area. It severely damages steel-framed buildings and can pick up cars and carry them some distance before dropping them. These extreme events can also produce freakish effects. There have been accounts of houses lifted from the ground and then set down again—in one case a house was carried for 2 miles (3.2 km)—and a roof was blown 12 miles (19 km). In 1958, at El Dorado, Texas, a woman survived being blown through the window of her home and carried 60 feet (18 m).

In the United States, 69 percent of all tornadoes are weak, 29 percent are strong, and 2 percent are violent. On average, only one F-5 tornado strikes the United States each year. In April 1998, an F-5 tornado killed 33 people in Alabama.

Rating	Wind speed		Damage
Weak	mph	km h^{-1}	
F-0	40–72	64–116	Slight
F-1	73–112	117–180	Moderate
Strong			
F-2	113–157	182–253	Considerable
F-3	158–206	254–331	Severe
Violent			
F-4	207–260	333–418	Devastating
F-5	261–318	420–512	Incredible

Fujiwara effect A phenomenon that occurs on average once every year and that was first described in 1921 by the Japanese meteorologist Sakuhei Fujiwara. If two TYPHOONS of approximately similar size approach to within about 900 miles (1,450 km) of each other they begin to interact. They start to turn about a point that lies about halfway between them. If one storm is much bigger than the other, they turn about a point that is closer to the larger storm. The big storm then absorbs the smaller one.

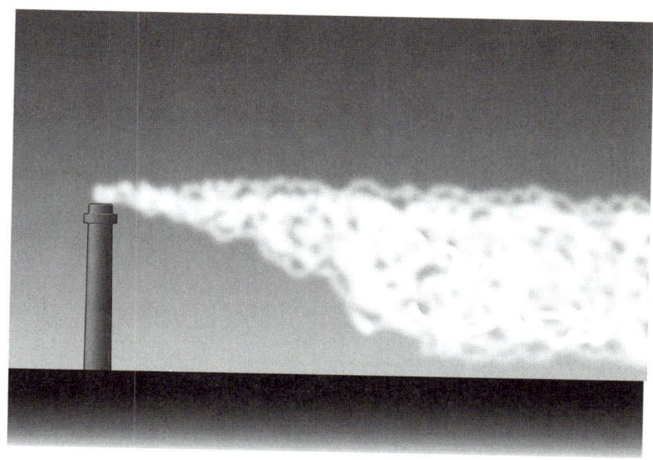

Fumigating. **The plume of gases and particles widens and sinks to ground level, causing serious pollution.**

fume A mass of solid particles, less than 0.00004 inch (1 μm) in diameter, that are suspended in the air and that result from the CONDENSATION of vapors, DEPOSITION, or chemical reactions. Fumes often contain metals or metallic compounds that may be harmful to health, and inhalation of the particles themselves may cause respiratory ailments.

fumigating One of the patterns a CHIMNEY PLUME may make as it moves away downwind. The plume widens and sinks with increasing distance from the smokestack. This carries the gases and particles to ground level, where they pollute the air. Fumigating occurs when the ENVIRONMENTAL LAPSE RATE is greater than the DRY ADIABATIC LAPSE RATE in the layer of air extending from the surface to the height of the smokestack and there is a strong INVERSION in the air above the stack.

fumulus A CLOUD LAYER that is so thin and tenuous as to be barely visible.

funnel cloud A CLOUD shaped like a funnel that develops inside a MESOCYCLONE and then descends through the base of a CUMULONIMBUS cloud. It hangs from the parent cloud, usually snaking erratically, and if it touches the ground it becomes a TORNADO. Air in the funnel rotates, almost always in a CYCLONIC direction (but there are rare exceptions). Its energy is derived mainly from the LATENT HEAT of condensation,

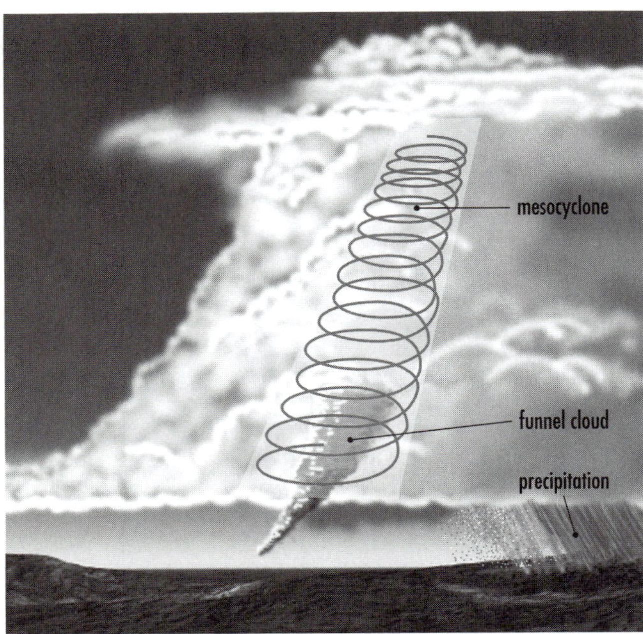

Funnel cloud. **The cloud begins to spin near the center of the mesocyclone, then extends downward until it protrudes beneath the cloud base.**

Funneling. **The sides of the valley, or buildings lining a street, constrain the air flowing parallel to the valley or street, causing it to pass through a smaller passageway in the same amount of time.**

and so it needs a constant supply of moist air to sustain it. The cloud is visible as a result of the condensed water droplets it contains. A funnel cloud is gray, and although most funnel clouds are wider at the top than at the bottom, some are wider at the base. Their width and length vary greatly.

funneling An acceleration of the wind that occurs when it is forced through a narrow passage. A funneling effect is felt when the wind direction is approximately parallel to the axis of a deep, narrow valley or a street lined by tall buildings. Although the wind is retarded to some extent by FRICTION, the fact that the rate at which air leaves the valley or street must be equal to the rate at which it enters means the flow must accelerate as it passes the constraint. At the same time, the air pressure decreases in proportion to the acceleration, as a result of the BERNOULLI EFFECT.

furiani A strong southwesterly wind that blows in the region near the Po River, Italy. It lasts for only a short time and is followed by a southerly or southeasterly gale.

furious fifties *See* POLAR WET CLIMATE.

fynbos *See* CHAPARRAL.

G

Gabriel *See* BENEDICT.

Gaia hypothesis The idea that on Earth, and by extension on any planet that supports life, the living organisms maintain conditions that are broadly favorable to them. In its weaker interpretation the hypothesis proposes that the totality of living organisms, called the *biota,* actively participates in the cycling of nutrient elements and in the regulation of climate and the salinity of seawater. The strong interpretation holds that Earth itself is a single living organism, which maintains a constant environment suitable to it.

JAMES LOVELOCK is the principal author of the hypothesis. In his first book on the subject (*Gaia: A New Look at Life on Earth* [Oxford: Oxford University Press, 1979]), he described Gaia as "a complex entity involving the Earth's biosphere, atmosphere, oceans, and soil; the totality constituting a feedback or cybernetic system which seeks an optimal physical and chemical environment for life on this planet."

Development of the hypothesis began when Lovelock was working as a consultant to the National Aeronautics and Space Administration (NASA) at the Jet Propulsion Laboratory at Pasadena, California. The Viking program was being prepared. This would place two landers on Mars, partly with the purpose of searching for life. Lovelock and his colleagues, including Dian Hitchcock, a philosopher employed to assess the logical consistency of planned experiments, discussed how it might be possible to discover whether a planet supports life, probably based on organisms utterly different from those on Earth. They reasoned that any living organism would need to absorb some materials and excrete others. This would produce chemical changes in its environment that should be detectable as a disequilibrium in the composition of the atmosphere. Later, Lovelock explored the idea further in collaboration with Lynn Margulis. The name *Gaia* was suggested later by Lovelock's friend and neighbor the novelist William Golding. Gaia (or Ge) represents the Earth in Greek mythology.

The chemical disequilibrium of Earth's atmosphere becomes evident when its composition is compared with those of Mars and Venus, both of which consist predominantly of carbon dioxide and are in chemical equilibrium. This means that no chemical reactions among its component gases are possible under the physical conditions that obtain. Earth's atmosphere is very different. It contains both methane (CH_4) and oxygen (O_2), for example, which react together to yield carbon dioxide (CO_2) and water (H_2O). Clearly, something must be constantly replenishing the CH_4 and the only reactions capable of this at the temperatures and pressures on Earth take place in bacterial cells.

Similarly, the atmosphere is predominantly (about 79 percent) nitrogen, yet it also contains oxygen, and in the world as a whole there are about 100 LIGHTNING flashes every second, or more than 8 million every day. Lightning supplies the energy needed to break the bonds holding nitrogen and oxygen atoms together in their molecules in the air close to the flash. Nitrogen (N) atoms then react with oxygen

(O) atoms to produce nitric oxide (NO) and then nitrate (NO_3). Nitrate is soluble, and so it is washed to the surface by rain. After some millions of years all the nitrogen should have been removed from the air. Something—in fact, denitrifying bacteria—is constantly returning it.

At one time Earth's atmosphere contained very much more CO_2 than the approximately 0.036 percent it contains today. There is no mystery about where the gas went. It is present as carbonate (CO_3) in limestone and chalk rocks. These are among the most abundant of surface rocks and represent a huge store of what was once atmospheric carbon. There is a set of inorganic chemical reactions that convert CO_2 to CO_3, but these proceed too slowly to account for the quantity of carbonate rocks that have been deposited over the time the planet has existed. Biological reactions are much faster. These involve combining CO_2 with calcium (Ca) to produce calcium carbonate ($CaCO_3$). This is insoluble in shallow water and many marine organisms use it to construct their shells. When the organisms die the shells accumulate as sediment on the seafloor and are eventually heated and compressed to form rock. This is the mechanism by which atmospheric carbon is removed and "buried."

CO_2 is a GREENHOUSE GAS, and removing it from the air has a climatic cooling effect. During the period since biological carbon burial began the Sun has grown about 30 percent hotter, yet the surface temperature on Earth has never varied far from its present average of 15° C (59° F). According to the Gaia hypothesis, this is one way in which living organisms have maintained a constant climate.

DIMETHYL SULFIDE, emitted by unicellular marine organisms, is the principal source of CLOUD CONDENSATION NUCLEI over the open ocean. Cloud formation helps to regulate the sea-surface temperature and so this is another example of a biological influence on climate.

The Gaia hypothesis has become widely known, but it has always been scientifically controversial. It is difficult to see how it can be reconciled with evolutionary biology, for example. Nevertheless, it has influenced scientific thinking on a number of practical issues connected with the biological response to environmental change. One of these has led to bioremediation, which is the use of biological organisms to clean up environmental pollutants. The organisms are usually bacteria that may have been genetically modified for the purpose.

(You can learn more about Gaia from James E. Lovelock, *Gaia: A New Look at Life on Earth,* [Oxford: Oxford University Press, 1979] and *The Ages of Gaia* [Oxford: Oxford University Press, 1989]; Michael Allaby, *A Guide to Gaia* [New York: E. P. Dutton, 1989]; Lawrence E. Joseph, *Gaia: The Growth of an Idea* [New York: St. Martin's Press, 1990] and Tyler Volk, *Gaia's Body: Towards a Physiology of Earth* [New York: Copernicus, 1998].)

gale A strong wind, ranging from one that exerts strong pressure on people walking into it to one that breaks and uproots trees. On the BEAUFORT WIND SCALE there are four categories of gale: moderate gale (force 7), fresh gale (force 8), strong gale (force 9), and whole gale (force 10). A wind stronger than a whole gale is called a *storm* and one weaker than a moderate gale is a *strong breeze.* The wind speeds of the four gales are moderate, 32–38 mph (51.4–61.1 km h^{-1}); fresh, 39–46 mph (62.7–74 km h^{-1}); strong, 47–54 mph (75.6–86.8 km h^{-1}); and whole, 55–63 mph (88.4–101.3 km h^{-1}). In the earliest version of the Beaufort wind scale, all the wind forces, including gales, were described in terms of their effect on sailing ships and speeds were not allotted to them.

gale warning A notification to shipping that winds of GALE force are expected imminently in designated sea areas. The warnings are prepared by national meteorological services and broadcast from coastal radio stations and attached to routine weather bulletins for shipping. A gale warning is issued when a fresh gale or wind gusts of 43–51 knots (49.5–58.6 mph; 80–94 km h^{-1}) are expected in part of a sea area, but not necessarily the whole of it. A typical gale warning for the seas around Great Britain might be, Gale warning Wednesday 1st March, 0150 GMT. Rockall, Malin, Hebrides, Bailey. Southwesterly gale force 8 imminent. *GMT* refers to GREENWICH MEAN TIME. Rockall, Malin, Hebrides and Bailey are the names of areas of sea around the British isles. Force 8 is a measure on the BEAUFORT WIND SCALE that refers to a fresh gale with a wind speed of 39–46 mph (62.7–74 km h^{-1}).

Galilei, Galileo (1564–1642) Italian *Physicist and astronomer* One of the most famous scientists who ever lived, Galileo is usually identified by his given name rather than by his family name, Galilei. His father was Vincenzio Galilei (c. 1520–91), a musician and mathematician. Galileo was born at Pisa on February 15, 1564.

He received his first lessons from a private tutor. Then, in 1574, the family moved to Vallombrosa, near Florence, and Galileo continued his education at a monastery there. In 1581 he enrolled to study medicine at the University of Pisa, but the family could not afford the expense, and in 1585 he returned home without having taken a degree.

While he was at the university his interest had turned toward mathematics and physics, and he had started to study these subjects by the time he had to leave. A popular story has it that Galileo once watched a lamp that was swinging in the cathedral at Pisa and noticed that no matter how large the range of its swing, the lamp always took the same amount of time to complete an oscillation. He timed the swings by counting his pulse. Later in life he confirmed this observation experimentally and suggested that the principle of the pendulum might be applied to the regulation of clocks.

After his return to Florence he obtained a post as a lecturer in mathematics and science at the Florentine Academy, at the same time continuing his studies of Euclid (c. 300 B.C.E) and Archimedes (287–212 B.C.E.). In 1586 he invented an improved version of a balance first devised by Archimedes that was used to measure specific gravity. At about this time his father was measuring the lengths and tensions in the strings of musical instruments that produce specific intervals between notes, and this may have helped convince Galileo that mathematical descriptions of phenomena could be tested by experiment.

Galileo became professor of mathematics at the University of Pisa in 1589. The appointment was an honor for him, but it was poorly paid and in 1592 he applied for and was awarded the better paid post of professor of mathematics at the University of Padua. He remained at Padua for the next 18 years and it is there that he did most of his best work.

As well as his experiments with gravity and motion and his astronomical observations and calculations, Galileo maintained the interest in the behavior of fluids that had begun with his studies of the work of Archimedes. In 1593 he invented the first THERMOMETER—called a *thermoscope*. It consisted of a bulb filled with air that was connected to a vertical tube containing a column of water. As the temperature rose and fell, the air in the bulb expanded and contracted, pushing the water up the tube or allowing it to fall. Unfortunately, the thermoscope was highly inaccurate, because no account was taken of changes in atmospheric pressure. Nevertheless, this was one of the earliest attempts to make an instrument for taking scientific measurements. Toward the end of his life, Galileo became interested in discovering whether air is a physical substance having mass. A young assistant, EVANGELISTA TORRICELLI, set to work on the problem, and the experimental apparatus he devised was the first BAROMETER.

Galileo Galilei. **The Italian physicist who invented the first thermometer.** ***(John Frederick Lewis Collection, Print and Picture Collection, The Free Library of Philadelphia)***

Galileo had little time for those who disagreed with his observations or the arguments he based on them. He was combative and could be sarcastic, but with good reason. He was convinced that natural phenomena can be described mathematically and that observation and experiment can then be used to validate the mathematical description. He was establishing what is now accepted as the basis of scientific procedure, but to do so he had to overthrow the prevailing verbal and nonmathematical approach that was derived from the work of ARISTOTLE.

His fame rests on three achievements. He was the first person to use a telescope to study the night sky. His observations provided evidence with which he supported the conclusion of Nicolaus Copernicus (1473–1543) that it is the Earth that orbits the Sun and not the reverse and therefore the Sun, not the Earth, is at the center of the universe. His studies of motion and gravitation outlined the principles Isaac Newton (1642–1727) later formalized as the laws of motion. His final achievement, and perhaps the most important, was his application of mathematics to the study of natural phenomena.

His support for the ideas of Copernicus led Galileo into conflict with the church, and in 1633 he was found guilty of heresy and sentenced to remain for the rest of his life in his villa at Arcetri, near Florence. He continued to study, experiment, and write, summarizing his early experiments and his thoughts on mechanics in *Discorsi e Dimonstrazione Matematiche Intorno a Due Nuove Scienze* (Discourses and mathematical discoveries concerning two new sciences), a book that was smuggled out of Italy and published in Leiden, the Netherlands, in 1638.

He became blind in 1637, but even this did not stop him working. He finally designed a pendulum-driven clock that was built in 1656 by Christian Huygens (1629–95) and directed the work of his assistants. He was still dictating to them when he fell ill with a fever toward the end of 1641. Galileo died at Arcetri on January 8, 1642.

gallego A cold northerly wind that blows across Spain and Portugal.

Galton, Francis (1822–1911) English *Scientist, inventor, and explorer* Sir Francis Galton was the first scientist to plot meteorological data onto a map and to attempt to produce a SYNOPTIC CHART showing the weather conditions over a large area. He played a large part in preparing the daily weather charts that were published by the *Times* of London from data supplied by the METEOROLOGICAL OFFICE. This was one of several contributions he made to the scientific study of weather.

He was born into a Quaker family on February 16, 1822, near Sparkbrook, now a suburb of Birmingham, England, and was the youngest of the nine children of a wealthy banker. He was also a first cousin of Charles Darwin (1809–82). Francis was able to read before he was three years old, and by the time he was four he was studying Latin.

In response to his father's wishes he studied medicine at Birmingham General Hospital and then at King's College, London, but interrupted his medical studies to study mathematics at Trinity College, Cambridge. After that he resumed studying medicine at St. George's Hospital, London, but he never completed the course. His father died and Francis inherited a fortune, so he left the hospital and for the rest of his life he pursued whatever topic interested him.

First he traveled through the Balkans, Egypt, Sudan, and the countries at the eastern end of the Mediterranean. He spent 1850 and 1851 on a journey of 1,700 miles (2,735 km) through southwestern Africa, after which he visited Spain. His observations in what was then a little known part of Africa led to the award of the Gold Medal of the Royal Geographical Society in 1854, and in 1856 he was elected a Fellow of the Royal Society.

His travels ended, in the 1860s Galton began to study weather. In particular, he wondered whether it might be possible to detect large-scale patterns in weather and, from these, to make forecasts. He circulated a detailed questionnaire to weather stations in different parts of the British Isles, asking for information about the weather conditions that had prevailed through the month of December 1861. When the replies arrived he plotted them on a map, using symbols he invented for the purpose. In 1862 he finally succeeded in compiling a detailed weather map. This showed a previously unsuspected relationship between atmospheric pressure and the speed and direction of wind.

Galton was familiar with the work of MATTHEW MAURY and Admiral ROBERT FITZROY. He also knew that the French astronomer Urbain Jean Joseph Leverrier (1811–77) issued daily weather charts of the North Atlantic, based on observations from ships and coastal stations, although these were so unevenly distributed that the charts included a great deal of guesswork. Maury, FitzRoy, and Leverrier had established the CYCLONIC circulation of air around a center of low pressure. Galton's questionnaire revealed its opposite: ANTICYCLONIC circulation around a center of high pressure. Galton coined the term ANTICYCLONE in a paper he submitted to the Royal Society. He published the results of his research in 1863 in the monograph *Meteorographica* and summarized them much later in his book *Memories of My Life,* published in 1908.

As well as preparing weather charts for publication, in the *Times* and in *Meteorographica,* Galton helped find a way to print them by using movable type. He had typefaces designed for the purpose and modified a drawing instrument called a *pantograph* so it used a drill to score curves and arrows in a soft material that could be used to make casts for printing.

Meteorologists were beginning to study the upper air by means of small balloons and kites to which instruments were attached. Galton had the idea of measuring the speed and direction of the wind at a specified location and time by means of the smoke emitted by an exploding shell. Galton was closely associated with the METEOROLOGICAL OFFICE and the shell was designed and fired experimentally under their auspices. The experiment was carried out over an area of the Irish coast where no ships might be damaged by falling debris and it was very successful. The shells exploded consistently at 9,000 feet (2,745 m), releasing a cloud of smoke Galton was easily able to track. On the suggestion of FitzRoy, Galton also invented the WIND ROSE.

In addition to his contributions to meteorology, Francis Galton was the first person to demonstrate the uniqueness of fingerprints and worked out a partial system for identifying them. He invented a teletype printer and the ultrasonic dog whistle and devised new techniques for statistical analysis, as well as a word-association test that was adopted by Sigmund Freud (1856–1939).

His main interest, stimulated by his cousin's book *On the Origin of Species by Means of Natural Selection,* lay in measuring the mental abilities of people and determining the extent to which these were inherited. From this he hoped it might be possible to improve them by means of selective breeding. This project was called *eugenics* and although it is now discredited Galton's contribution to it was of great importance to the scientific study of psychology.

Francis Galton was knighted in 1909. He died at Haslemere, Surrey, on January 17, 1911.

(You can learn more about Francis Galton from "Francis Galton: An Exploration in Intellectual Biography and History" at www.cimm.jcu.edu.au/hist/stats/galton/index.htm.)

Galveston A port and vacation resort in Texas that in 1900 suffered the worst hurricane disaster in American history. The TROPICAL CYCLONE formed in the Caribbean on August 27 and reached Galveston on September 8. The town of Galveston lies on Galveston Island, a barrier island that is nowhere more than 3 miles (5 km) wide and is an average 4.5 feet (1.4 m) above sea level. In 1900 it was a prosperous port with a population of nearly 40,000. Although the Weather Bureau had warned of the approach of the storm, most people ignored the warning. On the morning of September 8 the weather deteriorated rapidly and the sea level rose. By noon all the bridges linking Galveston to the mainland were flooded and impassable. Waves smashed buildings near the shore and the wind, 77 mph (124 km h^{-1}) gusting to 120 mph (193 km h^{-1}), destroyed those farther inland. The city was flooded to a depth of 4 feet (1.2 m). The hurricane began to move away from Galveston at about 10 P.M. and the wind abated. The following morning the damage became apparent. The city was largely reduced to rubble and smashed wood. About 6,000 people had been killed and 5,000 injured. More than 2,600 homes had been destroyed and about 10,000 people were homeless. The storm finally dissipated on September 15. Galveston is now protected by a strong sea wall that withstood an even stronger storm in September 1961.

(You can read an account of the Galveston disaster, and other hurricanes, in Michael Allaby *Dangerous Weather: Hurricanes* [New York: Facts On File, 1997].)

gamma rays Electromagnetic radiation that has a wavelength of 10^{-8} µm to 10^{-4} µm. Gamma radiation, often written with the Greek letter γ (gamma), has a shorter wavelength than X RAYS and therefore possesses more energy. Less than 1 percent of the radiation emitted by the Sun is at gamma wavelengths, and all the solar gamma radiation reaching the Earth is absorbed in the upper atmosphere. None reaches the surface.

GARP *See* GLOBAL ATMOSPHERIC RESEARCH PROGRAM.

garrigue *See* CHAPARRAL.

garúa (camanchaca, Peruvian dew) A wet mist or very fine drizzle that falls on the lower slopes of the Andes in Peru in winter. It sometimes lasts for weeks. On the low-lying coastal fringe the mist is clear of the ground and forms a low blanket of gray STRATUS cloud. The garúa carries cold, dismal weather, but it also provides some moisture for vegetation in the otherwise arid climate. When Charles Darwin (1809–82) visited the region in 1835, during his voyage on HMS *Beagle,* he recorded in his *Journal* that "a dull heavy bank of clouds constantly hung over the land, so that during the first sixteen days I had only one view of the Cordillera behind Lima. It is almost become a proverb that rain never falls in the lower part of Peru. Yet this can hardly be considered correct; for during almost every day of our visit there was a thick drizzling mist, which was sufficient to make the streets muddy and one's clothes damp; this the people are pleased to call Peruvian dew."

gas constant A value that is used when the GAS LAWS are combined into the EQUATION OF STATE. The universal gas constant applies to 1 MOLE of an IDEAL GAS and has a value of 8.314 J K^{-1} mol^{-1}. The specific gas constant (R) varies from one gas to another and has a value of $R = 10^3\ R*/M$, where $R*$ is the universal gas constant and M is the molecular weight of the gas (the value must be multiplied by 1,000 because moles are defined in grams and the unit of mass is the kilogram). Air is a mixture of gases. Constants for each can be added together, giving a specific gas constant for air with a relative molecular mass of 29.0 of $R = 287$ J kg^{-1} K^{-1} (17.3 cal lb^{-1} $°F^{-1}$). This is true for dry air, but only approximately true for moist air, because the presence of water vapor reduces the density of moist air to about 0.5 percent less than that of dry air at the same temperature and pressure. The difference becomes important in precise calculations of conditions inside clouds. For these, the use of the VIRTUAL TEMPERATURE gives moist air the same gas constant as dry air.

gas laws The physical laws by which the temperature, pressure, and volume of an IDEAL GAS are related. These were discovered by ROBERT BOYLE, EDMÉ MARIOTTE, JACQUES ALEXANDRE CÉSAR CHARLES, and JOSEPH GAY-LUSSAC.

In 1662, Boyle found that gases can be compressed. When he applied pressure to gas, by pouring mercury into the long, open end of a J-shaped tube that was sealed at the other, short end, he observed that the volume of the gas decreased. He found the volume was inversely proportional to the amount of pressure to which it was subjected—the greater the pressure, the smaller the volume. From this he calculated that pV = a constant, where p is the pressure and V the volume. In English-speaking countries this is known as *Boyle's law.*

About 15 years later, Mariotte independently reached the same conclusion, but with an important addition. Mariotte noticed that when a gas is heated, it expands, and it contracts when it is cooled. It follows, therefore, that pV = a constant is true only if the temperature remains constant. This improved version is known in French-speaking countries as *Mariotte's law,* and because his version is more complete there is a case for adopting the name in English-speaking countries as well.

In about 1787, Jacques Charles repeated experiments performed by GUILLAUME AMONTONS. Amontons had found that the volume of a gas changes with its temperature, provided the pressure under which it is held remains constant (allowing it to expand). Charles discovered the amount by which its volume changed with a given change in temperature. This is known as *Charles's law,* and is written as $V \div T$ = a constant, where T is the temperature. Charles did not publish it, however, and it was discovered again, independently, by Joseph Gay-Lussac, who did. For this reason it is sometimes known as *Gay-Lussac's law.*

Charles found that for every degree the temperature rises the volume of a gas increases by $1 \div 273$ of its volume at 0° C, and for every degree the temperature falls the volume decreases by the same amount. At a temperature of –273° C, therefore, the volume should reach zero and no lower temperature can exist (*see* ABSOLUTE ZERO).

A third law, known as the *pressure law,* is derived from the first two laws. It states that the pressure within a gas is directly proportional to its temperature provided the volume remains constant.

The three laws can be combined into a single equation of state, pV= a constant, from which a universal gas equation can be derived. This is $pV = nR*T$, where n is the amount of gas in moles and $R*$ is the gas constant (8.31434 J K^{-1} mol^{-1}). It is difficult to consider a volume of air in the atmosphere, because the air is not confined, but volume is equal to the mass (m) of the gas multiplied by its density (ρ). Also, the specific gas constant for air (R, *see* EQUATION OF STATE) can be substituted for the universal gas constant $R*$. The law can then be expressed as $p = \rho RT$.

Real gases obey the gas laws to only a limited extent. They come closest to obeying them at low pressures and high temperatures.

Gaussian distribution (**normal distribution**) In statistics, the way values of a variable quantity appear on a graph when there is an equal area beneath the graph curve to each side of the MEAN. The curve is shaped like a bell and its maximum height marks the mean. It is called ***normal*** because this is the distribution that occurs in the absence of any factor forcing it in one direction or the other. Pollutant particles inside a PLUME RISE are distributed in this way, for example, so the mathematics of Gaussian distribution can be used to calculate their position. The German mathematician, physicist, and astronomer Karl Friedrich Gauss (1777–1855) is usually credited with having devised the equations describing this distribution, but it is possible that they were discovered earlier by the French mathematician Abraham de Moivre (1667–1754).

Gaussian year *See* ORBIT PERIOD.

Gavin A CYCLONE that struck Fiji in March 1997. It killed at least 26 people, 10 of whom died when their fishing trawler sank.

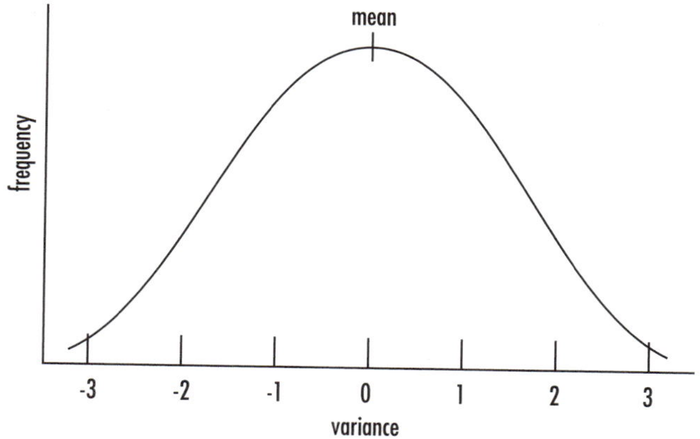

Gaussian distribution. **When the values of a variable quantity are plotted against the frequency of their occurrence, a Gaussia n distribution appears as a bell-shaped curve in which the value of the variables smaller than the mean is equal to that greater than the mean.**

Gay A TYPHOON that struck Thailand on November 4 and 5, 1989. It killed 365 people and destroyed or damaged 30,000 homes.

Gay-Lussac, Joseph Louis (1778–1850) French *Chemist and physicist* Joseph Gay-Lussac was born on December 6, 1778, at St. Léonard, Haute Vienne, in central France. His father was the judge Antoine Gay, who added *Lussac* to the name to prevent confusion with other families called *Gay*. Lussac was the name of an estate near St. Léonard. Antoine was arrested in 1793 for showing sympathy to the aristocrats.

Joseph began his education locally, and in 1797 he enrolled at the École Polytechnique; in 1799 he transferred to the École des Ponts et Chaussées (school of bridges and roads). He graduated in 1800. His interest was in engineering, but while still a student he began to assist the distinguished French chemist Claude Louis Berthollet (1748–1822), working at Berthollet's home at Arcueil, near Paris. Berthollet was very famous, and his home was a meeting place for many of the leading scientists of the day. For a time, Gay-Lussac worked alongside Berthollet's son in a factory where linen was bleached.

In 1802 Gay-Lussac was appointed a demonstrator in chemistry at the École Polytechnique. He spent 1805 and 1806 on an expedition to measure terrestrial magnetism led by ALEXANDER VON HUMBOLDT. In 1808 he married Geneviève Rojet, and on January 1, 1810, he became professor of chemistry at the École Polytechnique. He was also professor of physics in Paris at the Sorbonne University, a position he held from 1808 until 1832, when he resigned to take up the post of professor of chemistry at the Musée National d'Histoire Naturelle, in the Jardin des Plantes. In 1806 he was made a member of the Academy of Sciences.

Gay-Lussac was also a politician. In 1831 he was elected to the chamber of deputies to represent his home *département* of Haute-Vienne, and in 1839 he was made a peer and entered the chamber of peers, which was then the upper house of the French parliament.

He published his first research results in 1802. In collaboration with his friend Louis Jacques Thénard (1777–1857), Gay-Lussac had formulated a law stating that when the temperature is increased by a given amount, all gases expand by the same fraction of their volume. JACQUES ALEXANDRE CÉSAR CHARLES had discovered this law in 1787 and it is usually known as *CHARLES'S LAW*, but Charles had not published it. In 1804 Gay-Lussac and the physicist Jean Baptiste Biot (1774–1862) were commissioned by the Academy of Sciences to measure the Earth's magnetic field high above the surface. On August 24 they ascended by balloon from the garden of the Conservatoire des Arts and climbed to 13,120 feet (4,000 m). On September 16 Gay-Lussac made a solo ascent in which he reached a height of 23,012 feet (7,019 m) above sea level. This was higher than the tallest peak in the Alps, and it established an altitude record that stood for 50 years. Measurements made during the flights showed not only that the magnetic field remains constant with height, but that the chemical composition of the atmosphere also does. This discovery makes Gay-Lussac one of the founders of METEOROLOGY. The same year he read a paper describing research on a method of chemical analysis he had used in collaboration with von Humboldt. Using this method they had found (among a number of other discoveries) that the proportions of the volumes of hydrogen and oxygen in water were 2:1.

In 1809 Gay-Lussac published what may have been his most important discovery. He had found that when gases combine they do so in simple proportions by volume and that the products of their combination are related to the original volumes. This is known as *Gay-Lussac's law* and it is used in chemical equations. One of the examples Gay-Lussac used to illustrate it shows that when two molecules of carbon monoxide (CO) combine with one molecule of oxygen (O_2) the product is two molecules of carbon dioxide (CO_2):

$$2CO + O_2 \rightarrow 2CO_2$$

From about 1810, Gay-Lussac concentrated increasingly on pure chemistry. He made many important discoveries. These included improving the processes used to manufacture oxalic and sulfuric acids and devising ways to estimate the alkalinity of potash and soda and the amount of chlorine in bleaching powder. He developed volumetric analysis and in 1832 introduced a method for estimating the amount of silver in an alloy by using common salt.

His advice was constantly in demand and he held a number of official positions. In 1805 he was appointed to the consultative committee on arts and manufactures. In 1818 he was appointed to the department responsible for the manufacture of gunpowder, and in 1829 he became chief assayer to the Mint. Both of these positions were lucrative government appointments.

Joseph Gay-Lussac died in Paris on May 9, 1850.

Gay-Lussac's law *See* GAS LAWS and GAY-LUSSAC, JOSEPH LOUIS.

GCOS (Global Climate Observing System) *See* WORLD CLIMATE PROGRAM.

GEF *See* GLOBAL ENVIRONMENT FACILITY.

gegenschein A faint glow that is sometimes seen in the sky opposite the Sun as a circle or ellipse of light. The name is German for "reflection."

GEMS *See* GLOBAL ENVIRONMENTAL MONITORING SYSTEM.

gending A dry, southerly FÖHN WIND that blows across the northern plains of Java. The wind crosses the mountains near the south coast of the island and is funneled between the volcanoes.

general circulation The general circulation comprises all of the movements of the atmosphere by which heat is transported away from the equator and into higher latitudes, winds are generated, and CLOUDS and PRECIPITATION are produced. It includes all of the air motion that results in what we experience as weather.

People had always been interested in the weather and had sought explanations for meteorological phenomena. The phenomena occurred on a local scale, but general explanations could be applied to them. Whatever it is that causes a thunderstorm, gale, fog, or blizzard can cause these events anywhere. Explain one thunderstorm and you have explained all thunderstorms. Other events are not explained so easily. The monsoon, for example, is not the same as the many rainstorms it brings.

It was not until late in the 17th century that the idea of a global weather system began to develop and interest in it was triggered by the growth of world trade. Sailing ships carrying cargoes around the world encountered belts where winds were steady and reliable. The most reliable of these were the TRADE WINDS, and in attempting to explain the cause of the trade winds EDMUND HALLEY proposed a movement of air away from the equator at a high level and the return of air near the surface. GEORGE HADLEY improved on this explanation half a century later, but it was not until the 19th century that the American meteorologist WILLIAM FERREL completed the description. The result is known as the *THREE-CELL MODEL*.

The difference between the Halley and Hadley explanations centered on the eastward component of the trade winds, which Halley failed to account for and, as Ferrel discovered much later, Hadley accounted for incorrectly. Their general description was correct. Warm air rises over equatorial regions and moves away from the equator. Cold air subsides over the Poles. Cool air flows toward the equator.

In the early years of the 20th century VILHELM BJERKNES, TOR HAROLD PERCIVAL BERGERON, and their colleagues at the BERGEN GEOPHYSICAL INSTITUTE filled in many of the details of this very general outline. They introduced the concepts of the AIR MASS and the fronts that separate them (*see* FRONTOGENESIS).

The next major advance occurred with the launch of observation satellites that were able to view the Earth from above. They provided constant monitoring of cloud patterns, temperature, ocean waves, and changing vegetation over large areas and eventually over the entire planet. The introduction of powerful supercomputers made it possible to perform the millions of calculations needed to construct MODELS of the general circulation. It is these models that are used in estimating the climatic effects of various events, including the enhanced GREENHOUSE EFFECT.

Until recently, little was known about the STRATOSPHERE. Scientists now know that it has its own general circulation. This is quite distinct from the circulation in the TROPOSPHERE. Stratospheric air rises over the TROPICS, driven by rising tropospheric air, and subsides over the Poles. This produces variations in the height of the TROPOPAUSE, which is highest over the equator and lowest over the Poles, with sharp "steps" in its height at the subtropical front and POLAR FRONT. These are usually represented as being merged into a single step in each hemisphere. Within the stratosphere, air radiates warmth into space. This has a cooling effect. At the same time, however, the absorption of ULTRAVIOLET RADIATION by OZONE raises the temperature. Consequently, the temperature rises with increasing height, reaching a maximum over the Pole in the summer hemisphere and a minimum over the Pole in the winter hemisphere.

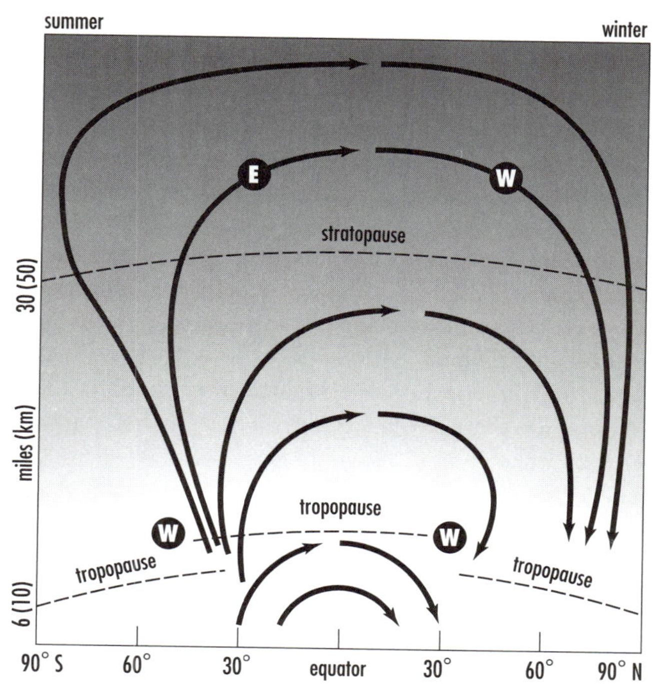

General circulation. **The general circulation in the stratosphere, showing the levels of the tropopause and stratopause. The letters *E* and *W* indicate the directions of the prevailing stratospheric winds in the middle latitudes of both hemispheres.**

general productivity model A climate MODEL that interprets climatic conditions to predict agricultural production throughout the major farming regions of the world. This makes it possible to estimate future commodity prices and, from that, to plan production in particular areas.

Geneva Convention on Long-Range Transboundary Air Pollution A legally binding international agreement that was drafted by scientists after a link had been established between sulfur emissions in Europe and the acidification of Scandinavian lakes (*see* ACID RAIN). The convention was drawn up under the auspices of the United Nations Economic Commission for Europe; was signed in Geneva, Switzerland, in 1979; and went into force in 1983. It lays down the general principles that form the basis for international cooperation to reduce the emission of air pollutants that drift across international frontiers and establishes an institutional framework for research and the development and implementation of policy. The Executive Body issues an annual report. Since it came into force, eight protocols have been added to the convention:

The 1984 Protocol on Long-term Financing of the Cooperative Programme for Monitoring and Evaluation of the Long-Range Transmission of Air Pollutants in Europe
The 1985 Protocol on the Reduction of Sulphur Emissions or their Transboundary Fluxes by at Least 30 Per Cent
The 1988 Protocol Concerning the Control of Nitrogen Oxides or Their Transboundary Fluxes
The 1991 Protocol Concerning the Control of Emissions of Volatile Organic Compounds or Their Transboundary Fluxes
The 1994 Protocol on Further Reduction of Sulphur Emissions
The 1998 Protocol on Heavy Metals
The 1998 Protocol on Persistent Organic Pollutants
The 1999 Protocol to Abate Acidification, Eutrophication, and Ground-level Ozone

(You can learn more about the convention and its protocols at www.unece.org/env/1rtap/env_eb1.htm.)

genitus The emergence of a new cloud from a MOTHER CLOUD, in which only one part of the mother cloud is affected by the change.

Genoa-type depression A type of DEPRESSION that is common in winter over the western Mediterranean, after a sudden drop in pressure that occurs around October 20 when the AZORES HIGH collapses. The sea-surface temperature in the Mediterranean is then about 2° C (3.6° F) higher than the mean air temperature, and when colder air crosses the sea it quickly becomes unstable. A Genoa-type depression develops over the Gulf of Genoa, in the LEE of the Alps and Pyrenees Mountains, in MARITIME POLAR AIR. Because the air in the WARM SECTOR is very unstable, it produces very intense precipitation along the WARM FRONT and heavy showers and THUNDERSTORMS to the rear of the COLD FRONT, often with CUMULIFORM clouds extending to a height of more than 20,000 feet (6,000 m). About 74 percent of winter depressions are of this type. About 9 percent are depressions that form over the Atlantic and about 17 percent are SAHARAN DEPRESSIONS.

gentle breeze In the BEAUFORT WIND SCALE, force 3, which is a wind that blows at 8–12 mph (13–19 km h^{-1}). In the original scale, devised for use at sea, a force 3 wind was defined as "or that in which a man-of-war with all sail set, and clean full would go in smooth water from". On land, a gentle breeze makes leaves rustle, small twigs move, and flags made from lightweight material stir gently.

geological time scale *See* Appendix III.

geometrical mean *See* MEAN.

geomorphology *See* CLIMATIC GEOMORPHOLOGY.

geopotential height A value for height above mean sea level that takes account of the increase in gravitational acceleration with latitude. The change in gravitational acceleration is due to the oblate shape of the Earth (the surface is closer to the center of the Earth at the Poles than it is at the equator) and to the changing balance between the gravitational force and the inertial force due to the Earth's rotation (*see* CENTRIPETAL ACCELERATION).

The gravitational force always acts directly toward the center of the Earth with a magnitude that varies according to the INVERSE SQUARE LAW. The inertial force acts away from the axis of the Earth's rotation with a magnitude that is directly proportional to the distance from that axis. Consequently, the inertial force is greatest at the equator and zero at the Poles. A component of the inertial force acts against the gravitational force, but its effect decreases with distance from the equator, and therefore the effect of the gravitational force increases with distance from the equator.

The gravitational potential, or geopotential, energy of a body is equal to the work that must be done to raise that body from mean sea level to a given height. Since this amount of work varies with the changes in the gravitational force, the geopotential energy can be used to designate height. The geopotential height is the height above mean sea level at which a body would have to be located in order to have the same geopotential if the gravitational acceleration were the same everywhere.

Geopotential heights are widely used by meteorologists, because particles that are free to move usually flow downhill, from regions of higher to lower potential energy. This means that if a PARCEL OF AIR is imagined as being located on a surface at a constant geometrical height (measured as vertical distance) it experiences a component of the gravitational force moving it toward the equator, where the geopotential is lower. If the location of the parcel is given as a surface at a constant geopotential height the forces acting on it are equal in all directions and so it remains stationary.

Geopotential height is measured in geopotential meters, which are units of energy per unit mass. One geopotential meter is equal to $1/9.8\ m^2\ s^{-2}$. If the values for gravitational acceleration are known, ordinary (geometrical) heights can be converted to geopotential heights without altering any of the equations that are used in calculating atmospheric behavior. This greatly simplifies the calculations.

Georges A HURRICANE, rated as category 3 on the SAFFIR/SIMPSON HURRICANE SCALE, that struck islands in the Caribbean and the Gulf Coast of the United States from September 21 to 28, 1998. It caused the deaths of about 250 people in the Dominican Republic, at least 27 in Haiti, about 20 in other islands, and 4 in the United States, where it caused damage in parts of Louisiana, Mississippi, Alabama, and Florida.

Geostationary Earth Radiation Budget (GERB) An instrument carried on MSG (*see* METEOSAT) satellites to observe and measure the radiation reflected and emitted by the Earth.

Geostationary Operational Environmental Satellite (GOES) A series of U.S. weather satellites that transmit data to the National Oceanic and Atmospheric Administration (NOAA) receiving station at Wallops, Virginia. Two GOES satellites are operational at any time, both in a GEOSTATIONARY ORBIT. *GOES-8* orbits at 75° W and observes eastern North America, the western Atlantic, and western South America. *GOES-10,* which replaced *GOES-9* at the end of its useful life in July 1998, orbits at 135° W and observes western North America, the eastern North Pacific as far as Hawaii, and the eastern South Pacific. The satellites transmit their data every half-hour.

(It is possible for amateurs to receive GOES data. Information on how to do so is at http://www.aa6g.org/weather/goes.html.)

Geostationary Operational Meteorological Satellite *See* ELEKTRO SATELLITE.

geostationary orbit (geosynchronous orbit, Clarke orbit) A satellite orbit at a height of about 22,370 miles (36,000 km), which is about 5.6 times the radius of the Earth, in which the satellite travels in the same direction as the Earth's rotation. At this height, the satellite completes a single orbit in precisely one SIDEREAL DAY; therefore, it remains permanently above the same point on the equator. Its field of view is almost an entire hemisphere. The satellite takes 20 minutes to complete a scan of its field of view and provides images with a resolution nearly as good as those from satellites in much lower POLAR ORBIT. The writer Sir Arthur C. Clarke first suggested the possibility of placing satel-

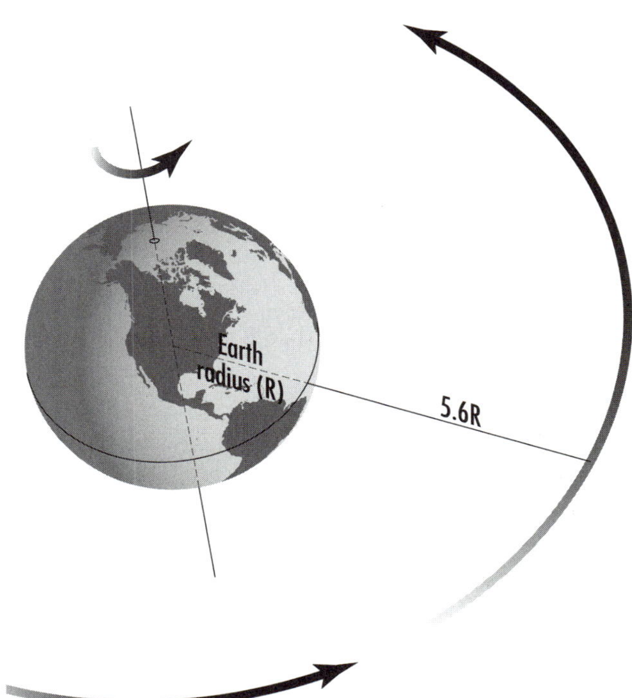

Geostationary orbit. **An orbit in which a satellite moves at the same rate as the Earth rotates, so it remains permanently above one point on the surface.**

lites into geostationary orbit and the orbit is sometimes named after him.

geostrophic balance The condition that exists in the wind above the BOUNDARY LAYER when the PRESSURE GRADIENT FORCE and CORIOLIS EFFECT are precisely equal. The wind then blows parallel to the ISOBARS.

geostrophic departure A difference between the wind speed that is observed and that of the GEOSTROPHIC WIND. Winds in the vicinity of the entrance region of the POLAR FRONT JET STREAM tend to be SUPERGEOSTROPHIC, because they are being accelerated in the direction of the wind. Accelerations that are at right angles to the wind direction also occur and cause geostrophic departures.

geostrophic equation The mathematical equation that is used to calculate the speed of the GEOSTROPHIC WIND:

$$V_g = (1/(2\Omega \sin \o\rho))(\delta p/\delta n)$$

where V_g is the geostrophic wind velocity, Ω is the ANGULAR VELOCITY of the Earth (= 15° h^{-1} = $2\pi/24$ rad h^{-1} = 7.29×10^{-5} rad s^{-1}), ø is the latitude, ρ is the air density, and $\delta p/\delta n$ is the horizontal PRESSURE GRADIENT. $2\Omega \sin ø$ is known as the CORIOLIS PARAMETER and sometimes designated by f, so the equation then becomes

$$V_g = (1/f\rho)(\delta p/\delta n)$$

geostrophic flux The transport of some substance or atmospheric property by the GEOSTROPHIC WIND.

geostrophic wind The wind that blows almost parallel to the ISOBARS. Close to the surface, friction slows the wind and this causes it to blow across the isobars at an angle of 10° to 30° depending on the speed of the wind and local topographical features. Friction has no effect above the BOUNDARY LAYER, where the winds are geostrophic. Suppose there is an area of low atmospheric pressure in air that is stationary. A PRESSURE GRADIENT exists across the isobars surrounding the low-pressure center and therefore there is a PRESSURE GRADIENT FORCE causing air to move toward the center. As soon as it starts to move the air becomes subject to the CORIOLIS EFFECT. This deflects it to the right in the Northern Hemisphere and to the left in the Southern Hemisphere, acting at right angles to the direction of flow. A component of the Coriolis effect acts in the same direction as the wind. This increases the wind speed, but the magnitude of the Coriolis effect varies in proportion to the wind speed, so as the wind accelerates the deflection increases. The process continues until the pressure gradient force and the Coriolis effect balance each other and the wind blows parallel to the isobars. Should the pressure gradient force increase, causing the wind to accelerate and turn toward the low-pressure center, the Coriolis effect also increases, correcting the wind. This is then called the *geostrophic wind,* from the Greek *ge* meaning "Earth" an *strepho* meaning "to turn", because the direction of the wind is turned by the Earth.

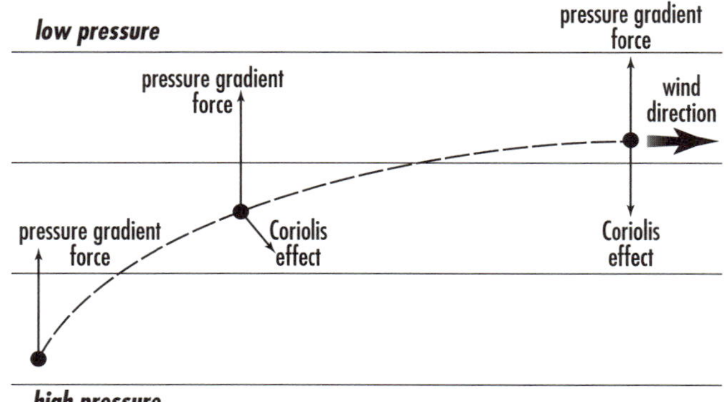

Geostrophic wind. **The wind is driven by the pressure gradient force (PGF) at right angles to the isobars (the horizontal lines) from the region of high pressure and toward the region of low pressure. The Coriolis effect (CorF) swings it to the right (in the Northern Hemisphere) and accelerates it, because a component of the CorF acts in the same direction as the wind. Acceleration increases the CorF until eventually the PGF and CorF are in balance, with the wind blowing parallel to the isobars.**

geostrophic wind level (gradient wind level) The lowest height at which the wind flow becomes GEOSTROPHIC. This is above the BOUNDARY LAYER, at an altitude of 1,650–3,300 feet (500–1,000 m).

geostrophical wind scale A diagram from which the GEOSTROPHIC WIND speed can be read. The diagram is drawn from solutions of the GEOSTROPHIC EQUATION.

geosynchronous orbit See GEOSTATIONARY ORBIT.

geothermal flux The flow of energy from the interior of the Earth to the surface that is due partly to the impacts of the bodies that collided to form the planet and its subsequent gravitational collapse, but mainly to the decay of radioactive elements in and beneath the crust. The geothermal flux heats the atmosphere where the energy reaches the surface at geysers, thermal springs, and volcanoes. This is the second largest source of energy affecting the atmosphere, but averaged over the entire Earth it amounts to no more than about 0.05 W m^{-2}. This is very small compared with proportion of the SOLAR CONSTANT that reaches the surface.

Geralda A CYCLONE, rated as category 5 on the SAFFIR/SIMPSON SCALE, that struck Madagascar from February 2 to 4, 1994. It generated winds of up to 220 mph (354 km h^{-1}) and destroyed 95 percent of the buildings in the port of Toamasina. It killed 70 people and rendered 500,000 homeless.

GERB *See* GEOSTATIONARY EARTH RADIATION BUDGET.

Gert A HURRICANE that struck Bermuda in September 1999. At its peak it generated winds of 145 mph (233 km h^{-1}) and was classed as category 4 on the SAFFIR/SIMPSON SCALE. By the time the edge of the hurricane reached Bermuda the wind speed had fallen to 110 mph (177 km h^{-1}). Gert generated a STORM SURGE of 5 feet (1.5 m) and waves 10 feet (3 m) high but caused no deaths or serious injuries.

gharbi A wind that blows across the Mediterranean from the Sahara Desert to the Adriatic and Aegean Seas. It sometimes reaches gale force, but it consists of warm air that has crossed the sea and it produces damp, CLOSE weather. It often produces FOG, heavy DEW, and heavy rain, especially on mountainous coasts. The rain is sometimes red because it carries desert dust.

gharra Severe SQUALLS, accompanied by heavy rain and THUNDERSTORMS, that cross Libya from the northeast. They are frequent and appear suddenly.

ghibli A hot, dry wind of the SIROCCO type that blows in northern Libya. It occurs when a DEPRESSION moves along the Mediterranean, drawing air from far to the south into the CYCLONIC flow.

GHOST *See* GLOBAL HORIZONTAL SOUNDING TECHNIQUE.

Gil *See* FAITH.

Gilbert A HURRICANE, rated category 5 on the SAFFIR/SIMPSON SCALE, that struck islands in the Caribbean and the Gulf Coast in the United States from September 12 to 17, 1988. It caused widespread damage in Jamaica before moving toward the Yucatán Peninsula, Mexico. It killed about 200 people and caused an estimated $10 billion of damage in Monterrey, Mexico. In Texas it killed at least 260 people and generated nearly 40 TORNADOES.

GIMMS *See* GLOBAL INVENTORY MONITORING AND MODELING SYSTEMS.

GISP *See* GREENLAND ICE SHEET PROJECT.

glacial anticyclone (glacial high) The semipermanent region of high AIR PRESSURE that covers the GREENLAND ICE SHEET and Antarctica.

glacial high *See* GLACIAL ANTICYCLONE.

glacial limit A line that marks the farthest point reached by a GLACIER at some time in the past. Geomorphologists identify glacial limits by the presence of terminal or lateral moraines; outwash plains, which are accumulations of rock deposited by water flowing from a glacier; the channels of rivers that once carried meltwater from the glacier; and lakes lying in depressions hollowed out by the ice.

glacial period (ice age) A prolonged time during which a substantial part of the surface of the Earth is covered by ice. In this sense, the term is somewhat vague, since *substantial* is not defined. The term is applied more precisely to a particular ice age. The most recent of these was the WISCONSINIAN GLACIAL, which was equivalent to the DEVENSIAN GLACIAL in Britain, the Weichselian glacial elsewhere in northern Europe, and the Würm glacial in the Alps.

glacial theory The theory that was first advanced during the 1830s and 1840s by a number of scientists, of whom LOUIS AGASSIZ was the best known, that at one time most of northern Europe, northern Asia, and North America had been covered by ICE SHEETS. The time during which this occurred later came to be called the PLEISTOCENE EPOCH. The theory was used to explain various geomorphological features and the extinction of certain animals, such as mammoths. Scientists have since expanded the theory to include a number of GLACIAL PERIODS rather than the single one that was originally supposed to have taken place.

glaciated cloud A cloud in which all the particles are ICE CRYSTALS.

glaciation The burying of an area beneath an ICE SHEET, as occurs during a GLACIAL PERIOD. The term is also used to describe the change of water droplets to ICE CRYSTALS in the upper part of a CONVECTIVE CLOUD.

glacier A large mass of ice that rests on the land surface or is attached to it but projects over the sea as an ICE SHELF. Most glaciers flow, but very slowly (the expression "glacial speed" is applied to events that happen extremely slowly). Large glaciers, like those in Antarctica, move between about 0.4 and 4 inches (0.01–0.1 m) a day and VALLEY GLACIERS at between 4 inches and 6.5 feet (0.1–2 m) a day, although they can occasionally move very much faster, at 165–330 feet (50–100 m) in a day. Glaciers can be classified in several ways. The most widely used classification is based on the mechanism by which the ice flows, which is related to the temperature at the base of the glacier. In this system glaciers fall into three categories: COLD GLACIERS, COMPOSITE GLACIERS, and TEMPERATE GLACIERS.

(You can learn more about glaciers from Michael Allaby, *Dangerous Weather: Blizzards* [New York: Facts On File, 1997], from Glaciers.net at www.glaciers./net/, and from the World Glacier Monitoring Service at www.geo.unizh.ch/wgms/index1.htm.)

glacier ice *See* FIRN.

glacier wind *See* FIRN WIND.

glacioisostasy The rise in the level of the land that takes place, very slowly, after an ICE SHEET has melted. During a GLACIAL PERIOD the weight of ice that accumulates on the surface depresses the crust. In Scandinavia, the ice depressed the rocks beneath it by about 3,300 feet (1,000 m). Since the ice melted the land has risen by about 1,700 feet (520 m), and eventually it will rise by a further 1,600 feet (480 m). Eastern Canada is also rising for the same reason and is expected to rise by a further 650 feet (200 m).

glaciology The scientific study of ice in the air, lakes, rivers, oceans, and below ground, but especially the study of GLACIERS.

glaciomarine sediment A sediment lying on the seabed that consists of materials that were transported by GLACIERS and ICEBERGS. These materials can be identified and dated. They provide evidence of glacial advances and HEINRICH EVENTS, which in turn provide evidence of past climates.

Gladys A TYPHOON that struck South Korea on August 23, 1991. It produced 16 inches (406 mm) of rain in Pusan and Ulsan, killing 72 people and rendering 2,000 homeless.

glaze (clear ice) Clear, solid ice that covers surfaces with a layer that is up to 1 inch (2.5 cm) thick. The weight of ice that forms in an ICE STORM can be enough to break branches from trees and knock down power lines. Glaze forms when liquid raindrops freeze on contact with a surface. This happens if the temperature of the surface is at or below freezing and the raindrops are either close to freezing or SUPERCOOLED.

Global Area Coverage Oceans Pathfinder Project (GAC) A study of sea-surface temperatures using data from measurements made by ADVANCED VERY HIGH RESOLUTION RADIOMETERS carried on the *NOAA 7*, *NOAA 9*, and subsequent odd-numbered National Oceanic and Atmospheric Administration (NOAA) satellites. The project is a collaboration of NOAA, the National Aeronautics and Space Administration (NASA), the University of Miami, and the University of Rhode Island. Its final results are checked at the Jet Propulsion Laboratory, a NASA facility located at the California Institute of Technology.

(Further information on Global Area Coverage Oceans Pathfinder Project can be found at http://podaac-www.jpl.nasa.sst.)

Global Atmospheric Research Program (GARP) A project that ran from 1968 until the early 1980s. Planning for it began in 1961, after President John F. Kennedy had proposed to the United Nations that orbiting space satellites should be used for peaceful purposes. Its aim was to observe the atmosphere from space over a long period in order to estimate its variability and the processes occurring within it, with a view to improving the quality of weather forecasts. GARP was organized by the WORLD METEOROLOGICAL ORGANIZATION and the International Council for Scientific Unions and its first leader was the American meteorologist Jule Gregory Charney (1917–81). The first GARP experiment, sometimes called the *Global Weather Experiment* (GWE), began on December 1, 1978, and ended on November 30, 1979. It involved scientists from more than 140 countries and was the biggest atmospheric experiment ever conducted up to that time. WORLD WEATHER WATCH ships and upper-air observations, weather buoys mainly in the oceans of the Southern Hemisphere, and commercial aircraft participated, as well as satellites. The data that were collected during the GARP years formed the basis of subsequent understanding of the dynamics of the atmosphere, and GARP served as a model of how to arrange and manage a large-scale international scientific collaboration.

(You can learn more about the first GARP experiment at neonet.nlr.nl/ceos-idn/campaigns/GARPFGGE.html.)

Global Change System for Analysis, Research and Training *See* INTERNATIONAL GEOSPHERE–BIOSPHERE PROGRAM.

Global Climate Observing System *See* WORLD CLIMATE PROGRAM.

Global Environmental Monitoring System (GEMS) An organization that was established in 1975 as part of the EARTHWATCH PROGRAM of the United Nations Environment Program (UNEP). It was first proposed by a meeting of experts that was convened in 1971 by the organizers of the United Nations Conference on the Human Environment, held in Stockholm, Sweden, in June 1972. The purpose of GEMS is to acquire data pertaining to the natural environment and to make them available to governments and other organizations that need them. The first part of GEMS to be inaugurated was GEMS/AIR, in 1975. It is managed jointly by UNEP and the World Health Organization (WHO).

(You can learn more about the history of environmental monitoring, including GEMS, at www.plas.bee.qut.edu.au/wwwjsc/psb320/320envmon01.htm.)

Global Environment Facility (GEF) An international organization that was established in 1990. It provides practical assistance to the environmental improvement programs of governments. The GEF is managed by the World Bank, which controls two-thirds of its funds. The United Nations Environment Program (UNEP) controls the remaining one-third.

Global Horizontal Sounding Technique (GHOST) A project that forms part of the WORLD WEATHER WATCH. GHOST uses balloons for the direct sensing of the atmosphere. The balloons are designed to float at various constant-density levels. Sensors carried beneath the balloons measure temperature, humidity, and pressure. The balloons are tracked by satellites in POLAR ORBIT, which receive data from the balloons and transmit them to receiving stations.

Global Inventory Monitoring and Modeling Systems (GIMMS) A set of data on global vegetation that are held at the National Aeronautics and Space Administration (NASA) Goddard Space Center. The data are used to produce the NORMALIZED DIFFERENCE VEGETATION INDEX (NDVI), with a resolution of about 4.7 miles (7.6 km).

Global Ozone Observing System *See* WORLD METEOROLOGICAL ORGANIZATION.

Global Resource Information Database (GRID) An international organization, based in Geneva, Switzerland, that was established in 1985 by the United Nations Environments Program (UNEP) and the Swiss government. It uses computers and software developed by the National Aeronautics and Space Administration (NASA) to analyze and integrate environmental information from different sources, including the World Health Organization and the Food and Agriculture Organization.

global warming The idea that the mean temperature of the atmosphere is increasing throughout the world. There are three sets of measurements from which trends in the mean temperature are estimated. The first are collected at the surface of the land and sea. These indicate general increase in temperature of 0.27° F (0.15° C) per decade. If sustained, by the year 2100 this would produce temperatures an average 2.7° F (1.5° C) warmer than those of today. These measurements are not very reliable, however. Land measurements are made from thermometers mounted in boxes and sited in easily accessible locations, principally at airports. There they are influenced by the HEAT ISLAND effect that has increased over the last century as a result of urban expansion. The measurements themselves are made accurately and provide a good geographical cover only in North America, Europe, Russia, China, and Japan. The system is much less comprehensive in Africa, South America, Australia, and throughout much of Southern Asia. Sea-surface readings are taken from buoys, piers, and ships that measure the temperature of water drawn onboard to cool their engines. These measurements do not cover the entire world.

Measurements are also made by RADIOSONDE balloons. These are released twice each day, at noon and midnight Greenwich Mean Time, from more than 1,000 sites. The launch sites are located predominantly in industrial nations. These measurements show the mean temperature is decreasing by 0.04° F (0.02° C) per decade. If this trend continues, by 2100 the global mean temperature will be 0.4° F (0.2° C) cooler than it is now.

The most widespread coverage is provided by temperature measurements from satellites. These are compiled as part of a joint project by the National Aeronautics and Space Administration (NASA) and the University of Alabama in Huntsville that has been monitoring temperatures since January 1979 from data gathered by MICROWAVE SOUNDING UNITS on nine TELEVISION AND INFRARED ORBITING SATELLITE (TIROS-N) satellites. These provide more than 30,000 readings every day from each satellite. When adjusted for the effect of a decay in the satellite orbits and their drift from east to west, the trend shows the temperature is cooling by 0.02° F (0.01° C) per decade, a trend that would produce an overall cooling of 0.2° F (0.1° C) by 2100.

Estimates from the INTERGOVERNMENTAL PANEL ON CLIMATE CHANGE (IPCC) indicate a warming at the surface of 0.3° F (0.18° C) per decade, or 3° F (1.8° C) by 2100. CLIMATE MODELS suggest this should mean weather balloons and satellites measure a temperature rise of 0.4° F (0.23° C) per decade. Clearly they do not. IPCC climatologists maintain that the surface warming is genuine, but accompanied by a cooling at higher levels, so there has been an unexplained steepening of the LAPSE RATE. It is uncertain whether global warming is occurring, but it is known that short-term variations, especially those generated by EL NIÑO and LA NIÑA, produce series of warm and cool years.

(For more information see John L. Daly, "Still Waiting for Greenhouse," at www.vision.net.au/~daly/index.htm and www.vision.net.au/~daly/peterson.htm; "Using Satellites to Monitor Global Climate Change" at www.atmos.uah.edu./essl/msu/background.html; and Roy Spencer, "Measuring the Temperature of Earth from Space" at science.nasa.gov/newhome/headlines/notebook/essd13aug98_1.htm.)

global warming potential (GWP) The amount of CLIMATIC FORCING that a particular GREENHOUSE GAS exerts. This is compared to the forcing exerted by CARBON DIOXIDE, which is given a value of 1. Water vapor, the gas with the strongest greenhouse effect, is not included, because the atmospheric content is highly variable and beyond our control. CFCs have a strong GWP, but these gases are now being phased out under the terms of the MONTREAL PROTOCOL, so their influence will decline. GWPs for the remaining greenhouse gases are under constant revision as the scientific techniques for estimating them improve.

In the table, hydrofluorocarbon gases (HFCs) were developed as alternatives to CFCs, because they have no effect on the OZONE LAYER. They are used mainly in refrigeration units and in the manufacture of semiconductors. Perfluorocarbons (PFCs) are also used as alternatives to CFCs in semiconductor manufacture, and they are a by-product of aluminum smelting and uranium enrichment. Sulfur hexafluoride is used as an industrial insulator and in the manufacture of cooling systems for electrical cables.

GWPs for Principal Greenhouse Gases

Gas	GWP
Carbon dioxide	1
Methane	21
Nitrous oxide	310
CFC-11	3,400
CFC-12	7,100
Perfluorocarbons	7,400
Hydrofluorocarbons	140–11,700
Sulfur hexafluoride	23,900

(For further information on GWPs see www.ncdc.noaa.gov/ogp/papers/solomon.html and www.state.gov/www/global/oes/fs_sixgas_cop.html.)

Global Weather Experiment *See* GLOBAL ATMOSPHERIC RESEARCH PROGRAM.

***Globigerina* ooze** A sediment, or ooze, that covers most of the floor of the western part of the Indian Ocean, the middle of the Atlantic Ocean, and the equatorial and South Pacific Ocean. It is less extensive elsewhere, but in total it covers about half of the entire ocean floor. At least 30 percent of the sediment consists of the shells (called *tests*) of tiny ameboid protozoa of the order Foraminiferida, and most of these belong to the genus *Globigerina*. *Globigerina* species drift at the ocean surface as part of the plankton, but they survive only within certain temperature limits. Each species has its own temperature requirement. Consequently, the species that are found in the sediment can be used to indicate the temperature of the water in which they lived. The presence of *Globigerina menardii* is taken to indicate warm water, for example, and *G. pachyderma* indicates cold water. The tests of some *G. truncatulinoides* coil to the right and those of others coil to the left. The direction of coiling is believed to indicate temperature differences: right coiling indicates warm water and left coiling indicates cold water.

gloom The condition in which daylight is markedly reduced by thick cloud or dense smoke, but horizontal VISIBILITY remains good. Gloom is not the same as ANTICYCLONIC GLOOM.

Gloria A TYPHOON that struck the Philippines on July 25, 1996. It killed at least 30 people. It was then downgraded to a TROPICAL STORM. It reached Taiwan and the southeastern coast of China on July 26, where it killed three people.

glory An optical phenomenon in which a shadow cast onto a layer of cloud appears surrounded by one or more circles of light. The light is faintly colored, with the colors of the RAINBOW and red on the outside of the circle, and if there is more than one circle the innermost one is the brightest. Glories are caused by the reflection and refraction of light by very small water droplets of fairly uniform size. The Sun, observer, and glory form a straight line, so the glory always surrounds the shadow of the observer. Glories are most often seen by people in aircraft flying above cloud, in which the glory surrounds the shadow of the airplane, but they also appear to people on the ground. The observer must be looking at a bank of cloud or fog with the Sun behind. The glory then forms around the observer's head, like a halo, and hence the name *glory*.

GMT *See* GREENWICH MEAN TIME.

goats According to weather folklore, if a goat grazes with its head facing into the wind the weather will be fine. If it grazes with its tail to the wind the weather will be wet.

GOES *See* GEOSTATIONARY OPERATIONAL ENVIRONMENTAL SATELLITE.

Gordon A TYPHOON that struck Luzon, Philippines, on July 16, 1989. At least 200 people were killed. TROPICAL STORM Gordon struck islands in the Caribbean, Florida, and South Carolina from November 13 to 19, 1994. It killed 537 people and caused damage estimated at at least $200 million.

gorge wind *See* MOUNTAIN-GAP WIND.

grab sampling A technique for obtaining a sample of air for analysis in which the air is collected very quickly, so that the time taken to obtain the sample is insignificant when compared to the duration of the process or rate of change that is being studied.

gradient flow The horizontal movement of air when there is no FRICTION and the ISOBARS and STREAMLINES coincide. In this situation the tangential ACCELERATION is zero throughout the system.

gradient wind The wind that flows parallel to the ISOBARS at a speed that results from the interplay of the PRESSURE GRADIENT FORCE, the CORIOLIS EFFECT, and CENTRIPETAL ACCELERATION. These are the three forces acting around a CYCLONE or ANTICYCLONE. The pressure gradient force makes air move out of an area of high pressure and into an area of low pressure. As it moves, the air is deflected by the Coriolis effect. Once it is following a curved path, the air is subject to centripetal acceleration, which is equal to the difference between the pressure gradient force and the Coriolis effect. Around a center of high pressure, where the tendency is for the air to move outward, the Coriolis effect exceeds the pressure gradient force. This accelerates the wind speed, which becomes SUPERGEOSTROPHIC. The effect is small, however, for two reasons. The first is that the PRESSURE GRADIENT around a center of high pressure is usually lower than that around a center of low pressure, so the winds are in any case lighter. The second is that the rotation of the Earth is CYCLONIC and this acts against ANTICYCLONIC flow, slowing it. Around a center of low pressure, where the tendency is for the air to move inward, the Coriolis effect is weaker than the pressure gradient force. This reduces the wind speed, which becomes SUBGEOSTROPHIC. The name *gradient wind* refers to the fact that the wind speed is proportional to the pressure gradient.

gradient wind level *See* GEOSTROPHIC WIND LEVEL.

granular snow *See* SNOW GRAINS.

grape belt A stretch of land that runs for about 60 miles (96 km) along the southern shore of Lake Erie. This area benefits from the LAKE EFFECT due to its proximity to the lake. Winters are less severe than in other parts of the United States at the same latitude, with frosts beginning late and ending early, and the fall is long and mild.

grasslands climate (prairie climate, subhumid climate) In the THORNTHWAITE CLIMATE CLASSIFICATION, a climate in HUMIDITY PROVINCE C, with a PRECIPITATION EFFICIENCY INDEX of 32–63.

grass minimum temperature The temperature that is registered by a MINIMUM THERMOMETER set in the open at the level of the tops of the blades of grass in short turf. The reading is of value to farmers and horticulturists.

grass temperature The temperature that is registered by thermometer set in the open with its bulb at the level of the tops of grass in short turf. This is the temperature to which crop plants are exposed and it is therefore of relevance to farmers and horticulturists.

graupel (soft hail) Precipitation in the form of ice pellets that are soft enough to flatten or smash into fragments when they hit a hard ground surface. The pellets are 0.1–0.2 inch (2–5 mm) in diameter but can also be less than 0.04 inch (1 mm) across, and most are spherical. They are white and opaque and form when RIME FROST accumulates on ice crystals, producing a very loose structure. *See* SNOW GRAINS and SNOW PELLETS.

gravitational force The attraction that exists between bodies with a magnitude that decreases with increasing distance between the bodies according to the INVERSE SQUARE LAW. Gravity is by far the weakest of the fundamental physical forces (the others are the strong and weak nuclear forces and the electromagnetic force), but it is felt on Earth because the Earth is very large.

Every body possesses mass. This is measured in pounds or tons and in SYSTÈME INTERNATIONAL D'UNITÉS (SI) units in kilograms. According to the first LAW OF MOTION, the gravitational force is equal to the mass (m) of the body multiplied by the ACCELERATION (a) it produces (ma). The Earth exerts a gravitational force on a PARCEL OF AIR and the parcel of air exerts an equal and opposite force on the Earth, although this has no observable effect.

The gravitational acceleration produced by the Earth is designated by g and has an average value of 9.8 m s^{-2} (32 ft s^{-2}). The gravitational force acts from the center of the Earth and therefore varies with altitude and latitude, because of the inverse square law. (It varies with latitude, because the Earth is not perfectly spherical and so the distance from the center of the Earth is not everywhere the same.) For most purposes the value of g can be taken as 10 m s^{-2} (33 ft s^{-2}).

gravitational settling The process by which solid particles fall from the air as a result of the force of gravity. The rate at which they do so depends on their size: large particles settle faster than small ones.

gravitational water *See* SOIL MOISTURE.

gravity wave *See* ATMOSPHERIC WAVE.

gravity wind *See* KATABATIC WIND.

graybody A body that absorbs electromagnetic radiation uniformly at all WAVELENGTHS. Consequently its absorptivity and EMISSIVITY are independent of the wavelength of the radiation. The emittance (E) of a graybody is given by

$$E = \epsilon\sigma T^4$$

where ϵ is the emissivity, σ is the STEFAN–BOLTZMANN CONSTANT, and T is the ABSOLUTE TEMPERATURE. Many natural bodies are graybodies across a wide range of wavelengths.

gray mist at dawn According to weather folklore, a sign that the summer day to follow will be fine. The rhyme is,

Gray mists at dawn,
The day will be warm.

Great interglacial *See* MINDEL–RISS INTERGLACIAL.

green flash (green ray) A bright flash of emerald green light that is seen just above the horizon immediately after the Sun has set or immediately before it rises. It lasts for 1–10 seconds, or sometimes longer. Light is refracted (*see* REFRACTION) slightly as it passes through the atmosphere, because of variations in atmospheric density. When the Sun is low in the sky, its light is refracted almost parallel to the surface of the Earth, so the real position of the Sun in the sky is about half a degree lower than its apparent position. Different wavelengths of light are refracted by different amounts. When the Sun is close to the horizon the atmosphere acts as a prism, breaking sunlight into its constituent (RAINBOW) colors, with red at the bottom. Sunlight is also scattered (*see* SCATTERING); blue is the color most affected. The atmosphere absorbs yellow light. When the red and orange parts of the solar spectrum are just out of sight below the horizon, the remainder of the spectrum is above it. The yellow light is absorbed and the blue light is scattered in all directions. This leaves the green part of the spectrum, which is neither scattered nor absorbed, and so there is a brief flash of brilliantly green light. For a green flash to appear the horizon must be absolutely flat and a long way away, so the effect is most often seen over the sea. The air must be stable, because instability causes variations in density that vary the degree of refraction. Dry air is usually more stable than moist air, and so green flashes are seen more often in dry climates than in humid ones.

greenhouse effect The warming of the atmosphere that is due to the absorption and reradiation of heat by the molecules of certain gases, known as *GREENHOUSE GASES*. Solar radiation is emitted at all wavelengths. Very short-wave, high energy gamma and X radiation is absorbed at the top of the atmosphere and does not penetrate it. The shorter wavelengths of ULTRAVIOLET RADIATION are absorbed by atmospheric oxygen (*see* OZONE LAYER).

The radiation that penetrates deeply into the atmosphere is predominantly at wavelengths between 0.2 μm and 4.0 μm, with a strong peak at 0.5 μm. Visible light is radiation at wavelengths between 0.4 μm (violet) and 0.7 μm (red); the 0.5-μm peak corresponds to blue–green light. Radiation between 0.7 μm and 4.0 μm corresponds to infrared radiation and heat. Of all the radiation Earth receives from the Sun, 9 percent is ultraviolet, 45 percent is visible light, and 46 percent infrared and heat.

Some of the incoming energy is reflected into space (*see* ALBEDO) from clouds and from the different surfaces of the land and sea. The remainder is absorbed. The absorbed radiation warms the surface, and as it warms the Earth starts to radiate as a BLACKBODY. This radiation is at much longer wavelengths, between about 5 μm and 50 μm, with a strong peak at about 12 μm.

The gases the atmosphere comprises are transparent to incoming solar radiation, but the atmosphere is partly opaque to outgoing long-wave radiation. Certain molecules absorb radiation at wavelengths determined by their own size and shape. Water vapor absorbs radiation at 5.3–7.7 μm and at wavelengths higher than 20 μm, for example; carbon dioxide absorbs at 13.1–16.9 μm; and ozone absorbs at 9.4–9.8 μm. No gas absorbs radiation at 8.5–13.0 μm.

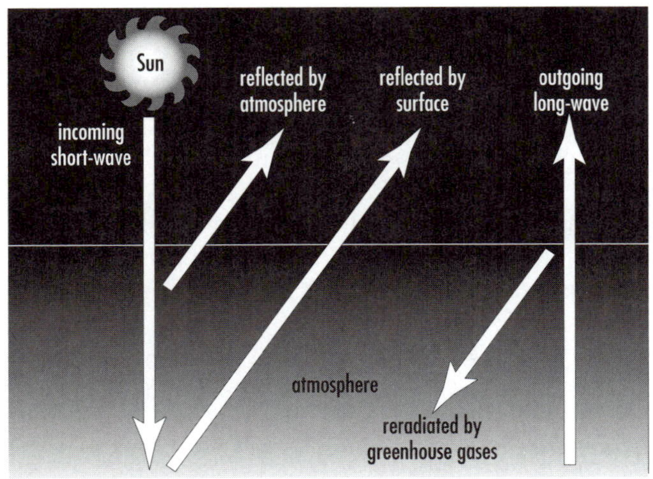

Greenhouse effect. **The warming of the atmosphere due to the absorption of long-wave radiation by certain gases that are naturally present in the air.**

This is the ATMOSPHERIC WINDOW through which radiation escapes into space.

Radiation that is absorbed by gases in the atmosphere is then reradiated by them. This radiation travels in all directions. Some returns to the surface and some is absorbed by other molecules. The overall effect is to retain in the atmosphere more than 90 percent of the heat radiated from the Earth's surface. Heat does not accumulate indefinitely, of course (*see* RADIATION BALANCE), but the net effect is to raise the temperature of the atmosphere.

The greenhouse effect makes the atmosphere 54°–72° F (30°–40° C) warmer than it would be if the atmosphere were completely transparent to radiation at all wavelengths. Instead of the global mean surface temperature's being 59° F (15° C) as it is now, it would be between about 5° F and –13° F (between –15° C and –25° C). This is the figure that is most often used in calculations involving the greenhouse effect.

Some scientists challenge this figure, however, on the ground that it makes an incorrect allowance for the planetary ALBEDO and that it fails to take account of the extent to which heat is absorbed by and reradiated from the uppermost layer of the ocean. Adjustment for the albedo reduces the warming effect to about 36° F (20° C), and when heat retention by the oceans is also taken into account the net warming effect is probably about 25° F (14° C). Without the greenhouse effect, therefore, the mean surface temperature of the atmosphere would be about 34° F (1° C).

This is what is known as the *greenhouse effect*. The comparison with a greenhouse is based on the notion that the glass of a greenhouse is transparent to incoming radiation but opaque to outgoing, long-wave radiation, causing the air inside the greenhouse to be warmer than the air outside. The phenomenon is not confined to Earth. It also occurs on all planets with an atmosphere containing greenhouse gases and is most marked in the atmosphere of VENUS.

Many scientists believe human activities that release greenhouse gases may be increasing the natural greenhouse effect and that the resulting enhanced greenhouse effect will induce GLOBAL WARMING. Should this prove to be correct, the amount of warming will be related to the calculated value for the natural greenhouse effect. The critics of this calculation hold that the magnitude of any global warming will be quite small.

greenhouse gas A gas that absorbs energy that is radiated from the surface or atmosphere of the Earth or another planet. The principal greenhouse gases are WATER VAPOR (H_2O), CARBON DIOXIDE (CO_2), NITROUS OXIDE (N_2O), METHANE (CH_4), OZONE (O_3), chlorofluorocarbons (CFCS), and hydrofluorocarbons. Each gas absorbs at particular wavelengths, and where the wave bands of two or more gases overlap the amount of absorption is shared by those gases. The amount of radiation absorbed by each gas varies greatly and is reported as its GLOBAL WARMING POTENTIAL. At certain wavelengths, known as the *ATMOSPHERIC WINDOW,* no outgoing radiation is absorbed.

It is individual gas molecules that absorb the radiation. It imparts energy to them that they then reradiate. Radiation can be absorbed only if it encounters an appropriate molecule in its passage through the atmosphere and into space. The amount of energy absorbed therefore depends on the concentration of those molecules: the higher the concentration, the greater the chance of an impact. It follows that as the atmospheric concentration of a particular greenhouse gas increases, so does the amount of radiation that gas absorbs. Once the concentration reaches a level at which all the radiation at the wavelengths absorbed by that gas is being absorbed, however, adding more of the gas will have no effect. This means there is a limit to the possible magnitude of any GREENHOUSE EFFECT.

greenhouse period A time when there were no GLACIERS or ICE SHEETS anywhere on Earth. Such periods have occurred many times in the history of the Earth, and evidence for some of them is found in seabed sediments. They appear to develop at intervals that are related to the MILANKOVICH CYCLES.

(For more information, see www.earthsky.com/1996/esmi960906.html.)

Greenland Icecore Project (GRIP) A drilling program sponsored by the European Science Foundation that was established to retrieve an ICE CORE 1.86 miles (3 km) long from Summit, the highest point on the GREENLAND ICE SHEET and 17.4 miles (28 km) to the east of the site of the GREENLAND ICE SHEET PROJECT. Analysis of the ice core and of the dust and air trapped in it provided information about the global climate over the past several hundred thousand years. Drilling began in January 1989 and ended in December 1995. By the end of 1990, the drill had reached a depth of 2,526 feet (770 m), where the ice is 3,840 years old. In 1991 the drill reached 8,270 feet (2,521 m) and ice 40,000 years old, and on August 12, 1992, the drill reached bedrock at a depth of 9,938 feet (3,029 m). Ice at that depth is 200,000 years old. The core is stored, in sections, at the University of Copenhagen, Denmark.

(You can learn more about GRIP at www.esf.org/life/lp/old/grip/lp_013a.htm.)

Greenland ice sheet The layer of ice, or glacier, that covers 80 percent of the area of Greenland. Beneath the ice sheet, the ground surface is close to sea level over most of central Greenland, but with mountain ranges around the center, parallel with the coast. The mountains contain the ice, except in a small area in the northwest, close to Qaanaaq (formerly Thule), where the ice reaches the sea, but without forming an ICE SHELF. Ice does reach the sea along valleys through the mountains; the biggest VALLEY GLACIER is the Ilulissat (formerly Jakobshavn) Glacier on the western coast. This flows at a rate of 66–72 feet (20–22 m) per day at its terminus and is the source of most of the ICEBERGS entering Arctic waters. The inland ice sheet is almost 1,490 miles (2,400 km) long in a north–south direction and is 680 miles (1,100 km) wide at its widest point, at 77° N. It covers an area of 670,272 square miles (1,736,095 km^2). Smaller ice caps and glaciers cover an additional 18,763 square miles (48,599 km^2). Two north–south domes or ridges, one in the north and the other in the south, are the location where the ice is thickest. The southern dome is about 9,845 feet (3,000 m) thick at 63°–65° N and at about 72° N the northern dome is about 10,795 feet (3,290 m) thick.

The climate over the ice sheet is cold and dry. Mean temperatures fall to –24° F (–31° C) on the north dome and –4° F (–20° C) on the south dome. Precipitation averages about 2.6 inches (67 mm) a year.

In 1993 and 1994, the National Aeronautics and Space Administration (NASA) surveyed the entire ice sheet by means of radar ALTIMETERS carried on aircraft. These measured the elevation of the ice surface very accurately. In 1998, the southern part of the ice sheet was surveyed again. This showed three areas where the ice was thickening by more than 4 inches (10 cm) a year, but also large areas of thinning at elevations up to 5,000 feet (1,500 m), especially in the east. The thinning exceeded 3 feet (1 m) a year at the mouth of some outlet glaciers. Above 6,500 feet (2,000 m), the thickness of the ice sheet was not changing, but overall the southern ice sheet was losing ice as a result of outlet glaciers that were flowing faster, rather than of warming.

(For more information about Greenland and its ice sheet see http://www.greenland-guide.dk/gt/green-10.htm and http://www.uwin.siu.edu/announce/press/atlas.html.)

Greenland Ice Sheet Project (GISP) A United States drilling program sponsored by the National Science Foundation (NSF) that retrieves ICE CORES from the GREENLAND ICE SHEET at a site 17.4 miles (28 km) to the west of the European GREENLAND ICECORE PROJECT. The cores are used to obtain information about past climates from analysis of the ice (*see* OXYGEN ISOTOPE RATIOS) and the dust and gases trapped in it. The first core reached bedrock at a depth of about 9,843 feet (3,000 m), and in 1988 the Office of Polar Programs of the NSF authorized the drilling of a second core, GISP2. The drilling of GISP2 was completed on July 1, 1993, when the drill penetrated 5 feet (1.55 m) into bedrock at a depth of 10,018.34 feet (3053.44 m). The ice at the base of the ice sheet is about 200,000 years old and samples of ice at any level can be dated, so the cores provide a continuous climate record that extends over a very long period. When the oldest ice fell as snow over Greenland, 200,000 years ago, the WOLSTONIAN GLACIAL was just beginning. The Wolstonian was the ice age before the most recent (DEVENSIAN or WISCONSINIAN GLACIAL) and separated from it by the IPSWICHIAN INTERGLACIAL.

(You can learn more about GISP at gust.sr.unh.edu/GISP2/.)

green ray *See* GREEN FLASH.

green taxes Taxation that is levied on the purchase of certain commodities in order to protect the environment by discouraging their use. Taxes are used in preference to banning the use of the commodity in question, because some people or industries have no alternative to using it. This clearly discriminates against certain groups of people and is therefore unfair, but it is argued that manipulating the price encourages more efficient use of the commodity and the development of alternatives. The carbon tax is the best known example. It aims to discourage the use of carbon-based fuels, principally petroleum, but in 2000 it triggered popular protests against the high fuel prices across Europe and in many other countries.

Greenwich Mean Time (GMT, Z) The mean time, calculated from the position of the Sun, at the 0° meridian. This passes through Greenwich, England, which was formerly the site of the Royal Observatory. The 0° meridian was agreed on at an international conference held in Washington, D.C., in 1884. At first, Greenwich Mean Time was calculated from noon. This was altered to midnight in 1925, but because of the possibility of confusion UNIVERSAL TIME was adopted in 1928.

gregale A strong northeasterly wind that blows across Malta and the adjacent parts of the Mediterranean region. It is associated with a large ANTICYCLONE over the Balkans and a CYLONE over North Africa. Its name means "wind from Greece" in the Maltese language. The gregale most often occurs in spring and fall. It produces very variable weather, sometimes with clear skies and at other times with mist and heavy rain. It also causes high seas to run into harbors, endangering ships.

GRID *See* GLOBAL RESOURCE INFORMATION DATABASE.

GRIP *See* GREENLAND ICECORE PROJECT.

Grosswetterlage A record of the general weather situation over a large area that is used by meteorologists preparing forecasts and by climatologists studying long-term weather patterns. The word is German and means "large-scale weather situation." It is used because the best known record, covering central Europe, is maintained on a daily basis by the Deutscher Wetterdienst (German weather service). Similar records are now compiled by the weather services of other European nations.

(You can learn more about the Deutscher Wetterdienst, in English, at www.dwd.de.)

ground fog Any FOG that covers less than 60 percent of the sky. Ground fogs are often very shallow, sometimes not extending as far as the head of a person walking through them.

ground frost The condition in which the air temperature is above freezing, but the temperature at ground level is below freezing. This occurs on still, clear nights, when the ground and plant surfaces lose heat by radiation but the air remains warm at a higher level. Extensive cloud cover prevents ground frost by reflecting the radiated heat and wind prevents it by constantly mixing the ground-level air with warmer air.

ground heat flux (soil heat flux) The flow of heat into the ground by day and upward from the ground at night. Its extent depends on the THERMAL CONDUCTIVITY of the material composing the ground and the vertical temperature gradient. If these are known, the FLUX can be calculated mathematically, but this is usually impractical because there are too many uncontrolled variables. The more practical alternative is to measure the flux directly by using a heat flux plate. This is a plate made from material with a known thermal conductivity, protected on both sides by thin metal sheets. It is equipped with a sensor to measure the temperature difference on each side. The plate is buried horizontally and the electrical output from its sensor is proportional to the temperature gradient across the plate. The thermal conductivity of the plate should not be greatly different from that of the ground, and the plate should not be so large as to require digging a big hole for it that disturbs the soil.

Groundhog Day February 2, the day on which an old tradition holds that the spring weather can be predicted. In parts of North America, this is the day on which the groundhog is said to emerge from hibernation to look for his shadow. If he sees it, he anticipates bad weather and returns to his hole for a further six weeks. If he cannot see his shadow, because the day is cloudy, he takes this as a good omen and remains above ground, anticipating fine weather. February 2 is also Candlemas Day (so named because it used to be celebrated with candles), and there is an ancient rhyme from which the Groundhog Day tradition may be derived:

If Candlemas be fair and bright,
Winter'll have another flight.
But if Candlemas Day be clouds and rain,
Winter is gone and will not come again.

ground inversion *See* SURFACE INVERSION.

ground speed *See* TAILWIND.

ground streamer At the start of a LIGHTNING stroke, a column of IONIZATION that rises from a point on the ground toward which a STEPPED LEADER is descending.

ground visibility Horizontal VISIBILITY that is measured at ground level, or at the height of an airfield control tower. This affects the movement of aircraft on the ground and during take-off and landing.

groundwater (phreatic water) Water that lies below the ground surface in the zone of saturation, or phreatic zone, where all the spaces between soil particles are filled with water. The word *phreatic* is from Greek *phreatos,* which means "well," and it is used because a hole dug from the surface into the phreatic zone will fill with water. Only about 5 percent of the water that reaches the ground from PRECIPITATION drains downward to join the groundwater. The upper boundary of the zone of saturation is called the *water table.* Above the water table there is a narrow layer of soil called the *capillary layer* or *capillary zone,* where water is drawn upward by CAPILLARITY. Above the capillary layer the soil is unsaturated. This is called the *zone of aeration.* Groundwater may flow through the soil. The material holding it is then an AQUIFER. The rate of flow varies according to the PERMEABILITY of the soil. In many soils the groundwater moves from a few feet a day to a few feet a month, but it can move much faster and much more slowly. Water travels at more than one mile per hour (1.6 km h^{-1}) through limestone that contains underground cavities, but less than one foot (30 cm) a century through clay or shale.

growing season That part of the year during which the weather allows agricultural and horticultural crops to grow. Temperature is the limiting factor for plant growth in middle and high latitudes; in the TROPICS and subtropics it is the availability of water. In North America, the growing season is sometimes defined as the period between the last KILLING FROST of spring and the first killing frost of autumn. This describes the conditions in which plants can survive, however, rather than those in which they can grow. Most plant growth ceases when the temperature falls below 43° F (6° C) and so the growing season is also defined as the number of days on which the mean temperature exceeds this value. This is the only definition that is used in Europe. The growing season lasts almost all year at sea level in southwestern England, but for an average 260 days a year at sea level in Scotland. The length of the growing season decreases with elevation above sea level, by about two days for every 100 feet (30 m) in New England, for example.

growler A small piece of an ICEBERG, less than about 33 feet (10 m) long. It floats low in the water and its surface is often pitted. Although they are small, growlers can cause considerable damage to any ship that collides with them.

guba A strong SQUALL, usually with heavy rain and thunder, that occurs at night along the southern coast of New Guinea during the summer northwest MONSOON. They are especially common and violent near the mouths of large valleys.

Guinea Current An ocean current that flows from west to east along the West African coast and into the Gulf of Guinea. It is part of the EQUATORIAL COUN-

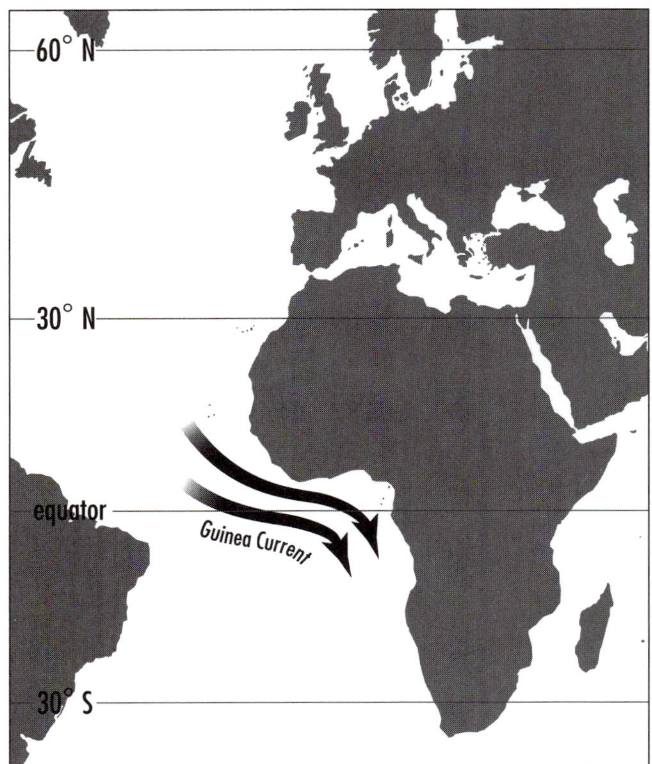

Guinea Current. **This warm ocean current flows past the West African coast into the Gulf of Guinea.**

TERCURRENT and carries warm water with a temperature that often exceeds 80° F (27° C). The current carries hot, humid weather to coastal regions in the west, but farther east, off the coast of western Nigeria, the current produces UPWELLINGS. These frequently produce FOG and a fairly low rainfall of about 30 inches (762 mm) a year, compared with the 100 inches (2,540 mm) a year along the coast to the east of Lagos.

Gulf Stream A system of ocean currents that convey warm water from the Gulf of Mexico to the center of the North Atlantic Ocean. It begins as the FLORIDA CURRENT where the NORTH EQUATORIAL CURRENT enters the gulf and ends in the latitude of Spain and Portugal, where it turns south, becoming the CANARY CURRENT and rejoining the North Equatorial Current. A branch from it becomes the NORTH ATLANTIC DRIFT. Its influence on climates makes it the most important current system in the Northern Hemisphere. It is clearly defined, as a belt of water at a fairly constant temperature of 64°–68° F (18°–20° C) and salinity of 36 parts per thousand (‰).

Gulf Current. **A warm ocean current that follows an approximately circular, clockwise path in the North Atlantic.**

gully squalls A name sailors have given to the violent SQUALLS that blow over the coastal waters of the eastern South Pacific from the mountain ravines of the Andes.

Gunung Agung *See* MOUNT AGUNG.

Günz glacial The earliest of the PLEISTOCENE GLACIAL PERIODS of the European Alps to be dated. It lasted from approximately 800,000 years ago until about 600,000 years ago. It is equivalent to the Menapian glacial of northern Europe and the NEBRASKAN GLACIAL of North America. It is named after an alpine river. The Günz was followed by the GÜNZ–MINDEL INTERGLACIAL.

Günz–Mindel interglacial An INTERGLACIAL period in the European Alps that lasted from approximately 600,000 years ago until about 350,000 years ago. It is equivalent to the CROMERIAN INTERGLACIAL of Britain and partly coincides with the AFTONIAN INTERGLACIAL of North America.

gust A sudden, violent increase in the wind speed that lasts for only a short time, after which the wind speed drops to its former level. Gusts are often caused by EDDIES that are due to TURBULENT FLOW. They are also generated as a GUST FRONT ahead of a storm. A gust is more short-lived than a SQUALL.

gust front (pressure jump line) A region of warm air that lies immediately ahead of an advancing storm. Descending cold air inside the storm cloud pushes forward beneath the warm air, lifting it. This makes the warm air unstable. It rises into the cloud and at the same time starts forming a new convective cell ahead of the cloud. Air being drawn into a large, vigorous cloud produces strong, gusty winds.

gustiness The quality of a wind that generates frequent GUSTS.

gustiness factor A measure that is used to describe the intensity of the GUSTS generated by a particular wind. The gustiness factor is calculated by measuring the wind speeds during gusts and during the lulls between gusts to determine the range of wind speeds. Then the mean wind speed is calculated across the period in question, including the speeds during gusts and lulls. The gustiness factor is then the ratio of the range of wind speeds to the mean wind speed. This value is taken into account when designing urban developments and forestry plantations.

gustnado A small TORNADO that forms in the GUST FRONT where downcurrents spill out from a CUMULONIMBUS cloud containing a SUPERCELL. Because they spin in diverging air (*see* DIVERGENCE), gustnadoes often rotate in an ANTICYCLONIC direction (clockwise in the Northern Hemisphere), unlike most tornadoes, which rotate in a CYCLONIC direction (counterclockwise in the Northern Hemisphere).

GWE See GLOBAL ATMOSPHERIC RESEARCH PROGRAM.

gymnosperm *See* POLLEN.

gyres The major ocean currents form approximately circular patterns that flow in an ANTICYCLONIC direction in both hemispheres. These circles of moving water, traveling clockwise in the Northern Hemisphere and counterclockwise in the Southern Hemisphere, are called *gyres*, which is a word derived from the Greek *guros*, which means "ring."

Surface ocean currents are driven by the wind. On each side of the equator, the PREVAILING WINDS are the TRADE WINDS, blowing from the northeast in the Northern Hemisphere and from the southeast in the Southern Hemisphere. The anticyclonic flow of air around the AZORES HIGH and Bermuda high also drives water in the North Atlantic.

The currents do not flow in the same direction as the wind, however, but form an EKMAN SPIRAL. At the surface, the water moves at an angle of about 45° to the wind, deviating to the right in the Northern Hemisphere and to the left in the Southern Hemisphere. The result is that the currents in both hemispheres flow parallel to the equator.

The movement of surface water generates differences in pressure between the upper and lower layers of the ocean. These differences cause water below the surface to move in the opposite direction to the surface current.

When they approach continents, the currents are deflected, and as they move farther from the equator their direction comes under the influence of the CORIOLIS EFFECT, which diverts them farther from the equator. The magnitude of the Coriolis effect increases with latitude, and as the currents leave the TROPICS and enter the middle latitudes they are driven by the prevailing westerly winds. The winds drive them toward the east and the Coriolis effect also deflects them to the east (in both hemispheres). When they approach the continents on the other sides of the oceans the currents turn back toward the equator.

The North Atlantic gyre begins as the NORTH EQUATORIAL CURRENT. Its name then changes according to

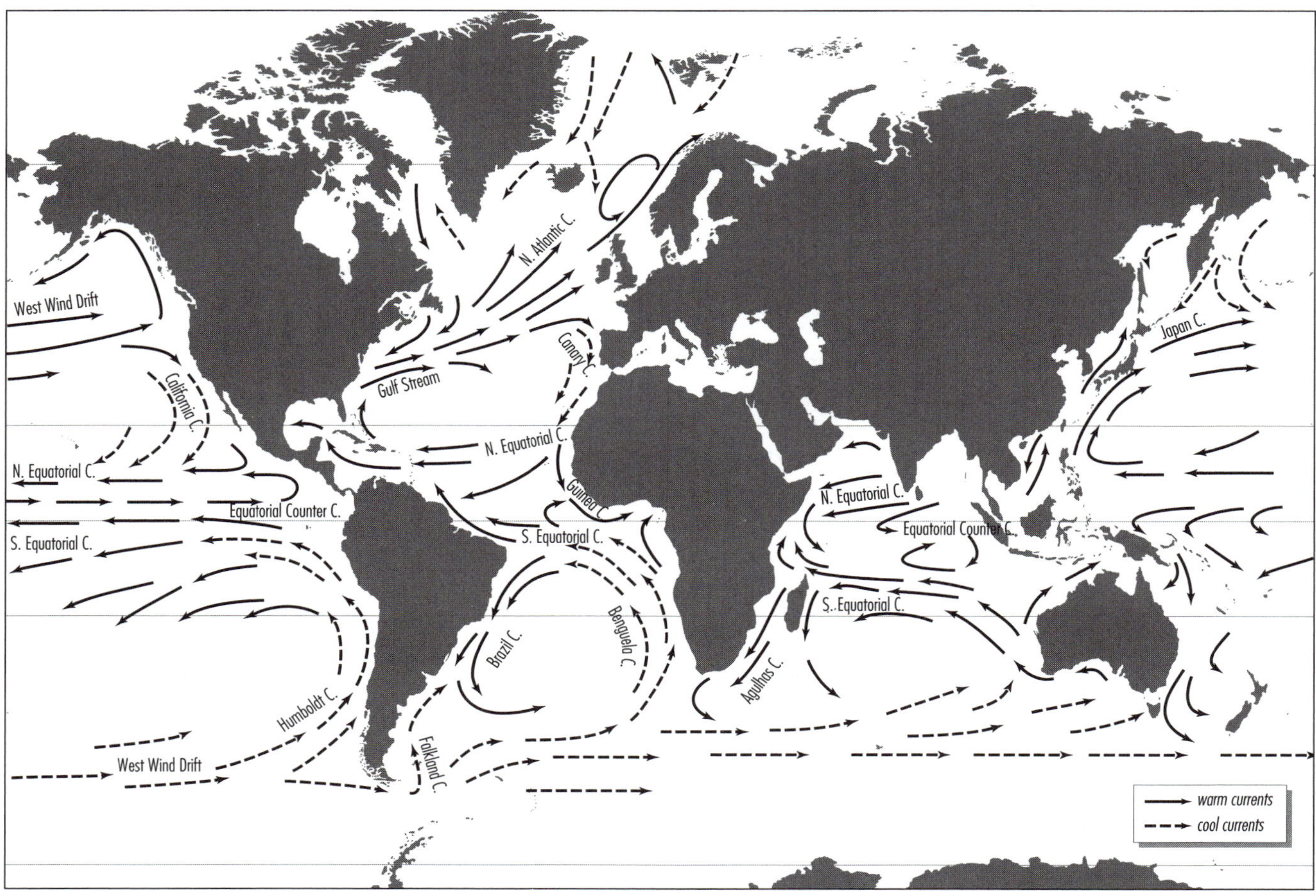

Gyres. **The major currents flow in circles, called *gyres,* around each of the oceans. They flow in a clockwise direction in the Northern Hemisphere and counterclockwise in the Southern Hemisphere.**

the region it traverses. It becomes the ANTILLES CURRENT and FLORIDA CURRENT before becoming the GULF STREAM. Its southward side is the CANARIES CURRENT.

The North Pacific gyre begins as the North Equatorial Current. It moves north as the KUROSHIO CURRENT and NORTH PACIFIC CURRENT and returns to the Tropics as the CALIFORNIA CURRENT.

The South Atlantic gyre flows past the eastern coast of South America as the BRAZIL CURRENT, crosses the Southern Ocean as the WEST WIND DRIFT, and returns to the Tropics as the BENGUELA CURRENT.

The South Pacific gyre begins as the SOUTH EQUATORIAL CURRENT. It flows southward past the coast of Australia as the EAST AUSTRALIAN CURRENT, then returns northward as the PERU CURRENT.

The Indian Ocean gyre begins as the South Equatorial Current, flows southward along the African coast as the AGULHAS CURRENT, becomes part of the West Wind Drift, and then returns northward as the WEST AUSTRALIAN CURRENT.

Ocean gyres are BOUNDARY CURRENTS that carry warm water along the western coasts of all the continents except Antarctica and cool water along the eastern coasts. The GULF STREAM appears to be an exception to this rule, but only because the warm side of the current divides into two branches, one of which, the NORTH ATLANTIC DRIFT, carries warm water to northwestern Europe. Cool water flows past the western coasts of southern Europe and Africa.

H

H *See* HENRY.

haar A cold SEA FOG is common along the eastern coast of Britain, especially in fine weather. The fine weather is associated with an ANTICYCLONE to the north. This draws warm air from continental Europe across the cold water of the North Sea, producing fog over a large area. The easterly winds then carry fog to the British coast. It disperses quickly as it moves inland and is warmed by contact with the sunlit ground, but it can persist all day along the coastal strip. The name may be from the Old Norse *hárr,* meaning "gray" (with age), and from which we also derive the word *hoary.*

haboob A severe DUST STORM that occurs in northern Sudan, most often late in the day in summer. It is associated with the northward advance of the INTERTROPICAL CONVERGENCE and develops when a strong SQUALL blows across a surface covered with loose sand and dust. Pillars of dust rise to a height of several thousand feet and merge to form a wall of dust about 15 miles (24 km) long that advances at about 35 mph (56 km h^{-1}). There is often thunder, but rain evaporates before reaching the ground. The storm lasts for about three hours.

Hadley cells The modern version of description of the GENERAL CIRCULATION of the atmosphere that was proposed in 1735 by the British meteorologist GEORGE HADLEY. Although the idea has been modified considerably, Hadley's account was largely correct.

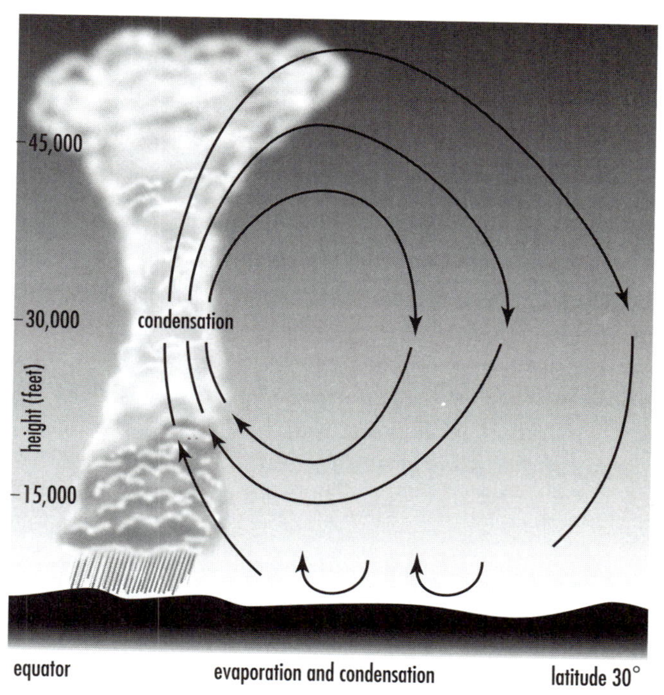

Hadley cells. **Air rises over the equator, cooling as it does so and producing clouds and high rainfall. The air moves away from the equator at a high level and subsides over the Tropics and subtropics.**

Hadley was seeking to explain the direction and reliability of the TRADE WINDS. EDMUND HALLEY had suggested that these represent air returning to the equator at a low level to replace air that had been warmed, risen, and moved away at high level. This explanation failed to account for the fact that the trade winds on each side of the equator blow not from the due north and south, but from the northeast and southeast. This was the missing component that Hadley included. He suggested that air in contact with the surface is heated more strongly at the equator than it is anywhere else. The warm air rises and flows away from the equator at a very high level. By the time it reaches the Poles it is very cold and subsides. It then returns to the equator at a low level, completing the circulation within the CONVECTION CELL. As the air moves across the surface it is deflected by the rotation of the Earth beneath it. It is this deflection that swings the airflow into a more westerly direction, so it arrives at the equator as northeasterly and southeasterly winds. Hadley imagined there was a single convection cell in each hemisphere.

It is not quite so simple. In the first place, the CORIOLIS EFFECT, which had not yet been discovered but was the cause of the deflection on which Hadley's account relied, is very small in the Tropics and zero at the equator. In 1856 WILLIAM FERREL pointed out that the real reason for the deflection is the conservation of ANGULAR MOMENTUM in the moving air.

Nor does the cell extend all the way from the equator to the Poles. Air rises at the equator and subsides in the subtropics, at about latitude 30° in both hemispheres. Subsiding air produces the SUBTROPICAL HIGHS. If there were a single convection cell, there would be a continuous subtropical region of high pressure associated with the subsiding side of the cell. In fact there are several subtropical highs separated by lows. This demonstrates that there is not one convection cell, but as many as there are subtropical highs. There are about five in winter and four in summer. However, the intensity of the convection cells and the subtropical highs appear to be unconnected.

It is only in spring and autumn that there are two sets of cells arranged symmetrically on each side of the equator. This is because the EQUATORIAL TROUGH, which is the center of the CONVERGENCE and rising air, moves north and south with the seasons. The most important cells lie on the winter side of the equator and air spilling out from them crosses the equator into the summer hemisphere. In addition, horizontal air movements are now known to contribute significantly to the transport of heat away from the equator. They are especially important in middle latitudes, where weather systems transport heat.

In the modern version of the Hadley cells, air rises by CONVECTION at the equatorial trough. The air is moist, and as it cools in an ADIABATIC process its water vapor condenses. Condensation releases LATENT HEAT, increasing the instability of the air (*see* STABILITY OF AIR) and producing giant CUMULONIMBUS clouds. This accounts for the heavy rainfall of the humid Tropics.

Air moves away from the equator at the level of the TROPOPAUSE. This air is extremely dry. As it moves away from the equator it loses heat by radiation. This increases its DENSITY. At the same time, the magnitude of the Coriolis effect increases, deflecting the air to the east. By the time it reaches the subtropics, much of the air is moving in an easterly direction. The high-level flow of air toward the Poles is concentrated mainly on the western sides of the subtropical highs. Air accumulates and this also increases its density. Consequently, the air subsides.

As it descends, the dry air becomes still drier. It warms adiabatically, and as its temperature rises its RELATIVE HUMIDITY decreases.

The tropical Hadley cells are now incorporated into a more complex circulation system known as the *THREE-CELL MODEL*. This provides a simplified description of the general circulation.

Hadley, George (1685–1768) English *Meteorologist* George Hadley was born in London on February 12, 1685. He studied law and qualified as a barrister (under the English legal system, a lawyer who is permitted to appear as an advocate in the higher courts).

Hadley became increasingly interested in physics, however, and in particular in the physics of the atmosphere. He was placed in charge of the meteorological observations that were prepared for the Royal Society, a task he performed for at least seven years.

In 1686 EDMUND HALLEY had proposed an explanation for the reliability of the TRADE WINDS (the name *trade* refers to the constancy of the winds and not to their importance to trade). His theory was plausible, but it failed to account for the direction of the trade winds, which blow from the northeast in the Northern Hemisphere and from the southeast in the Southern Hemisphere. George Hadley aimed to complete the theory by explaining the direction of the winds. He agreed with Halley that hot air rises over the equator and is replaced by cooler air flowing toward the equator from higher latitudes, but he noted that the Earth itself is rotating in an easterly direction at the same time. Consequently, the moving air is deflected in a westerly direction, resulting in winds that blow in the directions that are observed. Two centuries passed before this deflection was described in detail by GASPARD GUSTAVE DE CORIOLIS. It is now known as the *CORIOLIS EFFECT.* In 1735 Hadley presented his account of the atmospheric movements that produce the trade winds to the Royal Society in the paper "Concerning the Cause of the General Trade Winds."

Hadley had produced the first model for the GENERAL CIRCULATION of the atmosphere. This supposed one large CONVECTION CELL in each hemisphere. Warm air rises over the equator; moves to the Pole, where it subsides; and then returns to the equator. Although it explains the easterly component of the trade winds, however, this model fails to account for the westerly winds that prevail in middle latitudes. His mistake was to assume there is just a single convection cell in each hemisphere. In fact three sets of cells are included in the THREE-CELL MODEL of the atmospheric circulation. Also, meteorologists now know that the Coriolis effect is weak in low latitudes and does not exist at the equator, so it cannot be the full explanation for the direction of the trade winds. WILLIAM FERREL discovered the true cause of the deflection in the 19th century.

Despite these failings, George Hadley was one of the first people to develop a credible scientific description of the atmospheric circulation. His paper to the Royal Society aroused little interest at the time, and it was not until 1793, long after his death, that Hadley's importance as a meteorologist was recognized by JOHN DALTON. That contribution to meteorology is acknowledged to this day in the name of the HADLEY CELLS.

George Hadley died at Flitton, Bedfordshire, on June 28, 1768.

hail Precipitation in the form of HAILSTONES, which are hard, more or less spherical pieces of ice. They range in size from 0.2 inch (5 mm) to 2 inches (50 mm)—about the size of a golf ball—but they can be bigger. Hailstones have been known to attain the size of softballs, and one that fell at Coffeyville, Kansas, on September 3, 1970, measured 5.5 inches (14 cm) across and weighed 1.67 pounds (766 g). It is believed to have been traveling at more than 100 mph (160 km h^{-1}) when it hit the ground. Large hailstones can cause injury and damage to property, breaking windows, including car windows, and roofs. Intense hailstorms can flatten farm crops.

Hail is produced only in large CUMULONIMBUS clouds and is associated with violent storms. Storms that generate TORNADOES usually deliver intense falls of hail.

(For more information see www.chaseday.com/hail.htm.)

hail day A day on which hail falls.

hail region One of 13 regions into which the United States is divided according to the frequency and intensity of the HAIL they experience. A coded label for each hail region indicates the basic cause of HAIL DAYS, the average annual frequency of hail days, the peak season for hail, and the intensity of the hail that falls. The basic cause may be marine (B), macroscale (large) storms (M), or orographic (O)—meaning storms are caused when air crosses mountains and becomes unstable. The peak season may be early (E), late (L), autumn (A), summer (Su), spring (Sp), or winter (W). The intensity may be light (L), moderate (M), or heavy (H). A coded label appears in the form M/2–3/ESu/L. This means hail is associated with macroscale storms, falls on two or three days in an average year, is most likely in early summer, and hail falls are light.

hailshaft A column of falling HAILSTONES that is visible beneath the CLOUD BASE.

hailstone An ice pellet, approximately spherical in shape, that falls from a CUMULONIMBUS cloud. Hailstones vary greatly in size (*see* HAIL), but most are 0.2–2 inches (5–50 mm) in diameter.

Many hailstones consist of alternate layers of clear and opaque ice. An individual hailstone begins as a raindrop at a low level in the cloud that is carried upward by a vertical air current. The air temperature inside the cloud decreases with height, and as it rises the raindrop enters a region of the cloud where the air is below freezing temperature. The raindrop freezes and as an ice pellet it continues to be carried up and down by air currents.

As it travels, it encounters SUPERCOOLED water droplets. The temperature of some of these is only slightly below freezing. They form a layer of water on the surface of the pellet and then freeze. This produces a layer of clear ice. Other droplets are much colder and freeze instantly on contact with the pellet, forming tiny ice crystals with air trapped between them. This produces a layer of spongy, white, opaque ice.

Eventually, the path of the hailstone carries it into a region of the cloud where downcurrents predominate and it is swept downward and out of the cloud. Its size depends on the number of times it has been carried up and down, growing with each cycle by the accumulation of additional layers, and on the vigor of the upcurrents. The hailstone falls from the cloud when these are no longer strong enough to support its weight. In both cases, therefore, the size of the hailstones reflects the size and vigor of the cloud that produced them.

hailstreak A strip of ground that is completely covered by fallen HAILSTONES. Hailstreaks are seen inside HAILSWATHS.

hailswath An area of ground that is partly covered in fallen HAILSTONES.

hair hygrometer (hygroscope) An instrument that measures atmospheric HUMIDITY and gives a direct reading of the RELATIVE HUMIDITY. It is the most common type of hygrometer for domestic use, but atmospheric scientists usually prefer the PSYCHROMETER. The hair hygrometer exploits the fact that the length of a human hair varies according to the humidity. As the humidity increases, hairs lengthen, and they shrink as the humidity falls. This is a characteristic derived from the physical structure of the hair and remains constant for a particular hair, at 2 percent to 2.5 percent for a change from RH 0 percent to RH 100 percent. A hair hygrometer contains a bundle of hairs that are wound around a drum and linked mechanically to a pointer on a dial. Each instrument is calibrated individually to take account of the particular characteristics of the hairs used.

halcyon days A period of calm, peaceful weather, especially in winter. In Greek mythology, the halcyon was a bird that laid and incubated its eggs around the time of the winter SOLSTICE in a nest that floated upon the sea. The halcyon charmed the wind and waves to make the sea calm. The mythical bird is sometimes identified with the kingfisher, and *halcyon* is sometimes used as a poetic name for this bird. The scientific name of the white-breasted kingfisher of North America is *Halcyon smyrnensis.*

Hale cycle The approximately 22-year cycle during which the magnetic polarity of the Sun reverses. Climate records for the last 300 years show that droughts in the western United States are most severe in the 2–5 years that follow one of the SUNSPOT minima that occur every 11 years, and therefore twice in the course of the Hale cycle. The English solar astronomer EDWARD MAUNDER found a link between sunspot minima and extremes of climate (*see* MAUNDER MINIMUM). The Hale cycle was discovered by the American astronomer George Ellery Hale (1868–1938). Professor Hale was also famous for his development of large astronomical telescopes, including the 200-inch Palomar telescope and the 60- and 100-inch reflecting telescopes at the Mount Wilson Observatory, as well as for his research into solar physics.

half-arc angle The angle bisecting the arc that extends from the point directly above the head of an observer (the zenith for that observer) to the horizon seen by the same observer.

half-life The time that elapses while a given quantity is reduced to half its original value. The term can be applied to atmospheric pollutants, but it is most commonly used to describe rates of radioactive decay. When a radioactive element decays, only the nucleus of the atom is affected. This means the decay occurs independently of outside chemical or physical conditions and is EXPONENTIAL. The probability that a particular atom will decay in a unit time is called the *decay constant* and is usually expressed in units of 10^{10} atoms per year. The decay constant for uranium-235 (^{235}U) is 9.72, for example, and that for ^{238}U is 1.54. It is impossible to predict the lifetime of an individual atom, but simple to calculate how long it takes for half the atoms in a very large sample to decay. This is the half-life for that element and it is given by $T = 0.693/\lambda$, where T is the half-life and λ is the decay constant.

Halley, Edmund (1656–1742) English *Astronomer* Edmund (or Edmond) Halley was born at Haggerston, Shoreditch, which was then a village near London, on October 29 according to the Julian calendar that was then in use, or on November 8 according to the Gregorian calendar that is in use today, in the year 1656. He said he was born in the year 1656, but there is some doubt about this, which arises partly from the fact that in 17th-century England the year began on March 25, and not January 1 as it does today. The family was from Derbyshire, and Edmund's father, who was also called Edmund, had grown wealthy by making soap at a time when the use of soap was increasing throughout Europe. He was murdered in 1684.

Edmund senior could afford a good education for his son and after employing a tutor to teach him at home, sent the boy to St. Paul's School. Young Edmund excelled in Latin and Greek and in mathematics, and he displayed a talent for devising and making scientific instruments. In 1673 he entered Queen's College at the University of Oxford, where he wrote a book on the laws of Johannes Kepler. This drew him to the attention of John Flamsteed (1646–1719), the Astronomer Royal. Halley left Oxford without taking a degree, a practice that was not unusual. Flamsteed employed him as an assistant and then helped him launch his career by spending two years on the island of Saint Helena, charting the stars of the Southern Hemisphere. The project was financed partly by Halley's father and partly by King Charles II. In 1678, after his return from Saint Helena, Halley was elected to the Royal Society.

His reputation grew rapidly. On December 3, 1678, Oxford University awarded him a degree without requiring him to take the examination, and in 1679 the Royal Society sent him to Danzig (now Gdansk) to resolve a disagreement between ROBERT HOOKE and the German astronomer Johannes Hevelius (1611–87). Hevelius, a close friend of Halley's, had made certain astronomical observations without using telescopic sights and Hooke claimed that the observations could not be accurate.

Halley became friendly with Isaac Newton (1642–1727) and financed the publication of Newton's major work, *Philosophae Naturalis Principia Mathematica* (Mathematical principles of natural philosophy). His success alienated Flamsteed, who became increasingly hostile to him.

In about 1695 Halley began making a careful study of the orbits of comets. He calculated that the comet that had appeared in 1682 was the same object

Edmund Halley. **The English astronomer who proposed the first explanation of the trade winds.** ***(John Frederick Lewis Collection, Print and Picture Collection, The Free Library of Philadelphia)***

that had been seen in 1531 and 1607, and in 1705 he predicted that it would be seen again in December 1758. When it appeared on December 25, 1758, it was given his name. It is still known as *Halley's comet.*

Although astronomy was his principal interest, Halley also studied tides, winds, and weather phenomena. In 1686 he attempted an explanation of the TRADE WINDS by proposing that air is heated more strongly at the equator than elsewhere. The warm air rises and draws in cooler air from higher latitudes, flowing toward the equator.

For a time he commanded the *Paramore Pink,* a warship, exploring the coasts on both sides of the Atlantic, charting variations in compass readings, and investigating the tides along the English coast. He also inspected harbors in southern Europe on behalf of Queen Anne.

In 1704 Halley was appointed Savilian professor of geometry at Oxford, and in 1720 he succeeded Flamsteed as Astronomer Royal. Flamsteed's widow was so angry at this that she arranged for all of her husband's instruments to be removed from the Royal Observatory and sold, to prevent Halley from using them.

Edmund Halley died in Greenwich, London, on January 14, 1742, by the Julian calendar and January 25, 1742, by the Gregorian calendar. His tombstone bears an inscription stating that he died in 1741, but this is also correct. In 17th-century England the year began on March 25, so January 25 would have fallen in 1741 by the reckoning of the time, but in 1742 by our reckoning.

(You can learn more about Edmund Halley in Lisa Jardine, *Ingenious Pursuits* [London: Little, Brown and Company, 1999]; at es.rice.edu/ES/humsoc/Galileo/Catalog/Files/halley.html, www.astro.uni-bonn.de/~pbrosche/persons/pers_halley.html; and at www-groups.dcs.st-and.ac.uk/~history/Mathematicians/Halley. html.)

halo An optical phenomenon in which a circle of white light surrounds the Sun or Moon. It is seen when the Sun or Moon is behind a layer of CIRROSTRATUS cloud, so it appears as a white disk. Most halos have a radius subtending an angle of 22° to the eye of the observer. Less commonly, the radius may subtend an angle of 46°. The halo is caused by the REFRACTION of light through ice crystals in the cloud. The difference in size is due to the path taken by light through the crystals. If light enters through one side of the crystals and exits through another side it produces a 22° halo. If it enters through a side and exits through the top or bottom it produces a 46° halo. Halos often mean rain is on its way and a halo surrounding the Moon (a ring around the Moon) means rain is likely by morning.

HALOE *See* HALOGEN OCCULTATION EXPERIMENT.

halogen A chemical element that belongs to group VII of the periodic table. The halogens are fluorine, chlorine, bromine, iodine, and astatine. When halogens enter the STRATOSPHERE they are able to engage in chemical reactions that deplete the OZONE LAYER. Halogens are emitted naturally during volcanic eruptions, and the amount and intensity of volcanic activity partly determine the amount of OZONE present in the stratosphere. Certain halogens are also used in industrially manufactured compounds. Chlorine and fluorine are used in chlorofluorocarbons (*see* CFCs), and chlorine, fluorine, and bromine are used in HALONS. CFCs and halons are implicated in the depletion of the ozone layer and their use is being phased out.

Halogen Occultation Experiment (HALOE) An experiment that is carried on the UPPER ATMOSPHERE RESEARCH SATELLITE. It measures the absorption of infrared radiation, but only at sunrises and sunsets (occultations). This means it acquires data at only two latitudes each day, but since the satellite is in an INCLINED ORBIT the area it views changes so that eventually it completes a global coverage of the STRATOSPHERE. It measures concentrations of a range of gases, including ozone (O_3), nitric oxide (NO), nitrogen dioxide (N_2O), water vapor (H_2O), methane (CH_4), hydrochloric acid (HCl), and hydrogen fluoride (HF).

halon The commercial name for one of the bromofluorocarbons, which are chemical compounds that are used in fire extinguishers. Halon-1211 is a compound of carbon, chlorine, fluorine, and bromine ($CClF_2Br$) that is used in portable fire extinguishers. Halon-1301 is a compound of carbon, fluorine, and bromine (CF_3Br) used in fire extinguishers that are installed in fixed positions and work by flooding the area adjacent to them. Once it is released into the atmosphere halon-1211 survives for 16 years before being removed by natural processes and halon-1301 survives for 65 years. When halons reach the STRATOSPHERE they are broken down by exposure to ULTRAVIOLET RADIATION and then engage in a series of reactions that deplete the OZONE LAYER. Halon-1211 has an OZONE DEPLETION POTENTIAL of 3.0 and that of halon-1301 is 10.0.

halophyte A land plant that is adapted to grow in saline soil or salt-laden air. Many coastal plants are halophytes. Glassworts (*Salicornia* species) are halophytes found near seacoasts and in other salt habitats throughout the world, except in Australia.

hard frost *See* BLACK FROST.

hard UV *See* ULTRAVIOLET RADIATION.

harmattan A moderate or strong hot, dry, dusty wind that blows across West Africa to the south of the Sahara, from the north on the eastern side of the desert and from the northeast on the western side. It occurs at all times of year but only during the day; at night the wind dies down. It is the TRADE WIND that becomes hotter and drier as it crosses the desert and is usually trapped beneath a temperature INVERSION that prevents the air from rising by CONVECTION. It is so dry it hardens leather, warps wood, and in some places reduces the average RELATIVE HUMIDITY to below 25 percent. In the Tropics, however, it produces welcome relief from the high humidity and is sometimes called the "doctor."

Harvey A TROPICAL STORM, with 60-mph (96-km h^{-1}) winds, that produced more than 10 inches (254 mm) of rain in southwestern Florida in September 1999.

Hawaiian eruption *See* VOLCANO.

haze A reduction in VISIBILITY that occurs when AEROSOLS absorb and scatter (*see* SCATTERING) sunlight. Horizontal visibility in a haze does not fall below 1.2 miles (2 km). Usually, the aerosols are large enough to scatter all wavelengths of light. This makes the sky appear white. The amount of scattering depends on the OPTICAL DEPTH of the layer containing the aerosols, which depends in turn on the height of the Sun above the horizon. This is why haze is more common early and late in the day and why an early morning haze often disappears later. People say the haze has "burned off," but in fact the Sun has risen higher into the sky, reducing the optical depth of the haze layer. Hazes also tend to be thicker when the Sun is low, because the aerosols reflect back to the observer sunlight that was reflected from the ground.

Some aerosols absorb certain wavelengths of sunlight but not others. This can give the haze a yellow or brown color. There are also blue hazes, caused by hydrocarbon aerosols emitted by trees and other plants. The particles are so small they scatter blue light but not light at longer wavelengths. Blue hazes have given the Smoky Mountains and the Blue Mountains of Australia their names. The burning of fuels and vegetation also releases particles that are small enough to produce haze, and the air over large cities is often hazy. In dry climates, fine dust particles raised by CONVECTION can cause haze. All types of haze are more likely to develop in the still air beneath a temperature INVERSION, where particles accumulate.

If the RELATIVE HUMIDITY is high, water vapor condenses onto aerosols. The aerosols absorb the water, which increases their size and makes the haze thicker.

A heat haze is different. This is a shimmer caused by a change in the refractive index of air at different densities and is a type of MIRAGE.

haze droplet A liquid droplet that forms by CONDENSATION onto a HYGROSCOPIC NUCLEUS in air with a RELATIVE HUMIDITY (RH) greater than about 80 percent. The droplets grow as the RH rises above 90 percent and begin to reduce VISIBILITY. The droplets are smaller than 1 μm (0.0004 inch) in diameter and remain suspended in the air. Their effect on visibility is too small to be seen in a small volume of air, but they often form a wet HAZE beneath CUMULUS clouds that is clearly visible to observers on mountains or in aircraft, who have an approximately horizontal line of sight through the haze. If the air reaches SUPERSATURATION the haze droplets begin to grow very rapidly. This process removes water molecules from the air, sometimes reducing the RH to below supersaturation. Alternatively, the droplets may grow into CLOUD DROPLETS.

haze factor The ratio of the brightness of an object that is seen through HAZE or FOG to the brightness of the same object seen through clear air. It is a measure of the extent to which the haze or fog reduces VISIBILITY.

haze horizon The top of a HAZE layer when seen from above against a clear sky. In the distance the upper surface of the haze resembles the true horizon. Despite the resemblance, the haze horizon may not be horizontal.

Hazel A HURRICANE that struck islands in the Caribbean and the eastern United States in October 1954. It formed in the Lesser Antilles; then struck Haiti, South Carolina, North Carolina, Virginia, Maryland, Pennsylvania, New Jersey, and New York State; then continued into Canada. On October 18, the hurricane moved out into the Atlantic; eventually it caused strong winds and heavy rain in Scandinavia. It killed an estimated 1,000 people in Haiti, 95 in the United States, and 80 in Canada and caused $250 million of damage in the United States and $100 million in Canada.

haze layer A band of HAZE that is usually seen beneath a temperature INVERSION and that extends to the surface.

haze line The boundary between the upper surface of a HAZE LAYER and the clear air above it.

headwind *See* TAILWIND.

heat capacity (thermal capacity) The amount of heat energy that must be supplied to a substance in order to raise the temperature of that substance. It is measured in relation either to a unit mass of the substance, when it is known as the *specific heat capacity* (symbol c), or to a unit amount of the substance, when it is known as the *molar heat capacity* (symbol C_m). The heat capacities of gases are given either at constant volume (c_v) or at constant pressure (C_v). In the SYSTÈME INTERNATIONAL D'UNITÉS (SI) units that are used scientifically heat capacity is reported in joules per gram per kelvin (J g^{-1} K^{-1}). In the C.G.S. SYSTEM it is reported in calories per gram per degree C (cal g^{-1} °C^{-1}): 1 J g^{-1} K^{-1} = 0.239 cal g^{-1} °C^{-1}; 1 cal g^{-1} °C^{-1} = 4.187 J g^{-1} K^{-1}.

Heat capacity varies widely from one substance to another and water has a higher heat capacity than most common substances, including rock and sand. In the c.g.s. system the specific heat capacity of water is 1.0 cal g^{-1} °C^{-1} (4.187 J g^{-1} K^{-1}). This is an average value, because heat capacity varies slightly with temperature. When stating the heat capacity for a particular substance it is usual to mention the temperature at which this value applies. For example, at freezing temperature the specific heat capacity of water is 4.2174 J g^{-1} K^{-1} (1.007 cal g^{-1} °C^{-1}), at 50° C (122° F) it is 4.1804 J g^{-1} K^{-1} (0.9985 cal g^{-1} °C^{-1}), and at 90° C (194° F) it is 4.2048 J g^{-1} K^{-1} (1.004 cal g^{-1} °C^{-1}). The table gives the specific heat capacities of various substances at specified temperatures.

As a result of its high heat capacity water absorbs a large amount of heat with very little change in its temperature. This is why the sea and large lakes warm only slowly during the spring and early summer, despite the warmth of the sunshine and of the air. In winter, however, large bodies of water also cool very slowly, and as they do so they release a great deal of heat. If a layer of seawater 3.3 feet (1 m) thick cools by 1.8° F (1° C), the heat that is released is sufficient to warm the air above the sea to a height of 33 feet (30 m) by 18° F (10° C).

The huge volume of water that is contained in the oceans acts as a major heat store. The oceans absorb heat during the spring and summer and then release it slowly during the autumn and winter. Consequently, AIR MASSES that cross the ocean are warmed in winter and cooled in summer. The high LATENT HEAT of water also contributes to the summer cooling, because of the large amount of heat that is absorbed from the air in order to evaporate water. It is the contrast between the heat capacities of water and of the rock and sand on dry land that produces the marked differences between continental and maritime climates (*see* CONTINENTALITY and OCEANICITY).

Substance	**Temp.**		***c***	
	°C	**°F**	**J g^{-1} K^{-1}**	**cal g^{-1} °C^{-1}**
freshwater	15	59	4.19	1.00
seawater	17	62.6	3.93	0.94
ice	–21 to –1	–5.8–30.2	2.0–2.1	0.48– 0.50
dry air	20	68	1.006	0.2403
basalt	20–100	68–212	0.84–1.00	0.20–0.24
granite	20–100	68–212	0.80–0.84	0.19–0.20
white marble	18	64.4	0.88–0.92	0.21–0.22
quartz	0	32	0.73	0.17
sand	20–100	68–212	0.84	0.20

Water mixes with sand and other mineral particles, and this means the heat capacity of soils varies with their moisture content. Wet soil warms and cools more slowly than dry soil. This is why wet soil or sand feels cold on a warm day, and dry sand can feel very hot. Crop seeds do not germinate below a certain temperature and for this reason a wet spring delays sowing, because farmers must wait until the soil is warm enough.

heat index *See* TEMPERATURE–HUMIDITY INDEX.

heating degree-days A version of DEGREE-DAYS that is used by heating engineers in their calculations of fuel consumption. One heating degree-day is equal to one degree by which the mean daily temperature falls below a base level. In the United States the base level is 65° F or 19° C. The sum of the heating degree-days over a month, season, or year indicates the amount of heating that will be needed, although the calculation takes no account of cooling by radiation, evaporation, or wind.

heat island An area that is markedly warmer than its surroundings and that is therefore like an island of warmth in a sea of cool air. This effect was first described by LUKE HOWARD in *The Climate of London,* published in 1818–19, and it has since been observed in many large cities. It affects all urban areas, but large cities most of all. It is patchy: more extreme in some parts of a city than in others, depending on the kind of activity pursued—industrial or residential, for example—and on the configuration of the buildings and streets. The accumulation of warmth modifies the city atmosphere in a number of ways and produces a distinctive URBAN CLIMATE.

The temperature in the city rises rapidly during the day and falls again at night. Sometimes the city center is cooler than the countryside at night, but this is unusual. More commonly the heat-island effect reaches a maximum during the first part of the night, when the difference between the temperature at ground level in the city center and in the surrounding countryside is often 5°–6° C (9°–11° F) and can reach 8° C (14° F). The temperature decreases fairly steadily from the city center to the outlying suburbs along a temperature gradient. Once in the countryside the temperature remains constant.

When the air is otherwise calm, the heat island often produces a COUNTRY BREEZE. This is similar to a sea breeze (*see* LAND AND SEA BREEZES).

The heat island results from several causes. Buildings reduce wind speed. The wind slows the more deeply it penetrates the city, and this reduces its capacity for exchanging air. Consequently, city air that is warmed is replaced less frequently by cooler air. Air is exchanged much more frequently in the open countryside.

The removal of most plants reduces the amount of TRANSPIRATION. Where there is an extensive vegetation cover, transpiration absorbs a considerable amount of LATENT HEAT, which cools the air. Trees that line city streets transpire too little moisture to make much difference, but transpiration by the plants in large city parks produces smaller islands of cool temperatures inside the larger heat island.

EVAPORATION is also reduced, because rain and melted snow are carried away rapidly through storm drains, rather than being left to dry by evaporating from exposed water surfaces. In the countryside, some surface water evaporates immediately and some soaks into the ground, but much of the water that sinks from the surface either is drawn upward again by CAPILLARITY or enters plants and is returned to the air by transpiration. Evaporation also absorbs latent heat from the layer of air close to the ground.

Bricks, concrete, stone, and asphalt absorb heat better than the soil, water, and plants of the open countryside (*see* HEAT CAPACITY). They have a much lower ALBEDO than farmland and a skyscraper absorbs up to six times more heat than a field per unit of surface area. This means city streets and buildings warm quickly during the day and lose their heat just as rapidly at night. The warm surfaces warm the air in contact with them. On a warm summer day the temperature in the city can rise by as much as 17° C (31° F) between dawn and the middle of the afternoon.

Cities also generate heat. Buildings are heated in winter and in warm climates they are cooled by air conditioners in summer. Heated buildings lose heat through windows and doors to the outside air and air conditioners work by expelling heated air from the building and replacing it with chilled air. Machines of all kinds warm the air around them and city streets are usually filled with vehicles, all pumping out heated air. A very large urban area, such as Los Angeles or the

Boston–Washington megalopolis, generates an amount of heat that is equal to half the solar radiation that falls on a horizontal surface in winter and about 15 percent of the amount that falls in summer.

There are now so many very large cities that heat islands are clearly visible in INFRARED satellite photographs. Many weather stations are located at city airports that have become surrounded by urban development as cities have expanded. They are now within the boundaries of heat islands. Since the weather stations are the source of data from which changes in surface temperatures are calculated, some climatologists suspect that, although corrections are made to allow for it, at least some of the reported GLOBAL WARMING may be due to the heat island effect.

heat lightning *See* THUNDER.

heat low *See* THERMAL LOW.

heat stress, index of *See* COMFORT ZONE.

heat thunderstorm A popular name for the type of THUNDERSTORM that sometimes develops on a warm, humid afternoon, when the ground has been strongly heated and has made the air above it unstable (*see* STABILITY OF AIR).

heat wave A period of at least one day, but more usually lasting several days or weeks, during which the weather is unusually hot for the time of year. Heat waves in middle latitudes are often caused by BLOCKING. In North America, summer heat waves occur when the belt of prevailing westerly winds is shifted to the north. The SUBTROPICAL HIGH then expands and cT air replaces mT air (*see* CONTINENTAL AIR, MARITIME AIR, and TROPICAL AIR.)

During a heat wave affecting much of the central United States in 1936 the temperature over parts of the Great Plains exceeded 120° F (49° C), and several eastern states reached 109° F (43° C). Nearly 15,000 people died as a result of that heat wave. A heat wave in 1980 affecting Missouri, Georgia, Tennessee, and Texas produced temperatures so high that asphalt roads became plastic, concrete road surfaces expanded until they cracked and buckled, and sometimes they exploded violently. More than 1,200 people died. The highest temperatures were recorded in Memphis (108° F [42.2° C]), Augusta (107° F [42° C]), and Atlanta (105° F [41° C]). A heat wave is especially dangerous if the humidity rises as well as the temperature. Advice is available to help people protect themselves from the risks of high temperature and humidity (*see* COMFORT ZONE and TEMPERATURE–HUMIDITY INDEX).

(You can learn more about coping with a heat wave from the American Red Cross at www.redcross.org/disaster/safety/heat.h.)

heavenly cross A SUN PILLAR that is crossed by a horizontal bar.

heiligenschein An optical phenomenon in which the shadow of an observer on ground covered with vegetation has a bright, white light surrounding the head. It is seen when the Sun is shining and the surface vegetation is covered with DEW, provided the dewdrops are of the right size and in the right position to act as lenses, focusing an image of the Sun onto the plant surface beneath. The observer sees this image of the Sun repeated innumerable times and magnified by the dewdrops through which it is seen. *Heiligenschein* is the German for "halo."

Heinrich event The relatively sudden calving of a large swarm of ICEBERGS from North America into the North Atlantic. This is believed to have occurred six times between 70,000 and 16,000 years ago, during the most recent GLACIATION. The icebergs, from the LAURENTIDE ICE SHEET, carried rock rubble scraped from the land surface. They drifted eastward, and as they slowly melted the rubble sank to the ocean floor. Fragments of it were found and their Canadian origin determined by the German oceanographer Hartmut Heinrich, and it was he who proposed, in 1988, that icebergs had transported them to the northeastern Atlantic, where he found them. Various causes have been proposed for these events. Some scientists suggest the calving resulted from warming, perhaps linked to one of the MILANKOVICH CYCLES. Others believe it was due to the extent to which the ice sheet thickened during the period of the events. Still others maintain earthquakes fractured the sheet. As the icebergs melted, they formed a layer of fresh water floating above the seawater. This is believed to have suppressed the formation of NORTH ATLANTIC DEEP WATER and thus altered or shut down the ATLANTIC CONVEYOR.

(There is more information about Heinrich events and the earthquake theory at http://www.dukenews.duke.edu/Research/HEINRICH.HTM.)

heliotropic wind A slight change in the wind direction that takes place during the course of the day and is caused by the east-to-west progression of the surface area that is most intensely warmed by the Sun.

helium (He) An odorless, colorless, gaseous, nonmetallic element that constitutes 0.0005 percent by volume of the atmosphere below a height of 15.5 miles (25 km). Helium is completely chemically inert and forms no known compounds. It is released into the atmosphere through the ALPHA DECAY of radioactive potassium and uranium that are present naturally in rocks. Radioactive potassium (^{40}K) also occurs in all organic matter. Alpha particles, comprising two protons and two neutrons, are the nuclei of helium atoms. Because the air is constantly being mixed, the proportion of helium it contains remains fairly constant throughout most of the atmosphere. The EXOSPHERE, however, consists of oxygen, hydrogen, and helium atoms. About 1 percent of the helium atoms in the exosphere are ionized (*see* IONIZATION).

helm bar *See* HELM WIND.

Helmont, Jan Baptista van (1577–1644) Flemish *Physician and alchemist* Jan van Helmont discovered the existence of gases and claimed to have coined the word GAS. He also identified the gas we know as carbon dioxide. He was born in Brussels (his date of birth is not known, but it was in the year 1577) and was educated at Louvain. He studied several sciences and finally concentrated on medicine, in which he graduated in 1599. In 1609 he moved to Vilvorde, near Brussels, where he spent the rest of his life practicing medicine and conducting chemical experiments.

In his work and ideas van Helmont bridged the medieval and modern worlds. He was a mystic, interested in the supernatural, and an alchemist who believed in the philosopher's stone—the object alchemists believed would change base metals such as lead into gold. He even claimed to have seen it used. He believed in spontaneous generation, by which living organisms develop from nonliving materials, and asserted that mice are produced from dirty wheat. At the same time he was in touch with the scientific developments of his day and his experiments were performed carefully and accurately. In one experiment he grew a willow tree in a measured quantity of soil to which he added only water. At the end of five years he found the tree had gained 164 pounds in weight, but the soil had lost only two ounces. He concluded that the tree was converting water into its own tissues. This was incorrect, but it was an early example of a carefully quantified experiment in biology, and it did prove that plants do not draw their principal nourishment from the soil.

Some of his experiments produced vapors, and he was the first person to recognize that these are distinct substances, each with its own properties. Unlike liquids and solids, vapors immediately fill any space they enter. He thought this meant they existed in a state of chaos, a word he spelled *gas,* which is the way it sounded when spoken by a Flemish speaker.

Charcoal gives off a gas when it is burned and because the gas is from wood, for which the Latin word is *silva,* he called it *gas sylvestre.* He discovered that the same gas is given off when malted barley is fermented to make beer. It is the gas we call *carbon dioxide.*

Jan van Helmont continued experimenting and working as a physician until his death, on December 30, 1644.

helm wind A cold, northeasterly wind that blows along the valley of the River Eden in Cumbria, northern England, especially during late winter and spring. It derives its name from the distinctive cloud, shaped like a helmet (or helm), that is seen when the wind is blowing over the high ground of the western side of the Crossfell Range. The cloud itself is known as a *helm bar* and comprises a thick bank of cloud along the top of the hills, with a narrow band of almost motionless cloud protruding from it away from the hills.

hemispheric wave number *See* ANGULAR WAVE NUMBER.

Hengelo interstade (**Hoboken interstade**) An INTERSTADE that occurred in northern Europe around 40,000 years ago, during the middle of the DEVENSIAN GLACIAL. July temperatures averaged about 50° F (10° C). Together, the Moershoofd (when summer temperatures were 43°–45° F [6°–7° C]), DENEKAMP, and Hengelo interstades are equivalent to the UPTON WARREN INTERSTADE.

henry (H) The derived SYSTÈME INTERNATIONAL D'UNITÉS (SI) UNIT OF INDUCTANCE, which is equal to the inductance of a closed circuit that varies uniformly at one AMPERE per SECOND and produces an electromotive force of one VOLT. The name of the unit was adopted in 1893 in honor of the American physicist JOSEPH HENRY.

Henry, Joseph (1797–1878) American *Physicist* As a boy, Joseph Henry showed little interest in school. He was born in Albany, New York, on December 17, 1797. His family was poor, and at 13 Joseph left school and was apprenticed to a watchmaker. He might have remained a watchmaker, but for a curious event that happened when he was 16, spending a vacation on a farm owned by a relative. A rabbit he was chasing ran beneath a church and Joseph crawled under the building to follow it. Some of the floorboards above him were missing, so he climbed through the gap and into the church, where he came across a shelf of books. One, called *Lectures on Experimental Philosophy,* attracted him. He started to read it and was inspired to return to school.

Henry enrolled at Albany Academy to study chemistry, anatomy, and physiology, paying his way by teaching in country schools and tutoring. He graduated, hoping to practice medicine, but in 1825 he obtained a job surveying a route for a new road in New York State. This aroused his interest in engineering, but in 1826 he returned to Albany Academy to teach mathematics and science—which was then known as *natural philosophy.*

Electricity and magnetism were research topics of great interest in Europe and Henry began experimenting with electromagnets. He was the first American since BENJAMIN FRANKLIN to undertake important experimental work with electricity. He was the first to insulate wire for the magnetic coil, using silk from one of his wife's petticoats, and he invented spool winding. In 1829 he made the first electromagnetic motor, and by 1831 he was able to make a range of electromagnets, from very delicate ones to those that could lift 750 pounds (340 kg) and another that lifted more than one ton.

In 1832 he was appointed a professor at the College of New Jersey (now Princeton University). The same year Henry discovered self-induction—the phenomenon in which an electric current flowing through a wire coil induces a second current in the coil, so the current becomes the original and induced currents combined. Henry had read a preliminary account of work on induction by the British physicist and chemist Michael Faraday (1791–1867). Henry was the first to perform the key experiments, but Faraday was the first to publish his results. Nevertheless, at a meeting in Chicago in 1893, the Congress of Electricians named the henry as the unit of inductance. It is now the SI unit of inductance (*see* SYSTÈME INTERNATIONAL D'UNITÉS).

This work gave Henry another idea. Suppose there were a very long wire with a battery at one end, an electromagnet at the other, and a key that would open and close the circuit. Every time the key closed the circuit a current would reach the electromagnet, which would attract a small iron bar, causing it to move. When the key was released, opening the circuit, the attraction would cease and a small spring would pull the iron bar back to its original position. When the iron bar moved it would make an audible click, so if the key were pressed repeatedly to make a pattern of taps, the same pattern would be heard as clicks from the iron bar. By 1831, Henry had made this device and it was working. He had invented the telegraph, and in 1835 he invented the relay. This was a series of similar circuits in which each circuit activated the next. It overcame the diminution in the signal passing through a length of wire that is caused by resistance in the wire itself.

Henry believed that scientific discoveries should benefit everyone. He did not patent his inventions and he was happy to describe them in detail. Consequently it was SAMUEL MORSE who made the first practical telegraph line and who patented the telegraph.

In December 1846 Joseph Henry was elected the first secretary of the Smithsonian Institution, which had only recently been founded. An excellent administrator, he turned the institution into a major clearinghouse for scientific information.

Henry mobilized scientific effort during the American Civil War. He was elected the second president of the National Academy of Sciences and was active in organizing the American Association for the Advancement of Science and the Philosophical Society of Washington.

Joseph Henry. **The American physicist who contributed to the invention and development of the telegraph and who established a network of volunteer weather observers throughout the United States.** ***(John Frederick Lewis Collection, Picture and Print Collection, The Free Library of Philadelphia)***

One of his first projects at the institution reflected another of his interests: meteorology. Henry established a corps of voluntary observers, located all over the United States, who used the telegraph to send reports to a central office at the institution. This was the first use of the telegraph for a scientific purpose. For the next 30 years Henry supported and encouraged this volunteer corps. When the U.S. WEATHER BUREAU was established in 1891 it used the system for collecting data that Henry had devised.

Joseph Henry died in Washington on May 13, 1878. His funeral was attended by many government officials and by Rutherford B. Hayes, president of the United States.

Herb A TYPHOON that struck Taiwan on July 31, 1996. It produced 44 inches (1118 mm) of rain in 24 hours in Mount Ali, Taiwan, flooding thousands of homes. It killed at least 41 people. Herb then crossed the Taiwan Strait, reaching Pingtan, in Fujian Province, China, on August 1.

hertz (Hz) The derived SYSTÈME INTERNATIONAL D'UNITÉS (SI) UNIT OF FREQUENCY, which is equal to one cycle per SECOND. The name of the unit was adopted in 1933, in honor of the German physicist Heinrich Rudolph Hertz (1857–94).

heterogeneous nucleation The freezing of SUPERCOOLED water droplets onto FREEZING NUCLEI. *Heterogeneous* refers to the fact that two different substances—a solid particle and water—are involved.

heterosphere That part of the atmosphere, above a height of about 50 miles (80 km), that has a heterogeneous (nonuniform) composition. In the lower region, molecular nitrogen (N_2) is predominant, together with oxygen in both molecular (O_2) and atomic (O) forms. Above that, atomic oxygen predominates, with some molecular and atomic (N) nitrogen. Higher still the most abundant (but extremely tenuous) gas is atomic helium (He), and above that there is atomic hydrogen (H).

hibernal Pertaining to the winter. The word is derived from *hibernus,* the Latin word for "wintry."

hibernation The method by which some mammals survive the winter, when food is very scarce. Most animals that hibernate are no larger than a bat or mouse. Marmots (*Marmota* species) are the largest. They weigh an average 11 lb (5 kg).

Prior to hibernation, an animal eats voraciously to lay down layers of body fat that will be metabolized gradually through the winter to supply the energy its body needs. The animal then finds a place where it will be safe from predators and where its own body temperature is unlikely to fall below freezing. It makes a nest there and stocks the nest with food that it will need if it should wake up during the winter.

As it enters hibernation, the animal falls into a deep sleep. Its metabolic rate then slows to the minimum needed for survival, typically about 1 percent of the rate when the animal is active. Its rate of breathing slows, its heartbeat slows to about half its nonhibernating rate, its blood vessels constrict, and its body temperature falls, usually to about 40° F (4.5° C). The temperature is maintained at this level throughout the winter, and if it threatens to fall lower the animal awakens and shivers to warm itself. It replaces the energy by feeding on food it stored before entering hibernation.

The change on entering and leaving hibernation is dramatic. In ground squirrels, for example, the first step toward entering hibernation is a drop in the pulse rate from about 200 beats per minute to about 10 beats per minute. Then the temperature falls from 95° F (35° C) to about 39° F (4° C). Then the breathing slows from about 100 breaths per minute to about 4 breaths per minute.

When the weather starts to grow warmer the animal begins to arouse itself. Its heartbeat accelerates, it shivers violently, its breathing accelerates, its blood vessels dilate, and its body temperature rises from 40° F (4.5° C) to 95° F (35° C).

(You can read more about hibernation in Michael Allaby, *Ecosystem: Deserts* [New York: Facts On File, 2001].)

high An area in which the surface atmospheric pressure is higher than the pressure in the surrounding area. The term refers only to horizontal comparisons of pressure and not to differences in pressure at varying heights above each other, and it most commonly is applied to conditions at sea level. Comparisons can be made at any height above the surface, however, provided the pressures being compared are measured at the same height. Thus it would be correct to speak of a high at a particular height, although this use of the term is unusual. It also refers to measurements taken at the same time and not to variations in pressure in the same place but at different times. Air circulates in an ANTICYCLONIC manner around a high and *high* is often used interchangeably with ANTICYCLONE, but the two words are not synonymous (*see also* LOW).

high-altitude station A weather station that is located at a sufficiently high elevation for the conditions it records to be significantly different from those at sea level. High-altitude stations are sited no lower than about 6,500 feet (2,000 m) above sea level.

high cloud Cloud that has a base at above 6,000 m (20,000 feet) altitude. CIRRUS, CIRROCUMULUS, and CIRROSTRATUS are classified as high clouds.

high fog FOG that develops on the upper slopes of a mountain and sometimes extends as STRATUS cloud over the valley on the LEE side.

high föhn (stable-air föhn) The situation that occurs when cold, stable air lies beneath a layer of potentially warmer air. It develops when cold, stable air moves against a mountain range. In stable air (*see* STABILITY OF AIR) the upper layer is at a higher POTENTIAL TEMPERATURE than the lower layer. The upper layer of air spills over the mountains and descends, warming in an ADIABATIC manner as it does so. A high föhn can also develop when subsiding air in an ANTICYCLONE is chilled by a cold land or sea surface.

high index A high value of the ZONAL INDEX, which means that the JET STREAM is strongly developed and blowing steadily in an approximately easterly direction (*see* INDEX CYCLE). Weather systems move at a fairly steady pace in the same direction as the jet stream.

highland climate In the climate classification devised by A. N. Strahler, a climate in his group 3, comprising climates controlled by polar and arctic air

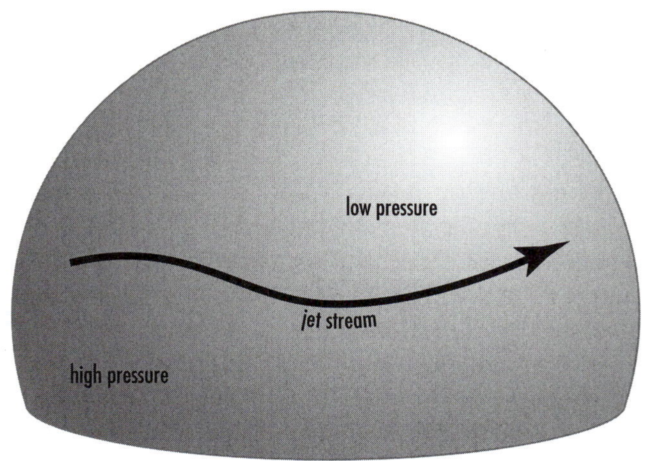

High index. **The wave pattern in the jet stream is only weakly developed and the jet stream is aligned approximately from west to east.**

masses. Highland climates occur at high altitudes in mountain ranges throughout the world. They are cool and moist, but local in extent.

high-level inversion A temperature INVERSION that forms 1,000 feet (300 m) or more above the surface. It is caused by the cooling of subsiding air in an ANTICYCLONE that occurs when the surface is much colder than the air. Above the chilled layer the air is warmer.

high-resolution Doppler interferometer (HRDI) An INTERFEROMETER that is carried on the UPPER ATMOSPHERE RESEARCH SATELLITE. It measures the DOPPLER EFFECT on sunlight that is scattered in the STRATOSPHERE and on AIRGLOW emissions in the MESOSPHERE. From this it provides data on the temperature and winds throughout much of the stratosphere (but with a gap in the upper stratosphere) and most of the mesosphere.

high-resolution picture transmission (HRPT) A method used by satellites to transmit images with a resolution of 0.7 mile (1.1 km), allowing them to depict features 0.5 mile (0.8 km) across. HRPT transmits two visible and three infrared channels.

hill fog (upslope fog) FOG that forms when moist air is made to rise, causing it to cool in an ADIABATIC manner. If it cools to its DEW POINT TEMPERATURE, water vapor condenses. This is the only type of fog caused by adiabatic cooling. Hill fog occurs on hillsides, but also on the Great Plains of North America, when moist air from the Gulf of Mexico moves westward across the continent, up a gentle but long gradient toward the Rocky Mountains.

historical analog model A CLIMATE MODEL that is based on comparisons with weather conditions that obtained at some time in the past. It is an ANALOG MODEL because it uses comparisons, rather than calculations of physical values for pressure, temperature, and precipitation. When estimating the possible consequences of increasing the atmospheric concentration of CARBON DIOXIDE, for example, modelers might study the global distribution of temperature and precipitation during periods in the past when the CO_2 concentration is estimated to have been higher than it is today. Alternatively, future changes can be estimated from the difference in the conditions that prevailed during the 5 coldest and 5 warmest years over a 50-year period.

hoar frost Ice that forms as a thin layer of white crystals on grass, other herbs, shrubs, trees, spider webs, and other exposed surfaces. It is seen in the morning after a night in which heat radiated from the surfaces of objects has reduced their temperature to below freezing. Air in the layer immediately adjacent to the cold surfaces has been chilled to below its frost point. This raised its relative humidity to above 100 percent, causing water vapor to change directly to ice (SUBLIMATION).

Hoboken interstade *See* HENGELO INTERSTADE.

Holocene epoch (Post Glacial, Recent) The geological time (*see* GEOLOGICAL TIME SCALE) in which we are now living. It commenced about 10,000 years ago, when the most recent (WISCONSINIAN) GLACIAL ended. The Holocene follows the PLEISTOCENE EPOCH, and the two together compose the Quaternary subera of the Cenozoic era.

The history of the Holocene has been reconstructed by the study of POLLEN found in layers of soil that can be dated (*see* POLLEN ANALYSIS and POLLEN DIAGRAM) and is divided into five ages, the PRE-BOREAL, BOREAL, ATLANTIC, SUB-BOREAL, and SUB-ATLANTIC PERIODS.

Holocene climates are warmer than those of the Pleistocene, but the present epoch has continued for only about the length of one of the Pleistocene INTERGLACIALS. Despite concerns about GLOBAL WARMING, it is likely that we are living in an interglacial that will be followed by a new ice age. If that is so, it may be that we are really living in a Pleistocene interglacial and the use of the term *Holocene* should be abandoned.

Holocene interglacial *See* FLANDRIAN INTERGLACIAL.

Holsteinian interglacial *See* HOXNIAN INTERGLACIAL.

homogeneous atmosphere A hypothetical atmosphere in which the density of the air remains constant at all heights.

homogeneous nucleation The spontaneous freezing of SUPERCOOLED water droplets that occurs at very low temperatures in the absence of FREEZING NUCLEI. Because only one substance—water—is involved, the process is said to be homogeneous.

homosphere That part of the atmosphere in which the chemical composition is homogenous (uniform). This is the lower part of the atmosphere, extending from the surface to a height of about 50 miles (80 km) and comprising the TROPOSPHERE and STRATOSPHERE. *See* ATMOSPHERIC COMPOSITION.

Honorinnia A CYCLONE that struck Madagascar on March 17, 1985. It destroyed 80 percent of the buildings in Toamasina and left 32 people dead and 20,000 homeless.

Hooke, Robert (1635–1703) English *Physicist* One of the most ingenious experimenters and instrument makers who ever lived, Robert Hooke was born on July 18, 1635, at Freshwater, in the Isle of Wight, an island off the southern coast of England. His father, John Hooke, was a clergyman, employed as the curate at Freshwater. A curate is an assistant to the parish priest. It is a poorly paid occupation.

John wanted his son to enter the church, but as a child Robert was often ill and he was not strong enough to undertake the necessary studies. He spent much of his time alone, amusing himself by making mechanical toys. Later he attended school in Oxford, and after the death of his father in 1648 Robert was sent to London. At first he seems to have embarked on an apprenticeship, for which he paid the £100 he inherited from his father. Then he enrolled at Westminster School. He learned Latin and Greek, although he never wrote in either, but his best subject was mathematics, in which he excelled. It took him only a week to master the geometry of Euclid. In 1653 he entered Christ Church College, at the University of Oxford. At first he was a chorister and then he became a servitor, an undergraduate student who was assisted financially from the college funds, in return for which he performed certain menial duties. Robert was already a highly skilled instrument maker, and he earned a living by selling ideas for modifications and improvements to the owners of the professional workshops where scientific instruments were made.

He did not take a degree, but while he was at Oxford he joined a group of brilliant scientists, one of whom was ROBERT BOYLE. Boyle was a wealthy man and employed Hooke as an assistant, paying him generously. The two became close lifelong friends, and although Hooke ceased to be employed by Boyle in 1662, Boyle continued to pay him until 1664, when Hooke was in a better financial position. Hooke's first task, in 1658, was to help design a pneumatic pump that would remove air from a vessel into which animals and scientific instruments could be placed and the effect on them observed. Hooke succeeded and Boyle used the pump for his experiments on gases.

In 1659 the group of friends began to separate. Most of them, including Boyle and Hooke, moved to London, where, in 1660, they formed a scientific society. In 1662 this became the Royal Society of London. Hooke was appointed curator of experiments, a position that required him to demonstrate new experiments at each weekly meeting. He was paid £30 a year and was provided with accommodation in Gresham College, in London. In 1663 he was elected a Fellow of the Society and in the same year he was awarded an M.A. degree from Oxford. In 1664 he was made a lecturer in mechanics at the Royal Society, for which he received £50 a year, and in 1665 he was appointed professor of geometry at Gresham College, also at a salary of £50 a year. He remained in these two posts for the rest of his life. From 1677 until 1683 he was secretary to the Royal Society.

His work for the society, and especially his job as curator of experiments, involved him in every branch of science and allowed him to develop to the full his mechanical skills. It also meant, however, that although he originated many ideas and improved other people's inventions, he left many projects unfinished.

Despite that, his achievements were considerable. He found that the stress placed on an elastic body is proportional to the strain it produces. This is known as *Hooke's law.* He invented an anchor escapement mechanism for a watch and claimed it was the first, although that assertion led to a dispute over priority with Christopher Huygens (1629–95). He also claimed to have discovered gravitation and the INVERSE SQUARE LAW before Isaac Newton (1642–1727). He greatly improved the design of microscopes and his book *Micrographia,* published in 1665, contained the first important set of drawings of microscopic observations.

He insisted that fossils were the remains of plants and animals that lived long ago.

Hooke invented the wheel barometer, which indicated the pressure by means of a needle on a dial. He suggested ways to apply barometer readings to weather forecasting. It was Hooke who first labeled a barometer with the words *change, rain, much rain, stormy, fair, set fair,* and *very dry.* He designed, but did not make, a weather clock to record air temperature, pressure, rainfall, humidity, and wind speed on a rotating drum. He suggested that the freezing point of water be used as the zero reference point on thermometers.

Hooke died in London on March 3, 1703.

(You can read more about the world in which Robert Hooke lived in Lisa Jardine, *Ingenious Pursuits* [London: Little, Brown and Co., 1999].)

hook pattern A distinctive shape that is often visible in radar images taken from directly above the clouds associated with a SUPERCELL storm. The hook pattern usually occurs at the edge of a MESOCYCLONE and therefore indicates an immediate risk of TORNADOES. The hook pattern is not entirely reliable, however, since not all mesocyclones produce one and not all mesocyclones lead to tornadoes. DOPPLER RADAR provides more reliable information on conditions inside a storm cloud.

horologion *See* TOWER OF THE WINDS.

horse latitudes The latitudes of the SUBTROPICAL HIGHS, at approximately 30° in both hemispheres, where air that has risen over the equator is subsiding and diverging on the poleward side of the HADLEY CELLS. These cells are not continuous, but where air from them is sinking to the surface the winds are light and variable and often the air is calm. Sailing ships were sometimes becalmed in these latitudes. The name *horse latitudes* refers to the fact that ships often carried cargoes of horses. When the ships were becalmed supplies of water sometimes ran low and horses died and were thrown overboard.

Hortense A TROPICAL STORM that struck Martinique on September 7, 1996, and then headed for the British Virgin Islands, Puerto Rico, and the Dominican Republic. By September 10 it had strengthened to a category 1 HURRICANE on the SAFFIR/SIMPSON SCALE. It then moved to the Bahamas and Turks and Caicos Islands. It killed 16 people in Puerto Rico and the Dominican Republic.

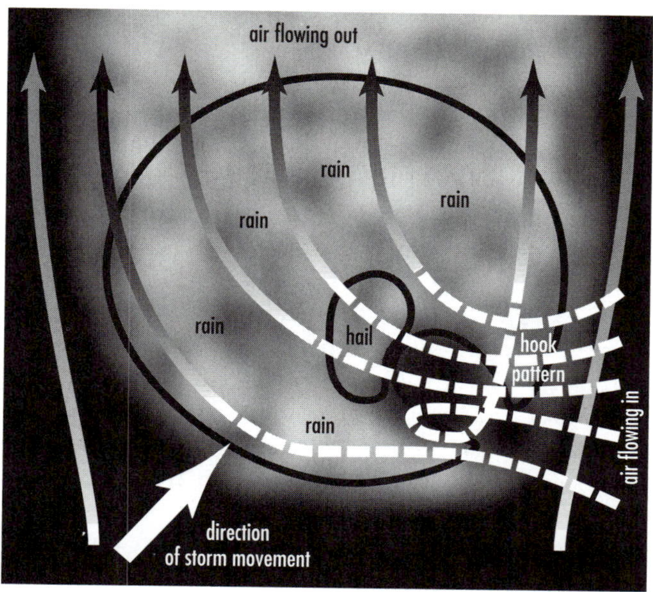

Hook pattern. **A hooklike shape visible from above in the radar image of a cloud that indicates the imminent risk of tornadoes. The broken lines indicate air entering the storm at low level and climbing. The solid lines indicate air at high level leaving the storm.**

hot belt The belt around the Earth within which the mean annual temperature exceeds 68° F (20° C).

hot lightning LIGHTNING that ignites forest fires, in contrast to COLD LIGHTNING. The current carried by the lightning stroke is sustained for a fraction of a second until the RETURN STROKE. This generates enough heat to ignite dry material.

hot tower A narrow column of air that is rising rapidly by CONVECTION and that is enclosed by a much larger volume of air that is rising very little or that is sinking. Hot towers occur widely in the Tropics and the equatorial air that rises in the HADLEY CELLS does so in the form of many hot towers.

Howard, Luke (1772–1864) English *Chemist, pharmacist, and meteorologist* Luke Howard devised a method for classifying clouds that formed the basis of the system that was adopted internationally and that remains in use to this day. His was not the only attempt at classification. At about the same time, the chevalier de Lamarck also proposed a scheme (*see* CLOUD CLASSIFICATION) and developed it over several years. Lamarck was a leading scientific figure and an authority on biological classification, but his cloud classification was based on rather vague definitions. Whether for that reason or some other, his scheme was not adopted, and Howard's was.

Luke Howard was born in London into a Quaker family and was educated at a Quaker school at Burford, near Oxford. He was educated as a pharmacist (druggist), and when he grew up he earned his living as a manufacturer of chemicals. He was a businessman and chemist but was never a professional scientist.

His interest in meteorology began in the summer of 1783, when he was 11 years old. That year there were two major volcanic eruptions, one in Iceland and the other in Japan. Volcanic dust formed a haze around the world that produced spectacular skies. Then, on August 18, Luke witnessed a dramatic meteor blaze across the sky. He began to keep a record of his meteorological observations and maintained the habit for more than 30 years.

Howard was a founder member of the Askesian Society, which was one of the many philosophical societies (*philosophy* was the name given to what we now call science) formed in the late 18th and early 19th centuries. At a meeting of the society in 1800 he presented a paper, "The Average Barometer," and in 1802 he presented another paper, "Theories of Rain."

Howard became interested in biological classification, which was then being strongly influenced by the work of the Swedish naturalist Linnaeus (Carl Linné, 1707–78). In 1735, Linnaeus had published a book called *Systema Naturae,* in which he introduced a system based on a binomial nomenclature, in which every species was given two names, one of its genus and the other of the species itself. In 1803 (although some historians say it was in December 1802), Howard presented a paper to the Askesian Society, "On the Modification of Clouds." By *modification* he meant what we would call "classification," and in this paper he followed Linnaeus in allotting Latin names to cloud types and proposed ways of combining the names in a Linnaean, binomial fashion.

Howard proposed that all clouds belong to one of three groups: *cumulus, stratus,* and *cirrus.* A fourth group, called *nimbus,* denoted a cloud that was producing rain, hail, or snow. *Cumulus,* the Latin word for "heap," Howard described as "convex or conical heaps, increasing upward from a horizontal base—Wool bag clouds." *Stratus,* the past participle of the Latin verb *sternere,* "to strew," and meaning "layer," he described as "a widely extended horizontal sheet, increasing from below." *Cirrus,* the Latin word for "curl," he described as "parallel, flexuous fibers extensible by increase in any or all directions." *Nimbus,* Latin for "cloud," but to which Howard attached the meaning "rain," he described as "a rain cloud—a cloud or system of clouds from which rain is falling."

He maintained that rain could not fall from cumulus, stratus, or cirrus as long as they "retain their primitive forms." Clouds can alter their forms, however. Cumulus clouds could fill the sky to become cumulostratus, which is "cirro-stratus blended with cumulus." Similarly, cirrus could become cirrocumulus and cirrostratus.

His classification attracted widespread attention and Howard became a celebrity; his fame increased more when all his meteorological papers up to that time were collected by Thomas Forster and published in 1813 as *Research about Atmospheric Phaenomenae.* Wolfgang von Goethe even dedicated four poems to Howard.

In 1806, Howard began the publication *Meteorological Register,* which appeared regularly over several years in the *Athenaeum Magazine.* In 1818–19, he published the first book ever written on urban climate, *The Climate of London,* in two volumes, and in 1833 published an expanded second edition in three volumes. In it he made what is believed to be the first reference to what is now called a *heat island,* with temperature records to support it. A heat island is an urban area that is warmer than the surrounding countryside. A series of lectures he delivered in 1817 were published in 1837 as *Seven Lectures in Meteorology* and became the first textbook on the subject. His last book, *Barometrographia,* appeared in 1847.

In recognition of his contributions to meteorology Howard was elected in 1821 a Fellow of the Royal Society of London, the highest honor British scientists can bestow on one of their colleagues.

Luke Howard remained a devout Quaker throughout his life. He died at a great age in London in 1864.

Luke Howard. **The English meteorologist who devised the first successful system for classifying clouds.** ***(Copyright, Royal Meteorological Society, UK)***

(There is more information about Luke Howard in Keith C. Heidorn, "Luke Howard, the Man Who Named the Clouds," at www.islandnet.com/~see/weather/history/howard.htm and "The Godfather of the Clouds" at cloudman.com/luke1/luke_howard1.htm.)

Hoxnian interglacial An INTERGLACIAL period in Britain that followed the ANGLIAN GLACIAL and preceded the WOLSTONIAN GLACIAL. It is named after the village of Hoxne, Suffolk. The Hoxnian is equivalent to the Holsteinian interglacial of northern Europe and the Mindel–Riss interglacial of the European Alps. It partly coincides with the YARMOUTHIAN INTERGLACIAL of North America. Dating is uncertain, but the Hoxnian probably began about 250,000 years ago and ended about 200,000 years ago. Early humans (*Homo erectus*) were making stone hand axes in Britain at this time. The climate during the Hoxnian was strongly oceanic (*see* OCEANICITY), with high rainfall, mild winters, and warm but not hot summers.

HRDI *See* HIGH-RESOLUTION DOPPLER INTERFEROMETER.

HRPT *See* HIGH-RESOLUTION PICTURE TRANSMISSION.

Hugo A HURRICANE, rated category 5 on the SAFFIR/SIMPSON SCALE, that struck islands in the Caribbean and the eastern coast of the United States from September 17 to 21, 1989, generating winds of up to 140 mph (224 km h^{-1}). On September 11, Hugo reached Guadeloupe, where it killed 11 people. It reached Dominica in the Lesser Antilles (Leeward Islands) on September 17. Then it moved to Saint Croix, Saint John, Saint Thomas, and the smaller islands of the U.S. Virgin Islands and Puerto Rico, arriving on September 19. It killed 10 people on Montserrat, 6 in the Virgin Islands, and 12 in Puerto Rico. Almost the entire population of Montserrat was made homeless, as well as 90 percent of the population of Saint Croix and 80 percent of the people of Puerto Rico, and 99 percent of all homes were destroyed in Antigua. Hugo then turned north and weakened to category 4. On September 21, it struck Charleston, South Carolina, where one person died when a house collapsed. At Folly Beach, South Carolina, 80 percent of the population was left homeless. The following day Hugo reached Charlotte, North Carolina, where one child was killed. It crossed the Blue Ridge Mountains the same day and in the afternoon moved across Virginia, where two people died and winds of 81 mph (130 km h^{-1}) were recorded. There was a STORM SURGE at Awendaw, South Carolina, and the storm triggered several TORNADOES that caused damage on several islands and on the mainland in North Carolina. In all, the hurricane caused damage in the United States estimated at $10.5 billion.

hum *See* HUMILIS.

Humboldt Current *See* PERU CURRENT.

Humboldt, Friedrich Heinrich Alexander, Baron von (1769–1859) German *Geologist, geophysicist, geographer* Alexander von Humboldt was born in Berlin on September 4, 1769. Berlin was then the capital of Prussia, and von Humboldt's father was a Prussian officer who served as an official at the court of the king, Frederick II (Frederick the Great) and who wanted his son to pursue a political career. Alexander was more interested in science, however, and after the death of his father in 1779 he was educated privately before enrolling at the University of Göttingen in 1789 to study science.

While there he met Georg Forster, who had accompanied James Cook on the second of his voyages of exploration. The two became firm friends and Humboldt spent only one year at Göttingen before he and Forster set off on a journey through the Netherlands and England, where he met many leading scientists. On his return to Prussia, Humboldt realized he would need a formal qualification if he were to make any useful contribution to science. In 1791 he became a student at the Freiburg Bergakademie (School of Mining). He spent two years there before graduating in geology, and while studying mining he became fascinated by the plants that grow in and around mines.

After graduation from the Bergakademie, Humboldt was appointed assessor of mines and later director of mines in the Prussian principality of Bayreuth. He founded a school of mining, improved conditions for the miners, and also conducted his own research into the magnetic declination of the rocks in the area. He spent the years from 1792 to 1797 on a diplomatic mission that took him to the salt-mining regions of several central European countries. In the course of these travels he met more of the most senior scientists of the day.

Humboldt's mother died in 1796, and Alexander inherited a share of the family fortune. This meant he no longer needed to earn a living and could indulge his passion for travel. He went first to Paris and from there to Marseilles, accompanied by the French botanist Aimé Bonpland (1773–1858). They planned to travel to Egypt, where they hoped to join Napoleon, but instead they went to Madrid, where the prime minister, Mariano de Urquijo, became their patron. With his support they changed their plans and determined to visit the Spanish colonies in South America.

They sailed from Spain in 1799, landed in New Andalusia (modern Venezuela), and early in 1800 started on a four-month expedition through Latin America. They explored the course of the Orinoco and confirmed that the headwaters of the Orinoco were linked to those of the Amazon. By the time they returned to their base on the coast, at Cumaná, they had traveled 1,725 miles (2,775 km). They then sailed to Cuba, stayed there for several months, and returned to South America in March 1801, arriving at Cartagena, Colombia.

They the embarked on a second expedition, this time on a route that crossed the Andes. As they climbed, Humboldt noted the changes in vegetation at different elevations and recorded the decrease in air temperature with height. He also made many geophysical observations of the alignment of volcanoes and of the Earth's magnetic field. When they reached the Pacific coast Humboldt measured the temperature of the water offshore and discovered the existence of the cold current that is sometimes named after him (*see* PERU CURRENT).

Humboldt and Bonpland left South America in February 1803, spent a year in Mexico, visited the United States, and sailed for Europe on June 30, 1804. In the course of their explorations the two men had covered about 6,000 miles (9,600 km).

Humboldt spent the following years in Berlin and Paris arranging the vast amount of material he had collected during his travels—including 60,000 plant specimens, many of which were new to science—and writing accounts of his experiences and discoveries. His major work, *Voyage de Humboldt et Bonpland,* appeared in 30 volumes between 1805 and 1834. Most of his fortune had now been spent and in order to secure an income Humboldt agreed to serve as a Prussian diplomat in Paris.

Humboldt was one of the founders of biogeography, the study of the geographical distribution of plants and animals. He was the first scientist to measure the decrease of temperature with altitude, and he investigated the cause of TROPICAL STORMS. This supplied information other scientists used later to determine the processes involved in the weather systems of middle latitudes.

His many discoveries and his liberal opinions had made Humboldt a celebrity. He was said to be the second most famous man in Europe, after Napoleon. He died in Berlin on May 6, 1859, at the age of 89 and was given a state funeral.

humid climate In the THORNTHWAITE CLIMATE CLASSIFICATION, a climate in which the MOISTURE INDEX is between 20 and 100 and the POTENTIAL EVAPOTRANSPIRATION is 22.4–44.9 inches (57–115 cm). There are four subdivisions (B_1 to B_4). In terms of

Alexander von Humboldt. **The German geologist, geophysicist, and geographer who was the first to measure the decrease of temperature with altitude and who investigated the cause of tropical storms.** ***(From the collections of the Library of Congress)***

THERMAL EFFICIENCY, this is a mesothermal climate (B'_1 to B'_4).

humid continental climate In the CLIMATE CLASSIFICATION devised by A. N. STRAHLER, a climate in his group 2, comprising climates controlled by both tropical and polar AIR MASSES. Humid continental climates occur in latitudes 35°–60° N in the central and eastern regions of continents, where polar and tropical air masses meet. The weather is very variable and there are strong contrasts between summer and winter. Precipitation is abundant and increases in summer, when maritime tropical air masses enter. Winters are dominated by continental polar air masses that invade frequently from the north. This climate type includes the cold, snowy forest, or humid microthermal, climate. It is designated *Dfa* in the KÖPPEN CLASSIFICATION if it is wet throughout the year and the summers are hot and DFB if they are warm, *Dwa* if winters are dry and summers hot, and *Dwb* if winters are dry and summers warm.

humidification The deliberate addition of moisture to the air in order to increase its HUMIDITY, often with the aim of reducing the accumulation of STATIC ELECTRICITY.

humidifier A device that injects moisture into the air. It works by putting the air in contact with water, with the air and water both at the same temperature. If necessary, the air is warmed or cooled to change it to the temperature of the water. Humidifiers are often fitted to air-conditioning or central heating systems. The operation of a humidifier is controlled by a humidistat, which monitors the HUMIDITY of the air, bringing air and water into contact only when the humidity is low. Some humidifiers incorporate a dehumidifier, which removes moisture by putting the air in contact with very cold water, which causes some water vapor to condense, or by passing the moist air over a bed of hygroscopic crystals.

humidistat *See* HUMIDIFIER.

humidity The amount of water vapor that is present in the air—the "wetness" or "dryness" of the air. The term refers only to water that is present as a gas; clouds, which are composed of ice crystals or liquid droplets of water, and precipitation falling through the air do not count in the measurement of humidity.

Humidity can be measured in several ways: as the MIXING RATIO, SPECIFIC HUMIDITY, ABSOLUTE HUMIDITY, and the most widely used measure of all, as RELATIVE HUMIDITY. When weather reports state the humidity, relative humidity is usually what they mean. The instruments that measure humidity are known as *HYGROMETERS*.

The amount of water vapor air can hold varies according to the temperature—the warmer the air the more water vapor it is able to hold. At 59° F (15° C), for example, which is the mean temperature over the entire surface of the Earth, air can hold as much water vapor as produces 17 mb of pressure (*see* VAPOR PRESSURE). When the temperature rises to 95° F (35° C), the amount of water vapor the air can contain increases to a SATURATION VAPOR PRESSURE of 50 mb. At 14° F (–10° C) it can hold only 2.6 mb. As proportions of sea-level atmospheric pressure, these are 1.7 percent, 4.9 percent, and 0.26 percent, respectively. This demonstrates that water vapor is never more than a minor constituent of the air and that the amount is very variable. It is not included in lists of the ingredients making up the air, because the amount varies from place to place and from hour to hour.

Air temperature decreases with height; therefore, the moisture-holding capacity of the air also decreases with height. Because of this, most of the water vapor in the atmosphere is contained in the air below about 18,000 feet (5,500 m).

Water vapor can also be measured as *precipitable water*, which is the depth to which an area of the ground surface would be covered by rain water if all the water vapor in the column above that area were converted to rain. The amount is surprisingly small. Over southern Asia, during the MONSOON, the air holds no more than about 2.5 inches (64 mm) of rain, accounting for about 7 percent of the air by weight. Air over a low-latitude desert holds about 0.5 inch (10 mm), and air over central Antarctica often holds no more than one-tenth of that amount (by weight about 2 percent and 0.2 percent, respectively).

Water enters the air by EVAPORATION and TRANSPIRATION. Evaporation reaches a maximum during the day but may continue at a reduced rate through the night. This means the movement of water vapor in the air is predominantly upward and moist air is carried aloft mainly by eddies. As it rises, drier air descends from higher levels, diluting it. Moisture returns to the surface by PRECIPITATION.

humidity coefficient A measure of the effectiveness for plant growth of the PRECIPITATION falling over a region during a specified period. It relates the amount of precipitation to the temperature and is calculated as $P/1.07^t$, where P is the amount of precipitation in centimeters and t is the mean temperature for the period in question in degrees Celsius.

humidity index A measure of the extent to which the amount of water available to plants exceeds the amount needed for healthy growth. The term was introduced by C. W. THORNTHWAITE and is calculated as $100W_s/PE$, where W_s is the WATER SURPLUS and PE is the POTENTIAL EVAPOTRANSPIRATION.

humidity province One of the five categories into which climates are divided on the basis of their PRECIPITATION EFFICIENCY INDEX (P–E) value in the THORNTHWAITE CLIMATE CLASSIFICATION. The five provinces are labeled A, B, C, D, and E. Province A, with a P–E index greater than 127, is the rain forest climate. B, with a P–E index of 64–127, is the forest climate. C, with a P–E index of 32–63, is the grasslands climate. D, with a P–E index of 16–31, is the steppe climate. E, with a P–E index of less than 16, is the desert climate.

humid mesothermal climate *See* TROPICAL WET–DRY CLIMATE, HUMID SUBTROPICAL CLIMATE, MARINE WEST-COAST CLIMATE, and MEDITERRANEAN CLIMATE.

humid microthermal climate *See* HUMID CONTINENTAL CLIMATE and CONTINENTAL SUBARCTIC CLIMATE.

humid subtropical climate In the CLIMATE CLASSIFICATION devised by A. N. STRAHLER, a climate in his group 2, comprising climates controlled by both tropical and polar AIR MASSES. Humid subtropical climates occur in latitudes 20°–35° in both hemispheres along the eastern edges of continents exposed to moist maritime tropical air masses that move from the western side of oceanic high-pressure cells. Summers are hot, with heavy rain. Winters are cool and often affected by polar air masses, and there are frequent storms produced by FRONTAL SYSTEMS. This climate includes the temperate rainy, or humid mesothermal, climate and is designated *Cfa* in the KÖPPEN CLASSIFICATION.

humilis (hum) A species of CUMULUS clouds (*see* CLOUD CLASSIFICATION) that have flat bases and little vertical development. They represent cumulus that has failed to grow. *Humilis* is a Latin word that means "lowly."

hurricane A TROPICAL CYCLONE that forms in the North Atlantic or Caribbean. It moves in a westerly direction, carried by the easterly airflow of the Tropics, beginning as a TROPICAL DEPRESSION associated with an EASTERLY WAVE, intensifying to a TROPICAL STORM, and usually reaching hurricane force as it enters the Caribbean. Its westward track across the Caribbean may carry it across several inhabited islands. Then its VORTICITY causes it to turn onto a more northerly track that may carry it toward the United States, where it may reach the Gulf Coast, Florida, or the Carolinas. It continues to turn until it is on an easterly track that carries it back over the Atlantic, occasionally as far as northwestern Europe, where it may still have enough strength to cause considerable damage.

Hurricanes occur only in the North Atlantic and never to the south of the equator; this is because the EQUATORIAL TROUGH never moves south of 5° S, which is too far to the north for the CORIOLIS EFFECT to start moving air turning about its own vertical axis.

hurricane-force wind A wind of force 12 on the BEAUFORT WIND SCALE, with a wind speed in excess of 75 mph (121 km h^{-1}). This wind speed defines a category 1 hurricane on the SAFFIR/SIMPSON HURRICANE SCALE.

hurricane monitoring buoy A free-floating instrument package that detects the approach of a tropical cyclone and is designed to be expendable.

hurricane warning A notification of the approach of a HURRICANE to an inhabited area of the United States that is issued by the National Hurricane Center, in Miami, Florida, and broadcast on local radio, television, and WEATHER RADIO. A hurricane warning is broadcast when the hurricane is expected to arrive within 24 hours or less. Persons receiving a hurricane warning should prepare for the imminent arrival of the hurricane and should leave a radio or television switched on and tuned to the local station. Updated information and safety instructions will be broadcast and should be obeyed promptly.

hurricane watch A preliminary notification of the approach of a HURRICANE to an inhabited area of the United States that is issued by the National Hurricane Center, in Miami, Florida, and broadcast on local radio, television, and WEATHER RADIO. A hurricane watch, based on observations of the advancing storm and predictions of its future track, is issued one or two days before the expected arrival of the storm. It is given to people residing in a belt of coastline and its hinterland centered on the point where it is anticipated that the STORM CENTER will cross the coast. The affected area extends for a distance equal to about three times the radius of the hurricane to each side of this point. On receipt of a hurricane watch people should prepare for the arrival of the hurricane. This includes preparation to evacuate.

Hyacinthe A CYCLONE that struck the island of Réunion, in the Indian Ocean, in January 1980. At least 20 people were killed.

hydraulic conductivity *See* PERMEABILITY.

hydrodynamical equations *See* NUMERICAL FORECASTING.

hydrogen (H) The lightest element, and the most abundant in the universe. Its atom comprises one proton and one electron. Its atomic number is 1, relative atomic mass 1.00794, melting point –434.45° F (–259.14° C), and boiling point –423.17° F (–252.87° C). It is a colorless, odorless gas. Hydrogen occurs naturally as two ISOTOPES. Hydrogen-1 accounts for 99.985 percent of hydrogen. The remainder is deuterium, or heavy hydrogen (relative atomic mass 2.0144), in which the nucleus comprises one proton and one neutron. Deuterium oxide is heavy water. A third isotope, tritium, is made artificially. Hydrogen is present in water and in all organic compounds. It burns readily and is often proposed as an alternative fuel to gasoline for driving vehicles, because the only combustion product is water ($2H_2 + O_2 \rightarrow 2H_2O$ + heat).

hydrogen bond An attractive force that links molecules in which hydrogen is bonded to nitrogen, oxygen, or fluorine. These are POLAR MOLECULES, and the electrostatic attraction is between poles with opposite charge. In ammonia (NH_3), for example, there is a bond between the hydrogen (H^+) of one molecule and the strongly negative nitrogen (N^-) of an adjacent molecule. Oxygen atoms are also strongly negative, and water molecules are linked by hydrogen bonds between the hydrogen of one molecule and the oxygen of its neighbor. It differs from other types of chemical bonding in that it bonds molecules, not atoms. Hydrogen bonding in water causes the rearrangement of molecules on FREEZING that causes the density of ice to be less than that of liquid water.

hydrological drought A DROUGHT during which the GROUNDWATER table falls markedly. This reduces the flow of streams and rivers. It can be caused either by a prolonged period without rain over the watershed supplying the groundwater or by an unusually low accumulation of winter snow in a mountainous region, leading to a reduced flow of melt water in spring.

hydrological cycle (**water cycle**) The constant circulation of water from the oceans to the atmosphere and back to the surface as PRECIPITATION. There is a total of about 0.33 billion cubic mile (1.36 km³) of water on Earth, of which about 3 percent exists as FRESHWATER, and about 75 percent of all the freshwater on Earth is held in the polar icecaps and GLACIERS. The table sets out the way the freshwater is distributed:

Location	Percentage of total freshwater
icecaps and glaciers	75
groundwater	22
upper soil	1.75
lakes and inland seas	0.6
rivers	0.003

The small amount of freshwater that exists in the liquid phase must satisfy all the needs of organisms that live on land and in freshwater lakes and rivers. This water is transported through the atmosphere and the amount present in the atmosphere at any given time is approximately 3,120 cubic miles (13,000 km³), which is about 0.03 percent of the total amount of freshwater and about 0.001 percent of all the water on the planet.

Each year about 77,000 cubic miles (321,000 km³) of water evaporates from the oceans and about 15,000 mi³ (62,500 km³) leaves the land by EVAPOTRANSPIRATION, so a total of 92,000 mi³ (383,500 km³) of water enters the atmosphere. Of the total amount evaporating from the oceans, about 7,000 mi³ (29,000 km³) is carried by the air from the oceans to the land. The remainder, of about 70,000 mi³ (291,600 km³), falls back into the ocean as precipitation. About 21,000 mi³ (87,500 km³) falls as precipitation over land. This leaves a balance of 6,000 mi³ (25,000 km³), which is the amount that is carried by rivers from the land back to the sea.

It is EVAPORATION that injects water vapor into the atmosphere, the CONDENSATION of water vapor that produces CLOUDS, and clouds that produce precipitation. Although the amount of water present in the air as vapor, liquid droplets, hailstones, ice crystals, and snowflakes represents only a tiny proportion of the total amount of water on the Earth, it is nevertheless involved in most of the processes that generate our weather. The hydrological cycle, or water cycle, is what produces most of our weather.

hydrological model A MODEL that is used to simulate the behavior of real hydrological systems, such as drainage basins and river flow. The model may be a small-scale physical device that mimics the real system, a computer simulation, or a sequence of mathematical calculations.

hydrometeor A drop of water, ICE CRYSTAL, or SNOWFLAKE that has condensed or frozen from water vapor in the atmosphere and that is deposited on a surface. The particles that constitute every form of PRECIPITATION (DRIZZLE, FOG, RAINDROPS, GRAUPEL, HAIL, SLEET, and SNOW) are hydrometeors and so are DEW, FROST, and VIRGA.

hydrometeorology The branch of METEOROLOGY that specializes in the study of PRECIPITATION. Types of precipitation are sometimes called *hydrometeors*. This is the aspect of meteorology that is of most direct concern to farmers and growers, irrigation engineers, flood-control engineers, designers and managers of hydroelectric schemes, and others with a particular interest in surface waters.

hydrosphere Water that is present at or close to the surface of the Earth. The term includes the oceans, icecaps, lakes, and rivers, and also GROUNDWATER. It is contrasted with the atmosphere, which is the gaseous envelope surrounding the Earth, and the lithosphere, which is the solid surface of the Earth.

hydrostatic approximation The assumption that the atmosphere is in HYDROSTATIC EQUILIBRIUM.

hydrostatic equation A mathematical equation that relates the weight of a column of air to the pressure exerted on it from above, at height p_2, and below, at height p_1. The cross-sectional area of the column is taken to be 1. The weight of the air is equal to its mass (M) multiplied by its gravitational acceleration (g). The equation is then

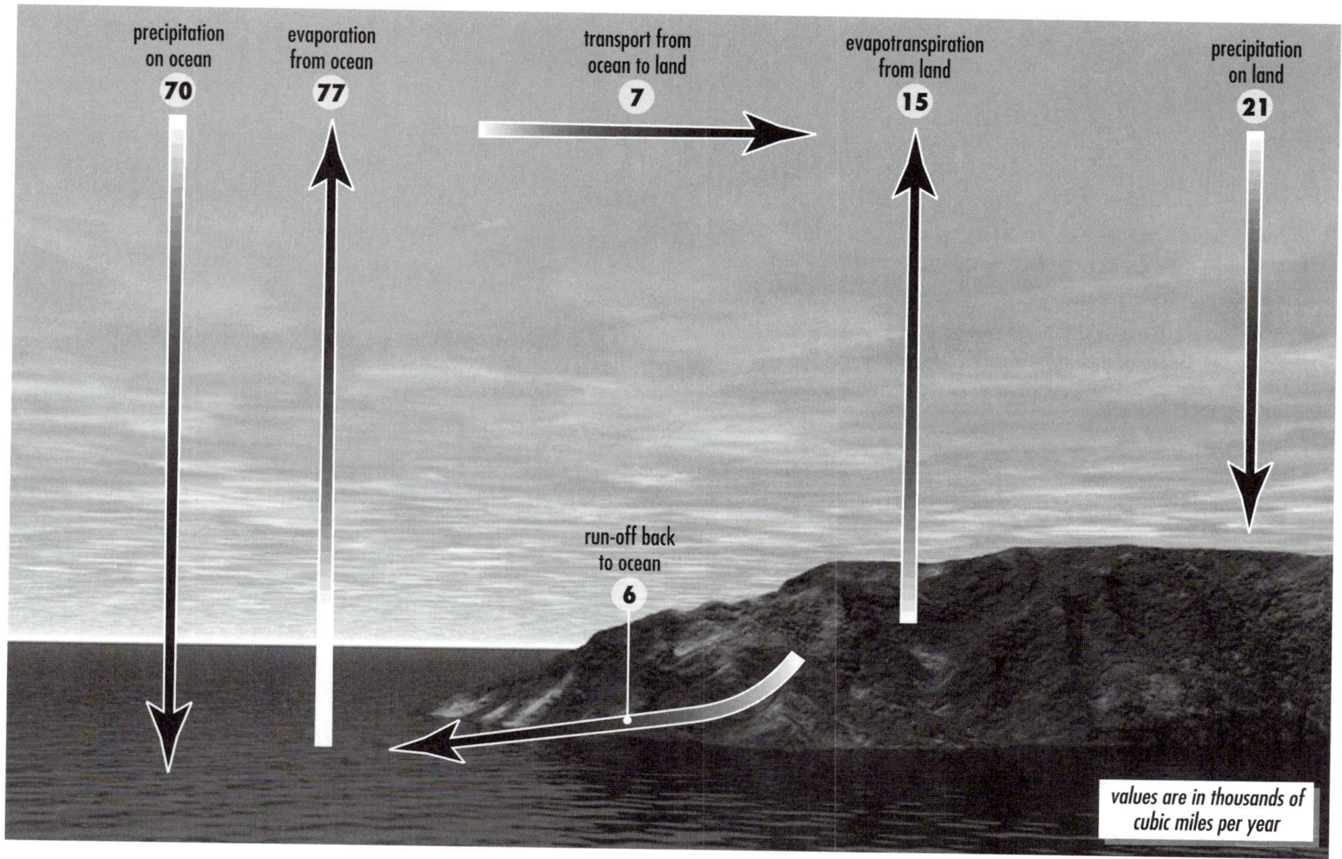

Hydrological cycle. **Water enters the air by evaporation and evapotranspiration and returns to the surface as precipitation. A small amount is transported from the oceans to the land through the air, and rivers carry the same amount from the land back to the sea. (Numbers do not add up precisely because of rounding.)**

$$M g = p_1 - p_2$$

This assumes that the density of the air remains constant throughout the column and that the gravitational acceleration is a constant. The density remains almost constant for a shallow column, but for a deeper column allowance must be made for the vertical pressure gradient. This produces a differential equation:

$$\delta p/\delta z = -g\rho$$

where p is pressure, z is height, ρ is density, and the minus sign indicates that density decreases with height. Gravitational acceleration decreases with height according to the INVERSE SQUARE LAW, but within the TROPOSPHERE the effect is so small it may be ignored. Gravitational acceleration also changes with latitude, and by an amount large enough to be significant in some calculations. To allow for this, GEOPOTENTIAL HEIGHT is used rather than vertical distance (geometrical height).

hydrostatic equilibrium The condition of the atmosphere in which the weight of a PARCEL OF AIR is exactly balanced by an upward pressure at its base. If a parcel of air sinks under its own weight, it compresses the air beneath it. This increases the upward pressure exerted by the air beneath the parcel until that upward pressure is equal to the weight of the parcel. Its descent is then halted and it is in hydrostatic equilibrium. The circumstances leading to hydrostatic equilibrium are described by the HYDROSTATIC EQUATION.

hydroxyl A hydroxide molecule, which consists of one oxygen atom bonded to one hydrogen atom (OH). It is produced mainly by the action of ULTRAVIOLET

RADIATION on OZONE in the presence of water molecules.

$$O_3 \rightarrow O_2 + O$$

$$O + H_2O \rightarrow 2OH$$

Hydroxyl carries a negative charge and is known as a *free radical.* It is extremely reactive. METHANE and many chemical pollutants are removed from the air by reactions with hydroxyl that yield harmless or soluble products. Methane, for example, reacts to form carbon dioxide and water.

hyetogram A chart that records the amount and duration of rainfall at a particular place.

hyetograph An instrument that measures and automatically records rainfall. It consists of a reservoir in which rainwater collects. The reservoir also contains a float that is connected mechanically to a pen. As the water accumulates, the float rises, moving the pen, which traces a line on a chart fixed to a rotating drum.

hyetography The scientific study of the annual geographical distribution of precipitation and variations in it.

hygrogram A chart that shows a continuous record of the RELATIVE HUMIDITY of the air over a period, usually of one week, as this is measured by the HYGROGRAPH.

hygrograph A type of HYGROMETER in which changes in the RELATIVE HUMIDITY of the air cause a pen to rise or fall, tracing a line on a chart that is fixed to a rotating drum. The resulting record is known as a *hygrogram.*

hygrometer An instrument that measures the HUMIDITY of the air. There are several types. The HAIR HYGROMETER is the one most often used domestically. It is the cheapest and gives a direct reading without requiring calculations. The PSYCHROMETER is often used by meteorologists. There are also the dew cell, which measures DEW POINT TEMPERATURE, and the RESISTANCE HYGROMETER, which is the instrument most often used in RADIOSONDES and weather satellites.

hygrophyte A plant that grows in wet places or in very humid climates.

hygroscope *See* HAIR HYGROMETER.

hygroscopic moisture Water that is absorbed from the atmosphere by soil particles. It is held very tightly by the particles and cannot enter the root hairs of plants.

hygroscopic nucleus An airborne particle that acts as a CLOUD CONDENSATION NUCLEUS because it possesses the property of *wettability*—it is made from a substance that absorbs water and swells in size as it does so, eventually dissolving into a concentrated solution. Common salt (sodium chloride, NaCl) is hygroscopic. Crystals of salt that are left exposed to the air gradually clump together and eventually turn into solution, using water they have absorbed from the air. Salt crystals occur naturally in the air over the sea. They remain airborne after the water has evaporated from drops of seawater that are thrown into the air as spray. Other naturally occurring hygroscopic nuclei include particles of dust, smoke, sulfate (SO_4), and sulfur dioxide (SO_2).

hygrothermograph An instrument that provides a continuous record of both RELATIVE HUMIDITY and temperature. It consists of a THERMOGRAPH and HYGROGRAPH, each connected to a separate pen.

hyperosmotic *See* OSMOSIS.

hypoosmotic *See* OSMOSIS.

hypsometer (hypsometric barometer) A BAROMETER that calculates AIR PRESSURE by measuring the temperature at which a liquid boils. This is possible because the SATURATION VAPOR PRESSURE of a boiling liquid is equal to the atmospheric pressure acting on the surface of the liquid. The saturation vapor pressure of distilled water at 212° F (100° C) is 14.7 pounds per square inch (1,013.25 millibars), at 122° F (50° C) it is 1.79 lb in^{-2} (123.45 mb), and at 77° F (25° C) it is 0.46 lb in^{-2} (31.69 mb). Consequently, if the air pressure is 1.79 lb in^{-2}, distilled water boils at 122° F. The hypsometer supplies heat to boil liquid contained in its reservoir and measures the temperature immediately above the liquid surface. Hypsometers are used to measure pressure at high altitudes, where other types of barometer are less accurate (*see* ALTIMETER). At very high altitudes, where the pressure is 0.03–0.7 lb in^{-2} (0.02–0.5 mb), carbon disulfide (CS_2) is sometimes used instead of water, because at sea-level pressure it boils at 114.8° F (46° C).

hypsometric barometer *See* HYPSOMETER.

hythergraph A diagram that allows the climates of two or more places to be compared at a glance. Mean temperature is plotted against one axis and mean precipitation against the other (it does not matter which is the *x* and which the *y* axis). The values are plotted for each month and joined by straight lines, ending with the point for December linking to that for January, thus producing a closed figure of irregular shape. The shape and dimensions of the figure indicate the range of temperature and precipitation experienced through the year, and the location of the figure indicates the overall temperature and precipitation. A hythergraph for New York City and Chicago, for example, reveals that both cities enjoy about the same summer temperatures, Chicago has slightly colder winters, and New York has the wetter climate of the two. The word is

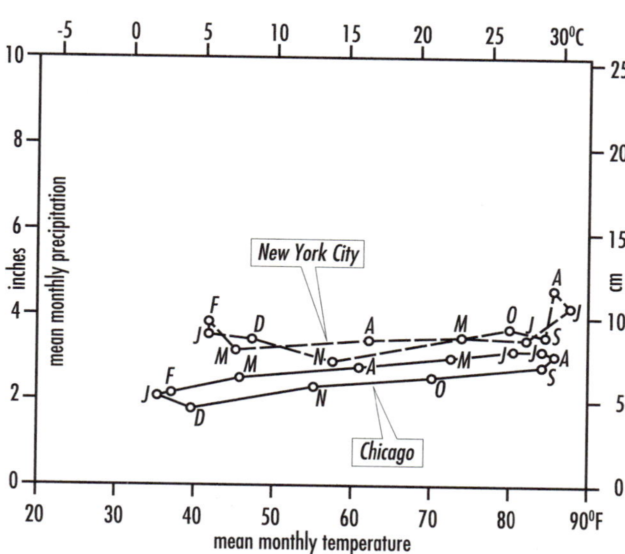

Hythergraph. **A diagram that allows the climates of different places to be compared easily. In this example New York and Chicago are compared. The position of the closed loop produced for each city by joining the points for each month shows whether one city is colder or wetter than the other.**

from the Greek words *hyetos,* meaning "rain," and *thermos,* meaning "temperature."

Hz *See* HERTZ.

I

ice age *See* GLACIAL PERIOD.

iceberg A large block of ice, floating in the sea, that has broken away from the edge of an ICE SHELF or the outlet of a GLACIER. Approximately 10,000 icebergs with a total volume of 67 cubic miles (280 km^3) are released each year into Arctic waters. Far more, with a total volume of about 430 cubic miles (1,800 km^3) enter Antarctic waters.

Most Arctic icebergs originate in western Greenland and some in eastern Greenland and Franz Josef Land, where valley glaciers enter the sea. They are often about 150 feet (46 m) tall and 55 feet (180 m) long; smaller ones, about the size of a house, are called *growlers*. Because they are derived from glaciers that have crossed land from which they acquire sediment and stones, Arctic icebergs are usually dark.

Antarctic icebergs break away (calve) from the edge of the ice shelves. They are of similar average height to Arctic icebergs, but often 5 miles (8 km) or more—sometimes much more—in length. In 1998, an iceberg 92 miles (148 km) long and 30 miles (48 km) wide broke away from the Ronne-Filchner Ice Shelf in the Weddell Sea; its area, of 2,751 square miles (7,125 km^2), made it slightly bigger than Delaware (2,044 square miles [5,294 km^2]). One of the largest, seen in 1927, was more than 87 miles (140 km) long; the largest of all known icebergs, measured by the icebreaker U.S.S. *Glacier* in 1956, was 207 miles (333 km) long and 62 miles (100 km) wide. Antarctic icebergs have not contacted the land surface and are white or blue.

Most icebergs have a specific gravity of 0.9, so about 85 percent of their volume lies below the sea surface.

ice blink A white gleam that is visible above the horizon when PACK ICE is present in the distance. The gleam is caused by the reflection of light from the ice.

icebow An optical phenomenon in which an arc of light is seen. It resembles a RAINBOW and is formed in the same way, but by the REFLECTION and REFRACTION of light through ice crystals rather than raindrops. The much smaller size of the crystals does not allow the colors to diverge, so an icebow is always white.

icecap climate In the KÖPPEN CLIMATE CLASSIFICATION, the type of climate that is found over the GREENLAND ICE SHEET and the Antarctic ICE SHEET. The temperature never rises above freezing, the high elevation of the ice sheets intensifies the cold, and precipitation is very low. At the South Pole, where the ice surface is about 9,200 feet (2,800 m) above sea level, the mean temperature is about –58° F (–50° C) and the annual precipitation is about 1 inch (25 mm). Over the Greenland ice sheet, at an elevation of 9,900 feet (3,020 m), the mean temperature is about –22° F (–30° C) and annual precipitation is about 2.6 inches (67 mm).

In the STRAHLER CLIMATE CLASSIFICATION an icecap climate is categorized as group 3, comprising climates controlled by polar and arctic AIR MASSES. An icecap climate occurs over the interior of Greenland

and Antarctica. These are SOURCE REGIONS for Arctic and Antarctic air masses, which are extremely cold; the low temperature is intensified by the high elevation of the Greenland and Antarctic ice plateaus. The climate is the coldest in the world and temperatures never rise above freezing. It is also classed as a polar or perpetual frost climate and designated *EF* in the Köppen classification.

ice concentration The term that is used to indicate the percentage of an area of ocean surface that is covered by ice. Ice concentration 0 percent means there is no ice; ice concentration 50 percent means half the area is covered; ice concentration 100 percent means the area is fully covered.

ice core A sample of ice for scientific study that is obtained by drilling vertically into a glacier and extracting a long cylinder. Glacial ice forms by the compression of snow, and each year's snowfall produces a recognizable band in the ice core. By counting the bands, starting from the top, it is possible to date the ice at each level. Cores are taken from the polar ice sheets, which are thick enough to yield records many thousands of years old. Analysis of the ice then reveals much valuable information. Air trapped in small bubbles in the ice shows the composition of the atmosphere at a particular time—the content of carbon dioxide and methane, for example. Dust mixed with the ice indicates the relative amount of rainfall—the more dust there is the drier the climate was, because rain quickly washes dust from the air—and dry conditions are associated with cold episodes. The OXYGEN ISOTOPE RATIOS in the ice also reveal the temperature.

ice-crystal cloud Cloud that consists entirely of ICE CRYSTALS because the whole of it is above the FREEZING LEVEL.

ice-crystal haze A HAZE that consists entirely of ICE CRYSTALS that are usually being precipitated from a cloud.

ice crystals When WATER freezes it forms solid crystals, the shape of which is determined by the shape of the water molecules from which they are made. A water molecule consists of two atoms of hydrogen (H + H) and one of oxygen (O) to yield H_2O. These form an isosceles triangle with an apex angle, at the center of the oxygen atom, of 104.5° and one hydrogen atom at the tip of each leg of the triangle. The length of each leg is 1.00 Å. Molecules are linked by hydrogen bonds, each of which is 1.76 Å long. The linkage can be described as O—H · · · O, where the three dots represent the hydrogen bond.

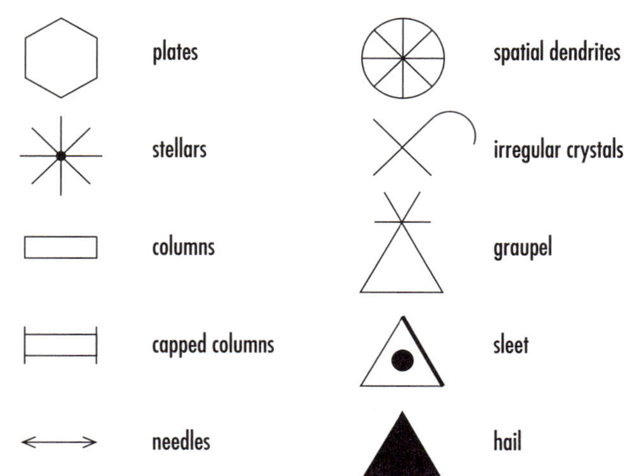

Ice crystals. **The standard symbols that are used to designate the 10 recognized types of ice crystals.**

As the temperature falls, the molecules move closer together. Each molecule bonds to four neighboring molecules. Its two hydrogens bond to the oxygens of two adjacent molecules, and its oxygen bonds above and below to one hydrogen of each of two adjacent molecules. This results in a hexagonal structure when seen from above and crystals that consist of stacks of puckered hexagonal rings. These rings join together, side by side, to form puckered hexagonal layers at right angles to the axes of the hexagons.

Each molecule is bonded to four others, but because of the shape of the molecule its hydrogen atoms can face toward any two of those four molecules. This allows each molecule to be oriented in one of six different ways. Each of the six orientations is equally probable, but whichever orientation occurs requires that the orientation of molecules nearby must be such as to maintain all the hydrogen bonds.

A consequence of this variability in orientation is that although all ice crystals are hexagonal, they occur in a variety of shapes. This affects the shapes of SNOWFLAKES, which are aggregations of ice crystals.

In 1951, an international system was adopted for classifying the shapes of ice crystals and the objects made from them. The classification recognizes 10 types; a symbol is assigned to each to help in reporting them:

Plates are flat, hexagonal rings.
Stellars are rings with six points.
Columns are cylinders that are hexagonal in cross section and columns are sometimes joined together.
Capped columns are columns that have a bar at each end. They may also be joined together and when they are the bars remain intact.
Needles are fine, resembling splinters, and they may be joined together.
Spatial dendrites are crystals with many fine branches, resembling fern fronds. These are what make the familiar frost patterns on windowpanes.
Irregular crystals are clumped together and have no regular shape. They are chaotic, as their name suggests.

In addition to these, there are three more categories for GRAUPEL, SLEET, and HAIL.

ice day A day on which the AIR TEMPERATURE does not rise above freezing and when ice on the surface of water does not thaw.

ice desert An area that is permanently covered with ice or snow and supports no vegetation of any kind, other than single-celled algae that arrive attached to snowflakes.

ice evaporation level The height at which ice crystals entering dry air change directly into water vapor by SUBLIMATION. Sublimation occurs if the temperature of the air is lower than –40° F (–40° C).

ice fog FOG that forms when relatively warm water is suddenly released into very cold air. This happens when sluices are opened in a dam. These are located low on the dam wall, and the water they release is from deep below the lake surface, where the water is warmer than the surface water. As this water encounters the very cold air—the temperature must be lower than –22° F (–30° C)—evaporation is rapid, because the VAPOR PRESSURE at the water surface is much greater than it is in the cold, dry air. Only a small amount of water vapor is required to cause SATURATION, however, and so water vapor sublimates (*see* SUBLIMATION), rapidly producing a fog of ice crystals.

icehouse period A period when GLACIERS and ICE SHEETS reached their maximum extent. It is possible that there have been times when the entire surface of the Earth was covered by ice. Less extreme icehouse periods have occurred many times in the history of the Earth, and evidence for some of them is found in seabed sediments. They appear to develop at intervals that are related to the MILANKOVICH CYCLES.

(For more information, see www.earthsky.com/1996/esmi960906.html.)

ice island A fragment of an ICEBERG that has an irregular surface, has a surface area of up to 200 square miles (520 km^2) or more, and is 50–150 feet (15–45 m) thick.

Icelandic low One of the two semipermanent areas of low pressure in the Northern Hemisphere (the other is the ALEUTIAN LOW). The Icelandic low is centered between Iceland and Greenland, between about 60° N and 65° N. It is described as semipermanent because although it forms, dissipates, and reforms, it is present for most of the year and moves very little. It is farther north than the Aleutian low because of the warm water carried northward by the GULF STREAM, and it covers a much larger area in winter, when the sea is warmer than the adjacent continents, than it does in summer. The intensity of the low varies; pressure is lowest in winter, when the atmospheric circulation is strong. The average pressure at the center in January is 996 mb and in July 1008 mb. The Icelandic low generates many storms. These are carried eastward along the POLAR FRONT by the prevailing westerly winds of middle latitudes and account for the storminess of the North Atlantic, especially in winter. The Icelandic low is also involved in the NORTH ATLANTIC OSCILLATION, which is a periodic fluctuation in the distribution of pressure that has a major effect on weather in Europe.

Iceland Scotland Overflow Water *See* NORTH ATLANTIC DEEP WATER.

ice nucleus Any small particle onto which supercooled water (*see* SUPERCOOLING) freezes. Both FREEZING NUCLEI and SUBLIMATION NUCLEI are classed as ice nuclei.

ice pellets Precipitation in the form of ice particles that are transparent or translucent, spherical or irregular in shape, and less than 0.2 inch (5 mm) in diameter. They are snowflakes that have partly melted and then refrozen, frozen raindrops or DRIZZLE droplets, or SNOW PELLETS that are enclosed in a thin coating of ice.

ice period The time that elapses between the first fall of snow in winter and the melting of the last patches of snow in spring.

ice prisms (diamond dust) ICE CRYSTALS so tiny they are barely visible. They seem to hang suspended in the air, twinkling as they catch and reflect the sunlight. They are shaped as needles, columns, or plates and are seen only in extremely cold weather. They may fall from STRATUS, NIMBOSTRATUS, or STRATOCUMULUS cloud but sometimes occur when the sky is cloudless, when they are a type of ICE FOG.

ice sheet A layer of ice that covers an extensive area of land and forms by the compression of snow that does not melt in summer. Snow that falls each year remains lying above the snow that fell in previous years, and as the layers accumulate their combined weight compresses the lower layers, packing the ice crystals together until they form solid ice. At the base of the ice sheet the weight causes ice to flow outward. Since flowing ice is known as a *GLACIER*, an ice sheet can also be called a glacier.

The GREENLAND ICE SHEET is the largest in the Northern Hemisphere, but the Antarctic ice sheet is very much bigger. It covers about 5.4 million square miles (13.9 million km^2) to an average depth of 6,900 feet (2,100 m) and contains more than 7 million cubic miles (29 million km^3) of ice. It has three parts. Ice on the Antarctic Peninsula comprises local ice caps, glaciers, and ICE SHELVES. The main part of Antarctica is divided in two by the Transantarctic Mountains. West Antarctica is largely covered by ice that flows toward the sea, where it forms large ice shelves. East Antarctica is much bigger, with an area of about 3.9 million square miles (10.2 million km^2). Its ice sheet is securely bonded to the underlying rock.

ice shelf A sheet of ice that extends from an ICE SHEET and covers an area of sea. Ice sheets flow outward under the pressure exerted by the weight of ice at the center. Where they reach the coast the sheets continue to advance into the shallow water, still in contact with the solid surface of the seabed. As they enter deeper water, the ice loses contact with a solid surface and the vertical movement of the water beneath the ice causes sections to break free, as ICEBERGS. An ice shelf forms where an ice sheet crosses a coast and enters a wide, but partly enclosed bay. It remains secured to land at each side, but there is water beneath its center. There are a few small ice shelves in the Arctic, but there the GREENLAND ICE SHEET is contained by mountain ranges. The really big ice shelves are found off the coast of Antarctica. The Ronne–Filchner, which has a large island near its center and for that reason is sometimes considered to be two distinct ice shelves, covers part of the southern Weddell Sea and the Ross Ice Shelf covers part of the southern Ross Sea. Each has an area of more than 154,400 square miles (400,000 km^2).

ice splinter A fragment of ice that becomes detached from an ICE CRYSTAL as the crystal grows inside a cloud. Vertical air currents detach the splinters, which then act as fresh FREEZING NUCLEI. Splinters also form in CUMULONIMBUS storm clouds, where there are very vigorous vertical currents and also supercooled (*see* SUPERCOOLING) water droplets. The splinters, released when supercooled droplets freeze, are so small that the air currents carry them to the top of the cloud, where they accumulate, because they weigh too little to fall. They carry a positive electric charge, and their accumulation high in the cloud is a major factor in the separation of charges that eventually breaks down in sparks of LIGHTNING.

ice storm A storm of FREEZING RAIN that deposits thick layers of ice onto structures such as radio masts, the rigging of ships, trees, and telephone wires. It is sometimes called *SILVER THAW,* because it is often followed by warmer conditions, or *BLACK ICE,* because it is transparent and therefore almost invisible as a coating on roads and sidewalks.

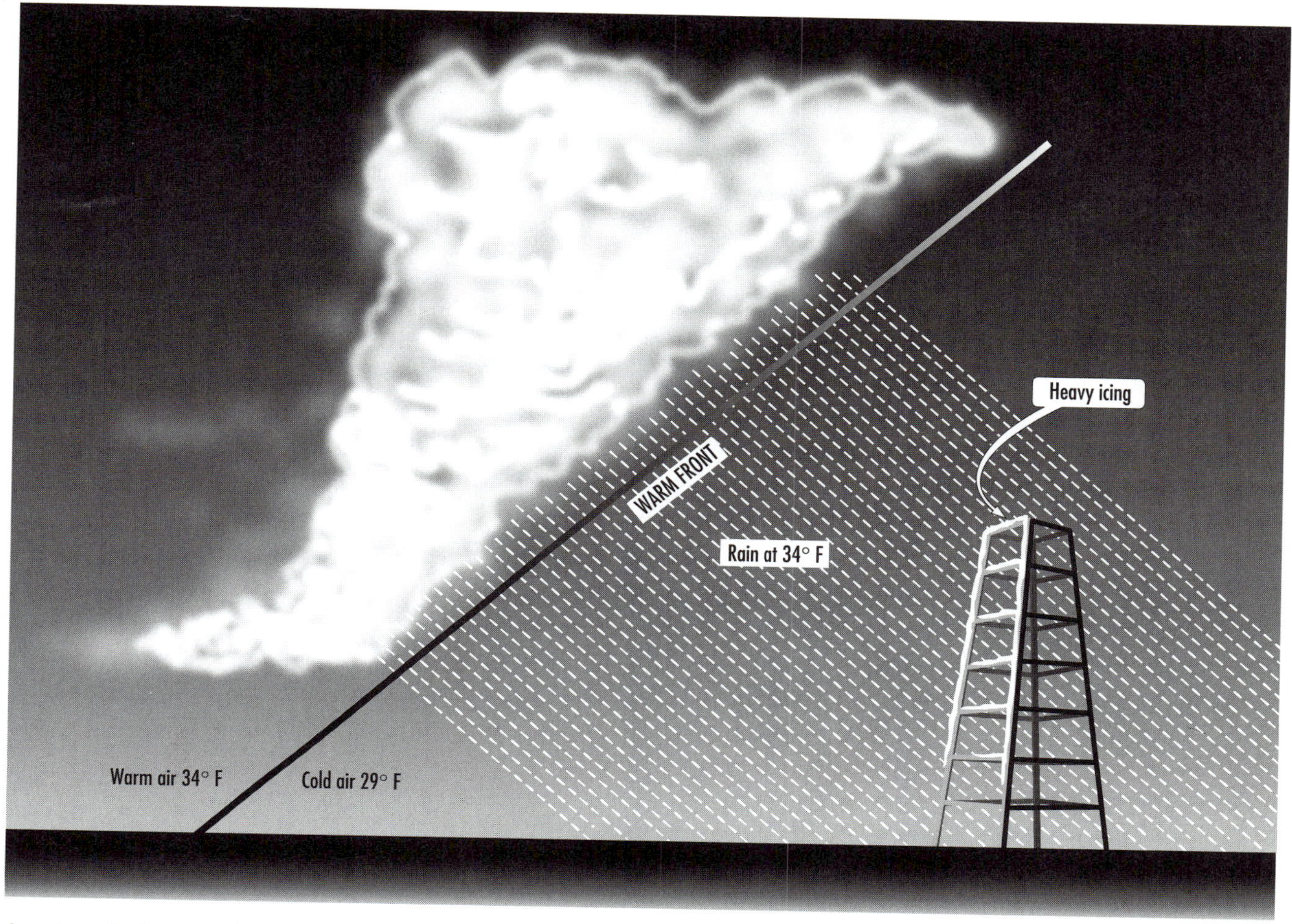

Ice storm. **Ice forms on the mast because rain, at a temperature just above freezing, cools as it falls through air that is below freezing temperature and freezes on contact with the chilled surface of the mast.**

Ice storms usually occur just ahead of a WARM FRONT, where the temperature of the air in the lower few hundred feet of the COLD SECTOR ahead of the front is well below freezing and surfaces in the cold air are at the same temperature. In the warm air behind the front, the temperature must be a degree or two above freezing, and water droplets in the frontal cloud must be at the same temperature. Finally, there must be a strong wind to drive the rain so it falls at an angle to the vertical. These conditions are crucial. A difference of just a few degrees determines whether the precipitation is in the form of SNOW or cold rain and whether it causes an ice storm.

Suppose the warm air and water droplets falling from it are at 34° F (4° C) and the cold air and surfaces below the front are at 29° F (–5° C). Rain falling from the cloud enters colder air. The droplets cool as they fall until, by the time they strike a surface, they are at or just below freezing. The temperature of the surface is also below freezing.

Even then, the ice storm soon abates unless the supply of cold air is constantly replenished. This is because the freezing of water droplets releases LATENT HEAT, which warms the surrounding air.

As a droplet strikes a surface, the water that first makes contact with the surface freezes instantly. The remainder of the droplet flows over the newly formed ice and also freezes as it encounters the cold surface. As more droplets fall, the layer of ice grows steadily thicker, and, because a strong wind is driving the rain, the ice layer thickens on the sides of objects that are facing into the wind much more than it does on the opposite, more sheltered sides. It is not only solid objects like radio masts and trees that accumulate ice. People walking into the wind are also coated, although their movement and the flexibility of their clothes break the ice into fragments before the layer can become very thick.

In a severe ice storm the ice can form a very thick layer and it can be extremely disruptive. It is dangerous for aircraft to land and take off when runways are thickly covered with ice, so air traffic is subject to delays. Railroads can also be immobilized when the rails are coated with ice, and road accidents increase when conditions are icy.

Ice is heavy and this weight causes other hazards. In January 1940, an ice storm in southern England snapped telephone wires that later were found to be loaded with up to 1,000 lb (450 kg) of ice. Telephone poles snapped when their load of ice exceeded about 11 tons (10 tonnes). Ice can form layers several inches thick on radio masts and the superstructure of ships. This is enough to cause structural damage and workers must struggle to remove the ice before it can do serious harm.

Roosting birds have frozen to tree branches during an ice storm and died of starvation because they were unable to free themselves. Ground-dwelling birds have had their wings coated with ice, which kept them grounded, although perhaps they were safe from predators, because cats caught in severe ice storms have been known to be frozen to the ground by their paws.

Ice storms can now be predicted from the temperatures and DEW POINT TEMPERATURES in the air at the height where the pressure is 850 mb (about 5,000 ft [1,500 m]), and the THICKNESS of the layers between 850 mb and 1,000 mb, and between 500 mb and 1,000 mb. The mean temperature of the air in these layers can be calculated from the thickness of the layers. In the United States, the NATIONAL WEATHER SERVICE issues warnings of impending ice storms.

icing The formation of ice on the surfaces of the leading edges of the wings, tailplanes, and tail fins of aircraft. It occurs when an airplane flies through air containing SUPERCOOLED water droplets and its own exposed surfaces are below freezing temperature. The droplets then freeze instantly on impact and ice can accumulate extremely rapidly. This is dangerous, because the shape of the ice coating can significantly alter the aerodynamic properties of aerofoil surfaces, so the airplane loses lift, and the weight of the ice can overload the mainspar that supports the wings. Most large aircraft are equipped to prevent icing by heating the affected surfaces or by using deicing "boots." These are flexible tubes running along the leading edges of vulnerable surfaces and covered by a flexible outer skin. Air is pumped through the tubes to expand and contract them one at a time; the movement breaks the ice, which is then swept away by the flow of air.

icing level The lowest altitude at which an aircraft is expected to experience ICING in a particular locality under the prevailing weather conditions.

ideal gas A gas composed of very small molecules in which there are no forces acting between the molecules. Such a gas obeys the EQUATION OF STATE under all conditions. The equation of state is $pV = nR^*T$, where p is the pressure, V is the volume, n is the number of MOLES, T is the ABSOLUTE TEMPERATURE, and R^* is the universal GAS CONSTANT (8.31434 J K^{-1} mol^{-1}). Real gases become more like an ideal gas as their pressure is reduced.

ideal gas law *See* EQUATION OF STATE.

IFR *See* INSTRUMENT FLYING.

IFR terminal minima The conditions of minimum surface visibility and CLOUD BASE under which aircraft are permitted by law to approach an airfield and land under instrument flight rules. These conditions vary according to the ground-based guidance systems available at the airfield, the type of aircraft, and the level of qualification in INSTRUMENT FLYING held by the pilot.

IGBP *See* INTERNATIONAL GEOSPHERE–BIOSPHERE PROGRAM.

IJPS *See* INITIAL JOINT POLAR SYSTEM.

Ike A TYPHOON that struck the Philippines on September 2 and 3, 1984. It killed more than 1,300 people and left 1.12 million homeless. On September 6, it struck the coast of Guangxi Zhuang, China, where it caused widespread damage and killed 13 fishermen whose boats were lost at sea.

Illinoian glacial A GLACIAL PERIOD in North America that began about 170,000 years ago and ended about 120,000 years ago. Average temperatures during the Illinoian were 3.6°–5.4° F (2°–3° C) cooler than those of today. The Illinoian is approximately equivalent to the MINDEL and RISS glacials of the Alps. It was preceded by the YARMOUTHIAN INTERGLACIAL and followed by the SANGAMONIAN INTERGLACIAL. The name is sometimes spelled *Illinoisan.*

immission The receipt of a substance, such as an atmospheric pollutant, from a distant source. It is the opposite of emission.

impaction The removal of solid particles from the air when they collide with surfaces and adhere to them. Dust and smoke particles are deposited on such surfaces as leaves, buildings, and the ground. Particles are also removed from the air by FALLOUT, RAINOUT, and WASHOUT.

impingement Placement of two or more substances or objects in contact with each other. For example, DUST is made to impinge onto a DUST COLLECTOR.

improved stratospheric and mesospheric sounder (ISAMS) An instrument carried on the UPPER ATMOSPHERE RESEARCH SATELLITE that measures the temperature in the STRATOSPHERE and MESOSPHERE. It also measures concentrations of the atmospheric gases: ozone (O_3), nitric oxide (NO), nitrogen dioxide (NO_2), nitrous oxide (N_2O), nitric acid (HNO_3), dinitrogen pentoxide (N_2O_5), water vapor (H_2O), methane (CH_4), and carbon monoxide (CO).

inactive front (passive front) A front or part of a front that has very little cloud or precipitation associated with it. People on the ground may notice its passing only by the change in temperature as one AIR MASS replaces another.

inclination The angle between the orbital plane of a body and a reference plane centered on the body about which it is orbiting. In the case of the Earth, the reference plane is the PLANE OF THE ECLIPTIC. In the case of satellites it is the equatorial plane of the planet about which they orbit. The angle between the rotational axis of a body and its orbital plane is also known as the *inclination.*

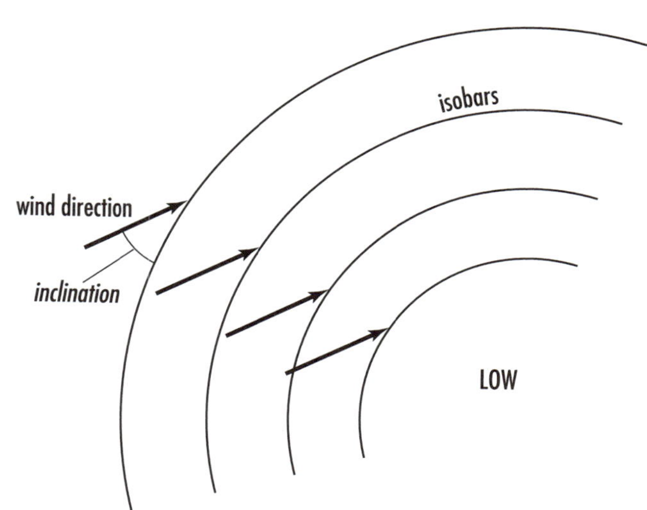

Inclination of the wind. **The inclination is the angle that is made between the direction of the wind and the isobars.**

inclination of the wind The angle that is shown on a weather chart between the wind direction and the isobars. This is an indication of the amount of FRICTION that is affecting the wind and also of the rate at which a CYCLONE is filling or an ANTICYCLONE weakening (*see* PRESSURE GRADIENT).

inclined orbit An orbit that falls between the GEOSTATIONARY and POLAR ORBITS. A geostationary orbit has an inclination of 0° and a polar orbit an inclination of 90°. The angle of inclination for the orbit is chosen to allow the greatest observation to be of a particular region of interest, but because an inclined orbit is not SUN-SYNCHRONOUS, that area is observed at a different time each day.

incus (anvil) A supplementary cloud feature (*see* CLOUD CLASSIFICATION) that constitutes the mass of CIRRIFORM cloud at the top of a CUMULONIMBUS cloud. This is often swept by the wind into the shape of an anvil. *Incus* is the Latin word for "anvil," and an incus is often called an *anvil.*

The incus is formed from ICE CRYSTALS. As the high-level wind sweeps them away from the top of the parent cloud they begin to fall. Outside the cloud, the RELATIVE HUMIDITY of the air is below 100 percent and so ice crystals falling into it vaporize by SUBLIMATION. As the remaining crystals are carried farther, they also eventually reach air into which they vaporize. It is this process that produces the characteristic anvil shape.

indefinite ceiling The condition in which the vertical VISIBILITY cannot be measured precisely, because it is determined not by the CLOUD BASE, but by FOG, HAZE, blowing snow, sand, or dust. PRECIPITATION does not produce an indefinite ceiling.

index cycle A progressive change in the circulation of air in the middle-latitudes upper TROPOSPHERE of the Northern Hemisphere that typically lasts between three and eight weeks, at the end of which the original circulation resumes. The change usually moves westward (from east to west) at a rate of about 60° of longitude a week, but it is very irregular both in the duration of the full cycle and in the speed with which it moves westward. It is especially common in February and March.

The "index" is the ZONAL INDEX. This is a number that represents the difference in pressure between two latitudes, usually 33° N and 55° N. These latitudes mark the boundaries that contain the POLAR FRONT and POLAR FRONT JET STREAM. When the pressure difference is great, the index is said to be *high* (*see* HIGH INDEX), and when it is small the index is said to be *low* (*see* LOW INDEX).

When the cycle commences the zonal index is high. The polar front and its jet stream are aligned approximately from west to east, some distance to the north of their mean positions. POLAR AIR lies to the north of the front and TROPICAL AIR to the south of it and there is very little mixing of them. A small number of ROSSBY WAVES lie along the front and jet stream. The AMPLITUDE of the waves is small and their WAVELENGTH is long.

Where the Rossby waves carry the flow northward, that part of the jet stream experiences a stronger CORIOLIS EFFECT (CorF), because the magnitude of CorF increases with latitude. In order to conserve the ABSOLUTE VORTICITY of the flow, the jet stream develops an equivalent negative vorticity. This turns the jet

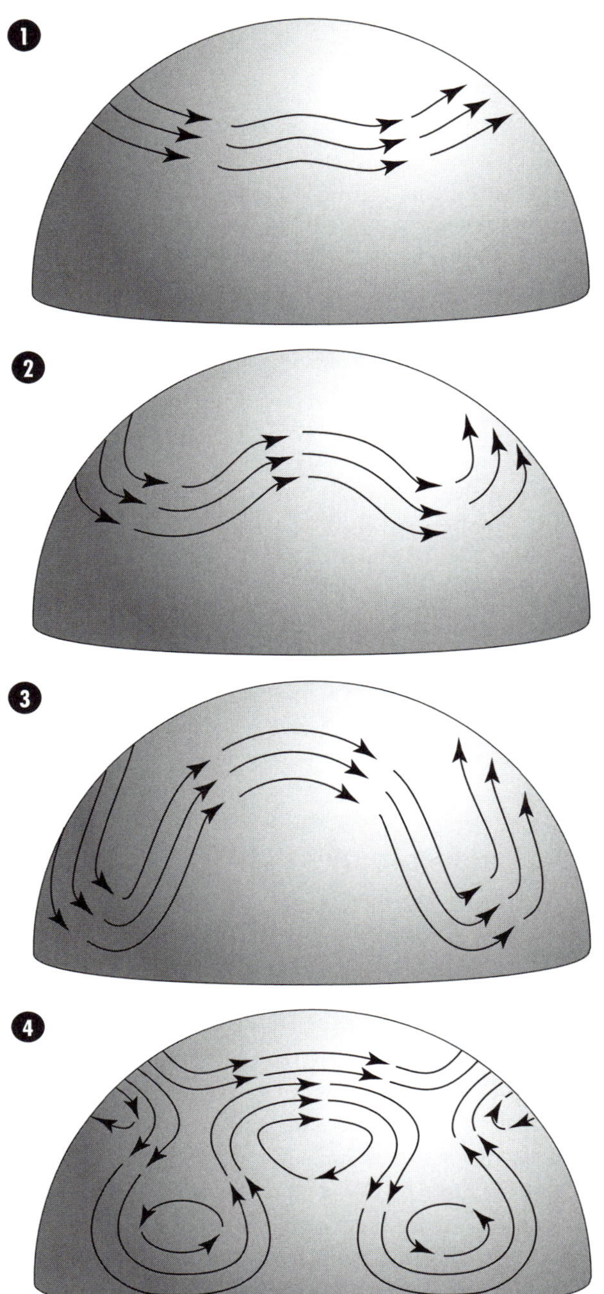

Index cycle. **Four stages in the development of the index cycle, during which the pressure difference to each side of the polar front weakens: (1) the initial condition, with a high zonal index; (2) the amplitude of the Rossby waves increases and their wavelength decreases; (3) the undulations in the waves become more and more extreme, carrying polar air a long distance south and tropical air a long distance north; (4) the flow breaks down into a series of cells, with anticyclones in the north and cyclones in the south.**

stream so that it curves in a CYCLONIC direction. It is then moving back to its original latitude, but it overshoots and, to compensate, develops positive vorticity that makes it curve in an ANTICYCLONIC direction.

This pattern can remain stable for long periods, with low-amplitude, long-wavelength Rossby waves. During an index cycle, however, each overshoot is slightly greater than the one preceding it. Both the cyclonic and anticyclonic curvatures continue longer, and so the wave amplitude increases and the wavelength decreases. The undulations in the jet stream and polar front become more and more extreme. The original zonal (east-to-west) flow is now much more meridional (north to south and south to north).

Finally, the pattern breaks down altogether. The flow of air joins on each side of the RIDGES and TROUGHS, forming isolated cells. Cells that form from the troughs lie to the south. The flow around them is cyclonic and they mark areas of low surface pressure. Cells that form from the ridges are located to the north and the flow around them is anticyclonic. They mark areas of high surface pressure.

At this stage the pattern may temporarily stabilize. This produces BLOCKING, resulting in prolonged periods of weather associated with low pressure on the southern side of the index, at about 33° N (about the latitude of Dallas, Texas, and Little Rock, Arkansas) and high pressure on the northern side, at about 55° N (a little to the north of Edmonton, Alberta, and Belfast, Northern Ireland).

As the anticyclones weaken and the cyclones fill, the cellular pattern dissipates. The original flow then reestablishes itself, with a high zonal index.

Indian summer A period of warm weather with clear skies that occurs in late September and October in the northeastern United States and especially in New England. FRONTAL SYSTEMS usually dominate the weather in early September, but late in the month these move southward and ANTICYCLONIC conditions become established. As the ANTICYCLONE extends southward its airflow produces fine, cold weather that is followed by warm, dry air from the southwest. It is this air that produces the fine weather of the Indian summer. It is late in the year, however, and the clear skies allow the surface to cool rapidly at night by radiation, so nights are cold. There is not an Indian summer every year and sometimes more than one occurs in a single year. A period of cool weather, with a KILLING FROST, must precede the warm weather for the change to be sufficiently marked to qualify as an Indian summer.

Other parts of the world also experience Indian summers. Anticyclones often become stationary over Britain in October and November, for example. They produce warm sunshine, but do not heat the ground strongly enough to produce vigorous CONVECTION that results in CUMULIFORM clouds and rain.

The origin of the name *Indian* is uncertain. It was first used in America in the 1790s and may reflect the idea that the fine weather is from a part of the country that was then inhabited by Native Americans. The use of the term in other English-speaking countries is derived from the American use.

indirect cell One of the cells that form part of the GENERAL CIRCULATION of the atmosphere and that are not driven by CONVECTION. The FERREL CELLS are indirect cells, so called because they are driven by the DIRECT CELLS to the north and south of them.

INDOEX The Indian Ocean Experiment, which is an international meteorological field experiment using data transmitted from *METEOSAT-5* that studies air pollution, clouds, interactions between clouds, and solar radiation in the INTERTROPICAL CONVERGENCE zone over the Indian Ocean. The experiment is supported by agencies and institutes in France, Germany, India, the Netherlands, Sweden, the United Kingdom, and the United States.

(More information about INDOEX, including images from it, can be found at http://www.eumetsat.de.)

inductance A property that is possessed by electric circuits. It takes two forms: self-inductance and mutual inductance. Self-induction is the production of an electromotive force when the current flowing through the circuit changes. Mutual inductance is the production of an electromotive force when the current flowing through a neighboring circuit changes. The unit of inductance is the HENRY.

industrial climatology The application of climatological studies (*see* CLIMATOLOGY) to industry in order to determine the influence of the climate on a particular industrial operation. If the industry imports materials or exports products by sea, for example, its proximity to a port that remains ice-free in winter is likely to be important. The direction of the PREVAILING WIND indicates areas that are most affected by factory emissions. Extreme temperatures make certain operations difficult. Chocolate factories sometimes close in very hot weather, because the molten chocolate does not set, and so it is important for the factories to be located in places where summer heat waves are uncommon. These are among the climatological factors industrial planners must take into account when alternative sites for a new factory are being considered.

industrial melanism An evolutionary adaptation to industrial pollution in which a species of animal adjusts its body color to that of its surroundings when these have been discolored by soot. More than 100 species of European and North American moths and butterflies have responded in this way; the most famous is the peppered moth (*Biston betularia*), which was described in 1973 by the British ecologist H. B. D. Kettlewell, who studied the phenomenon in England.

Variations in shape, size, or color are found among individuals of many species. The phenomenon is known as *polymorphism,* which means many (*poly-*) forms (*morphs*). The peppered moth, an insect about 2 inches (5 cm) across when its wings are extended, is ordinarily ivory-colored with dark speckles and blotches. It rests on tree bark and fences that are covered in lichens, against which its coloration makes it almost invisible. There is also a dark form of the moth, known as the *melanic* form because melanin is the pigment that gives animals a dark color. In melanic peppered moths the dark spots cover much more of the wings, and the most extreme form, called the *carbonaria* subspecies (*B. betularia carbonaria*), is almost completely black. A single gene that is dominant to the pale gene causes this color pattern.

The first *carbonaria* specimen was recorded in 1848, near Manchester. Smoke and sulfur dioxide from factories had killed most of the lichens and blackened tree bark and fences with soot. The ordinary, pale moth stood out strongly against this background, but the melanic form was well camouflaged. It was then found that the melanic form was much more common than the pale form in industrial areas and the pale form was more common in unpolluted rural areas.

Kettlewell bred several thousand moths and liberated both pale and melanic forms, some in the industrial Midlands and Northwest of England and others in the rural south. Then he watched with binoculars to see what happened. In polluted areas, where the background was dark, melanic forms were better hidden from birds and more of them survived. In unpolluted areas the pale forms survived better.

Melanic forms had become more common than the pale forms in polluted parts of the country, but after pollution controls were introduced the situation started to change. Gradually the lichens returned, and, as they did so, the proportion of pale moths increased. Prior to 1962, when controls were imposed, there was a 41–55 percent chance of a pale moth's being caught. After 1963 this fell to 21–22 percent.

The change from the pale to the melanic form in response to industrial pollution, *industrial melanism,* is a clear example of natural selection in action.

industrial meteorology The study of the ways in which day-to-day weather conditions affect a particular industry.

inert gases *See* NOBLE GASES.

inertial reference frame A location that is not accelerating (*see* ACCELERATION) and from which the acceleration of other bodies can be observed accurately. The second of Newton's LAWS OF MOTION states that the rate at which a body accelerates is proportional to the force exerted upon it. Newton assumed, however, that the acceleration is measured from a platform that is not itself accelerating, that is, from an inertial reference frame. Unless such a frame is used measurements of acceleration are inaccurate, the second law may appear to be flouted, and mysterious countervailing forces may manifest themselves.

The pilot of a fast jet airplane is pushed hard into the seat when the aircraft pulls out of a dive. If it were possible to measure his weight at that moment it would be greater than his true weight, because it is measured while his body is accelerating rather than while he is at rest and in an inertial reference frame. A passenger riding in a car that takes a corner at speed feels pushed outward. This is the supposed "centrifugal force." It is unreal, but it appears to be real because the passenger is attached to the car, which is accelerating, and is therefore in the same reference frame. An observer in an inertial reference frame, outside the car, could see that the car tended to continue moving in a straight line (the first law of motion) but that a force drawing it toward the center of the turn (the CENTRIPETAL FORCE) caused it to follow a curved path. The passenger is being pulled inward, not pushed outward, and there is no centrifugal force.

All meteorological observations are made from a reference frame that is fixed to the Earth, which is accelerating because of its rotation. That is why moving air appears to be deflected by the CORIOLIS EFFECT, which is sometimes incorrectly called a *force*. In order to understand the way air moves, the actual reference frame used to observe it must always be converted into an inertial reference frame.

inferior image See MIRAGE.

infrared radiation Electromagnetic radiation that has a wavelength from 0.7 μm to 1 mm. This is longer than visible red light and shorter than radio waves. Certain atoms and molecules vibrate at frequencies within this wave band, and because of this they absorb infrared radiation at characteristic wavelengths (*see* ABSORPTION OF RADIATION). Certain substances can be identified by the infrared wavelength at which they absorb, and absorption by some atmospheric gases produces the GREENHOUSE EFFECT. In a real greenhouse, the glass absorbs and is therefore opaque to infrared radiation with a wavelength greater than 2 μm. All warm bodies emit infrared radiation. Although infrared is invisible to the human eye, human skin glows at infrared wavelengths.

initial condition One of the values that is used as the base from which later values are calculated. For example, a weather forecast represents a set of calculations that aim to predict the way the weather will change from one state to another. The change is calculated from a set of measured or estimated values for a range of factors including temperature, pressure, humidity, and wind at various heights. These values are the initial conditions. Very small errors in the initial conditions tend to become increasingly exaggerated as the weather develops and the calculated values diverge from them. Because such discrepancies are too small to be noticed, the reliability of a forecast decreases with time. The acute sensitivity of a developing system to its initial conditions makes the development chaotic (*see* CHAOS).

Initial Joint Polar System (**IJPS**) A system of meteorological satellites in polar orbits that will be established and operated jointly between the NATIONAL OCEANIC AND ATMOSPHERIC ADMINISTRATION and EUMETSAT. The system will come into operation in 2003 and will enhance and supplement climate monitoring and NUMERICAL FORECASTING.

insolation The amount of solar radiation that reaches the surface over a unit area of the surface of the Earth. The word *insolation* is a contraction of *in*coming *sol*ar radi*ation*. This varies with the SEASONS and according to weather conditions. Knowing the insolation at a particular place is useful for agricultural and horticultural planning, and the insolation is also botanically important. Insolation is calculated for a whole day, then averaged over a number of days, chosen to include both cloudy and sunny conditions. Insolation is measured in megajoules per square meter (MJ m^{-2}). Sunshine is most intense at noon and the average total insolation during the hours of daylight is equal to $2/\pi$ multiplied by the noon maximum. A typical value for insolation in middle latitudes is about 25 MJ m^{-2}.

instability The tendency of air that has begun to rise to continue to rise. Air is absolutely unstable when the ENVIRONMENTAL LAPSE RATE is greater than the DRY ADIABATIC LAPSE RATE (*see* ABSOLUTE INSTABILITY) and conditionally unstable (*see* CONDITIONAL INSTABILITY) when the environmental lapse rate lies between the dry adiabatic lapse rate and the SATURATED ADIABATIC LAPSE RATE. *See* STABILITY OF AIR.

instability line Any line that is not associated with a FRONTAL SYSTEM and along which the air is subject to vigorous CONVECTION and is therefore unstable (*see* STABILITY OF AIR). If the INSTABILITY leads to the formation of active CUMULONIMBUS clouds and THUNDERSTORMS, the line is known as a *SQUALL LINE*.

instrument flight rules *See* INSTRUMENT FLYING.

instrument flying The operation of an aircraft when the pilot cannot see the horizon or the ground and is consequently unable to judge the attitude of the aircraft (whether or not it is flying straight and level) visually. Under these conditions the pilot must rely wholly on information supplied by the flight instruments. The aircraft is then subject to instrument flight rules (IFR), which specify the minimal conditions under which a pilot who has a particular level of qualification may fly an aircraft of a particular type. If an aircraft is to take off under IFR from one airfield and land under IFR at another, the pilot must file an IFR flight plan. This sets out the route to be followed, the height at which the aircraft will fly, and the estimated time of arrival at the destination airfield. A copy of the flight plan is sent to the ground controllers at the destination airfield.

insular climate The climate of an oceanic island or a coastal region, where the influence of the ocean is greater than that of the nearest large landmass. Although insular climates vary considerably, they are generally moister than continental climates and experience a smaller TEMPERATURE RANGE.

Integrated Hydrometeorological Services Core The branch of the United States NATIONAL WEATHER SERVICE that combines all of the forecast and warning programs into a single program. The aim is to ensure consistency in the management and operation of each of the services. The core includes the Aviation Weather Program (*see* AVIATION WEATHER FORECAST), Fire Weather Program (*see* FIRE WEATHER), Public Weather Program, and TROPICAL CYCLONE PROGRAM.

(You can learn more about the core at www.nws.noaa.gov/om/hydro.htm.)

intensification In SYNOPTIC METEOROLOGY, an increase in the PRESSURE GRADIENT that takes place over hours or days, leading to a strengthening of the winds.

interception The process by which an exposed plant surface catches PRECIPITATION, or the proportion of the total precipitation that falls on particular surfaces, such as the leaves of a tree. Sunshine is also subject to interception by tall objects that cast shadows, such as hills and buildings, and by gases or particles present in the air.

interfacial tension *See* SURFACE TENSION.

interference The effect of imposing one set of waves upon another. This produces a new wave with a different form and characteristics from either of the waves from which it is made. If the two interfering sets of waves are in phase, so their wave crests and troughs are aligned, the amplitude of the resultant wave is the sum of the amplitudes of the two waves. If the waves are directly out of phase, so the crests of one are aligned with the troughs of the other, the amplitude of the resultant wave is equal to the difference between the two amplitudes. This is called *destructive interference,* and if the two waves are of similar amplitude they cancel each other. When light is diffracted (*see* DIFFRACTION) some colors (wavelengths) become out of phase. These are subtracted from the light, causing the light to become colored by the wavelengths that are in phase.

interferometer An instrument that is used to measure the wavelength of radiation. It works by measuring the fringes that form between wavelengths that interfere with each other.

interglacial A prolonged period of warmer climates that separates two GLACIAL PERIODS. As an interglacial progresses in midlatitudes, POLLEN ANALYSIS reveals that tundra vegetation is replaced by a variety of herbaceous plants, which are replaced in their turn first by coniferous and then by broad-leaved, deciduous forest. In higher latitudes the progression may proceed no further than the establishment of coniferous forest, and in continental interiors, where the climate is too dry for forest, grassland may become the predominant vegetation type. The progression to more warmth-loving species is reversed as an interglacial nears its end and the climate enters a new glacial period. An interglacial typically lasts for about 10,000 years. At present we are living in an interglacial known as the *FLANDRIAN.*

Intergovernmental Panel on Climate Change (IPCC) An organization that was established in 1988 by the WORLD METEOROLOGICAL ORGANIZATION (WMO) and the United Nations Environment Program (UNEP). It is open to all members of WMO and UNEP; its task is to assess the scientific, technical, and socioeconomic information that is relevant to changes in climate that are caused by human activity and their consequences. The IPCC has three working groups.

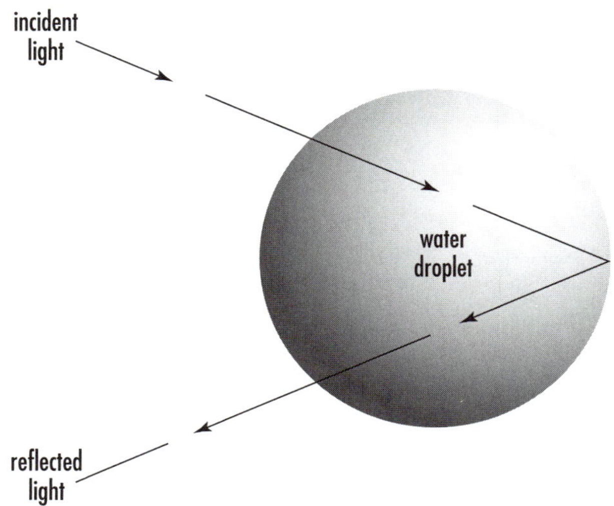

Internal reflection. **Incident light rays pass through the water droplet to its far side; from there they are reflected.**

Working Group I assesses the science, Working Group II considers the effects of climate change on socioeconomic and natural systems and the extent to which they may adapt, and Working Group III assesses the options for mitigating climate change, in particular by reducing the emission of GREENHOUSE GASES. The IPCC meets in plenary session approximately once each year to discuss and accept the reports from its Working Groups.

(For more information see www.ipcc.ch/about/about.htm.)

internal reflection The REFLECTION of light that has passed through a transparent body from the inside surface of the body on the far side. Internal reflection from water droplets is partly responsible for such optical phenomena as RAINBOWS.

International Cloud Atlas A book, published by the WORLD METEOROLOGICAL ORGANIZATION, that contains the pictures and definitions through which clouds are classified (*see* CLOUD CLASSIFICATION). The complete *Atlas* is in two volumes and there is also a single-volume, abridged version. Volume I of the complete *Atlas* is a loose-leaf manual describing the way clouds and other atmospheric phenomena should be observed. Volume II contains 196 pages of photographs, 161 of them in color, together with concise yet detailed captions defining cloud types. The abridged version contains 72 photographs, some in color and some in black and white, of all the cloud types and a variety of other phenomena, such as FOG and WATERSPOUTS, together with a text describing cloud observation.

The first edition of the *Atlas* was published in Paris in 1896 "by order of the International Meteorological Committee." The plan to produce a standard classification of clouds was agreed to at the WORLD METEOROLOGICAL CONGRESS held in 1874. The project drew on observations made by meteorologists in many countries and the *Atlas* was compiled by the Cloud Commission established by the committee. The observations contributed by American scientists were made at the Blue Hill Observatory, south of Boston, Massachusetts. The classification that was used was based on the system devised by LUKE HOWARD and published in 1803. The *Atlas* divides clouds into 10 major types, called *genera*. These are subdivided into 14 species and nine varieties. There are also ACCESSORY CLOUD features, such as PILEUS, TUBA, and VELUM.

The *International Cloud Atlas* has been revised several times since it was first published. The most recent edition of Volume I was published in 1956 and revised in 1975, the latest edition of Volume II was published in 1987, and the latest edition of the abridged version was published in 1956. The text is available in English, French, and Spanish editions.

(You can see some of the pictures from the first edition at orpheus.ucsd.edu/speccoll/weather/27.htm and order the latest edition of the *Atlas* at www.wmo.ch/web/catalogue/New%20HTML/frame/engfil/407.html.)

International Council of Scientific Unions *See* INTERNATIONAL GEOSPHERE–BIOSPHERE PROGRAM.

International Geosphere–Biosphere Program (IGBP) A project that was launched in 1986 by the International Council of Scientific Unions, which is the body that coordinates the activities of national scientific societies and academies. The aim of the IGBP is to "describe and understand the interactive physical, chemical and biological processes that regulate the total Earth system, the unique environment that it provides for life, the changes that are occurring, and the manner in which changes are influenced by human actions." In order to achieve this aim the IGBP seeks to establish a scientific basis from which to detect and measure changes, including climate change. This task is divided into six components: atmospheric chemistry, terrestrial ecosystems, biological drivers of the water cycle, coastal land–ocean interactions, ocean circulation, and past global changes. Scientists of more than 100 nations are engaged in studying these areas.

The IGBP has also established a network of regional centers where the local environment is studied, scientists are trained, and the regional relevance of change is estimated. This program is the Global Change SysTem for Analysis, Research and Training (START). START has also established 14 regional research networks in less industrialized countries.

(You can learn more about the IGBP at www.ciesin.org/TG/HDP/igbp.html.)

international index numbers A system devised by the WORLD METEOROLOGICAL ORGANIZATION that identifies each meteorological observing station by a number. Areas of the world are divided into blocks, each of which is given a two-digit number; a further three-digit number identifies each station within each block.

International Meteorological Committee *See* INTERNATIONAL METEOROLOGICAL CONGRESS, WORLD METEOROLOGICAL ORGANIZATION, *INTERNATIONAL CLOUD ATLAS*.

International Meteorological Congress A meeting that was first held in 1874 under the auspices of the International Meteorological Committee, a nongovernmental body consisting of the directors of the national weather services of a number of countries. The committee had been established the previous year with the aim of establishing regular communication among meteorologists throughout the world and developing standards for weather observation and recording. The most important decision made at the congress was to compile and publish the *INTERNATIONAL CLOUD ATLAS*. The name of the committee was later changed to the *International Meteorological Organization* and in 1947 it became the WORLD METEOROLOGICAL ORGANIZATION (WMO) of the United Nations. The congress, now called the *World Meteorological Congress*, still meets at intervals of at least four years and sets the policy for the WMO. Congresses were held in 1995 and 1999.

International Phenological Gardens (IPG) A network of gardens in Europe that covers an area from Macedonia (42° N) to Scandinavia (69° N) and from Ireland (10° W) to Finland (27° E). All the gardens grow genetically identical clones of trees and shrubs. Records are kept of the dates of phenological events, such as the unfolding of leaves, growth of shoots, flowering, leaf coloring, and leaf fall.

International Polar Year The year from August 1882 until August 1883, during which scientists conducted the first international collaboration to investigate the environment of the Arctic, and especially arctic meteorology, AURORAS, and the Earth's magnetic field. The participating nations financed the establishment of 12 observing stations in high latitudes, 6 of which were sited on the barren surfaces of the islands to the north of Eurasia and North America. Several of these stations remained occupied through the winter, and almost all of them recorded meteorological observations. The longest continuous weather records collected from this region during the 19th century covered a period of 1,000 days, from 1895 to 1897, and were made at Cape Flora, Franz Josef Land.

Although the International Polar Year focused attention on the Arctic, the observing stations were not maintained, and the record of meteorological data from the Arctic is much more sparse than that from Antarctica, which has been studied more thoroughly. A few permanent stations were established around the edges of the Arctic in about 1915, but there has never been one at the North Pole.

A Second International Polar Year was held from 1932 to 1933. Many more stations were established during that year around the continental coasts and on islands. The Soviet Union opened a series of them. Many of the stations opened during the second year have produced a constant meteorological record over many years. A few years later, in 1937, the Soviet Union deployed the first meteorological station on the surface of floating ice. It began about 30 miles (50 km) from the North Pole and survived for 18 months, drifting south with the EAST GREENLAND CURRENT. The United States began using similar floating observation stations in 1946. Many more observing stations were established during the International Geophysical Year, from 1957 to 1958.

International Satellite Cloud Climatology Project (ISCCP) A program that was established in 1982. It was the first project of the World Climate Research Program, which is part of the WORLD CLIMATE PROGRAM. It began collecting data on July 1, 1983. The ISCCP gathers a wide range of satellite radiation measurements at wavelengths covering the whole of the visible and infrared spectra, with spatial resolutions from 33 feet (10 m) to 1,240 miles (2,000 km) and at time intervals of 1/2 hour to 30 days. The images are transmitted by satellites in POLAR ORBITS and others in GEOSTATIONARY ORBITS. The data are analyzed to obtain information about the location, extent, and types of cloud over the entire surface of the Earth. ISCCP has its own system for cloud classification, based on the atmospheric pressure at the tops of the clouds, measured in millibars, and their OPTICAL DEPTH, which is a ratio and has no units.

International Satellite Land Surface Climatology Project (ISLSCP) A project in which satellite images are used to map vegetation. Data from the NORMALIZED DIFFERENCE VEGETATION INDEX are used to build illustrations based on a 1° latitude by 1° longitude grid that are then stored on CD-ROM and released to scientists as sets of CD-ROMs.

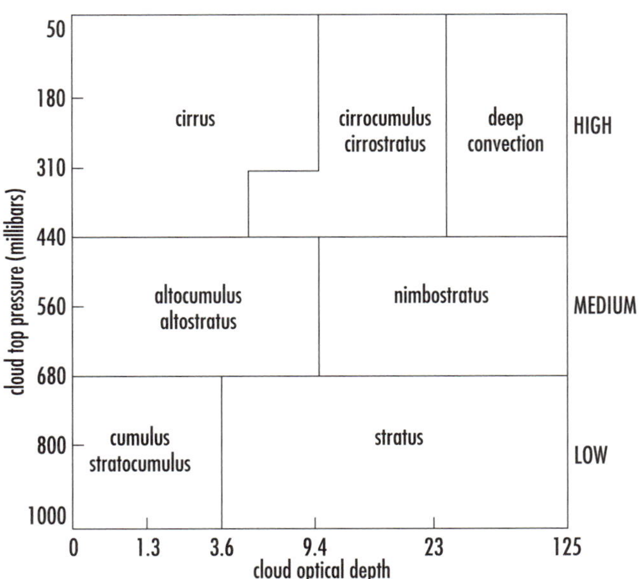

International Satellite Cloud Climatology Project. **The system of cloud classification that is used in the International Satellite Cloud Climatology Project is based on the atmospheric pressure at the cloud tops and the optical depth of the clouds.**

international synoptic code A code that is approved by the WORLD METEOROLOGICAL ORGANIZATION for the transmission of meteorological data. Each element of the data is encoded as a series of five-digit numerals.

interpluvial A period during which the climate was drier than in the PLUVIAL periods prior to or succeeding it.

interstade (interstadial) A time of warmer weather that occurs during a GLACIAL PERIOD. An interstade lasts for about 1,000–2,000 years. It is of much shorter duration than an INTERGLACIAL and may also be cooler, although this is uncertain. The vegetation present during an interstade, identified by POLLEN ANALYSIS, is typical of a climate cooler than that of an interglacial, but this may be because an interstade does not continue long enough for warmth-loving species to colonize the area.

interstadial *See* INTERSTADE.

Intertropical Convergence Zone (ITCZ) A belt surrounding the Earth and close to the equator, where the TRADE WINDS of the Northern and Southern Hemispheres meet. CONVERGENCE of air from the northeast and southeast causes the air to rise. When the rising air reaches the TROPOPAUSE, at 39,000–49,000 feet (12–15 km), it moves away from the equator and descends again in the subtropics to rejoin the trade winds. This is the HADLEY CELL circulation. Convergence also produces low atmospheric pressure at the surface and the ITCZ is a region of permanently low surface pressure. Winds are often light and variable within the ITCZ. It is where the DOLDRUMS are found.

Oceans cover most of the equatorial belt. Consequently, the trade winds gather moisture and the air rising in the ITCZ is moist. Its water vapor condenses, producing towering clouds and heavy precipitation. In satellite photographs, a line of cloud in the equatorial belt marks the position of the ITCZ. The cloud is not continuous but occurs in masses of cloud about 60 miles (100 km) across. These often appear as a series of waves, 1,200–2,500 miles (2,000–4,000 km) apart, which move slowly from east to west. They are known as *EASTERLY WAVES* and are implicated in the development of TROPICAL STORMS and TROPICAL CYCLONES.

Over the ocean, but not over the continental landmasses, the ITCZ also marks the region of highest temperature. This does not coincide with the geographic equator (*see* THERMAL EQUATOR). The difference arises because the location of the ITCZ depends strongly on the HUMIDITY of the air. Air has sufficient BUOYANCY to rise all the way to the tropopause only if it gains energy from the release of LATENT HEAT through the CONDENSATION of its water vapor. Air over land is less buoyant, and although low-latitude deserts, such as the Sahara and Australian Deserts, are much hotter than the ocean, the ITCZ lies on the side of them closer to the equator.

The position of the ITCZ changes with the SEASONS. After each EQUINOX, as the line joining places where the noonday Sun is directly overhead moves away from the equator, the ITCZ follows it. The largest seasonal migration occurs over land. Over South America and Africa, in January the ITCZ lies at about 15° S. In July it lies at about 15° N over Africa and at about 25° N over Asia. There is much less movement over the oceans. In January the ITCZ is close to the equator over all the oceans, and in July it lies at about 5°–10° N over the oceans. During the Northern Hemisphere summer, increased warmth causes an increase in EVAPORATION and cloud formation in the ITCZ. In winter, cooler temperatures produce a reduction in the amount of cloud. It is because the ITCZ never crosses the equator over the eastern and central Atlantic that HURRICANES never occur in the South Atlantic.

In addition to its seasonal migrations, the position of the ITCZ varies from year to year, following changes in sea-surface temperatures. If the sea is warmer than average the ITCZ is displaced toward the warm area. This happens during ENSO events.

intertropical front The name that was formerly given to the INTERTROPICAL CONVERGENCE ZONE (ITCZ). A front exists only where there is a marked contrast between two AIR MASSES. Such contrasts occur in some places within the ITCZ, but while the ITCZ remains close to the equator the characteristics of the air masses in the Northern and Southern Hemispheres are similar and there is no clearly marked front. A front develops as the ITCZ moves away from the equator into the summer hemisphere and it is most marked around the time of the summer SOLSTICE. In June and July, during the summer MONSOON, an intertropical front does exist over southern Asia and West Africa. Despite the name, however, even where it does exist the intertropical front produces little distinctive weather and is quite unlike the FRONTAL SYSTEMS of middle latitudes.

intortus A variety of CIRRUS cloud (*see* CLOUD CLASSIFICATION) in which the cloud consists of filaments that curve irregularly and appear to be entangled haphazardly. The name is derived from the Latin word *tortus,* which means "crooked."

inverse square law A law stating that the magnitude of a physical quantity is inversely proportional to the square of the distance between the source of that quantity and the place where it is experienced. The inverse square law determines the proportion of the total amount of energy emitted by the Sun that reaches each body in the solar system. Imagine that the Sun is surrounded by a series of concentric spheres, each having a radius equal to the distance between a planet and the Sun. Mercury "sits on" the innermost sphere, Venus on the next one, then Earth, Mars, and so forth. The whole of the energy emitted by the Sun reaches each of the spheres in turn. There is no diminution in the amount of energy as it travels. At each sphere, the solar energy is distributed evenly over the surface area of the sphere, which is equal to $4\pi r^2$, where r is the radius of the sphere. Since the area of the sphere is proportional to the square of its radius—the distance between the surface (or planet) and the Sun—the amount of energy arriving at any part of the surface, such as a planet, must be inversely proportional to the square of the radius ($1/r^2$). Consequently, the amount of solar radiation reaching each planet decreases rapidly with increasing distance from the Sun.

inversion The condition in which the air temperature increases with height, rather than decreasing. The layer of air in which the temperature increases with height is called an *inversion layer.*

Inversions can occur near the surface and are especially common in some areas that are surrounded by mountains. The shelter of the mountains restricts the movement and mixing of air, creating conditions in which an inversion can develop. Inversions over the ocean are also common in the Tropics.

The TRADE WIND INVERSION is a fairly permanent feature of the outer edges of the HADLEY CELLS at a height of 5,000–6,500 feet (1,500–2,000 m), increasing toward the equator. The lower STRATOSPHERE is also an inversion layer, and the TROPOPAUSE, marking the boundary between the TROPOSPHERE and stratosphere, is an inversion.

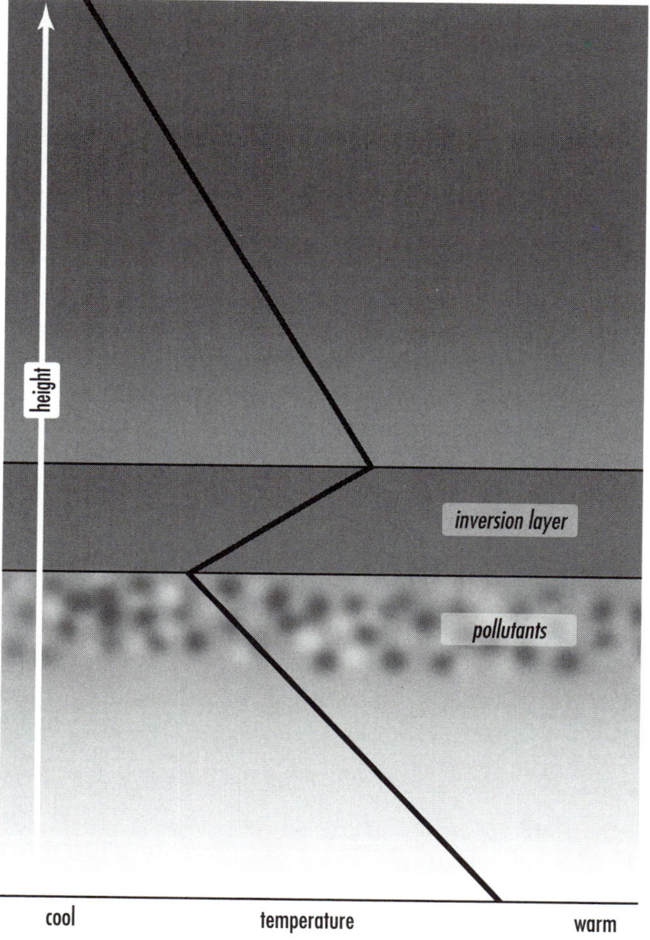

Inversion. **Ordinarily, air temperature decreases with height. Sometimes the situation is reversed in a layer of air where the temperature increases with height. This is an inversion layer and air beneath it is trapped. Consequently, pollutants accumulate.**

There are several ways in which an inversion can develop. On clear nights the ground cools by radiating the warmth it accumulated during the day. The dry air is almost transparent to outgoing INFRARED RADIATION, so the cooling is rapid. Air that is in contact with the ground is chilled, and if there is a little wind to produce a TURBULENT FLOW that mixes the air, all the air to a height of a few hundred feet grows cooler. Radiative cooling has little effect above this height, however, and so air above the mixed layer remains at its daytime temperature. By dawn this can produce a condition in which the lowest layer of air is cooler than the air above it. As the Sun rises and the ground begins to warm again, the temperature of the lower air starts to rise and the air expands upward into the inversion layer. The overnight inversion then breaks down and the usual LAPSE RATE is established.

It is this type of inversion that develops over the tropical oceans. Here, though, it is not the surface that cools by radiation, but the air above it. Because of its high HEAT CAPACITY the ocean surface does not cool at night. The air immediately above it is very moist and WATER VAPOR emits infrared radiation. This cools the air to a height of about 1,700 feet (500 m), and although the air is being warmed by contact with the sea surface, the rate of cooling exceeds the rate at which the air is warmed. Above that height the air is drier and so it loses less heat, forming an inversion.

Many inversions are caused by the ADVECTION of warm, stable air over a layer of cool, less stable air (*see* STABILITY OF AIR). Because the warm air is stable, it tends to return to its former level if it is made to rise. There is little mixing of the air and the warm air lies like a blanket on top of the cooler low-level air. This is called a FRONTAL INVERSION.

The most persistent inversions occur at the centers of ANTICYCLONEs, where air is subsiding. As the air sinks it is compressed and warms in an ADIABATIC manner. The subsiding air is unable to sink all the way to the surface, however, because it is prevented by turbulence in the lower air. This allows warm, subsiding air to form an inversion layer above the mixed air near the surface.

The trade wind inversions are of this type, and they form in the same areas of the Tropics as the low-level radiative inversions. Consequently, there are often two inversions, one at about 1,700 feet (500 m) and another above it, at about 5,000–6,500 feet

(1,500–2,000 m). There are a layer of mixed air beneath the lower inversion and another layer of mixed air between the two inversions. Mixing in the upper layer is driven by water vapor that rises by CONVECTION from the top of the lower inversion layer. Water vapor then condenses, releasing LATENT HEAT that increases the INSTABILITY and mixing and forming CUMULUS clouds.

The inversions that are so common over Los Angeles are usually caused by subsiding air at the eastern edge of the SUBTROPICAL HIGH that lies over the Pacific Ocean. Cool air moving from the ocean over the land intensifies these inversions by lifting the warmer air and forming a cool layer beneath it. The mountains to the east of Los Angeles prevent the pool of cool air from moving farther inland and weakening the inversion.

An inversion forms a cap over the air below it. Surface air cannot penetrate it to rise higher and any pollutants carried by the air are trapped. If the inversion persists for any length of time over a large city, pollutants accumulate in the air and AIR QUALITY deteriorates, sometimes to a level that is harmful to health. The PHOTOCHEMICAL SMOG that afflicts many cities, including Athens, Mexico City, and Los Angeles, forms in air trapped beneath an inversion. It was smoke held beneath an inversion that mixed with FOG that caused the "pea-soupers" for which London and other industrial cities were once notorious and that culminated in the LONDON SMOG INCIDENTS.

inversion layer *See* INVERSION.

invisible drought One of the types into which DROUGHTS are formally classified. It is a drought in which precipitation falls, but when the amount of water lost by evaporation and transpiration is deducted the amount of precipitation is insufficient to recharge AQUIFERS. Consequently, river levels and water tables remain low and plants continue to suffer stress. It is the kind of drought that often follows a more obvious drought, when there is little or no precipitation. Because there is precipitation, it is usually difficult to persuade people of the need to continue economizing in their use of water.

ion An atom or molecule that carries an electromagnetic charge because it has either lost or gained one or more electrons. An atom comprises a nucleus composed of protons, which carry positive charge, and neutrons, which carry no charge. The nucleus is surrounded by a cloud of electrons, each of which carries negative charge that is equal to the charge on a proton. Ordinarily, the negative charge on the electron precisely balances the positive charge on the proton, so the atom has no net charge. Should the atom gain one or more electrons, however, it would have a net negative charge and would be a negatively charged ion, or anion, its charge indicated by as many superscripted minus signs as are needed to describe its charge. Were it to lose one or more electrons it would become a positive ion, or cation, and its charge would be indicated by as many superscripted plus signs as necessary. Atoms that do carry charge usually occur naturally in compounds formed by IONIC BONDS between positive and negative. Chlorine (Cl^-), for example, bonds with sodium (NA^+) to form common salt (NaCl) and with hydrogen (H^+) to form hydrochloric acid (HCl). An ion is an atom with a number of electrons that is different from the number of protons in its nucleus.

ionic bond An attractive force that holds two or more atoms together and that is based on the exchange of electrons between the atoms. In the case of sodium chloride (NaCl), for example, sodium (Na) donates an outer electron to chlorine (Cl). This results in a more stable electron configuration for both atoms but leaves them charged, as Na^+ and Cl^-. The electrostatic attraction between positive and negative then bonds the two atoms together.

ionization The addition of electrons to an atom or the removal of electrons from it. The addition of electrons imparts a negative electric charge to the atom, and the removal of electrons leaves the atom with a positive charge. An atom carrying an electrical charge is known as an *ION*.

ionizing radiation Radiation that is capable of causing the IONIZATION of atoms. In the atmosphere, ionizing radiation strips electrons from nitrogen and oxygen atoms, thereby forming the IONOSPHERE. The radiation primarily responsible for ionizing atmospheric gases is solar radiation in the short-wave ULTRAVIOLET and long-wave X-ray wavelengths, of about 0.4–0.001 µm. COSMIC RADIATION also contributes to ionization.

ionosphere A layer of the atmosphere in which PHOTOIONIZATION causes the separation of positively charged IONS and negatively charged electrons. Because IONIZATION is maintained by solar radiation, the thickness of the ionosphere and the density of ions within it vary daily, seasonally, and with latitude. Generally, the ionosphere extends from a height of about 37 miles (60 km) to about 620 miles (1,000 km), although the ion density is greatest from about 50 miles (80 km) to 250 miles (400 km). The ionosphere is subdivided into four layers, designated *D, E,* F_1, and F_2, in ascending order of height. The D and E layers disappear at night, except in very high latitudes, as ionization ceases and collisions between ions cause them to recombine. At greater heights, where the atmosphere is so rarefied that collisions are less common, ions survive at night. Ions survive in very high latitudes because of the ionizing effect of the interaction of the SOLAR WIND and MAGNETOSPHERE. In these latitudes AURORAE are produced in the ionosphere. The ionosphere reflects radio waves. Waves at the lowest frequencies are reflected in the D and E layers, higher frequencies in one of the F layers.

IPCC *See* INTERGOVERNMENTAL PANEL ON CLIMATE CHANGE.

IPG *See* INTERNATIONAL PHENOLOGICAL GARDENS.

Ipswichian interglacial The INTERGLACIAL period in Britain that followed the WOLSTONIAN GLACIAL and preceded the DEVENSIAN GLACIAL. It lasted from about 130,000 years ago until about 72,000 years ago and is named for glacial deposits found near the town of Ipswich, Suffolk. Deposits dated from the very early Ipswichian and taken from the River Thames in London have given this interglacial the alternative name *Trafalgar Square interglacial.* Those deposits include remains of plants that are no longer found so far north as Britain, and of hippopotamus, elephant (*Elephas antiquus*), and rhinoceros, suggesting that average summer temperatures during the Ipswichian were 3.6°–5.4° F (2°–3° C) warmer than those of today. The Ipswichian is equivalent to the EEMIAN INTERGLACIAL of northern Europe and approximately equivalent to the SANGAMONIAN INTERGLACIAL of North America.

Irene A HURRICANE that struck Caribbean islands and then moved to Florida and North Carolina in October 1999. It formed as a TROPICAL STORM in the northwestern Caribbean and reached hurricane force, category 1 on the SAFFIR/SIMPSON SCALE, as it approached Cuba. Irene reached Florida, moved from there to North Carolina, and then traveled out into the Atlantic. Although its winds were not strong, Irene produced very heavy rain and triggered several TORNADOES. Two people were killed in Cuba, five in Florida, and in North Carolina one person died in a traffic accident caused by the hurricane.

iridescence *See* IRIDESCENT CLOUD.

iridescent cloud Cloud that is partly brightly colored, most often with patches of red and green but sometimes with violet, blue, or yellow. The color is caused by the DIFFRACTION of sunlight or moonlight by small water droplets or ice crystals. For iridescence (also called *irisation*) to occur the particles must be of approximately uniform size and the cloud must be in the same part of the sky as the Sun or Moon. Iridescent clouds are most often seen when the Sun or Moon is behind cloud or some barrier; when it is in full view its light is so intense as to make the colors invisible. Which colors are seen depends on the size of the droplets or crystals and the angles of the Sun, cloud, and observer.

irisation The appearance of colors around the edges of clouds. It is caused by the DIFFRACTION of sunlight by SUPERCOOLED water droplets or ICE CRYSTALS. The colors that are produced vary according to the size of the droplets and the angular distance from the Sun. For the phenomenon to occur the cloud must be in the same part of the sky as the Sun, and it is often at its best when the Sun is behind the cloud. Irisation often occurs with ALTOCUMULUS. *See also* IRIDESCENT CLOUD.

Irma Two TYPHOONS, the first of which, rated category 5 on the SAFFIR/SIMPSON SCALE, struck the Philippines on November 24, 1981. It generated winds of 140 mph (225 km h^{-1}) and caused great destruction in the coastal towns of Garchitorena and Caramoan. More than 270 people were killed and 250,000 were rendered homeless. The damage was estimated at $10 million. The second Typhoon Irma struck Japan on July 1, 1985. It caused the deaths of 19 people and extensive damage in Numazu and Tokyo.

Irminger Current An ocean current that flows past the southern coast of Iceland and continues past the southern cape of Greenland. It is a branch of the GULF STREAM that breaks away from the NORTH ATLANTIC DRIFT in about latitude 50° N and carries warm water northwestward. As it passes Greenland its water mixes with cold water from Baffin Bay, but it can still be detected by its higher salinity as far as 65° N.

irradiance The rate at which radiant solar energy flows through a unit area that is perpendicular to the radiation beam. Since the Sun emits radiation equally in all directions, illuminating a sphere around itself, irradiance decreases with distance from the Sun in accordance with the INVERSE SQUARE LAW. The irradiance at any point (I) is given by the equation $I = I_s(R_s/R)^2$, where I_s is the amount of radiation being emitted by the Sun, R_s is the radius of the PHOTOSPHERE, and R is the distance of the point from the Sun. The average annual irradiance at the top of the atmosphere is known as the *SOLAR CONSTANT.*

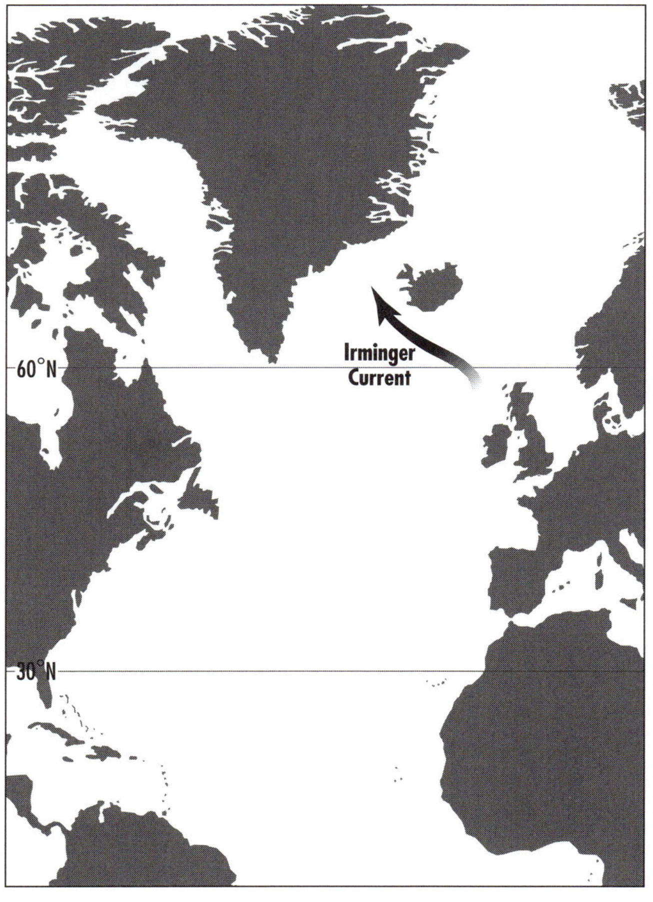

Irminger Current. **The current breaks away from the Gulf Stream and flows past the southern coast of Iceland and Greenland.**

irregular crystal A particle of SNOW that consists of a number of very small crystals that have formed erratically, so the resulting particle has an irregular shape. Irregular crystals are sometimes covered with a coating of RIME FROST.

Irving A TYPHOON that struck Thanh Hoa Province, Vietnam, on July 24, 1989. At least 200 people were killed.

isallobar A line that is sometimes drawn on a SYNOPTIC CHART to link places where the atmospheric pressure has changed by an equal amount during a specified period of time. It is usually based on data for the change in pressure over the preceding three hours that are included in every report from a weather station (*see* STATION MODEL).

isallobaric An adjective that describes a constant or equal change in atmospheric pressure over a spatial distance or over a specified time.

isallobaric high *See* PRESSURE-RISE CENTER.

isallobaric low *See* PRESSURE-FALL CENTER.

isallotherm A line that is sometimes drawn on a SYNOPTIC CHART to link places where the temperature has changed by the same amount over a specified period.

ISAMS *See* IMPROVED STRATOSPHERIC AND MESOSPHERIC SOUNDER.

isanomal *See* ISANOMALOUS LINE.

isanomolous line (**isanomal**) A line that is drawn on a chart to link points where the value of a particular meteorological quantity varies from the regional average by the same amount, that is, to join points of equal anomaly. The anomalies are adjusted for the effects of latitude and elevation. If the anomalous value is greater than the average the anomaly is said to be *positive* and if it is lower than the average it is said to be *negative*. A place that is usually warmer than the surrounding area has a positive temperature anomaly; if the rainfall is usually lower than that of the surrounding area it has a negative rainfall anomaly.

ISCCP See INTERNATIONAL SATELLITE CLOUD CLIMATOLOGY PROJECT.

isentrope *See* ADIABAT.

isentropic analysis A procedure in which winds, pressures, temperatures, and humidities across several ISENTROPIC SURFACES are extracted from RADIOSONDE data.

isentropic chart A SYNOPTIC CHART on which the elements of the weather, such as pressure, temperature, humidity, and wind, are plotted on a surface of equal POTENTIAL TEMPERATURE. This surface is equivalent to an ISENTROPIC SURFACE.

isentropic mixing Mixing of air that takes place across an ISENTROPIC SURFACE. Isentropic mixing results from ADIABATIC processes, in which ENTROPY is conserved.

isentropic surface A surface over which ENTROPY is everywhere the same. Entropy is conserved in ADIABATIC processes, and so an isentropic surface is a surface of equal POTENTIAL TEMPERATURE.

isentropic thickness chart A chart that shows the THICKNESS of an atmospheric layer that is bounded above and below by ISENTROPIC SURFACES. The thickness of such a layer is directly proportional to the CONVECTIVE INSTABILITY of the air within it.

isentropic weight chart A chart that shows the difference in pressure between two ISENTROPIC SURFACES.

ISLSCP *See* INTERNATIONAL SATELLITE LAND SURFACE CLIMATOLOGY PROJECT.

Ismael A HURRICANE that struck Mexico on September 14, 1995. It killed at least 107 people in the northwestern states, many of them fishermen who were lost at sea.

isobar A line that is drawn on a map to join points where the atmospheric pressure is the same. Isobars resemble the contours on a physical map and are said to depict an ISOBARIC SURFACE. Those drawn on a map that also shows the land surface are usually corrected to their sea-level value. A map showing the actual surface pressure would reflect the elevation of the surface and would be very difficult to interpret meteorologically. Isobars do not necessarily refer to the sea surface, however, and isobaric charts can be drawn for any altitude.

isobaric equivalent temperature *See* EQUIVALENT TEMPERATURE.

isobaric map A map that shows the distribution of atmospheric pressure at any given height above sea level.

isobaric slope *See* PRESSURE GRADIENT.

isobaric surface *See* CONSTANT-PRESSURE SURFACE.

isobront A line that is drawn on a map to link places at which THUNDERSTORMS reached the same stage of development at the same time.

isoceraun *See* ISOKERAUN.

isodrosotherm A line that is sometimes drawn on a SYNOPTIC CHART to link places where the DEW POINT TEMPERATURE is the same.

isogradient A line that is drawn on a map to link places where the horizontal gradient of pressure or temperature is the same.

isohel A line drawn on a map that joins places that experience equal numbers of hours of sunshine.

isohume A line that is drawn on a map to link places of equal HUMIDITY across a specific surface. The measure used may be the RELATIVE HUMIDITY, SPECIFIC HUMIDITY, or MIXING RATIO.

isohyet A line drawn on a map that joins places that receive the same amounts of rainfall.

isokeraun (**isoceraun**) A line that is drawn on a map to link places that experience the same frequency or intensity of THUNDERSTORMS.

isoneph A line that is drawn on a map to link places that are equally cloudy.

isonif A line that is drawn on a map to link places that received equal amounts of snowfall.

isopectic A line that is drawn on a map to link places where winter ice begins to form at the same time.

isophene A line that is drawn on a map to connect places at which a particular stage in plant development (*see* PHENOLOGY) occurred on the same date. An isophene map for a certain event, such as the commencement of the wheat harvest, thus shows the rate at which that event advanced across a region.

isopleth A line or surface that is drawn on a map or chart to connect points that are equal in respect to some quantity, such as temperature or pressure. ISOBARS and ISOTHERMS are isopleths. Because of the stratification of the atmosphere, most isopleths are approximately horizontal.

isoprene A volatile hydrocarbon compound (C_5H_8) that forms the structural base for many compounds synthesized by plants. Isoprene molecules form units that combine to form the larger molecules of a class of compounds called *terpenes*. Isoprene and terpenes are emitted by plants and especially deciduous trees. These compounds are highly reactive chemically and play an important part in the formation of OZONE in the lower atmosphere. In urban areas isoprene makes an important contribution to the formation of PHOTOCHEMICAL SMOG, and in rural areas it is the predominant hydrocarbon involved.

isopycnal A line or surface that is drawn on a map or chart to connect points of equal air density. Density is a more fundamental factor than temperature in determining the behavior of air, and a BAROCLINIC atmosphere is usually defined by the intersection of ISOBARS and isopycnals.

isoryme A line that is drawn on a map to link places where the incidence of FROST is the same.

isosteric surface A surface across which the density of the air remains constant.

isotach A line drawn on a map that joins places that experience winds of the same speed.

isothere A line that is drawn on a map to link places where average summer temperatures are the same.

isotherm A line drawn on a map or TEPHIGRAM that joins points that are at the same temperature.

isothermal equilibrium (**conductive equilibrium**) The condition of a large body of air that is at the same temperature throughout. This situation develops if the air remains unaffected by external conditions long enough for the temperature to equalize by CONDUCTION.

isothermal layer The layer of air composing the lower part of the STRATOSPHERE, in which the temperature remains constant with height.

isotope One of two or more varieties of atom that all belong to the same element. The atoms possess identical chemical properties but differ in their relative atomic masses. The chemical properties of an element are determined by the number of protons in the nuclei of its atoms. Atoms of all the isotopes of an element have the same number of protons in their nuclei but differ because their nuclei contain different numbers of neutrons, which affect only the mass of the atom. Most elements comprise two or more isotopes that occur naturally in constant proportions. Although all the isotopes of an element are chemically identical, the difference in their masses may affect aspects of their physical behavior. The OXYGEN ISOTOPE water molecules contain affects the evaporation of water, for example, and OXYGEN ISOTOPE RATIOS are used in studies of past climates.

isotropic Having properties that change independently of direction. The properties of an isotropic substance are the same regardless of the direction from which the substance is approached. On a small scale, of distances up to about 3 feet (1 m), the atmosphere is fairly isotropic, but on a large scale it is strongly ANISOTROPIC.

ISOW *See* NORTH ATLANTIC DEEP WATER.

ITCZ *See* INTERTROPICAL CONVERGENCE ZONE.

J

J *See* JOULE.

January thaw (January spring) A period of mild weather that sometimes occurs in late January in parts of the northeastern United States and in Britain. There is an English saying, "A January spring is worth nothing."

jauk A FÖHN WIND that occurs in the Klagenfurt Basin, Austria.

JDOP *See* DOPPLER RADAR.

jet-effect wind A wind that is accelerated as a consequence of FUNNELING. *See* MOUNTAIN-GAP WIND.

jet streak A region within the JET STREAM where the wind speed is higher than it is elsewhere.

jet stream A winding ribbon of strong wind that is found close to the TROPOPAUSE, in either the upper TROPOSPHERE or the lower STRATOSPHERE. Typically, it is thousands of miles long, hundreds of miles wide, and several miles deep. Its existence was discovered during World War II. The navigators of American aircraft flying at high altitude across the Pacific and of German aircraft flying in the Mediterranean region found their journey times sometimes varied greatly from those they had calculated using the winds predicted by meteorologists. At their cruising altitudes they encountered winds with speeds comparable to those of their aircraft. If they approached the jet stream from above or below they found the wind speed increased by about 37–73 mph for every 1,000 feet of altitude (18–36 km h^{-1} per 1,000 m). If they approached from the side it increased by the same amount for every 60 miles (100 km) of distance from the core of the jet stream. At the center of the stream the wind speed averages about 65 mph (105 km h^{-1}), but it sometimes reaches 310 mph (500 km h^{-1}).

There are several jet streams. With one exception, they all blow from west to east in both hemispheres. In the Northern Hemisphere, the POLAR FRONT JET STREAM is located between about 30° N and 40° N in winter and about 40° N and 50° N in summer. The SUBTROPICAL JET STREAM is located at about 30° N throughout the year. In summer there is also an EASTERLY JET at about 20° N extending across Asia and southern Arabia and into northeastern Africa. This is the jet stream that blows from east to west. The Southern Hemisphere polar front jet stream is at about 45° S in summer, with two branches that spiral into it. One of these begins off eastern South America at about 32° S and the other starts in the South Pacific at about 30° S 150° W. In winter, the two branches start from about 20° S, one from South America and the other from about 170° W, and the jet stream is at about 50° S. The Southern Hemisphere subtropical jet stream is at about 30° S.

Jet streams are THERMAL WINDS. That is to say, they are produced by the large horizontal temperature gradients associated with baroclinicity (*see* BAROCLINIC). These gradients reach a maximum across

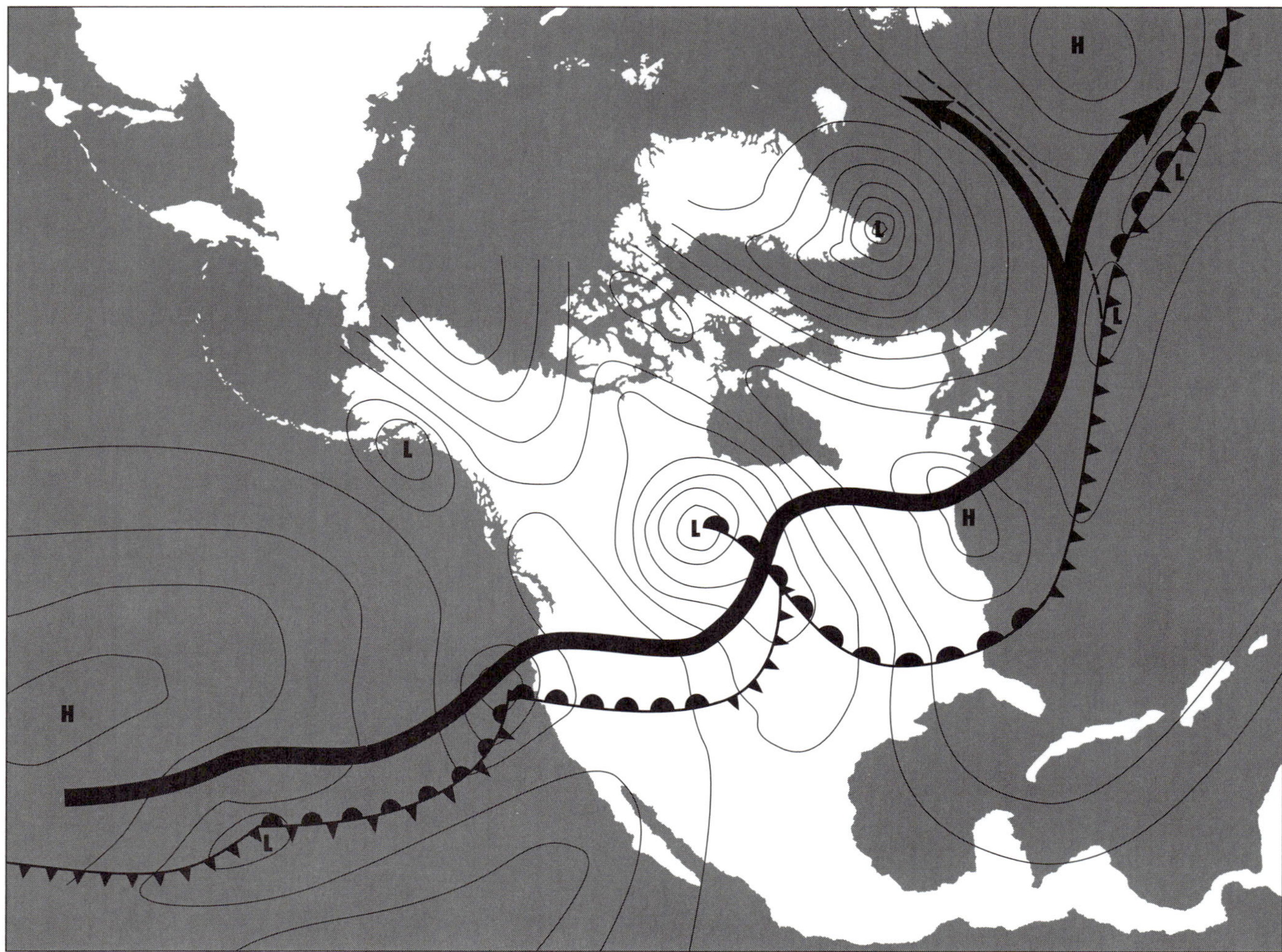

Jet stream. **The position of the polar front jet stream across North America and the distribution of pressure and fronts associated with it. Entrance regions are linked to anticyclones on the northern side and depressions on the southern side. Exit regions have depressions on the northern side and anticyclones on the southern side.**

major fronts, and especially the polar and subtropical fronts, and they increase with height. The steepest thermal gradient occurs at the top of the polar and subtropical fronts, close to the tropopause, where the jet streams are located. The precise altitude of the jet streams is determined by the air pressure. The polar front jet stream occurs at about the 300-mb level and the subtropical jet stream at about the 200-mb level. The core of a jet stream lies on the warm side of the front, and the wind blows with the cold air on its left in the Northern Hemisphere and on its right in the Southern Hemisphere, resulting in a westerly wind in both hemispheres.

As air is drawn into the jet stream from the colder side it accelerates into a region of higher positive (CYCLONIC) VORTICITY. This causes it to become narrower horizontally, producing CONVERGENCE, and it also sinks. High-level convergence produces a region of high surface pressure. The air then travels along the core and is expelled from the jet stream ahead of the core. Air leaving the core on the cold-air side enters a region of lower positive vorticity, and there is high-level DIVERGENCE with corresponding low-level low pressure. The pattern is reversed for air entering and leaving the core on the warm-air side, with divergence where the air is entering and convergence where it is leaving. It is in the core that wind speeds reach their maximum. Some of the air drawn into the core from the cold side is drawn down from the stratosphere.

High-level divergence and convergence produce areas of high and low surface pressure. These generate FRONTAL SYSTEMS and their associated DEPRESSIONS, which travel with the generally westerly movement of midlatitude air. Waves in the fronts are reflected in the jet stream; that is why the wind direction is not constant. ROSSBY WAVES develop along the jet stream, causing it to change direction according to an INDEX CYCLE.

Jevons effect The effect on the measurement of rainfall that is caused by the RAIN GAUGE itself. In 1861 the English economist, logician, and scientist William Stanley Jevons (1835–82) discovered that the gauge disturbs the flow of air past it. Some of the air is deflected, carrying some raindrops with it, so the amount of rain captured by the gauge is smaller than the amount that falls on ground a short distance from the gauge. The effect is too small to be of importance.

JNWP *See* NUMERICAL FORECASTING.

Joan A HURRICANE that struck the Caribbean coast of Central and South America October 22 to 27, 1988. It caused severe damage and killed at least 111 people in Nicaragua, Costa Rica, Panama, Colombia, and Venezuela. It then weakened and was renamed TROPICAL STORM Miriam. Miriam struck El Salvador, where it rendered 3,000 people homeless.

Joe A TYPHOON that struck northern Vietnam on July 23, 1980. More than 130 people were killed and about 3 million were made homeless.

John A CYCLONE from the Indian Ocean that struck the sparsely populated northwestern coast of Australia on December 15, 1999. At its full strength it had sustained winds of 130 mph (209 km h^{-1}) and gusts of up to 185 mph (298 km h^{-1}), making it the most powerful storm ever recorded in Australia. It crossed the coast near Whim Creek (estimated population 12), Western Australia, weakening as it did so, but still with sustained winds of 106 mph (170 km h^{-1}). The cyclone created STORM SURGES of up to 20 feet (6 m).

Joint Doppler Operational Project *See* DOPPLER RADAR.

Joint Numerical Weather Prediction Unit *See* NUMERICAL FORECASTING.

joran *See* JURAN.

Jose A HURRICANE, rated category 2 on the SAFFIR/SIMPSON SCALE, with winds of 100 mph (160 km h^{-1}) but later downgraded to a TROPICAL STORM with 65-mph (105-km h^{-1}) winds that struck Caribbean islands in October 1999. At its maximum extent its diameter was about 300 miles (480 km). It caused heavy rain, but no deaths and few serious injuries.

Joseph effect The tendency for a particular type of weather to persist and repeat itself. HEAT WAVES, DROUGHTS, and COLD WAVES are examples of persistence. This is used in weather forecasting, when the forecaster predicts that the weather over the next few hours or days will remain as it is at present. A forecast of this kind is often accurate, because in middle latitudes, where the weather is more changeable than it is elsewhere in the world, weather systems last for up to seven days. The tendency is known as the *Joseph effect* because of the dream that Pharaoh described to Joseph in the Book of Genesis: "Behold, there come seven years of great plenty throughout all the land of Egypt: And there shall arise after them seven years of famine" (Genesis 41: 29–30).

joule (J) The derived SYSTÈME INTERNATIONAL D'UNITÉS (SI) unit of energy, work, or quantity of heat, which is equal to the work that is done when a force of 1 NEWTON moves a distance of 1 METER, or the energy that is expended by 1 WATT in 1 SECOND. The unit was adopted in 1889, and in 1948 it was adopted as the unit of heat. It is named in honor of the English physicist James Prescott Joule (1818–89): 1 J = 10^7 ergs = 0.2388 calorie.

Judy Three TYPHOONS, the first of which struck South Korea on August 25 and 26, 1979. It caused severe flooding in which 60 people died and 20,000 were rendered homeless. The second Typhoon Judy, rated category 3 on the SAFFIR/SIMPSON SCALE, struck

Japan on September 11 and 12, 1982, killed 26 people, and caused extensive damage. The third Typhoon Judy struck South Korea in July 1989 and killed at least 17 people.

junk wind The name that is given in Thailand, Vietnam, China, and Japan to the southerly or southeasterly MONSOON wind. The name refers to the fact that this wind is favorable to the sailors of junks (traditional flat-bottomed sailing ships).

junta A MOUNTAIN-GAP WIND that blows through passes in the Andes. It sometimes reaches hurricane force (*see* BEAUFORT WIND SCALE).

juran (joran) A cold northwesterly wind that blows through the Jura Mountains, near Geneva, Switzerland. The juran often produces snow and is sometimes very blustery.

Justine *See* BENEDICT.

juvenile water Water that is formed by physical and chemical processes in the magma, which is the hot, plastic rock beneath the Earth's solid crust. Juvenile water has never been in the atmosphere or near the surface of the Earth, but it is released during volcanic eruptions. A very large amount of juvenile water is present in the magma. A layer of magma with a volume of about 4 cubic miles (10 km^3) and a density of 2.5 (that is, 2.5 times the density of water) may contain about 2 billion (2×10^9) cubic feet (1.25×10^9 m^3) of water. Water that falls from the sky is called *METEORIC WATER,* and water that was trapped when sediments were deposited and remains held inside sedimentary rocks is called *connate water*.

K *See* KELVIN and KELVIN SCALE.

K *See* STABILITY INDICES.

kachchan A hot, dry, westerly or southwesterly FÖHN WIND that blows from the mountains of central Sri Lanka in June and July, during the southwest MONSOON. It is especially strong at Batticaloa, on the eastern coast directly northeast of the mountains. There the kachchan wind is often powerful enough to overcome the afternoon sea breeze and produce temperatures approaching 100° F (38° C).

Kainozoic *See* CENOZOIC.

Kai Tak *See* KIROGI.

kal baisakhi A short-lived squall that occurs in Bengal at the start of the southwest MONSOON. The weather then is dry, and so the squall raises considerable amounts of dust.

Kamchatka Current *See* OYASHIO CURRENT.

kamikaze A "divine wind" (in Japanese) that saved Japan from invasion in 1281. The Mongols, who ruled China and Korea, had ordered the Japanese to submit to them. When the Japanese refused, a Mongol invasion force sailed in Korean ships for the southernmost Japanese island of Kyushu. Before landing, the invaders had to overcome small Japanese garrisons on the offshore islands of Tsushima and Iki. They did this without difficulty and advance parties of Mongolian troops landed in several places on Kyushu but were contained by the Japanese warriors. Had the main Mongol force landed, it is quite possible that it would have overwhelmed the Japanese forces and Japan would have been put under Mongol rule. Before this could happen, however, a TYPHOON destroyed most of the Korean ships and the Mongol soldiers drowned. The Japanese believed the typhoon had been sent by the gods to save them from being conquered by foreigners.

Kansan glacial A GLACIAL PERIOD in North America that began about 480,000 years ago and ended about 230,000 years ago. It is approximately equivalent to the GÜNZ GLACIAL of the European Alps. The Kansan was preceded by the AFTONIAN INTERGLACIAL and followed by the YARMOUTHIAN INTERGLACIAL.

karaburan A hot, dry wind of up to gale force that blows from the east-northeast across the deserts of Central Asia from early spring to the end of summer. Its name means "black storm" because of the large amount of desert dust it carries, darkening the sky. It is caused by the rapid heating of the Central Asian landmass.

karema A strong easterly wind that blows across Lake Tanganyika, in East Africa.

Karman vortex street. **A series of vortices that develop downwind of an obstruction when the wind is strong.**

karif (kharif) A strong, southwesterly wind, often reaching gale force (*see* BEAUFORT WIND SCALE), that blows across the coast of Somalia on the southern side of the Gulf of Aden during the southwest MONSOON. The wind is fiercely hot and heavily laden with sand.

Karman vortex street A series of vortices (*see* VORTEX) that develops when a fast-moving flow of air passes an obstruction, such as a building. As the moving air approaches the building it divides into two streams. These pass to each side of the building and both accelerate as a result of the VENTURI EFFECT. If the air is moving gently and fairly steadily, the two streams decelerate and rejoin a short distance downstream from the obstruction. If the air is moving rapidly, however, the two streams remain separated and a series of vortices develop between them. They rotate alternately clockwise and counterclockwise at a slower speed than the mean speed of the main airflow. The vortices are constantly forming and disappearing and greatly increase the TURBULENT FLOW of air. This phenomenon was first described by the Hungarian-born American aerodynamicist Theodore von Kármán (1881–1963).

katabaric (katallobaric) An adjective that is applied to a phenomenon associated with a fall in atmospheric pressure.

katabaric center *See* PRESSURE-FALL CENTER.

katabatic wind (drainage wind, fall wind, gravity wind) A cold wind that blows downhill across sloping ground. The word is from the Greek *katabatikos*, which means "going down." It usually develops on a still, cold night when the ground cools by radiating the warmth it absorbed by day and the air is cooled from below by contact with the ground. A layer of cold air may then lie trapped in an INVERSION beneath a layer of warmer air. The cold air then flows down any slopes and accumulates in hollows and valleys (*see* FROST HOLLOW). Katabatic winds also occur over the polar ice sheets, which are domed, with the highest elevation at the center. Along the coast of Antarctica, the average katabatic wind speed is about 45 mph (72 km h^{-1}) and BLIZZARDS of blowing snow are common.

katafront A WARM FRONT or COLD FRONT at which air in the WARM SECTOR is subsiding relative to the cold

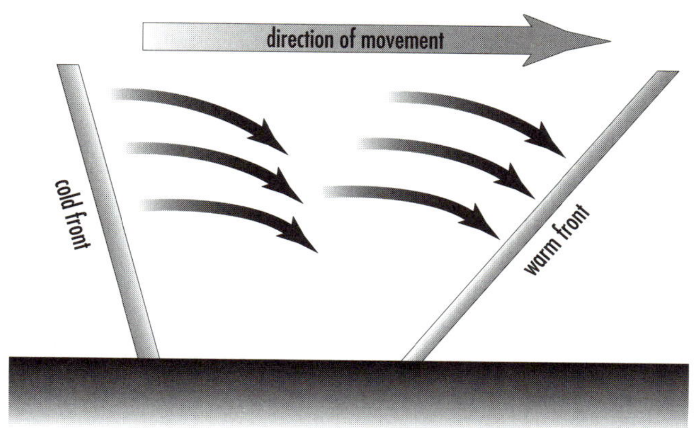

Katafront. **A front against which air is subsiding.**

air on each side of the fronts. Often, warm-sector air is trapped beneath a temperature INVERSION at about 10,000 feet (3 km). Although there is usually a complete cloud cover and cloud is especially thick in the vicinity of the fronts, the vertical extent of the cloud is limited by the inversion. Precipitation at a warm katafront consists of a belt of DRIZZLE or light rain and that at the cold front consists of showers.

katallobaric *See* KATABARIC.

Kate A HURRICANE that struck Cuba and Florida November 19 to 21, 1985. It killed at least 24 people.

kaus (cowshee, sharki) A southeasterly wind that blows over the Persian Gulf ahead of a DEPRESSION. It can reach gale force (*see* BEAUFORT WIND SCALE) and is usually associated with cloudy weather and squalls. After the depression has passed the kaus may be followed by a SUHAILI.

Keeler, James Edward *See* LANGLEY, SAMUEL PIERPONT.

Keith A HURRICANE that struck Central America in 2000. On September 30 it killed one person in El Salvador and the following day it struck Nicaragua, where a 16-year-old boy was swept away by a swollen river as the hurricane reached category 3 on the SAFFIR/SIMPSON HURRICANE SCALE, with winds of 135 mph (217 km h^{-1}). Keith then stuck Belize and the Yucatan Peninsula of Mexico, its winds dropping to 90 mph (145 km h^{-1}). It weakened to a TROPICAL STORM as it moved out over the Gulf of Mexico but then strengthened to category 2 as it headed back toward the Mexican coast. By the time it died, Keith had caused at least 12 deaths and had dropped 22 inches (559 mm) of rain on Belize.

Kelly A TYPHOON that struck the Philippines on July 1, 1981. It caused floods and landslides in which about 140 people died.

kelvin (K) The SYSTÈME INTERNATIONAL D'UNITÉS (SI) unit of thermodynamic temperature, which is defined as being 1/273.16 of the TRIPLE POINT of water. 1 K = 1° C = 1.8° F. The unit is named in honor of the Scottish physicist and electrical engineer William Thomson, the first baron Kelvin of Largs (1824–1907).

Kelvin–Helmholtz instability *See* SHEAR INSTABILITY.

Kelvin scale (K) The temperature scale that is used in the SYSTÈME INTERNATIONAL D'UNITÉS (SI) and that was devised by William Thomson (1824–1907), who later became Lord Kelvin. He called it the *absolute* scale. One kelvin, written as 1 K with no degree sign (not as 0° K), is equal to 1° C (1.8° F), and 0 K, or ABSOLUTE ZERO, is equal to –273.16° C (–459.67° F).

Kelvin waves Ocean waves that occur only in equatorial waters. Their amplitude is measured in tens of meters at the THERMOCLINE and their wavelength in thousands of meters. They take about two months to cross the Pacific and because of the CORIOLIS EFFECT are higher on the side closer to the equator. They are associated with EL NIÑO events. The waves are named after the Scottish physicist Lord Kelvin (William Thomson, 1824–1907), who discovered them.

Kew barometer A type of mercurial BAROMETER that allows the AIR PRESSURE to be read directly from a scale at the top of its tube; no adjustment needs to be made to the level of mercury in its reservoir. It is based on the principle that if both the reservoir and the tube above it are perfectly cylindrical but of different diameters, then a change in the level of mercury in the reservoir will be a definite fraction of the change in the level in the tube. The central part of the tube is usually narrower than the remainder of the tube. Kew barometers are often used at sea, because they can be made in such a way as to minimize the oscillation in the height of the mercury caused by the movement of a ship. Oscillation, or pumping, of the mercury level makes it impractical to use other types of mercurial barometers at sea. Kew barometers can also be transported. They are tilted until mercury fills the tube, then inverted and carried with the reservoir at the top. They can also be carried horizontally.

kg *See* KILOGRAM.

khamsin A hot, dry wind that blows from the southeast across Egypt and the Sudan. Its name is the Arabic word for "fifty" and refers to the fact that the wind usually blows on 50 days every year in late winter and early spring. It is a wind of the SIROCCO type associated with a low-pressure system over the Sudan that extends to the northeast and a simultaneous shallow DEPRESSION over the northern Sahara. These combine to draw in air from Arabia and possibly from as far away as the Arabian Sea. The air is hot, its temperature often exceeding 100° F (38° C), and extremely dry; it carries so much fine dust that visibility is seriously reduced and cars must use their headlights in the middle of the day. The wind lasts two or three days.

kharif *See* KARIF.

killing frost A drop in temperature that kills plants or prevents them from reproducing. As the falling temperature approaches freezing, small ICE CRYSTALS start to form in the spaces between cells. This reduces the amount of liquid in the intercellular space and water flows from the cells by OSMOSIS. If there is then a slow thaw, the ice melts and the cells reabsorb the water and recover, but if the temperature rises rapidly, the water is lost from the plant and cells die of dehydration. Dehydration also occurs if the temperature remains below freezing for a prolonged period, because water is lost from the plant by SUBLIMATION. If flowers or fruits that have not yet set seed are damaged by frost, the resulting dehydration prevents them from producing viable seeds. Seeds themselves are usually tolerant of frost because they contain little moisture.

kilogram (kg) The SYSTÈME INTERNATIONAL D'UNITÉS (SI) unit of mass, which is equal to the mass of a prototype that is kept at the International Bureau of Weights and Measures at Sèvres, near Paris, France; 1 kg = 2.20462 pounds.

Kina A CYCLONE, rated as category 4 on the SAFFIR/SIMPSON SCALE, that struck Fiji on January 2 and 3, 1993. It generated winds of up to 115 mph (185 km h^{-1}) and killed 12 people.

K Index *See* STABILITY INDICES.

kinematic viscosity The COEFFICIENT OF VISCOSITY of a fluid divided by its density. It is measured in square meters per second (m^2s^{-1}) in the SYSTÈME INTERNATIONAL D'UNITÉS (SI) system and square centimeters per second in the C.G.S. SYSTEM.

kinetic energy The energy of motion, which is usually defined in terms of the amount of work a moving body could do if it were brought to rest. It is equal to $mv^2 \div 2$, where m is the mass of the moving body and v is its speed. For a rotating body, such as the air in a TROPICAL CYCLONE or TORNADO, the kinetic energy is equal to $I\omega^2 \div 2$, where ω is the ANGULAR VELOCITY and I is the MOMENT OF INERTIA. Kinetic energy is one of the two forms of energy; the other is POTENTIAL ENERGY.

Kirchhoff's law A law stating that the amount of radiation of a particular wavelength that is emitted by a body is equal to the amount absorbed by that body at the same wavelength. This is expressed as

$$\epsilon\lambda = \alpha\lambda f(\lambda, T)$$

where $\epsilon\lambda$ is the emissivity at a stated wavelength, $\alpha\lambda$ is the absorptivity at the same wavelength, and $f(\lambda, T)$ is a function that varies with the wavelength (λ) and surface temperature (T), but not with the material of which the surface is composed. The law means that if a body is a good or poor absorber of radiation, it is an equally good or poor emitter of radiation at the same wavelength.

The law was first described in 1859 by the German physicist Gustav Robert Kirchhoff (1824–87). From this law he went on to derive the concept of a perfect BLACKBODY. Although the Scottish physicist Balfour Stewart (1828–87) had reached a similar conclusion in 1858, Kirchhoff presented it more persuasively and is usually given credit for its discovery.

Kirk A TYPHOON that crossed southwestern Honshu, Japan, on August 14, 1996, with winds of 130 mph (209 km h^{-1}) and up to 12 inches (300 mm) of rain. It returned to the northern part of the island on August 15, but by then it had weakened. From there it moved to northeastern China where it caused floods that inundated 845 villages along the Yellow River.

Kirogi A TYPHOON that struck the Philippines, Japan, and Taiwan in July 2000. Kirogi was closely followed by another typhoon, Kai Tak. Together the two typhoons killed 42 people in the Philippines, 5 in Japan, and 1 in Taiwan. Kirogi generated sustained winds of 89 mph (143 km h^{-1}) and Kai Tak winds of 93 mph (150 km h^{-1}). Both typhoons delivered heavy rain.

kloof wind A cold, southwesterly wind that blows across Simons Bay, South Africa.

knik wind A strong, southeasterly wind that blows in the area around Palmer, Alaska. It can occur at any time of year but is most frequent in winter.

knot A unit of velocity that was devised originally for use at sea but that is also widely used by aircraft. It is a velocity of one NAUTICAL MILE per hour and, by international agreement, equal to 1.852 km h^{-1}. The United States adopted the international knot in 1954, but British ships and aircraft continue to use a knot equal to 1.00064 international knots. The knot came into use as a unit in the late 16th century. Then, the speed of a ship was measured by dropping over the side a float attached by a knotted rope. The knots were spaced 7 fathoms apart (a fathom is 6 feet, or 1.8 m) and the sailor measuring the speed counted the number of knots that passed in 30 seconds. This gave the speed of the ship in nautical miles per hour. Subsequently, the length of the nautical mile was changed, as were the distance between knots and the period of time used for counting, until the unit became standardized.

koembang A dry, warm southerly or southeasterly wind of the föhn type (*see* FÖHN WINDS) that blows in Java, Indonesia. The wind is caused by the MONSOON and is accelerated by FUNNELING as it passes through the gaps in the mountains before descending on the LEE side in the area around Tegal, on the northern side of the island.

kona A southwesterly wind that blows across Hawaii about five times a year. It carries storms and heavy rain to the southwestern slopes of the mountains, which are ordinarily on the LEE side of the northeasterly TRADE WINDS.

kona cyclone (kona storm) A persistent, slow-moving SUBTROPICAL CYCLONE that delivers most of the winter rainfall in Hawaii.

kona storm *See* KONA CYCLONE.

Köppen climate classification A system for categorizing climates generically according to their temperatures and aridity that was devised by WLADIMIR PETER KÖPPEN. Of all the schemes for CLIMATE CLASSIFICATION, Köppen's is the one most widely used by geographers.

Köppen began by relating climate to the plant growth in a region and the type of vegetation present there. He was not the first scientist to note that particular plants grow only in certain places, and the distribution of vegetation types had already been mapped and described. Köppen worked from the map prepared by the Swiss botanist Alphonse Louis Pierre Pyramus de Candolle (1806–93), professor of natural history at the University of Geneva and possibly the most eminent botanist of his day. (His father had held the same position and was equally distinguished, mainly for his classification of plants.) Candolle published *Géographie Botanique Raisonée* (Botanical geography classified) in 1855 and *La Phytographie* in 1880 (phytogeography, to give it its English name, is the scientific study of the geography of plant distribution). In 1874, Candolle produced a world vegetation map in which plant distribution was linked to the physiological structure of the plants. Köppen used it to define climates according to the zones occupied by particular vegetation types and produced his first climate map in 1884. He then developed this into a complete classification scheme for climates, and published the first version in 1900 and the final version in 1936. Toward the end of his life, Köppen collaborated with the German climatologist Rudolf Geiger, who continued to work on modifications to the classification after Köppen's death.

The scheme is fairly complex and uses a code of letters to designate climatic details. Climates are grouped in six major categories, identified as A to F, all but one of which (B) are identified by temperature:

A. Tropical rainy climates in which the temperature in the coldest month does not fall below 64.4° F (18° C)
B. Dry climates
C. Warm temperate rainy climates in which temperatures in the coldest month are between 26.6° F (–3° C) and 64.4° F (18° C) and in the warmest month are higher than 50° F (10° C)
D. Cold boreal forest climates in which temperatures in the coldest month are below 26.6° F (–3° C) and temperatures in the warmest month are above 50° F (10° C)
E. Tundra climate in which temperatures in the warmest month are between 32° F (0° C) and 50° F (10° C)
F. Perpetual frost climate in which temperatures in the warmest month remain below 32° F (0° C)

The temperatures were chosen for particular reasons. Trees do not grow where the summer temperature is below 50° F (10° C), for example, and some plants do not grow where the winter temperature falls below 64.4° F (18° C). A mean temperature of 26.6° F (–3° C) implies frost and probably some snow.

These principal categories are then divided further, mainly according to the amount and distribution of precipitation they receive:

Af climates are hot and rainy throughout the year.
Am climates are hot and excessively rainy in one season.
Aw climates are hot and in winter are dry.
BSh climates are semiarid and hot.
BSk climates are semiarid and cool or cold.
BWh climates are those of hot deserts.
BWk climates are those of cool or cold deserts.
Cfa climates are mild in winter, hot in summer, and moist throughout the year.
Cfb climates are mild in winter, warm in summer, and moist throughout the year.
Cfc climates are mild in winter; have a short, cool summer; and are moist throughout the year.
Cwa climates are mild in winter and hot in summer and have dry winters.
Cwb climates are mild in winter; have short, warm summers; and are dry in summer.
Csa climates are mild in winter and hot and dry in summer.
Csb climates are mild in winter and have a short, warm dry summer.
Dfa climates are very cold in winter; have a long, hot summer; and are moist throughout the year.
Dfb climates are very cold in winter; have a short, warm summer; and are moist throughout the year.
Dfc climates are very cold in winter; have a short, cool summer; and are moist throughout the year.
Dfd climates are extremely cold in winter, have a short summer, and are moist throughout the year.
Dwa climates have a very cold, dry winter and long, hot summer.
Dwb climates have a very cold, dry winter and a cool summer.
Dwc climates have a very cold, dry winter and a short, cool summer.
Dwd climates have an extremely cold winter and a short, moist summer.
ET is a polar climate with a very short summer.
EF is a climate of perpetual frost and snow.

In addition to these categories, additional letters are used to add further qualifications:

a Mean temperature in the warmest month is about 71.6° F (22° C).
b Mean temperature in the warmest month is below 71.6° F (22° C), and during at least four months the mean temperature is higher than 50° F (10° C).
c For one to four months the mean temperature is higher than 50° F (10° C), and the mean temperature does not fall below –36.4° F (–38° C) in the coldest month.
d The mean temperature in the coldest month is less than –36.4° F (–38° C).
f Precipitation is sufficient for healthy plant growth in all seasons.
g The temperature is highest prior to the summer rainy season.

h The mean annual temperature is higher than 64.4° F (18° C).

i The climate is *isothermal:* there is less than 14° F (5° C) difference in mean temperature between the warmest and coolest months.

k Winters are cold, the mean annual temperature is below 64.4° F (18° C), and the mean temperature in the warmest month is higher than 64.4° F (18° C).

k' This is similar to k, but in the warmest month the mean temperature is below 64.4° F (18° C).

l The mean temperature in all months is between 50° F (10° C) and 71.6° F (22° C).

m There are a short, dry season and heavy rain throughout the remainder of the year.

n Fog is frequent.

n' Fog is infrequent, but the humidity is high although rainfall is low; the mean summer temperature is below 74.2° F (24° C).

p Conditions are similar to those in n', but the mean temperature in one summer month is between 74.2° F (24° C) and 82.4° F (28° C).

p' Conditions are similar to those in n', but the mean summer temperature is higher than 82.4° F (28° C).

s There is a dry season in summer.

u The coldest month occurs after the summer solstice.

v The warmest month is in the autumn.

w Winters are dry and rainfall during the remainder of the year is not sufficient to make good the deficiency, so plant growth is restricted.

w' There is a rainy season in autumn.

w'' There are two distinct rainy seasons and two distinct dry seasons.

x Maximum rainfall occurs in spring or early summer and the late summer is dry.

x' Conditions are similar to those of x, but there are infrequent, heavy falls of rain in all seasons.

Köppen, Wladimir Peter (1846–1940) German *Meteorologist and climatologist* Wladimir Köppen was born in St. Petersburg, Russia, on September 25, 1846. His parents were German, and after attending school in the Crimea, he studied at the Universities of Heidelberg and Leipzig. It was while he was at school in Russia that Wladimir first became interested in the natural environment and especially in the interaction between plants and climate. His student dissertation, which he completed in 1870, dealt with the relationship between temperature and plant growth.

After his graduation, from 1872 to 1873, he was employed in the Russian meteorological service. In 1875, he returned to Germany to take up an appointment as chief of a new division of the Deutsche Seewarte, based in Hamburg. His task there was to establish a weather forecasting service covering northwestern Germany and the adjacent sea areas. His primary interest lay in fundamental research, however, and from 1879, once the meteorological service was functioning, he was able to devote himself to it.

He embarked on a systematic study of the climate over oceans and also investigated the upper air, using kites and balloons to obtain data. In 1884, he published the first version of his map of climatic zones (*see* KÖPPEN CLIMATE CLASSIFICATION). He plotted these on an imaginary continent he called *Köppen'sche Rübe* (Köppen's beet). His CLIMATE CLASSIFICATION appeared in full in 1918, and after several revisions the final version of it was published in 1936.

In addition to writing hundreds of articles and scientific papers, Köppen coauthored with ALFRED WEGENER *Die Klimate der Geologischen Vorzeit* (The climates of the geological past), published in 1924, and wrote *Grundriss der Klimakunde* (Outline of climate science), which was published in 1931. In 1927, he entered into collaboration with Rudolf Geiger to produce a five-volume work, *Handbuch der Klimatologie* (Handbook of climatology). This was never completed, but several parts, three of them by Köppen, were published.

Köppen had moved to Graz, Austria, to work on the *Handbuch der Klimatologie* and it was there that he died, on June 22, 1940.

kosava A strong wind that blows from the east or southeast through the Iron Gate gorge on the Danube River, on the border between Romania and Serbia. It continues through the Carpathian Mountains, over Belgrade, and onto the Hungarian plain. It is usually a cold wind associated with winter DEPRESSIONS.

Kr *See* KRYPTON.

Krakatau (Krakatoa) A volcanic island that lies in the Sunda Strait between Java and Sumatra, Indonesia. On May 20, 1883, the volcano became active, and on August 26 and 27 it erupted in a series of increasingly violent explosions. These were heard in Australia, 2.200 miles (3,540 km) to the southeast, and at

Rodriguez Island, 3,000 miles (4,800 km) to the southwest. The eruption threw about 5 cubic miles (21 km^3) of solid particles to a height of more than 19 miles (30 km). Stratospheric winds distributed the dust over most of the Earth, producing spectacular sunsets for the following three years, which was the time it took for the dust to settle. During those three years, the Montpelier Observatory, in France, recorded a 10 percent decrease in the intensity of solar radiation and there was a small but significant fall in average temperatures.

Krakatoa *See* KRAKATAU.

krypton (Kr) A colorless gas that is present in the air in trace amounts (*see* ATMOSPHERIC COMPOSITION). It is a NOBLE GAS with atomic number 36, relative atomic mass 83.80, density (at sea-level pressure and 32° F [0° C]) 0.0000021 ounce per cubic inch (0.0000037 g cm^{-3}). It melts at –250° F (–156.6° C) and boils at –242.1° F (–152.3° C).

Kuroshio Current An ocean current that flows northward from the Philippines, along the coast of Japan, and then eastward into the North Pacific Ocean, carrying warm water northward. It is a narrow current, less than 50 miles (80 km) wide, and flows rapidly, at up to 7 mph (11 km h^{-1}).

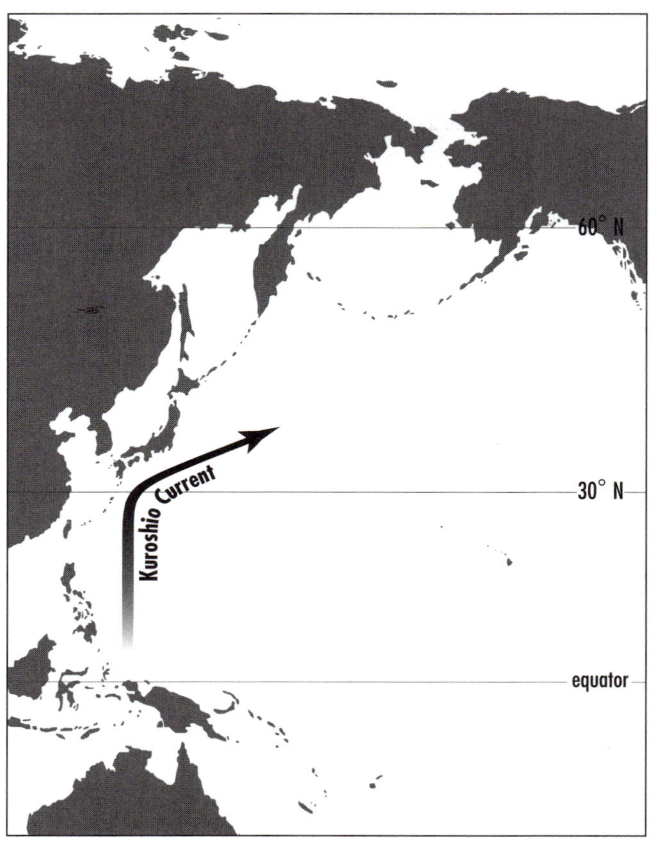

Kuroshio Current. **A warm ocean current that flows parallel to the eastern coast of Japan.**

Kyle A TYPHOON that struck Vietnam on November 23, 1993. It killed at least 45 people.

Kyoto Protocol An international agreement that was drawn up in 1997 under the auspices of the United Nations to provide guidelines for the implementation of the United Nations FRAMEWORK CONVENTION ON CLIMATE CHANGE. The protocol committed the industrialized nations to reducing their emissions of six GREENHOUSE GASES by an average of 5.2 percent (measured against their 1990 levels) by 2012. The European Union agreed to an 8 percent reduction, the United States to 7 percent, Japan to 6 percent, and 21 other countries to varying reductions. Less-developed countries were not required to make binding commitments. The protocol was adopted by about 170 nations, but not the United States, at Bonn, Germany, on July 23, 2001. The 5.2 percent target was retained, but countries were permitted to offset their emissions reductions by counting the absorption of carbon dioxide by forest planting, changes in forest management, and improved management of croplands and grassland. Nations emitting less than their target amounts were allowed to sell the surplus, up to 10 percent of their total emission entitlement, as credits to nations unable to meet their targets.

L

L *See* AVOGADRO CONSTANT.

Labor Day storm A HURRICANE that struck southern Florida on Labor Day, 1935. It was impossible to measure wind speeds accurately, but they were estimated at 150–200 mph (241–322 km h^{-1}) on some of the Florida Keys, making it a category 5 hurricane on the SAFFIR/SIMPSON HURRICANE SCALE. Surface atmospheric pressure in the eye of the hurricane was 892.4 millibars. This was the lowest surface pressure recorded in the Western Hemisphere until hurricane GILBERT in 1988. A total of 408 people died in the Labor Day storm.

Labrador Current An ocean current that conveys cold water from the Arctic Ocean into the North Atlantic Ocean. It flows in a southeasterly direction between the coasts of eastern Canada and western Greenland, often carrying ICEBERGS south into the North Atlantic. SEA FOGS are common off Newfoundland, where the cold water of the Labrador Current meets the warm water of the GULF STREAM.

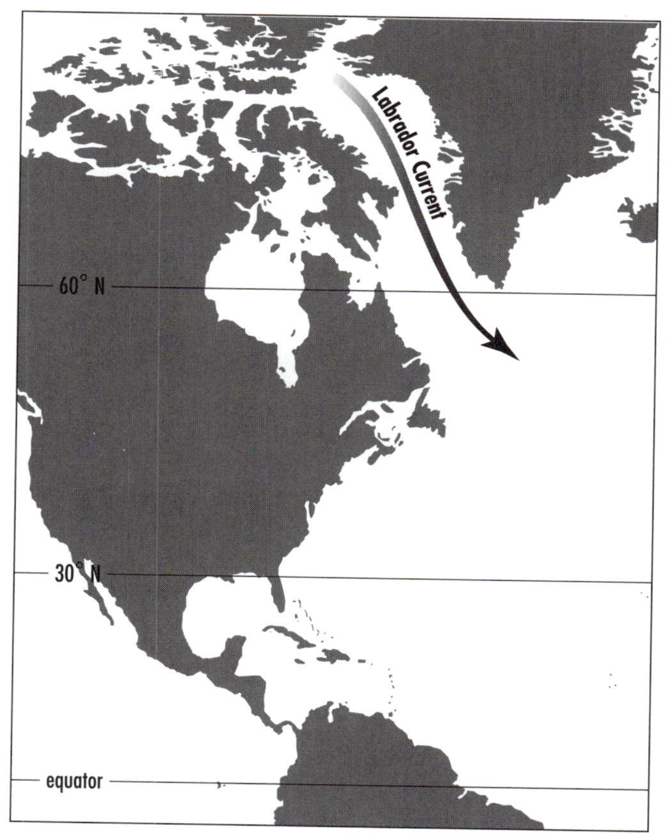

Labrador Current. **A cold ocean current that flows parallel to the northeastern coast of Canada.**

lacunosus A variety of clouds (*see* CLOUD CLASSIFICATION) that appear as patches, layers, or sheets of cloud that include approximately round holes distributed more or less evenly so the cloud and holes form a pattern reminiscent of a net. The cloud is usually thin and the holes often have fringes around the edges. Lacunosus occurs with the cloud genera CIRROCUMULUS and ALTOCUMULUS. The name *lacunosus* is derived from the Latin word *lacus,* which means "lake." This is the same source that gives us *lacuna,* meaning a hole or missing portion.

LAI *See* LEAF AREA INDEX.

Lake Bonneville A lake that formed in what is now Utah about 32,000 years ago and finally disappeared about 14,000 years ago, during the WISCONSINIAN GLACIAL period. At its greatest extent the lake was about 325 miles (523 km) long and 135 miles (217 km) wide and had a surface area of about 19,000 square miles (49,200 km^2). In places it was more than 1,000 feet (305 m) deep and what are now mountains projected above the surface as islands. During the Wisconsinian the average temperature was about 11° F (6° C) cooler than that of today and the climate was wetter. The lake was filled by water from precipitation, rivers, and the melting edges of GLACIERS. As the climate changed and most of the lake evaporated, the precipitation of salt formed the Bonneville salt flats. Modern Utah Lake is a freshwater remnant of Lake Bonneville, and the Great Salt Lake, Little Salt Lake, and Sevier Lake are saltwater remnants.

(You can learn more about Lake Bonneville from www.ugs.state.ut.us/pi-39/pi39pg1.htm.)

lake breeze A wind that is produced when air warms over land and rises by CONVECTION and its place is taken by cool, moist air drawn from over the surface of a lake. This can produce a change in wind direction and a sharp fall in temperature over land adjacent to the lake shore. The breeze develops during the afternoon, by which time the land has been warmed, and is similar to a sea breeze (*see* LAND AND SEA BREEZES). In summer, when the difference in land and lake-surface temperatures is most extreme, the movement of cool, moist air beneath warm, dry air produces a front that can trigger THUNDERSTORMS.

lake-breeze front *See* SEA-BREEZE FRONT.

lake effect A modification in the characteristics of air as it crosses a large expanse of water that is entirely enclosed by land. In summer, hot, dry CONTINENTAL AIR is cooled by contact with the open water surface and a large amount of water evaporates into it. Consequently, the weather on the lee side of the lake is somewhat cooler and more humid than that on the opposite side and the rainfall is greater over high ground.

lake-effect snow Snow that falls in winter on the LEE side of a large lake that is entirely enclosed by a large land mass. In most winters lake-effect snow falls very heavily to the east of the Great Lakes. Cold, dry CONTINENTAL AIR travels from west to east across North America. When it reaches the Great Lakes it makes contact with the surface of water, which remains liquid most of the winter and in some years does not freeze at all. The air is warmed a little by contact with the water and large amounts of water evaporate into it. When it reaches the other (eastern) side of the lakes the air crosses colder land. It is chilled, its water vapor condenses, and there are heavy falls of snow. Places in what is known as the *SNOW BELT,* up to about 50 miles (80 km) to the east of the lakes, receive much more winter snow than places to the west of the lakes and accumulations of 2 feet (50 cm) in 24 hours are not uncommon. When the lakes freeze over the effect ceases.

Lamb, Hubert Horace (1913–1997) English *Climatologist* One of the first scientists to draw attention to the variability of climates and the social and economic consequences of climate change, Hubert Lamb was possibly the greatest climatologist of the 20th century. In addition to studying climate change and the history of climates, Lamb was among the most skillful of weather forecasters.

Hubert Lamb was born at Bedford and educated at Oundle School, in Northamptonshire, and then at Trinity College, Cambridge, where he studied natural sciences and geography. He graduated in 1935 and received a master's degree in the same subjects in 1947. In 1981 he was awarded honorary doctorates from the Universities of Dundee (LL.D.) and East Anglia (D.Sc.) and in 1982 Cambridge University awarded him a doctorate of science.

In 1936, after his graduation from Cambridge, Lamb went to work at the British METEOROLOGICAL OFFICE. As war became increasingly probable, Lamb was asked to study the way clouds of poison gas would be carried by the wind. He refused to do this on moral grounds, and in 1940 he was transferred to the Irish Meteorological Office. In 1941 he was placed in charge of preparing forecasts for transatlantic flights. His forecasts were so accurate that during his period there transatlantic flights out of Ireland had a perfect safety record. Eventually he had a disagreement with the director, however, and in 1945 he returned to the Meteorological Office in Britain. He sailed as meteorologist on a Norwegian whaling expedition to Antarctica in 1946–47, and it was during this voyage that he came to realize the extent to which climate changes over time. He served as a weather forecaster in Germany from 1951 to 1952 and from 1952 until 1954 worked in Malta.

During all of this time, from 1945 until 1971, Lamb was employed by the Meteorological Office and devoted part of his time to long-range weather forecasting and studies of global weather and climate change. In 1954 he was placed in charge of the Climatology Division at the Meteorological Office. While there he undertook the first detailed study of the past climate records held at the office. He used these to trace ways the climate had changed since the middle of the 18th century and devised a classification system, known as *LAMB'S CLASSIFICATION*, for British weather.

He also came to realize that dust injected high into the air by volcanic eruptions can affect surface weather by reflecting incoming solar radiation back into space, thereby shading the surface. This led him to study every eruption since 1500 and estimate the amount of dust each had released and its effect on weather in the years following. From this he developed a method for estimating the climatic effect of volcanic dust, known as *LAMB'S DUST VEIL INDEX*.

In 1971 he left the Meteorological Office to establish the Climatic Research Unit at the University of East Anglia, in Norwich. He remained director of the unit until his retirement in 1977. After retiring he remained at the university as an emeritus professor. Under his direction the unit became one of the foremost centers for the study of climate change.

Hubert Lamb died in Norwich on June 28, 1997.

(You can learn more about Hubert Lamb from his autobiography, *Through All the Changing Scenes of Life* [ISBN 1 901470 02 4], [East Harling, U.K.: Taverner Publications, 1997]).

Lambert's law A law stating that the intensity of the radiation emitted in any direction from a unit surface area of a radiating body varies according to the cosine of the angle between the direction of the radiation and a line perpendicular to the radiating surface. The law was discovered by the German mathematician Johann Heinrich Lambert (1728–77), who was the first person to measure light intensities accurately and who coined the term *ALBEDO* (Latin for "whiteness"). A unit of luminance was named the *lambert* (L) in his honor. Equal to one lumen per square meter, it is no longer in use.

lambing storm The name given in parts of Britain to weather conditions that deliver a light fall of snow in the spring, when ewes are lambing.

Lamb's classification A catalog of the type of weather conditions experienced over the British Isles every day from January 1, 1861, until February 3, 1997. The catalog was compiled by Professor HUBERT HORACE LAMB and ended when his final illness made it impossible for him to continue working on it. The catalog describes the direction in which the weather systems are moving, broken into eight directions (N, NE, E, SE, S, SW, W, and NW), and classifies the systems as CYCLONIC, ANTICYCLONIC, or unclassifiable.

(You can learn more about Lamb's classification from www.cru.uea.ac.uk/~mikeh/datasets/uk/lamb.htm.)

Lamb's dust veil index (dust veil index, DVI) A set of numerical values that quantify the climatic effect of dust that is ejected into the atmosphere by a volcanic eruption. After an eruption, air movements disperse volcanic dust until it forms a thin veil covering a large area. This veil reflects a proportion of the incoming solar radiation and therefore exerts a climatic cooling effect. There is little exchange of air across the TROPOPAUSE, and if the dust is injected into the STRATOSPHERE the veil may endure for several years before finally dispersing. The index, compiled by Professor HUBERT HORACE LAMB, was based on his study of historical records from 1500 until 1983. He then calibrated the resulting values, relating them to the 1883 eruption of Krakatoa, to which he allotted a DVI value of 1,000. The full index lists the name of each erupting volcano, the year it erupted, its latitude and longitude, the maximum extent of the veil of dust it emitted, the duration of the dust veil, the DVI value for the entire world, the DVI value for the Northern Hemisphere, and the DVI value for the Southern Hemisphere. Lamb calculated the effect of the dust veil by comparing the recorded surface temperature with the average temperature for the same place at the same time of year. He also used estimates of the amount of material ejected during each eruption.

(You can learn more about Lamb's dust veil index in "Volcanic Loading: The Dust Veil Index (1985)" at cdiac.esd.ornl.gov/cdiac/ndps/ndp013.html and "Volcanic Activity and Environmental Change" at www.aber.ac.uk/~jpg/volcano/lecture2.html.)

laminar boundary layer The layer of air that is in direct contact with a surface and in which air molecules are able to move only slowly, as a result of VISCOSITY. Flow in this layer is laminar (*see* LAMINAR FLOW) and parallel to the surface.

laminar flow The motion of a fluid that occurs smoothly, the fluid (air or water) forming sheets that lie along STREAMLINES that are parallel to each other. Where the flow is laminar, the properties of the fluid at one level can be transferred to another level only through the random motion of molecules. Laminar flow changes to TURBULENT FLOW when the speed reaches a value that is directly proportional to the VISCOSITY of the fluid and inversely proportional to its DENSITY. This is because FRICTION increases with flow speed, increasing in turn the difference in speed from one laminar sheet to another and thus transferring the effect of friction throughout the flow. Consequently, the movement of air is almost always turbulent, except very close to the ground surface. There a layer of air that is never more than a few millimeters thick (1 mm = 0.04 inch) adheres to the surface. This is called the *LAMINAR BOUNDARY LAYER* and within it heat, momentum, and substances such as molecules of water vapor travel vertically only through random molecular motion.

Laminar flow. **Movement of a fluid in which all the stream lines are parallel to one another.**

land and sea breezes Winds that blow in coastal regions and along the shores of large lakes. In middle latitudes they are most common in fine weather in summer. In the Tropics, where seasonal temperature differences are less pronounced, coastal winds of this type can occur at any time of year.

During the morning the land is warmed by the Sun and its temperature rises. The Sun also shines on the water, but its temperature changes much more slowly. This is because water has a much higher HEAT CAPACITY than land. Air over the land is warmed by contact with the surface. It expands and rises, drawing cooler air from over the water to replace it. This establishes a CONVECTION CELL in which air rises over land, moves over the water at high level, sinks over the water, and flows back toward the land at low level. The cell is shallow, extending to no more than about 3,500 feet (1,070 m).

The wind flowing from the water to the land is a sea or LAKE BREEZE. It usually begins in the late morning and reaches a maximum in the middle of the afternoon, when the air temperature is highest. The wind speed can reach about 12 mph (20 km h^{-1}) in a sea breeze, but less in a lake breeze. In the case of sea breezes, a SEA-BREEZE FRONT may develop.

Sea breezes are gusty and carry cool, moist air over the coast, reducing the temperature by up to 18° F (10° C). In the Tropics the cool coastal belt this produces extends up to 100 miles (160 km) inland. In middle latitudes its effect is seldom felt more than about 20 miles (32 km) inland from seacoasts and a shorter distance from lakes. Moist air crossing warm land can become unstable, leading to the formation of CUMULIFORM cloud and showers.

At night the opposite effect occurs. The land cools more rapidly than the water, chilling the air in contact with it. This causes air to subside over the land, drawing air from over the sea to replace it at high level. Cool air flows from the land toward the sea at low level. This is the land breeze. The cell producing it is shallower than the one that produces a sea or lake breeze and the land breeze is weaker. It does not develop at all unless the land cools to below the temperature of the water surface.

Land and sea breezes. **During the day, warm air rises over the land and cool air flows from the sea to replace it. This is the sea breeze. At night the land cools. If its temperature falls below that of the sea surface, air flows from land to sea as a land breeze.**

The frequent movement of FRONTAL SYSTEMS often produces winds that obscure land and sea breezes in middle latitudes.

land blink A yellow glow that is seen in polar regions over an extensive snow-covered surface.

land contamination In satellite imaging, the effect of the footprint, about 30 miles (50 km) in diameter, that occurs near coasts. It is due to the fact that some of the radiation received at the satellite is from land and some from the ocean, making the data imprecise near coasts. This is especially important where the land is covered with ice but the sea is not, because it can make it appear that ice covers coastal waters.

Landsat A satellite observation program that was designed by the National Aeronautics and Space Administration (NASA), in collaboration with the Department of Agriculture, Department of Commerce, and an Environmental Sciences Group within the Environmental Sciences Services Administration (which subsequently became the NATIONAL OCEANIC AND ATMOSPHERIC ADMINISTRATION). The U.S. government funds the program and the satellites are operated by the Department of the Interior and NASA. The aim of the program is the REMOTE SENSING of the Earth's resources. This includes studies of weather systems and climates.

Planning commenced in 1965 and the first satellite was launched on July 23, 1972. It was called the *Earth Resources Technology Satellite* (ERTS 1). The name of the program and the satellites was changed to Landsat in 1975. *Landsat 1* (ERTS 1) ended its service on June 1, 1978. *Landsat 2* was launched on January 22, 1975, and was taken out of service on February 25, 1982. *Landsat 3* operated from March 5, 1978, until March 31, 1983. *Landsat 4* was launched on July 16, 1982, and its active life ended when it ceased transmitting data in August 1993. *Landsat 5* was launched on March 1, 1984, and is still functioning. *Landsat 6* operated from October 5, 1993, but failed to reach orbit. *Landsat 7* was launched on April 16, 1999. It is planned to launch *Landsat 8* in 2004.

All the Landsat satellites operate in POLAR ORBITS. *Landsats 1–3* orbited at an altitude of 570 miles (917 km), passing over every point on the surface at intervals of 18 days. *Landsats 4–7* orbit at 438 miles (705 km) and overfly every point at intervals of 16 days.

(You can learn more about the history of the Landsat program at geo.arc.nasa.gov/sge/landsat/lpchron.html.)

land sky The dark color of the underside of a cloud that is seen in polar regions over a land surface that is free of snow. The cloud appears relatively dark because its base is not illuminated by light reflected from a snow-covered surface, but it is not so dark as a WATER SKY.

landspout (nonsupercell tornado) A TORNADO that develops in a relatively weak CUMULONIMBUS cloud containing no MESOCYCLONE or SUPERCELL. Landspouts have been observed and the lack of a mesocyclone confirmed by DOPPLER RADAR, but scientists have not yet discovered the mechanism that sets the interior of the cloud rotating.

langley (ly) A unit of solar radiation that was suggested in 1942 by the German physicist F. Linke (in *Handbuch der Geophysik*, 8, 30) and named for SAMUEL PIERPONT LANGLEY. It is defined in the C.G.S. SYSTEM as one CALORIE per square centimeter (using the 15° C calorie) per minute. The SOLAR CONSTANT is equal to 1.98 langleys.

Langley, Samuel Pierpont (1834–1906) American *Astronomer and physicist* Using an instrument of his own invention, Langley was the first scientist to calculate the amount of energy the Earth receives from the Sun and the proportion of that energy that is absorbed by the atmosphere. He was also the first person to explain clearly how birds are able to soar without flapping their wings and he came very close indeed to inventing the airplane.

Langley was born in Roxbury, Massachusetts, on August 22, 1834. He was educated at Boston Latin School and Boston High School and graduated in 1851; he did not attend college and was largely self-educated. From 1857 until 1864 he worked as a civil engineer and architect, mainly in Chicago and St. Louis. At the same time he studied astronomy, and by the time he returned to Boston in 1865 he had attained a sufficiently high standard to be offered a post as an assistant at the Harvard University Observatory. In 1866 he left to teach mathematics at the U.S. Naval Academy, at Annapolis, Maryland, and in 1867 he was appointed director of the Allegheny Observatory in Pennsylvania and professor of physics and astronomy in the Western University of Pennsylvania. In 1887 he became secretary and then director of the Smithsonian Institution, in Washington, D.C., a post he retained until his death. While there he established the Washington National Zoological Park and the Smithsonian Institution's astrophysical observatory at Wadesboro, North Carolina.

Samuel Langley.* The American astronomer and physicist who measured the amount of energy the Earth receives from the Sun and the proportion that is absorbed by the atmosphere. *(John Frederick Lewis Collection, Picture and Print Collection, The Free Library of Philadelphia)

Throughout his life, Langley was fascinated by all solar phenomena and especially interested in discovering the amount of solar radiation that reaches the Earth and provides the energy to drive the atmospheric and ocean circulations that produce the world's climates. In 1881 he climbed Mount Whitney, California, accompanied by the American astronomer James Edward Keeler (1857–1900). From the summit the two were able to measure the heat of the solar rays and compare this figure with the value they measured at sea level. The measurements were made with a BOLOMETER, an instrument Langley had invented for the purpose of studying the SOLAR SPECTRUM.

He was also interested in the way that air flows across surfaces, and he made a number of experiments that allowed him to calculate the forces operating on a body moving through the air at a constant speed. He showed how thin wings of a certain shape could support the weight of an airplane. In 1896 he made a steam-powered airplane. It carried no pilot but it flew across the Potomac River, a distance of 4,200 feet (1,281 m). Between 1897 and 1903 he made three trials, but he failed to achieve a successful flight carrying a pilot. The structural materials available to him were not strong enough and his engines were unreliable. These trials cost $50,000, paid from public funds, and after the third failure he was unable to raise more. The *New York Times* published an editorial berating Langley for wasting public money on a foolish dream, predicting that it would be a thousand years before humans achieved powered flight. The Wright brothers made their flight nine days later.

Langmuir, Irving *See* CLOUD SEEDING.

La Niña A strengthening of the southeasterly winds in the equatorial South Pacific that occurs at intervals of two to seven years. It is the opposite of an EL NIÑO and completes a full ENSO event. The strengthening of the winds accelerates the warm surface current flowing from east to west. The pool of warm water near Indonesia deepens and rainfall there becomes heavier. Off the South American coast the water is cooler, the prevailing winds blow from the land toward the sea, and the weather is extremely dry.

Laplace, Pierre Simon, marquis de (1749–1827) French *Mathematician and astronomer* A mathematical and scientific genius, Laplace rose rapidly from humble origins to become very famous and influential. He was born on March 28, 1749, at Beaumont-en-Auge, Normandy, where his father owned a small estate. Neighbors who were better off than the Laplace family recognized that the boy had talent and helped to pay for his education. When he was 16, Pierre went to the University of Caen, where his tutors recognized his genius. He graduated in mathematics after two years, and in 1767, aged 18, he traveled to Paris with a letter of introduction to the famous French mathematician Jean le Rond d'Alembert (1717–83). D'Alembert refused to see him, so the young man sent him a paper on mechanics. This was of such high quality that d'Alembert was delighted with it and sponsored Laplace for a professorship. Only 18 years old, Laplace was appointed professor of mathematics at the École Militaire.

Early in his career he collaborated with the French chemist Antoine Laurent Lavoisier (1743–94) in determining the specific heats of many substances. In 1780 they were able to show that the quantity of heat needed to decompose a compound into its constituent elements is equal to the amount that is released when those elements combine to form the compound. This discovery is regarded as the beginning of the branch of chemistry called THERMOCHEMISTRY. It also developed further the work of JOSEPH BLACK on LATENT HEAT and pointed the way to the concept of the conservation of energy. Sixty years later this was to become the first law of thermodynamics (*see* THERMODYNAMICS, LAWS OF).

Most of Laplace's work was in mathematics and in the mechanics of the solar system. Between 1799 and 1825 he published the five volumes of *Mécanique Céleste* (Celestial mechanics), for which he is sometimes known as the "French Newton." In 1812 Laplace published *Théorie Analytique des Probabilités* (Analytical theory of probabilities), which he developed further in *Essai Philosophique* (Philosophical essay), published in 1814. This work gave the theory of probability its modern form and, with it, helped establish statistics as a branch of mathematics.

In 1799 Napoleon appointed Laplace minister of the interior but dismissed him after only six weeks and promoted him to the rank of senator. Napoleon later made him a count, and when Louis XVIII was restored to the throne in 1814, far from being penalized for his association with Napoleon, Pierre Laplace was made a marquis.

Laplace was elected to the Academy of Sciences in 1785. In 1816 he was elected to the far more prestigious French Academy and in 1817 he became its president. He died in Paris on March 5, 1827.

lapse line A curve on a graph that shows the change of temperature with height in the FREE ATMOSPHERE.

lapse rates The rates at which a rising PARCEL OF AIR cools with increasing height. This cooling is said to be *ADIABATIC,* which means it involves no exchange of

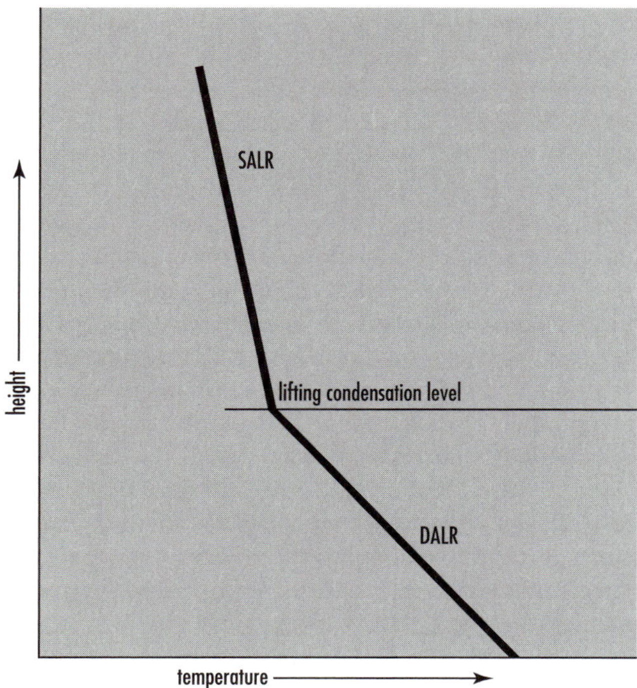

Lapse rates. **The lapse rate is the amount by which the temperature of a rising parcel of air decreases with altitude (and increases with descent). There are several lapse rates, of which the most widely used are the dry adiabatic lapse rate (DALR) and saturated adiabatic lapse rate (SALR). These separate at the lifting condensation level.**

heat with the air surrounding the parcel. Air can descend as well as rise. If it does so, it warms rather than cools and its rate of warming is equal to the lapse rate.

As a parcel of air rises, the atmospheric pressure exerted on it decreases. The parcel expands and the air cools, because the air molecules expend energy in pushing aside the molecules in the surrounding air. The rate at which it cools is proportional to the specific heat of air and to the change in pressure. The specific heat of air (the amount of heat that is required to change the temperature of a given volume by a given amount) is constant, and so is the change of pressure with height. Consequently, the rate of temperature change with height is also constant. This is the DRY ADIABATIC LAPSE RATE (DALR) and it is equal to 5.38° F for every 1,000 feet (9.8° C km^{-1}).

Most air contains some moisture, however, and when the parcel of air reaches the LIFTING CONDENSATION LEVEL the water vapor it carries starts to condense into liquid droplets. CONDENSATION releases LATENT HEAT. This warms the air and partially offsets the rate of cooling. While water vapor is condensing, therefore, the air cools at a slower rate of about 2.75° F for every 1,000 feet (5° C km^{-1}). This is the SATURATED ADIABATIC LAPSE RATE (SALR). Its value varies very slightly with height, because more latent heat is emitted and absorbed during condensation and EVAPORATION at high temperatures than at low temperatures. This is due to the fact that the lower the temperature the less water vapor the air can hold, and, therefore, the amount of vapor that condenses or evaporates decreases with temperature. Throughout the lower and middle TROPOSPHERE the value of 2.75° F per 1,000 ft (5° C km^{-1}) is sufficiently accurate for most purposes, but in the upper troposphere the SALR increases, approaching more closely the DALR.

The standard lapse rate, often simply called the *lapse rate,* is calculated from the average temperature at the surface and at the TROPOPAUSE, at a height of 36,000 feet (11 km). These temperatures are 59° F (15° C) and –70° F (–56.5° C), respectively, giving a lapse rate of about 3.6° F per 1,000 feet (6.5° C km^{-1}). This is very approximate, however, and varies with height, location, and season. The global average lapse rate is about 1.1° F per 1,000 feet (2° C km^{-1}) between the surface and about 6,500 feet (2 km), 3.3° F per 1,000 feet (6° C km^{-1}) between 13,000 feet and 20,000 feet (4–6 km), and 3.8° F per 1,000 feet (7° C km^{-1}) between 20,000 feet and 26,000 feet (6–8 km). The lapse rate is usually smaller in winter than in summer and in continental interiors the winter lapse rate is sometimes negative, and the air at the top of a mountain may be warmer than the air in the valley below. Over subtropical and tropical deserts the summer lapse rate can exceed 4.4° F per 1,000 feet (8° C km^{-1}).

Calculating the lifting condensation level is simple. The WET BULB DEPRESSION is obtained by subtracting the WET-BULB TEMPERATURE from the DRY-BULB TEMPERATURE and the DEW POINT TEMPERATURE can then be read from the table. Applying the DALR to the dry-bulb temperature then gives the lifting condensation level, which is the height of the CLOUD BASE.

An additional lapse rate must be taken into account, however. The rate of condensation and evaporation is affected by the atmospheric pressure, and consequently the dew point temperature decreases with increasing height by about 0.55° F for every 1,000 feet (1° C km^{-1}). This is known as the *dew point lapse rate.*

The actual lapse rate that is measured is called *ENVIRONMENTAL LAPSE RATE* (ELR), because it represents the temperature of the environment. The ELR is determined by the temperature of air in contact with the surface and consequently varies from place to place and with the time of day. On a fine, warm day in summer the ground is hot and so there is a large different between the temperature of air close to the ground and air that is high enough to escape this influence. The ELR is steep. At night the ground loses heat by radiation, chilling the air in contact with it. The ELR is then much shallower. Differences among the DALR, SALR, and ELR determine whether or not a parcel of air is stable (*see* STABILITY OF AIR).

large ion A charged particle of dust or other substance that exists as an AEROSOL. It usually consists of an ion that has attached itself to an AITKEN NUCLEUS. Charged particles can be moved through the air by applying an electric field. Large ions experience more drag than small ions and so they move more slowly. This provides a means for counting the proportion of large and small ions. Usually there are far fewer large ions than small ions, but because of their greater mass the large ions may account for more of the total mass. In 1950, the German atmospheric scientist Christian Junge discovered that the distribution of atmospheric particles is such that for every halving of the diameter of the particles their number increases approximately 10-fold.

laser altimeter *See* ALTIMETER.

last glacial maximum The time when the WISCONSINIAN GLACIAL, known in Britain as the *DEVENSIAN,* in northern Europe as the *Weichselian,* and in the Alps as the *Würm,* reached its greatest intensity. This was the most recent GLACIAL PERIOD and its maximum happened throughout North America and Europe about 21,000 years ago. The ICE SHEETS were at their greatest extent. The global mean temperature was about 9° F (5° C) cooler than that of today. Climates everywhere were generally drier. Over the ice sheets themselves, temperatures were much like those of present-day Greenland, rising above freezing for only a brief period during the summer.

late-glacial The period during which the WISCONSINIAN GLACIAL, known in Britain as the *DEVENSIAN,* in northern Europe as the *Weichselian,* and in the Alps as the *Würm,* was drawing to its close. This began about 14,000 years ago and continued until about 10,000 years ago, when the most recent GLACIAL PERIOD ended. At first, late-glacial temperatures rose sharply until climates were generally warmer than those of today. This warm period is known as the *BØLLING INTERSTADE* and it was followed by the OLDER DRYAS STADE. The Older Dryas gave way to the warmer ALLERØD INTERSTADE, which was followed by the long, cold YOUNGER DRYAS. The end of the Younger Dryas also marked the end of the Wisconsinian glacial, and after it temperatures rose to produce the climates of historical times.

latent heat The energy that is absorbed or released when a substance changes its phase between solid and liquid, liquid and gas, and directly between solid and gas. The absorption and release of latent heat do not alter the TEMPERATURE of the substance itself, but they do alter the temperature of the surrounding medium, because it is from that medium that absorbed latent heat is obtained and into the medium that released latent heat is introduced. The existence of latent heat was discovered, and the term *latent heat* coined, in about 1760 by the Scottish chemist JOSEPH BLACK. He called the heat latent because it cannot be measured directly with a THERMOMETER.

A solid consists of molecules that are tightly bound to each other by forces acting between them. Energy must be supplied in the form of heat to overcome these forces. As the bonding forces weaken, groups of molecules break free, and as more heat is supplied the solid changes into a liquid, in which small groups of molecules constantly break apart and reform and groups slide freely past each other. While this is happening, the temperature of the substance remains unchanged. All of the heat energy is used in loosening the bonds between molecules.

If still more energy is applied, it has the effect of making the groups of molecules move faster. When they move faster they strike harder against any object with which they come into contact. It is the speed of motion of molecules that a thermometer measures as temperature. Once a solid has melted to become a liquid, the application of additional heat raises the temperature of the liquid.

When a liquid is cooled, its molecules lose energy and move more slowly, and the temperature of the liquid falls. When their energy falls to a certain level the molecules start bonding together. This requires less energy than moving about, and so heat energy is released as the liquid solidifies. The temperature of the substance remains unchanged, but the surrounding medium is warmed by the release of energy.

The latent heat that is released when a liquid solidifies is exactly equal to the amount absorbed when the solid phase of the same substance melts. It is known as the *latent heat of fusion.*

As the temperature of a liquid rises, its molecules move faster and faster. A point is reached at which the absorption of heat energy breaks the bonds holding groups of molecules together. Instead of raising the temperature of the liquid, this absorption of energy is used to break the bonds between molecules. Individual molecules break free and the liquid becomes a gas. When a gas is cooled, its molecules move more slowly until a point is reached at which molecules bond together into small groups and the gas condenses into a liquid. The amount of energy that is released when a gas condenses is exactly equal to the amount that is absorbed when a liquid vaporizes. It is called the *latent heat of vaporization.*

Water has many remarkable properties. One of them is that, because of the HYDROGEN BONDS that link its molecules in its solid and liquid phases, much more latent heat is absorbed and released when it changes phase than when other substances do. It is because of its high latent heat of vaporization that we are able to cool our bodies by allowing sweat to evaporate. The release of the latent heat of vaporization warms the air inside CUMULIFORM clouds. This causes them to continue growing upward and is a major factor in the formation of storms and TROPICAL CYCLONES. The HADLEY CELLS are driven partly by the absorption and release of latent heat of vaporization.

The latent heat of fusion of water at 32° F (0° C) is 334 joules per gram (80 cal g^{-1}). The latent heat of vaporization is 2,501 J g^{-1} (600 cal g^{-1}) at 32° F (0° C). It is possible for water vapor to change directly into ice and for ice to vaporize without passing through the liquid phase. The two processes are called *DEPOSITION* and *SUBLIMATION*, respectively, and they also involve the absorption and release of latent heat. The same amount of energy is involved in breaking and forming the molecular bonds, so the latent heat of deposition and sublimation is equal to the sum of the latent heats of fusion and vaporization: 2,835 J g^{-1} (680 cal g^{-1}) at 32° F (0° C).

It is necessary to specify the temperature, because molecules possess an amount of internal energy that increases as the temperature of ice or water rises, and so the higher the temperature the less latent heat they need to change phase. At 122° F (50° C), for example, the latent heat of vaporization is 2,382 J g^{-1} (569 cal g^{-1}), and at 194° F (90° C) it is 2,283 J g^{-1} (545 cal g^{-1}). The relationship continues at temperatures below freezing. At –4° F (–20° C) the latent heat of fusion is 289 J g^{-1} (68 cal g^{-1}).

latent instability The condition in which a PARCEL OF AIR that acquires sufficient KINETIC ENERGY to rise through a layer of stable air becomes unstable once it is above the LEVEL OF FREE CONVECTION. It is a type of CONDITIONAL INSTABILITY that develops only if a parcel of air rises to the critical level.

lateral acceleration The acceleration of air in a direction that is perpendicular to the wind direction. This happens when air is affected by the CENTRIPETAL FORCE as it flows around a center of high or low pressure. In order to maintain the centripetal acceleration the PRESSURE GRADIENT FORCE must exceed the CORIOLIS EFFECT, so there is a net acceleration at right angles to the direction of the GEOSTROPHIC WIND and the resulting wind is SUBGEOSTROPHIC.

Laurentide ice sheet The ICE SHEET that covered Canada and the northern United States during the WISCONSINIAN GLACIAL which reached its maximum intensity about 20,000 years ago. The Laurentide ice sheet may have been joined to the GREENLAND ICE SHEET, which was then larger than it is today. The ice sheet covered southern Alaska and all of North America to the north of a line from about the position of Seattle to Boston, passing to the south of the Great Lakes. The release of meltwater from the Laurentide ice sheet may have triggered the cold episode known as the *YOUNGER DRYAS.*

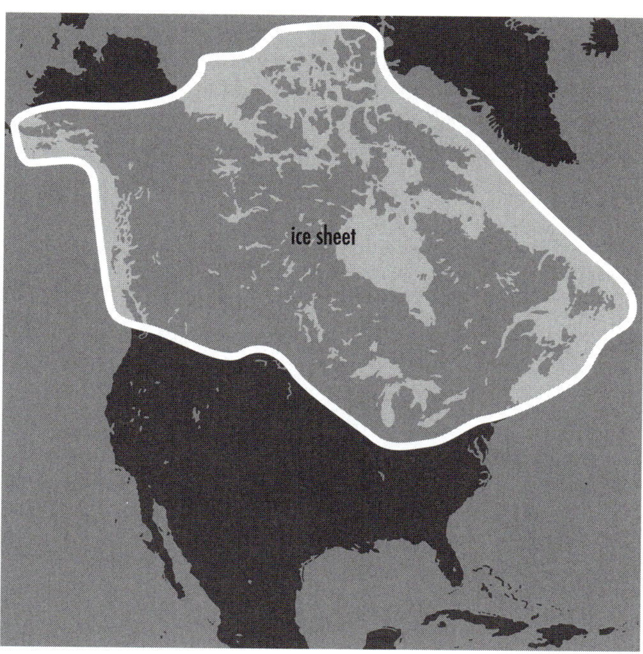

Laurentide ice sheet. **The line shows the boundaries of the Laurentide ice sheet at the time of the last glacial maximum, about 20,000 years ago. The sea was frozen to the north and northeast of the ice sheet.**

laws of motion The three laws that were proposed by Sir Isaac Newton (1642–1727) to describe the way bodies respond to the forces acting on them. The first law states that a body continues in a state of rest or uniform motion unless an external force acts on it. The second law states that the rate at which the MOMENTUM of a body changes is proportional to and in the same direction as the force acting on it. If the mass of the body remains constant, $F = ma$, where F is the force, m is the mass of the body, and a is the ACCELERATION. The third law states that if one body exerts a force upon another body, the second body exerts an equal force in the opposite direction, called a *reaction,* upon the first body. An equation derived from the second law and known as the *EQUATION OF MOTION* is widely used in meteorology.

layer cloud A STRATIFORM cloud that resembles a sheet and is of limited vertical extent.

leading-edge effect *See* ADVECTION.

lead pollution The injection of lead into the air, primarily in emissions from motor vehicles running on gasoline that contains tetraethyl lead. This compound is added to raise the octane number of the gasoline. High-octane fuel is less likely to ignite prematurely in the cylinder (this is called *knocking*), causing a loss in engine power and eventual damage to the engine. Lead is highly toxic in large doses and in small doses it is believed to harm the developing nervous system of young children. Lead also damages the catalytic converters that are fitted to cars to reduce the emission of other pollutants, especially CARBON MONOXIDE and NITROGEN OXIDES. The sale of gasoline containing tetraethyl lead is now forbidden in many countries, including the United States and the European Union. The United Nations Commission on Sustainable Development believes that by 2005 55 countries will have phased out the use of lead in gasoline.

(You can learn more about lead in gasoline at www.runet.edu/~wkovarik/papers/leadinfo.html and earthsummitwatch.org/gasoline.html.)

leaf area index (LAI) The ratio of the total area of plant leaves, counting one side only, to the area of ground. If there is 2 m^2 of leaves to every 1 m^2 of ground, for example, the LAI is 2 (the LAI is a DIMENSIONLESS NUMBER, so there are no units). In seasonal climates the LAI changes with the seasons, affecting the amount of solar energy that reaches the ground, the amount of gas exchange associated with PHOTOSYNTHESIS, and the amount of EVAPOTRANSPIRATION.

leaf margin analysis (LMA) The use of plant leaves to estimate past mean annual temperatures. The technique is based on the strong correlation that has been observed between the warmth of the climate and the likelihood that dicotyledon plants (plants with seeds that produce two seed leaves, or cotyledons) will have leaves with smooth edges. The technique is possible only if the leaves that fall are preserved in sediments. If enough leaves can be recovered, they can be used in the CLIMATE–LEAF ANALYSIS MULTIVARIATE PROGRAM.

lee The adjective that describes the side of a mountain or other obstacle that is sheltered from the wind. A lee shore is the shore on the lee side of a ship, so the wind tends to drive a ship toward the lee shore.

lee depression A DEPRESSION that forms in a westerly airflow on the LEE side of a range of mountains that is aligned north and south. As the air is forced to rise across the mountains, the layer of air is squeezed vertically. This causes the layer to contract vertically, which in turn produces a lateral expansion as the air spills out to the sides. The results are a tendency for DIVERGENCE and an ANTICYCLONIC movement of air above the mountain crest. On the other (lee) side the descending air expands vertically once more, producing a lateral contraction to compensate. On the lee side there is, therefore, a tendency for CONVERGENCE and CYCLONIC flow. Whether this will result in the closed circulation necessary for the full development of a depression depends on the size of the mountain barrier. Lee depressions are common in winter to the south of the Alps and Atlas Mountains. Where they do form, lee depressions remain close to the mountains for a time but may move away later. FRONTS may form in the depression, but this type of depression does not originate as a wave along a front. If a lee depression fails to develop there may nevertheless be a LEE TROUGH.

lee trough (dynamic trough) A TROUGH of low pressure that forms on the lee (downwind) side of a long, meridional (north–south) mountain barrier that lies across a zonal (west–east) airflow. It is most marked to the east of the Rocky Mountains.

Air approaching the mountains is forced to rise. This compresses the air into a thinner layer, causing the entire atmosphere to rise and air to accumulate on the upwind side of the barrier, but it also causes horizontal DIVERGENCE. This produces an ANTICYCLONIC airflow, diverting the air southward. Air is drawn inward on the lee side of the barrier, producing horizontal CONVERGENCE and a CYCLONIC airflow. Convergence on the downwind side cancels out divergence on the upwind side and would restore the circulation pattern to its original state, except that the anticyclonic flow turned the air southward. PLANETARY VORTICITY decreases with increasing proximity to the equator. This has the effect of increasing the strength of the cyclonic motion in the moving air, so instead of canceling out the air is left with a net cyclonic flow. This produces the trough. Beyond it, the circular flow carries the air northward, where planetary vorticity increases and the motion becomes anticyclonic. A series of long, horizontal waves develops, producing long-wave troughs and RIDGES.

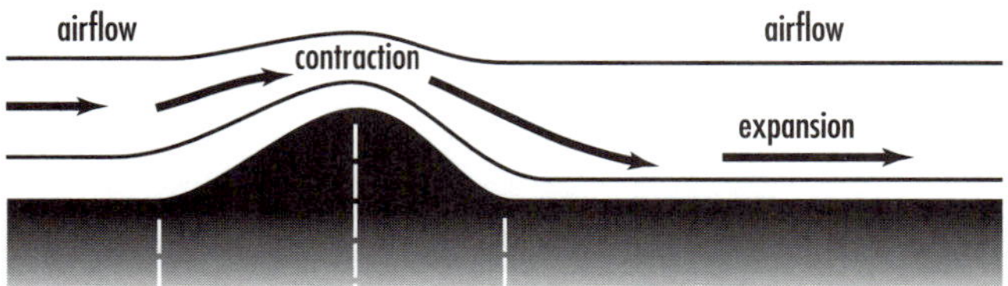

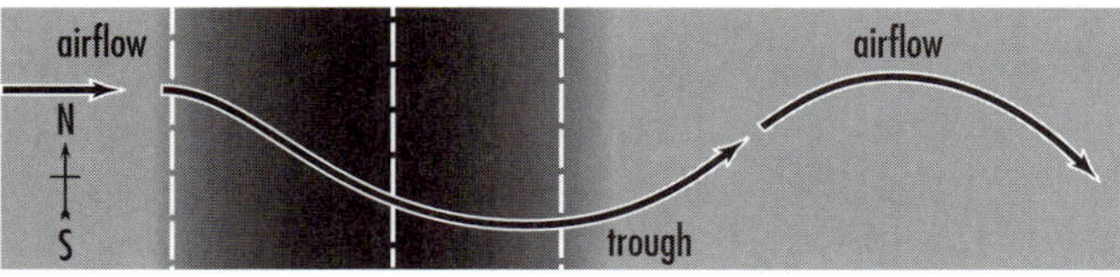

Lee trough. **The upper part of the figure shows a lateral view, with air contracting as it crosses a mountain barrier and expanding on the lee side. The lower part of the figure shows the same phenomenon in plan view, illustrating the way cyclonic airflow on the lee side of the barrier forms a trough.**

lee waves (standing waves) Waves that develop in stable air (*see* STABILITY OF AIR) on the lee (downwind) side of a mountain. The mountain need not be large, but it must be steep on both the upwind and downwind sides, so the air crossing it moves sharply upward and downward, and it must extend far enough across the path of the wind to prevent the air from simply flowing around it.

Air is forced to rise as it crosses the mountain. This can trigger instability if the air is conditionally unstable (*see* CONDITIONAL INSTABILITY), but stable air returns to its former level. If there is a layer of stable air at a high level, the vertical displacement of the air forms a series of waves. The waves remain stationary with the air moving through them. Their wavelength (*see* WAVE CHARACTERISTICS) is proportional to the wind speed; typical winds produce wavelengths of about 2–12 miles (3–20 km), and the more stable the air the shorter is the wavelength. The wavelength is not in any way related to the shape of the mountain or other high ground that triggers the formation of the lee waves. Usually the first wave crest is less than one wavelength downwind from the mountain peak. Wave amplitude decreases with height and with distance downwind.

Wavelength increases with increasing height and the crests and troughs are shifted progressively upwind.

Lee waves are made visible if the relative humidity is close enough to saturation for LENTICULAR CLOUDS to form at the crests. If the lee side of the mountain is very steep, a rotor, with a ROTOR CLOUD, may form beneath the crest of the first wave.

len *See* LENTICULARIS.

Lenny A HURRICANE that struck islands in the Caribbean in November 1999. It was upgraded from a TROPICAL STORM on November 14 and passed to the south of Jamaica with winds of 100 mph (160 km h^{-1}); its wind speeds increased to 125 mph (201 km h^{-1}) on November 17. The following day its winds increased to 150 mph (241 km h^{-1}) as it struck Saint Croix in the U.S. Virgin Islands, where the storm continued for 12 hours. Other islands in the region also suffered and a total of 13 people were killed. On November 19 it was downgraded to a tropical storm.

lenticels *See* TRANSPIRATION.

lenticular cloud A type of ALTOCUMULUS LENTICULARIS (Ac_{len}) that is often seen on the lee side of mountains. As stable air crossing the mountains is forced to rise and then descends to its former level, a wave motion becomes established. This produces a series of lens-shaped clouds extending downwind as far as the wave pattern. These are sometimes called *wave clouds.*

lenticularis (len) A species of cloud (*see* CLOUD CLASSIFICATION) that is most commonly found in association with the cloud genera CIRROCUMULUS, ALTOCUMULUS, and STRATOCUMULUS. The cloud has the shape of a lens, usually with well defined outlines, and sometimes has a very dramatic appearance. It can resemble a flying saucer!

Lenticularis forms when stable air moves over an obstruction, such as a hill or coastline. This forces the air to rise, but, being stable, it then sinks to its former level. If the air is moist, it may be lifted to its LIFTING CONDENSATION LEVEL, above which its water vapor starts to condense into droplets. As it sinks, the air moves below its lifting condensation level once more and condensation ceases. The vertical movement, with condensation occurring only at the top of the wave, is what produces the lens shape.

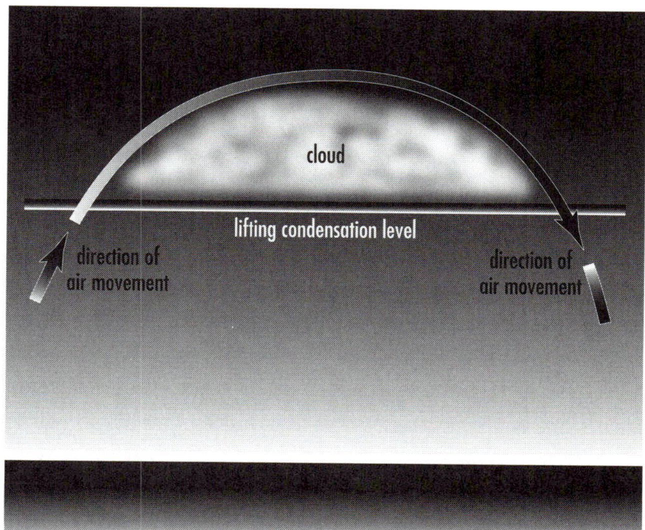

Lenticularis. **A cloud shaped like a lens. It forms where air is moving up and down with a vertical wave motion and the lifting condensation level lies below the wave crests.**

The name of the species is Latin and derived from *lenticula,* which means "lentil" or "lens" (from the similarity of shape).

leste A hot, dry easterly or southeasterly wind that blows from Africa to Madeira and the Canary Islands, lying off the northwestern coast of Africa. The wind carries much dust from the desert. Like the SIROCCO, it occurs ahead of a DEPRESSION that is moving toward the east.

levante *See* LEVANTER.

levanter (solano, levante, llevante) A strong easterly wind, sometimes of gale force, that blows across southern Spain and the Strait of Gibraltar. It carries air from the lands along the eastern coast of the Mediterranean, a region that was formerly known as the *Levant,* hence the name. The wind occurs at all times of year and produces moist weather, often with fog, that lasts for several days. It is caused by low-pressure systems.

levantera A persistent easterly wind that blows over the Adriatic Sea and usually produces cloudy weather.

leveche A hot, dry wind of the SIROCCO type that blows from the south over southeastern Spain bearing air from the Sahara. It is produced by depressions traveling eastward along the Mediterranean.

level of free convection The height at which a PARCEL OF AIR that is being forced to rise through a conditionally unstable atmosphere (*see* CONDITIONAL INSTABILITY) changes from being cooler than the surrounding air to being warmer than it. While it is near the surface, the parcel of air is moist, but not saturated. As it is forced to rise (*see* LIFTING) through air that is warmer than it it cools at the DRY ADIABATIC LAPSE RATE (DALR) until it reaches the LIFTING CONDENSATION LEVEL, beyond which it cools at the SATURATED ADIABATIC LAPSE RATE. Its temperature then decreases with height more slowly than that of the surrounding air, where the ENVIRONMENTAL LAPSE RATE is close to the DALR, until a level is reached at which it is warmer than the surrounding air. Beyond that point the parcel of air is unstable and continues to rise without being forced until it reaches a second level, where its temperature is once more lower than that of the surrounding air and it becomes stable.

level of nondivergence The height at which there is no CONVERGENCE or DIVERGENCE of air in a column of air that is rising or subsiding. Air that is converging at the surface is diverging at height, and vice versa. Consequently there must be a point in the column where convergence ceases and divergence begins. This is the level of nondivergence, which is usually at the height at which the air pressure is approximately 600 mb.

level of zero buoyancy *See* EQUILIBRIUM LEVEL.

LI *See* STABILITY INDICES.

libeccio A southwesterly wind that blows across Corsica and Italy at any time of year, but especially in winter, when it may carry storms.

lichen A composite organism that consists of a fungus and either an alga or a cyanobacterium living in close symbiotic association. Lichens are classified on

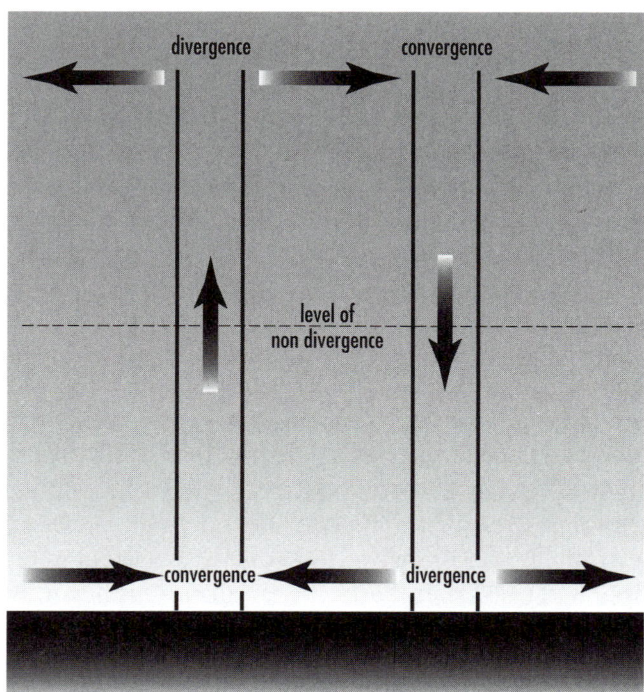

Level of nondivergence. **Nondivergence (or convergence) occurs at the level where convergence becomes divergence in a column of rising or sinking air.**

the basis of their fungal component. Many lichens are extremely sensitive to air pollution, especially by sulfur dioxide, and are widely used as pollution indicators.

LIDAR *See* LIGHT DETECTION AND RANGING.

life assemblage *See* POLLEN ANALYSIS.

Lifted Index *See* STABILITY INDICES.

lifting The forced ascent of an air mass that occurs as it crosses high ground or is undercut by denser air at a front. Lifting also occurs where AIRSTREAMS converge (*see* CONVERGENCE).

lifting condensation level The altitude at which the temperature of air that is rising and cooling in an ADIABATIC process falls to the DEW POINT TEMPERATURE. At this height the RELATIVE HUMIDITY reaches 100 percent and water vapor starts to condense into droplets. The lifting condensation level therefore marks the height of the cloud base. As the air continues to rise above this level its rate of cooling changes from the DRY ADIABATIC LAPSE RATE to the SATURATED ADIABATIC LAPSE RATE.

light air In the BEAUFORT WIND SCALE, force 1, which is a wind that blows at 1–3 mph (1.6–5 km h^{-1}). In the original scale, devised for use as sea, a force 1 wind is just sufficient to give steerage way to a sailing ship. On land, wind vanes and flags do not move, but rising smoke drifts.

light breeze In the BEAUFORT WIND SCALE, force 2, which is a wind that blows at 4–7 mph (6.4–11.3 km h^{-1}). In the original scale, devised for use at sea, a force 2 wind was defined as "or that in which a man-of-war with all sail set, and clean full would go in smooth water from." On land, a light breeze is just strong enough for drifting smoke to indicate the wind direction.

light detection and ranging (LIDAR) A ground-based technique that is used to measure the density of atmospheric AEROSOLS and OZONE, as well as temperature, in the stratosphere. A laser beam is directed upward into the atmosphere, and the desired information is calculated from an analysis of the intensity and wavelengths of the light reflected to the surface by aerosols or by atmospheric gases, depending on the wavelength. In the case of ozone measurements, for example, the laser transmits two beams at 308 and 351 nanometers (nm). Radiation at 308 nm is absorbed by ozone, but radiation at 351 nm is not. A comparison of the intensity of light received at the surface from each beam reveals the ozone concentration.

lightning An electrical discharge that neutralizes a charge separation that has accumulated within a cloud, between two clouds, or between a cloud and the ground. Charge separation occurs within large CUMULONIMBUS clouds and induces a charge on the ground surface beneath the cloud. Air is a good electrical insulator, but when the separation exceeds a critical threshold the insulation breaks down. There are several stages to a lightning flash.

It begins with a STEPPED LEADER. This carries negative charge from the cloud to the ground. The spark—in fact, a stream of electrons—travels through the electrical field that has become established and takes the path of least resistance. This carries it, very rapidly, to places where the field is weakest and accounts for its jagged path. As it travels, the lightning ionizes—imparts an electric charge to the molecules in—air around it. This creates a channel of ionized air, approximately 8 inches (20 cm) in diameter—the width of a lightning stroke. Near the accumulated positive charge it is met by a return stroke, carrying positive charge back to the cloud and traveling along the same path of ionization. This return stroke is the main part of the flash.

The leader and return stroke only partially neutralize the charge in the cloud. The first leader is neutralized by the return stroke, and the return stroke is neutralized by the charge in the cloud. This initiates a second lightning stroke, beginning from a point deeper inside the cloud. It is preceded by a DART LEADER, follows the same path through the ionized channel, and is met by a return stroke. Further flashes follow until the neutralization is complete; the entire process usually lasts about 0.2 second and consists of three or four flashes about 50 milliseconds apart. It is the release of energy by a lightning flash that causes THUNDER.

See also FORKED LIGHTNING and SHEET LIGHTNING.

lightning channel The path, about 8 inches (20 cm) across, along which a LIGHTNING stroke travels. It is made by the STEPPED LEADER and consists of air in which a significant proportion of the molecules have been ionized—an electric charge has been imparted to them—by the energy of the leader.

lightning conductor A metal rod that projects upward beyond the highest point of a building or other structure and that is connected to the ground by a metal strip or cable with a low resistance (less than 10 ohms). LIGHTNING discharges, traveling by the route of least electrical resistance, are attracted to the metal, which guides them to the ground, where they are earthed. A cone with an apex angle of 45° around the metal rod is protected from lightning strike. Copper is often used, because it is one of the best electrical conductors. The lightning conductor was invented by BENJAMIN FRANKLIN and soon proved its worth. By 1782 about 400 had been installed in Philadelphia alone. Eventually they were installed on ships of the British Royal

Navy, because lightning often killed sailors working aloft and made holes in the hulls below the masts. Lightning destroyed 70 British naval ships in one five-year period. At first the Admiralty would permit only portable conductors to be carried. These would be hoisted to the tops of the masts when thunderstorms were likely. Not surprisingly, this practice resulted in even more deaths among sailors, who were erecting lightning conductors during storms. Finally permanent conductors were allowed and ships became a great deal safer.

lightning stroke Any one of the components that together constitute a flash of LIGHTNING. The STEPPED LEADER, RETURN STROKE, and DART LEADER are all lightning strokes.

Linda A TYPHOON that struck Vietnam, Cambodia, and Thailand in November 1997, destroying thousands of homes and sinking hundreds of fishing vessels. It killed 464 people in Vietnam and more than 20 in Cambodia and Thailand.

linear acceleration Increase in speed in a straight line. It occurs in the ENTRANCE REGION of a JET STREAM and wherever air is accelerated in the direction of a wind. Winds entering the core region of a jet stream tend to be SUPERGEOSTROPHIC and angled slightly across the ISOBARS toward the region of lower pressure. This is because the CORIOLIS EFFECT exceeds the PRESSURE GRADIENT FORCE and acts at an angle to it, so there is a component of the forces acting on the air in the direction of the wind.

Linke scale A set of cards that is used to measure the blueness of the sky. Each card is a different shade of blue, and the cards are numbered from 2 to 26, using only the even numbers. The observer uses an odd number if the sky shade falls between those of two cards with even numbers. For example, sky shade 15 falls between shades 14 and 16, although there is no card bearing the number 15.

liquid The phase of matter that is between the solid and gaseous phases. In a liquid, the atoms or molecules do not occupy fixed positions in relation to each other as they do in a solid, but neither are they completely disorganized, as they are in a gas. The atoms or molecules of a liquid are joined into small groups and each group moves freely in relation to the other groups. This prevents the individual atoms or molecules from dispersing themselves evenly, so the volume of the substance increases to fill the space in which it is contained, but it does allow the liquid to flow and to adopt the shape of the container holding it, while maintaining a constant volume.

lithometeor Any solid particle that is suspended in or transported by the air. Airborne dust particles, sand grains, and smoke particles are lithometeors.

Little Ice Age A period of cold weather that began in the 16th century and lasted until the early 20th. It affected all of the Northern Hemisphere, but only parts of the Southern Hemisphere. Its cause is uncertain, but there is evidence linking it to a reduction in solar output (*see* MAUNDER MINIMUM). Although it is called "the" Little Ice Age, there have been other similar fluctuations in climate, although these are not so well documented.

In the first half of the 16th century the weather was generally mild, but there were three winters with temperatures low enough for the River Thames to freeze at London and summer temperatures were very variable. Snowfall increased in central Europe, with average winter temperatures about 2.3° F (1.3° C) lower in the period 1560–99 than in 1880–1930, and in Denmark temperatures were 2.7° F (1.5° C) lower in 1582–97 than in 1982–97. The weather began to grow warmer around 1600 but then cooled erratically to reach a minimum around the end of the 17th century. In Switzerland the mean temperature in 1687 was 7°–9° F (4°–5° C) cooler than it is today. In 1716 the Thames froze so firmly that in January a high spring tide that raised the river level by 13 feet (4 m) simply lifted the ice together with the fair being held on it. Yet the climate was much more variable than it is now. The summers of 1718 and 1719 were so hot and dry, for example, that there was DROUGHT over much of Europe, but 1740 was the coldest year in England since records began in 1659. The temperature rose again through the 18th century, only to fall to another minimum in the early 19th.

Recovery began in the middle of the 19th century, but it was erratic. The winter of 1878–79 was as cold in England as those of the 1690s. Between 1876 and

1879 there were droughts, failures of the MONSOON, and famines in India and China. Up to 18 million people are believed to have died in the Chinese famine of the 1870s. Although the river did not freeze over, there was a considerable amount of ice on the Thames in the winter of 1894–95. It was not until the 1920s and 1930s that the weather began to grow noticeably warmer. Even then there were cold winters. In 1924 people drove cars on the sea off Lund, Sweden, and the people of Malmö could walk all the way to Copenhagen.

Although the climate during the Little Ice Age was generally cooler than that of today, it was not consistently so. There were some very warm summers, and a shift in the ocean currents prevented the sea from freezing off Iceland from 1840 until 1855. The winter of 1845–46 was unusually mild in England. The summer of 1826 was even warmer than that of 1976, which caused severe drought over much of Europe.

PRECIPITATION was also very variable. The wet years of the 1840s contributed to the outbreak of late blight of potato that destroyed most of the British crop and led to the Irish Potato Famine.

(You can read more about the Little Ice Age in H. H. Lamb, *Climate, History and the Modern World,* [New York: Routledge, 1995].)

Liza A HURRICANE, rated category 4 on the SAFFIR/SIMPSON SCALE, that struck La Paz, Mexico, on October 1, 1976, killing at least 630 people and leaving tens of thousands homeless. It had winds of 130 mph (160 km h^{-1}) and produced 5.5 inches (140 mm) of rain. This destroyed an earth dam 30 feet (9 m) high, sending a wall of water 5 feet (1.5 m) high through a shantytown.

llebatjado A hot, gusty wind that blows down from the Pyrenees across Catalonia, in northern Spain.

llevantades A LEVANTER wind that produces especially stormy weather.

llevante *See* LEVANTER.

lm *See* LUMEN.

LMA *See* LEAF MARGIN ANALYSIS.

local climate The MICROCLIMATES found within an area that has a distinct type of surface. Taken together, the microclimates of an area such as a forest, farm, or city constitute a local climate. A local climate affects an area from 1 hectare (2.5 acres) to 100 km^2 (39 $miles^2$) to a height of about 100 m (330 feet) above the surface. In the widely used climatological classification introduced by the Japanese climatologist M. M. Yoshino, local climates are designated L_1 to L_7:

L_1 croplands
L_2 broad-leaved forest
L_3 city
L_4 coniferous forest
L_5 and L_6 mountain environments
L_7 intermontane grassland

local extra observation A weather observation that is taken at an airport at frequent intervals, often of 15 minutes, when conditions are close to the minima for taking off and landing and flight operations are imminent. The observation includes VISIBILITY, CEILING, CLOUD AMOUNT, and such other details as may be relevant.

local forecast A weather forecast that covers a small area and that is intended for the use of farmers, horticulturists, vacationers, and other people who need to know what conditions to expect in the next few hours or for up to about two days ahead. The local forecast is derived from the SHORT-RANGE FORECAST. This forecast is compiled for a large area, and local conditions may be strongly affected by topography, distance from a coast, or the amount of exposed water surface in the area. The local forecaster modifies the general forecast in the light of these influences.

local storm A STORM that affects only a small area. THUNDERSTORMS and SQUALLS are considered to be local storms.

local wind A wind that differs, in direction or strength, from the winds associated with the general distribution of pressure. The wind may be caused by such factors as a difference in temperature between two adjacent surfaces, producing ANTITRIPTIC WINDS, such as LAKE BREEZES and LAND AND SEA BREEZES, or a

KATABATIC WIND, or the FUNNELING of the wind through gaps in a mountain range, to produce a MOUNTAIN-GAP or JET-EFFECT WIND. The wind affects only a small area and usually only the lowest layer of the atmosphere. Many local winds have their own names.

Loch Lomond stadial *See* YOUNGER DRYAS.

loess *See* DUST.

lofting One of the patterns a CHIMNEY PLUME may make as it moves away downwind. The plume of gas and particles widens with increasing distance from the smokestack and rises gently as it disperses. Lofting causes little or no pollution of air close to ground level. It occurs when there is an INVERSION extending only as high as the top of the smokestack, and above the height of the stack the ENVIRONMENTAL LAPSE RATE is greater than the DRY ADIABATIC LAPSE RATE.

logarithmic wind profile The variation of wind VELOCITY with height throughout the PLANETARY BOUNDARY LAYER, but above the LAMINAR BOUNDARY LAYER. This is expressed by an equation:

$$\bar{u}_z = (u_*/k) \ln (z/z_0)$$

where $\bar{u}_z$ is the mean wind speed in meters per second at a height z, u_* is the FRICTION VELOCITY, k is the VON KARMAN CONSTANT, and z_0 is the roughness length (*see* AERODYNAMIC ROUGHNESS) in meters.

Lofting. **The plume of gas and particles rises gently as it travels downwind and disperses.**

London smog incidents Two episodes in London, England, each of which lasted for several days, during which very wet SMOG was trapped beneath a temperature INVERSION. The smog accumulated and thickened, causing illness and many deaths. In the first episode, in December 1952, more than 4,000 people died in the Greater London area as a direct consequence of the smog. Seven times more people died of bronchitis and pneumonia than in the same period in previous years. The 1952 smog was extremely acid, with a pH of about 1.6, making it very corrosive. This event stimulated the government to act and the Clean Air Act, imposing strict controls on the domestic burning of coal and wood in London and other cities, became law in 1956. The second smog incident occurred in December 1962. About 700 people died as a direct result of that smog. The reduction compared with the 1952 death toll was attributed to the Clean Air Act, which by then was starting to take effect. Although the number of deaths in the 1952 incident was dramatic and widely publicized, most of those who died were already in very poor health and it is doubtful whether they would have survived the winter even had there been no smog. Nor were these the first severe smog incidents. London "PEA SOUPERS" were well known. There were records of them in other years, and one similar to those of 1952 and 1962 lasted from December 27, 1813, until January 2, 1814.

long-day plants *See* PHOTOPERIOD.

long-range forecast A weather forecast that covers a period up to two weeks ahead and sometimes up to one month ahead. Forecasters compiling a long-range forecast cannot use the methods appropriate to SHORT-RANGE FORECASTS, because these describe atmospheric conditions with a lifetime of no more than about seven days. Instead, they rely on statistical methods in which current tendencies, such as BLOCKING, are projected into the future. This requires making allowance for the behavior of the JET STREAM and the stage that has been reached in the INDEX CYCLE. The influences of surface features, such as lying snow, are taken into account, and in coastal areas so is the SEA-SURFACE TEMPERA-

TURE. ANALOG MODELS provide an alternative approach. Although large amounts of data are available to forecasters and they have access to powerful MODELS, long-range forecasts are inherently unreliable. This is because weather systems behave in a chaotic fashion (*see* CHAOS), so variations in the initial conditions that are too small to detect can produce widely divergent outcomes.

long waves *See* ROSSBY WAVES.

loom A glow of light just below the horizon that is caused by the REFRACTION of light passing from cooler air aloft to warmer air below. It is a type of MIRAGE that does not involve an image of a physical object.

looming *See* MIRAGE.

looping One of the patterns a CHIMNEY PLUME may make as it moves away downwind. The gases and particles descend toward the ground and then rise again repeating this until the plume has dispersed. The plume causes pollution where its descent carries it close to ground level. Looping occurs when the wind is light and air is very unstable in the layer extending from the surface to the top of the highest loop. In this layer the ENVIRONMENTAL LAPSE RATE is greater than the DRY ADIABATIC LAPSE RATE.

Looping. **The plume of gas and particles from a factory smokestack moves downwind in looping spirals.**

James Ephraim Lovelock. **The British chemist who invented an instrument to detect extremely small concentrations of substances in the atmosphere and who composed the Gaia hypothesis.** ***(Copyright 2000 Sandy Lovelock)***

Lovelock, James Ephraim (1919–) English *Atmospheric chemist* James Lovelock was born on July 26, 1919, at Letchworth Garden City, Hertfordshire. His father was a keen gardener, and it was from him that James acquired a deep appreciation of the natural world.

The family was not wealthy, and James did not move directly from school to university. Instead he obtained a job in a laboratory. During the day he learned practical laboratory techniques and in the evenings he attended chemistry classes at night school. In time these gave him the qualifications he needed to become a full-time student. He enrolled at the University of Manchester and graduated in chemistry in 1941.

It was wartime and like all young people, Lovelock was recruited for war work. He joined the staff at the National Institute for Medical Research (NIMR) in London. After the war, in 1946, he remained with the NIMR and went to work at the Common Cold Research Unit, in Wiltshire, where he remained until 1951, taking part in the (ultimately fruitless) search for a cure for the common cold. In 1948 he received his Ph.D. in medicine, from the London School of Hygiene and Tropical Medicine. He received the degree of D.Sc. in biophysics in 1959, from the University of London. Still an employee of the NIMR, he was awarded a Rockefeller Traveling Fellowship in Medicine in 1954. He spent it at Harvard University Medical School, in Boston, and in 1958 spent a year working at Yale University as a visiting scientist. He resigned from the NIMR in 1961 in order to take up an appointment as professor of chemistry at Baylor University College of Medicine, in Houston, Texas.

Lovelock has a talent for devising extremely sensitive instruments. Eventually he decided it might be possible to earn his living in this way, and in 1964 he became a freelance research scientist. Many of his instruments helped to develop and refine the technique of gas chromatography, and in 1957 Lovelock invented the electron capture detector, which is still one of the most sensitive of all detectors. It revealed the presence of residues of organochlorine insecticides such as dichlorodiphenyltrichloroethane (DDT) throughout the natural environment, a discovery that contributed to the emergence of the popular environmental movement in the late 1960s. Later it registered the presence of minute concentrations of chlorofluorocarbon compounds (CFCs) in the atmosphere.

In the early 1960s, while living in Texas, Lovelock became a consultant at the Jet Propulsion Laboratory (JPL) of the California Institute of Technology, in Pasadena. He had already helped with some of the instruments that were used to analyze lunar soil, and he was asked to advise on various aspects of instrument design for the team of scientists planning the two Viking expeditions to Mars.

One aim of the Viking expeditions was the search for life on Mars, and although Lovelock was not directly involved in this, he began to speculate about how life on another planet might be detected and recognized. In discussions with Dian Hitchcock, a philosopher employed to assess the logical consistency of National Aeronautics and Space Administration (NASA) experiments, Lovelock reasoned that any living organism must alter its environment by removing substances (such as food) from it and adding substances (such as body wastes) to it. He thought such changes should be detectable and that the simplest place to look for them would be in the planet's atmosphere. Over the years that followed, this line of reasoning led him to formulate the GAIA HYPOTHESIS.

Although he has now handed over to younger scientists the task of exploring the implications of the Gaia concept, Lovelock retains a lively interest in environmental science and especially in atmospheric science. He lives in a converted mill in a remote part of southwestern England, surrounded by land he owns and has planted with trees, making it a haven for wildlife.

(You can read more about James Lovelock in his autobiography, *Homage to Gaia* [Oxford: Oxford University Press, 2000].)

low An area in which the surface atmospheric pressure is lower than the pressure in the surrounding area. The term refers only to horizontal comparisons of pressure and not to differences in pressure at varying heights above each other, and it is most commonly applied to the situation at sea level. Comparisons can be made at any height above the surface, however, provided the pressures being compared are measured at the same height. Thus it would be correct to speak of a low at a particular height, although this use of the term is unusual. It also refers to measurements taken at the same time and not to variations in pressure in the same place at different times. Air circulates in a CYCLONIC direction around a low and *low* is often used interchangeably with *cyclone,* but the two words are not synonymous (*see also* HIGH).

low cloud Cloud that forms with a base below 2,000 m (6,500 feet) altitude. STRATUS, STRATOCUMULUS, and NIMBOSTRATUS are classified as low clouds. CUMULUS and CUMULONIMBUS are sometimes counted as low clouds, because although they extend vertically, sometimes to a great height, their bases are often below 2,000 m (6,500 feet).

lower atmosphere The TROPOSPHERE, which is the part of the atmosphere lying beneath the TROPOPAUSE, including the PLANETARY BOUNDARY LAYER. It is the thinnest layer, but the one that contains about 80 percent of the mass of the atmosphere and in which almost all weather phenomena occur.

lowering The emergence of a mass of cloud beneath the base of a large CUMULONIMBUS cloud. The effect is

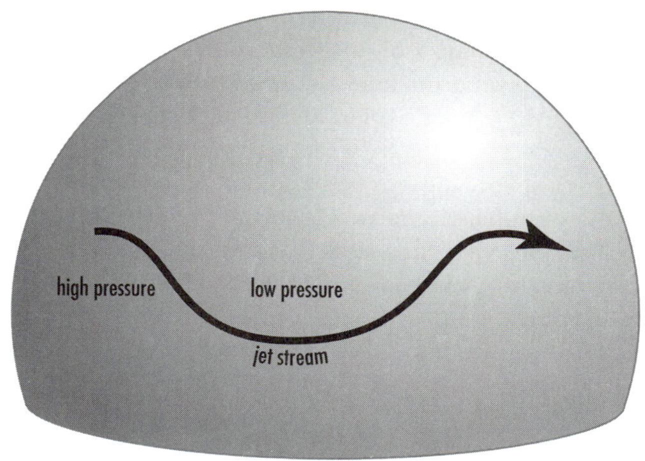

Low index. **The wave pattern in the jet stream is strongly developed, so the jet stream and polar front follow a very sinuous path.**

to lower the CLOUD BASE. Distinct rotational and vertical movement is sometimes visible in the lowered cloud. This is caused by air's being drawn into the base of the main cloud and it indicates that the storm has become tornadic (*see* TORNADO).

low index A low value of the ZONAL INDEX, which means that the ROSSBY WAVES in the JET STREAM and POLAR FRONT are well developed, so that the jet stream follows a sinuous path. Low-pressure POLAR AIR extends far to the south in some places, and in others TROPICAL AIR extends far to the north.

Luis A HURRICANE, rated as category 4 on the SAFFIR/SIMPSON SCALE, that struck Puerto Rico and the U.S. Virgin Islands September 4 to 6, 1995. Its winds gusted to more than 140 mph (225 km h^{-1}). At least 15 people were killed. Luis was shortly followed by MARILYN.

lull A temporary fall in the speed of the wind or cessation of PRECIPITATION.

lumen (lm) The derived SYSTÈME INTERNATIONAL D'UNITÉS (SI) UNIT of the rate of flow of light, known as *luminous flux*. It is equal to the rate at which light flows from a point emitting a uniform intensity of 1 CANDELA in a solid angle of 1 STERADIAN.

luminous meteor Any atmospheric phenomenon that appears as a pattern of light in the sky. Luminous meteors include AURORAE, CORONAS, HALOS, FOGBOWS, RAINBOWS, and similar phenomena, but not LIGHTNING, which is excluded.

lunar atmospheric tide That part of the ATMOSPHERIC TIDES that is caused by the gravitational attraction of the Moon. Its magnitude is very small and it can be detected only by careful analysis of records of atmospheric pressure made by readings taken at hourly intervals over a long period. The hourly readings must then be related to lunar time. The largest lunar tide detected in this way is a fluctuation of 0.003 lb in^{-2} (20 Pa, 0.003 mb).

lux (lx) The derived SYSTÈME INTERNATIONAL D'UNITÉS (SI) UNIT of illuminance, which is the amount of light energy that reaches a unit area of surface in a unit of time. The lux is equal to the illuminance produced by a lumous flux of 1 LUMEN distributed uniformly over an area of 1 square METER.

lx *See* LUX.

ly *See* LANGLEY.

Lynn A TYPHOON that struck Taiwan on October 24, 1987, and destroyed 200 homes.

lysimeter An instrument that is used to measure the rate of EVAPOTRANSPIRATION. It consists of a block of soil together with all the plants growing in it held within a waterproof container with only the upper surface open to the air. Strictly, the lysimeter is the block of soil, but the term is usually applied to the whole device. The container is usually circular and should be not less than 3.3 feet (1 m) deep and 3.3–20 feet (1–6 m) in diameter. A lysimeter installed in a forest may contain a mature tree, so the dimensions of its container must be sufficient to accommodate the root system. Water may be prevented from draining from the base of the soil block, or a drainage pipe may be fitted and water allowed to drain naturally into a sump, where the amount of drainage can be measured. The amount of water that enters the soil block from precipitation, including dew, and irrigation is mea-

Lysimeter. **The instrument measures changes in the amount of water in a block of soil, complete with its vegetation, held in a waterproof container. Drainage water is fed into a sump and measured. In this design changes in the mass of the soil block are registered as changes in pressure and monitored by a manometer. In other designs the soil block is weighed.**

sured. Any change in the mass of the soil block must be due to the entry of known amounts of water or to losses by evapotranspiration. Changes are measured either by a balance system, which measures the weight of the soil block, or by a manometer, which measures changes in pressure at the base. Lysimeters can produce very precise and reliable data, but they are large and require very careful installation and maintenance. Consequently, they are used mainly at agricultural research stations.

INDEX

Note: Page numbers in **boldface** indicate main entries; page numbers in *italic* refer to illustrations, maps, photographs, and tables.

C

D

I

U

X

Y

Z